Sustainable Development Goals Series

The **Sustainable Development Goals Series** is Springer Nature's inaugural cross-imprint book series that addresses and supports the United Nations' seventeen Sustainable Development Goals. The series fosters comprehensive research focused on these global targets and endeavours to address some of society's greatest grand challenges. The SDGs are inherently multidisciplinary, and they bring people working across different fields together and working towards a common goal. In this spirit, the Sustainable Development Goals series is the first at Springer Nature to publish books under both the Springer and Palgrave Macmillan imprints, bringing the strengths of our imprints together.

The Sustainable Development Goals Series is organized into eighteen subseries: one subseries based around each of the seventeen respective Sustainable Development Goals, and an eighteenth subseries, "Connecting the Goals," which serves as a home for volumes addressing multiple goals or studying the SDGs as a whole. Each subseries is guided by an expert Subseries Advisor with years or decades of experience studying and addressing core components of their respective Goal.

The SDG Series has a remit as broad as the SDGs themselves, and contributions are welcome from scientists, academics, policymakers, and researchers working in fields related to any of the seventeen goals. If you are interested in contributing a monograph or curated volume to the series, please contact the Publishers: Zachary Romano [Springer; zachary.romano@springer.com] and Rachael Ballard [Palgrave Macmillan; rachael.ballard@palgrave.com].

More information about this series at http://www.springer.com/series/15486

Yansui Liu

Urban-Rural Transformation Geography

Springer

Yansui Liu
Institute of Geographic Sciences
and Natural Resources Research
Chinese Academy of Sciences
Beijing, China

Key Laboratory of Regional Sustainable
Development Modeling
Chinese Academy of Sciences
Beijing, China

University of Chinese Academy
of Sciences
Beijing, China

ISSN 2523-3084 ISSN 2523-3092 (electronic)
Sustainable Development Goals Series
ISBN 978-981-16-4834-2 ISBN 978-981-16-4835-9 (eBook)
https://doi.org/10.1007/978-981-16-4835-9

This Springer imprint is published by the registered company Springer Nature Singapore Pte Ltd.
The registered company address is: 152 Beach Road, #21-01/04 Gateway East, Singapore 189721, Singapore

Foreword

Geography is a thriving discipline with a long history which has seen it evolve into a major integrative science of great relevance to the environmental and societal crises that the world currently faces. Geography has developed two prominent characteristics, namely regionality and comprehensiveness. The former is characterized by spatial differentiation of elements and the formation of regional patterns, while the latter encapsulates the diversity of geographical phenomena and their interactions. Moreover, geography is also an interdisciplinary subject, spanning both the natural and social sciences, as well as the humanities, and focuses on the pattern, process, and mechanisms of interactions between humans and the earth's environments. Geographical research not only involves consideration of the laws of physical geography, but also involves analysis of the laws of social economy. Due to its unique characteristics and research objects, geography has become a practical interdisciplinary discipline and provides a theoretical and scientific basis for recognizing, studying, and resolving various contemporary issues such as environmental change, resource overutilization, ecological degradation, and rural development, among others.

In terms of regional spatial differentiation, a region is typically characterized by human settlements of a range of sizes, from village to city. The urban–rural relationship emerges as the most basic social and economic relationship in the development of a region and, ultimately, of the country as a whole. From a global perspective, the "global village" is a complete and open huge urban–rural system, and there exist a range of urban–rural relationships at different scales in both developed and developing countries. Therefore, strengthening the cooperation and exchange between countries and regions, narrowing the gap between urban and rural areas, and realizing regional sustainable development have become common elements of human development as a whole.

In the process of regional development, urbanization and urban–rural relationships are of great significance to governments and academic researchers alike. In recent decades, the global urbanization rate has increased from 33.0% in 1960 to 56.2% in 2020, and rapid growth is particularly evident in Asia and Africa. However, problems such as the decline in investment in rural areas that has resulted in a decline in agriculture have led to loss of rural vitality and sustainability. In the absence of appropriate governance, some countries have developed what can be called the "urban disease," characterized by overcrowded and unhealthy housing, traffic congestion and environmental pollution, and a corresponding "rural disease" characterized by hollowing out, population aging, and economic poverty. The concurrence of these urban and rural "diseases" has become an important obstacle impeding integrated urban–rural development. Against this background, how to realize a transformation of the urban–rural relationship from isolation and opposition to coordination and integration is an important element of achieving the SDGs and a vital topic for the discipline of geography.

In the past two decades, research on urban–rural transformation has attracted considerable attention globally, and significant advances have indeed been forthcoming. In China, the acclaimed academic leader Prof. Yansui Liu steers the agricultural geography and rural development research team of the Institute of Geographical Research and Natural Resources Research in the Chinese Academy of Sciences to conduct high-quality, comprehensive research on rural geography, human–earth system science, and the theory and practice of urban–rural development. He has successively presided over and completed a substantial number of key projects of the National Natural Science Foundation of China, a major project of the National Social Sciences Foundation of China, and the National Key Technology Research and Development Program of China. In particular, he has completed three key projects of the National Natural Science Foundation of China, viz. Models of New Countryside Construction and Methods for Sustainable Development in Eastern Coastal China, The Impact of Urban-Rural Development Transformation on Resources, and The Environment and its Optimal Control and Evolution Processes of the Rural Human-Earth System and its Resource and Environmental Effects in China. The projects envisage human settlements as organisms that host a community of life, have innovated the research paradigm of processes, mechanisms, and patterns of rural development, and analyze the urban–rural relationship, urban–rural system and urban–rural integration from the aspects of element composition, structural organization, and functional state. In so doing, this research has facilitated the development of a deep and systematic understanding and scientific framework for the geography of urban–rural transformation.

These nine chapters of this important book, *Urban-Rural Transformation Geography*, present the latest and most significant results of Prof. Yansui Liu's group on the study of urban–rural transformation. This book systematically summarizes the main advances in theoretical research on urban–rural relationships of the past half century and analyzes deeply the basic laws and scientific components of urban–rural transformation. Innovative and novel concepts and perspectives such as "transformation tree," "transformation law," and "transformation threshold" and "fusion body" are proposed, and the basic theoretical methods of urban–rural transformation geography are explained. According to the characteristics of China's national conditions and development in the new era, this book systematically discusses the strategic objectives and regional orientation of China's urban–rural transformation, explains the technical methods and presents a comprehensive evaluation of models of China's urban–rural transformation research, explores the basic situation of urban–rural transformation at different spatial scales, analyzes the spatiotemporal pattern and mechanism of China's urban–rural transformation, and reveals the dynamic mechanism of spatial agglomeration and diffusion of population, land, capital, and other elements in the process of urban–rural transformation. In particular, from the comprehensive perspective of population, industrial, employment and land transformations, this book deconstructs the interactions within the transformation of urban–rural development, the evolution of regional industrial structure, and the change of rural resources and environment; analyzes the fixed effect of urban–rural transformation on rural resources and environment, the transmission effect of internal key elements, and the effect of rural resources consumption and environmental change. Finally, the countermeasures and strategic decisions for the optimization and regulation of urban–rural resources and environment are proposed. These diverse elements all embody the classical research characteristics of theoretical discussion, method construction, and practical application.

As an academic treatise, the book integrates theoretical innovation, disciplinary guidance, and practical application, with distinct and systematic disciplinary foresight, while at the same time offering insightful strategic guidance. It is not only excellent recognition of Prof. Liu's research and applied work on rural–urban transformation geography over the last 20 years, but also representative of the role of geography in responding to SDGs. The world has entered a critical phase and is facing many daunting challenges, among which are the pressing need to continue efforts to alleviate poverty and food insecurity to optimize the global urbanization network system and to slow the pace of anthropogenic climate change—all of this against the background of the COVID-19 pandemic. In highlighting the role of geography in servicing and supporting social and economic development and in promoting the construction of a sustainable urban–rural system, I firmly believe that the publication of this book is not only of great theoretical value but of considerable practical significance.

Michael E. Meadows
President of the International Geographical Union (IGU)

Fellow of the Royal Society of South Africa
University of Cape Town
Cape Town, South Africa

Preface

In 2015, the United Nations (UN) adopted the 2030 agenda for sustainable development and established 17 sustainable development goals (SDGs), aiming at completely solving the development problems in three dimensions of society, economy, and environment from 2015 to 2030 and achieving global sustainable development. In terms of the objects, the goal of "no poverty" and "zero hunger" is targeted at rural areas, whose development depends on the interaction between rural self-development ability and the driving force of industrialization and urbanization. Although the countryside breeds the city, there is a common problem of "attaching importance to the city and neglecting the countryside" in the process of promoting urbanization around the world, which leads to the long-term urban–rural separation and the widening gap between urban and rural areas and aggravates urban diseases and rural decline. Therefore, it is urgent to optimize the urban–rural relationship to promote urban–rural integrated development.

The urban–rural relationship is one of the most basic social and economic relations and is the frontier field of geography for UN SDGs and in-depth systematic research. As the largest developing country in the world, China is in the critical stage of transformation. However, due to unreasonable institutional arrangements, the long-standing contradictions of urban–rural segmentation, human–land separation, and land division have been increasingly intensifying. These further lead to the solidification of urban–rural dual structure, the widening urban–rural gaps, and the aggravation of rural diseases, which seriously hinder the equal exchange of urban–rural elements and its optimal allocation and is not conducive to socioeconomic sustainable development.

Urban–rural transformation is a comprehensive humanistic process, in which the urban–rural relationship transforms from isolation and opposition to coordinated and integrated development. And, it is embodied in the factors transfer, strategies change, and mechanism transformation of urban–rural regional system. As a coupled complex of urban–rural regional structure and spatial–temporal process, urban–rural system has all kinds of factors of urban and rural system. In general, the interaction of different subsystems in urban and rural areas leads to the formation of new relevance, which is reflected by the integrity and dynamic of urban–rural regional structure, morphology, and function.

Over the past 40 years of reform and opening-up, China has experienced a transition from "promoting urban development through rural reform," which is market-oriented, to the "driving rural development through polarization trickle-down effect of urban areas," which is guided by the strategy of balancing development of urban and rural areas, and gradually evolves to the "urban–rural integration" stage, and even to the advanced stage of "urban–rural equivalence." In the stage of "promoting urban development through rural reform," industrialization is the basic driving force for the development of central cities. Through administrative means to intervene the free operation of farmers and limit the flow of urban and rural population, the industrialization priority development strategy has been guaranteed. However, urban–rural dual system further aggravates the identity differences and regional differences between urban and rural residents, which to a certain extent becomes the root of "valuing the urban area while neglecting the rural area" and "urban rise while rural decline." Entering the twenty-first century, China has successfully joined the World Trade Organization (WTO), accelerated urbanization, focused on the implementation of the "five overall planning" strategy, and promoted the transformation of urban–rural relation to a new stage of "driving rural development through the polarization trickle-down effect of urban areas." In particular, the expansion of central cities and the rapid development of township enterprises promoted regional transformation of urbanization and agriculture and country and increased the flexibility of urban–rural dual structure, which has built a new platform and created a new environment to promote urban–rural interactive and coordinated development.

With the rapid development of urbanization and land conversion, the hollowing of rural areas and "rural disease" is becoming more and more serious. In 2011, China's urbanization rate exceeded 50% for the first time, reaching 51.3%, which marks that China has entered a new historical stage of urban society from a large peasant country. However, the urban–rural dual structure has not been fundamentally changed, and the dual structure within urban areas is also very prominent. To solve the problem of "double dual structure," it is necessary to speed up urban–rural coordinated development and strengthen the integration of urban and rural construction. The stage of "urban–rural integration" emphasizes the relationship of equality, coordination, and integration between urban and rural areas and is characterized by the coordinated development of industrialization, urbanization, informatization, and agricultural modernization. In this stage, the ultimate goal of social and economic development is to build a new pattern for urban–rural integrated development, rural revitalization, and then realize the equal development of urban and rural areas.

In terms of global urbanization development, some developed countries promote rural modernization through the development of urbanization and industrialization and ultimately achieve the urban–rural coordinated development. Due to the large number of people and the lack of land and the poor infrastructure, the human–land contradiction in rural China is prominent. And, the disorder resource allocation and vicious competition in rapid industrialization and urbanization result in serious destruction and waste of cultivated land and rapid non-agricultural transformation of rural elements.

As a result, the phenomenon of rural hollowing, population aging and weakening, and soil and water environment pollution are aggravating. Since the 13th five-year plan, economic development in China has entered a stage of the new normal and the contradictions and problems faced by urban–rural transformation are increasingly complex. Correspondingly, it is difficult to solve the problems in the process of transformation and seek scientific development strategies. In macro-level, the goal of "urban–rural integrated development" was clearly put forward in the 18th National Congress of the Communist Party of China (CPC), which emphasized promoting the equal exchange of urban–rural elements and the balanced allocation of public resources. Therefore, the systematic research on the pattern, process, and mechanism of urban–rural transformation needs to make substantial progress. Human–earth areal system is the core of geographic research. The evolution of urban–rural relationship, driving force of urbanization, and urban–rural morphology are essentially a complex natural, economic, and technological process. Paying attention to and strengthening the comprehensive researches of "process" and "pattern" has put forward new challenges and opportunities for the innovative research and development of geography.

Supported by the key project of National Natural Science Foundation of China (41931293, 41130748) and National Key Research and Development Program (2017YFC0504700), the author tried to introduce the theories of system science and engineering technology into geographic researches of urban–rural transformation, comprehensively integrated the geographical transect method, geo-statistics method, and geographical detector method, and applied the theories of system theory, economics, and engineering mechanics to explore urban–rural transformation theories, including "transformation tree," "transformation law," "transformation threshold," and "urban–rural integration," finally reveal the level–structure–function of urban–rural transformation, and promote the scientific researches on the process-pattern-mechanism of urban–rural transformation. Based on the systematic diagnosis of new-type urbanization, urban–rural land use, industrial structure transformation and public facilities allocation, this book explores the long-term mechanism, innovative model, and scientific way of urban and rural land optimal allocation and spatial reconstruction and develops the urban–rural transformation geography. Taking the process–mechanism–problem–regulation of urban–rural transformation as the main line, the study of urban–rural transformation geography is guided by solving urban–rural development problems and promoting urban–rural coordination and focuses on the spatial allocation, structural optimization, and function upgrading of urban–rural elements. Thus, it explores the process, pattern, mechanism, and response of urban–rural transformation, reveals the transformation characteristics, types, and regional differences of urban–rural regional system, and further puts forward countermeasures for deepening the reform of urban–rural integrated development strategy, system mechanism, and governance system according to the laws and principles of urban–rural economic and social transformation. All in all, this book starts from the perspective of geography, focuses on solving the realistic problems of unbalanced urban–rural development and inadequate rural development, promotes the transformation of urban–rural development and realizes the strategic goals of healthy urbanization and rural revitalization, and provides scientific path and successful cases for the construction of sustainable urban–rural system and urban–rural development model, contributing China's wisdom to the construction of a community with a shared future for mankind and the realization of 2030 SDGs.

Driven by environmental change, rapid urbanization, and technological progress, the urban–rural relationship has become increasingly complex and the change of urban–rural pattern has accelerated, which requires in-depth innovation of the dynamics, morphology, and typology of the evolution of rural regional system and urban–rural integrated system. Therefore, we sincerely hope to strengthen extensive cooperation and exchanges with professional scholars at home and abroad. In the professional research, practical exploration and typical demonstration of urban–rural transformation, we had studied and referred to the research results of many predecessors and professional peers, which laid a solid foundation and provided valuable inspiration for the project research. Team members, Hualou Long, Yifeng Zhang, Liming Wang, Yufu Chen, Jieyong Wang, Yurui Li, Maochao Yan, Yuheng Li, Yang Zhou, Yuanyuan Yang, Zhengjia Liu, Yongsheng Wang, Zhi Cao, and Yuanzhi Guo, as well as graduated doctoral students, Yangfen Chen, Guogang Wang, Ren Yang, Zhichao Hu, and Yanfei Wang, participated in the project research and demonstration practice, and special thanks to Jieyong Wang, Yurui Li, Yanfei Wang, and Yuanzhi Guo for their contribution in the writing and drawing. Additionally, this book is supported by the National Natural Science Foundation of China, University of Chinese Academy of Sciences, Institute of Geographic Sciences and Natural Resources Research of the Chinese Academy of Sciences, and Key Laboratory of Regional Sustainable Development Modeling of the Chinese Academy of Sciences and has received careful guidance and help from Michael E. Meadows, president of the International Geographical Union (IGU); Dadao Lu and Bojie Fu, former president of Chinese Geographical Society and academician of Chinese Academy of Sciences; Fahu Chen, president of Chinese Geographical Society and academician of Chinese Academy of Sciences; Huadong Guo, Chenghu Zhou, Mingan Shao, Caineng Zou, Peng Cui, Jun Xia, and Guirui Yu, academician of Chinese Academy of Sciences; Chunli Bai, Romain Murenzi, Shouyang Wang, Yonglong Lv, Jikun Huang, and Linxiu Zhang, Fellow of The World Academy of Sciences(TWAS); Changqing Song, director of the Faculty of Geographical Science of Beijing Normal University, and Springer's editors. Here, I would like to express my heartfelt thanks and high respect.

Beijing, China Yansui Liu

Contents

About the Author

Yansui Liu is Fellow of The World Academy of Sciences (TWAS) and the Distinguished Professor of Changjiang Scholars, Ministry of Education, PRC. Currently, he is Director General of Key Laboratory of Regional Sustainable Development Modeling, CAS; Director of the Center for Assessment and Research on Targeted Poverty Alleviation, CAS; Director of the Center for Regional Agriculture and Rural Development (CRARD) at Institute of Geographic Sciences and Natural Resources Research, CAS; Chairman of the Agriculture Geography and Land Engineering Committee of the International Geographical Union (IGU-AGLE); Chairman of the Alliance of Poverty Reduction and Development of the Alliance of International Science Organizations (ANSO-APRD); Advisory board member of Universal Scientific Education and Research Network (USERN). He is serving as Associate Editor of *Land Use Policy* and Member of Editorial Board or Guest Editor of *Journal of Geographical Sciences* and *Journal of Rural Studies*.

He is an internationally famous agricultural and rural geographer, and his research interests focus on human and economic geography, urbanization and urban–rural system, agricultural geography and rural development, as well as land use and human–earth system science. He has been in charge of more than 70 projects, including National Key Research and Development Program, National Key Technology Research and Development Program of China, Key Project of the National Natural Science Foundation of China, and Major Project of the National Social Sciences Foundation of China, etc. He has published 18 monographs and more than 480 papers, of which over 170 are in SCI/SSCI indexed journals and has been selected as Elsevier Chinese Most Cited Researchers in

Social Sciences since 2015, and was selected as one of Global Highly-Cited Researchers in 2018, 2019, and 2020. He was also awarded the TWAS Prize, National Science and Technology Progress Awards (2nd Class), National Excellent Science and Technology Work Award, Innovation Award of National Poverty Alleviation, the Technology Development Promotion Prize from CAS, and more than 20 provincial and ministerial science and technology awards.

List of Figures

List of Tables

1 Background and Value of Urban–Rural Transformation Research

Abstract

Urban–rural transformation geography is the systemic integration, comprehensive meta-synthesis and theoretical development of urban geography and rural geography. It is one of the innovations in the discipline of geography to meet the major national strategic needs of balancing urban and rural development and promoting urban–rural integration. As a subfield of geography, urban–rural transformation geography is a comprehensive and interdisciplinary subject, focuses on the urban–rural relationship regional system, aims at building urban–rural integration system to promote the optimization of urban–rural hierarchical structure and sustainable development, supports and serves the equal exchange of urban and rural elements, urban–rural integrated market construction, balanced allocation of urban and rural public resources and urban and rural spatial planning, systematically reveals the process, pattern, mechanism and effect of the evolution of urban–rural regional system.

Geography is an interdisciplinary subject with the responsibility of studying the human-earth areal system and its changing law. For a long time, it has focused on the thematic or comprehensive research of regional and watershed factor agglomeration, structural change, dynamic mechanism and its functional effect. At the discipline level, it shows relatively obvious and specific discipline boundaries and regional boundaries, including economic geography, urban geography, rural geography, arid basin geography and tropical geography. In terms of the urban–rural relationship, China's traditional urban–rural dual structure has resulted in the pattern of urban–rural division, land partition and human-land separation. Since the initiation of rapid urbanization development strategy in the mid-1990s, there has been extensive discussion on the urbanization model and pattern and process of urbanization transformation, but the disciplinary divergence of research scale still exists. Some experts proposed that rural geography should focus on the geographical space of the county, and urban geography should focus on settlement space above medium and small cities. In fact, urbanization is a compound process connecting rural and urban areas, and is an open, continuous and dynamic comprehensive human process, so it is the basic object of the study of the geography of urban–rural transformation.

The guideline of geography in China is to respond to the needs of national strategy and support national economic and social development. In 2003, the "Five Unified Arrangements" was proposed in the Third Plenary Session of the 16th Central Committee of the Communist Party of China (CPC), namely, coordinating urban–rural areas, regions, human and nature, social and

Y. Liu, *Urban-Rural Transformation Geography*, Sustainable Development Goals Series,
https://doi.org/10.1007/978-981-16-4835-9_1

economic development, and opening-up and internal-reform. In 2007, the Third Plenary Session of the 17th Central Committee of the CPC further put forward the strategy of urban–rural integration, promulgated and implemented the Urban–Rural Planning Law of the People's Republic of China (PRC). Since then, urban–rural coordination, urban–rural interaction and urban–rural integration have become the hot topics of geographic research, especially human-economic geography. However, the long-term situation of urban geography research on "urban area" and rural geography research on "rural area" is not competent and suitable for the requirements of the times of "urban–rural integration" theory and model innovation. Is it necessary to create urban–rural geography to promote the integration and systematic development of urban geography and rural geography? Is it necessary to create urban–rural transformation geography to meet the needs of the development of urban–rural relationship from isolation and opposition to coordination, integration and equivalence? Based on the regional system of urban–rural relationship, how to deepen the comprehensive geographical process of urban–rural elements exchange, urban–rural resource allocation and urban–rural coupling pattern? All these questions are being discussed widely by scientist groups. In 2011, the key project of "Research on the resource and environment effects of urban–rural transformation in China and its optimal regulation", supported by National Natural Science Foundation, was launched to answer these questions, and took the Bohai Rim region as an example to carry out in-depth research on the process-pattern-effect of urban–rural transformation. With five years' exploration and implementation, the exploratory research in this field had made gratifying progress and achievements, which laid a solid foundation for establishing and developing urban–rural transformation geography in terms of epistemology and methodology.

1.1 The Realistic Problems of Urban–Rural Transformation in China

1.1.1 Large Gap Between Urban and Rural Development

In recent four decades after the reform and opening-up, China has achieved great success in social-economic development, featured by constant improvements of the disposable incomes and wellbeing of urban and rural residents as well as optimization of economic structure. The social-economic achievements laid a solid foundation to realize domestic urban–rural equalization and integration. However, as China has long adhered to the strategy of giving priority to the development of capital-intensive heavy industry, the structural contradiction between supply and demand has become prominent, which has resulted in the relative decline of employment demand in urban sectors and rural residents cannot effectively transfer to urban areas. In particular, the strategy of "emphasizing city over countryside" and "emphasizing industry over agriculture" and many economic policies with urban bias continue to widen the gap between urban and rural development, which has gradually evolved into a complex social problem.

① The income gap between urban and rural residents is widening. The ratio of urban to rural incomes rose to 2.72:1 in 2016 from 2.57:1 in 1978, with a peak of 3.33:1 in 2009. If we consider that urban residents enjoy subsidies in housing, social security, public health and education, while the income of rural residents needs to pay taxes and invest in agricultural production, the actual ratio of urban to rural incomes would increase to 5:1–6:1. It was easy to find that the growth rate of urban residents' income is significantly faster than that of farmers' income (Fig. 1.1). ② There were huge gaps between urban and rural consumption. In 1990, the per capita consumption expenditure of urban

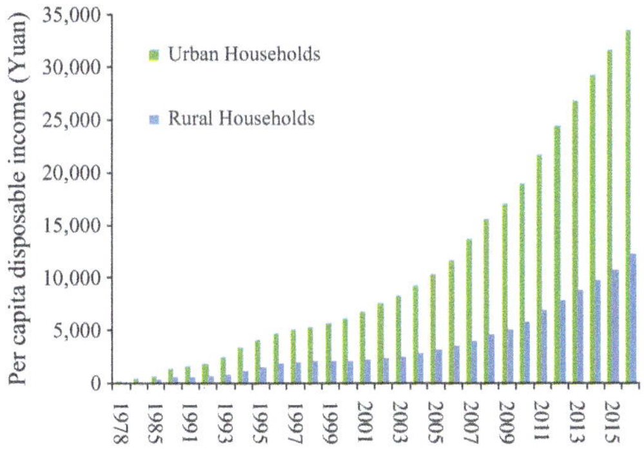

Fig. 1.1 Per capita disposable income of urban and rural residents in China from 1978 to 2016

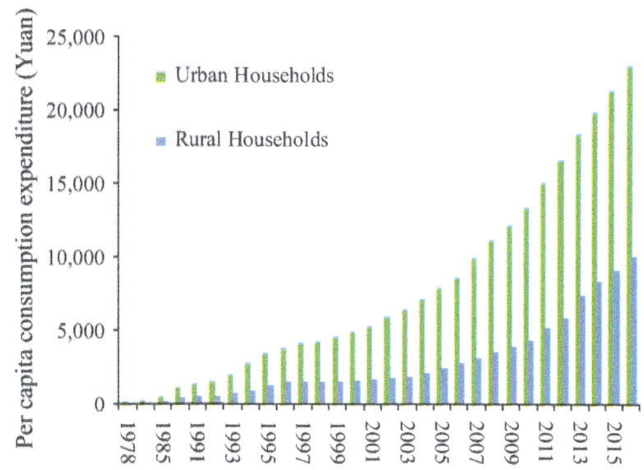

Fig. 1.2 Per capita consumption expenditure of urban and rural residents in China from 1978 to 2016

households was 1278.89 yuan, while that of rural households was 584.63 yuan, with a difference of 694.26 yuan, which was 2.19 times of that of rural households. In 2003, the per capita consumption expenditure of urban and rural residents was 6510.94 yuan and 1943.3 yuan respectively, which means that the urban residents were 3.35 times of the rural residents, and the gap between them was 4567.64 yuan. In 2016, the per capita consumption expenditure of urban and rural residents was 23,079 yuan and 10,130 yuan respectively, and the gap between them widened to 12,949 yuan. Obviously, the consumption gap between urban and rural residents is still very large, and the growth rate of consumption expenditure of urban residents is significantly higher than that of rural residents (Fig. 1.2). ③ Rural residents could not enjoy wellbeing and public services equivalent to those of urban residents because of the dualistic urban–rural citizenship. Residents with urban *hukou* enjoyed preferential treatment in employment, income, pension, medical-care and education. However, migrant workers with rural *hukou* were unable to enjoy the same welfare as urban citizens since a series of restrictions and a certain degree of discrimination.

With the continuous advancement of urbanization and industrialization in China, the income disparity among different regions, industries and groups has been enlarged considerably. As shown in a report of the National Bureau of Statistics, China's Gini coefficient has passed the international threshold of 0.4 in the past ten years, reaching a peak of 0.474 in 2012. The situation of urban–rural unbalanced development has lasted for a long time and intensified day by day, which has become an urgent problem to be solved at this stage. As documented by the China Migrant Population Development Report in 2017, the total volume of floating population in China has reached 245 million at the end of 2016, with the population mainly flowing to large cities and mega cities above prefecture level. Another report on the survey of rural migrants showed that the total volume of rural migrant workers has reached 282 million in 2016, among whom 30.5% and 19.7% were engaged in manufacturing and constructing industry, respectively. On the one hand, most rural migrant workers could hardly afford the social insurance and they could not obtain the same access to education, medical care, and social assistance etc. as urban citizens. On the other hand, the high cost of citizenization in big cities and megacities made it difficult for migrant workers to fully integrate into the city, which led to the long-term "urban–rural dual wandering". Although governments at all levels have attached great importance to the overall development of urban and rural areas since 2002, the urban–rural relationship was still dominated by urban department due to its strong economic power. Urban expansion was at the cost of loss of rural land resources, which brought about social and economic issues in rural areas because farmers lost their lands to make a living (Gu and Li 2013).

1.1.2 Rapid Non-agricultural Transformation of Land

Resource allocation is an eternal theme in the research domain of socioeconomic development. According to Lewis's dual-sector model, it is a general trend that surplus labor from the traditional agricultural sector will be transferred to the modern industrial sector to promote industrialization and stimulate sustained development. This labor transition law also applies to capital and land resource allocation in a developing economy. In China, rapid urbanization and industrialization have been accompanied by nonfarm utilization of rural land resources and nonfarm employment of rural labor (Liu et al. 2010; Long et al. 2009, 2012). Economic and social transformation and its land demand led to the continuous conversion of cultivated land to construction land, which has become the leading process of land conversion (Liu et al. 2013). As an important production factor, land transfer from rural sector to urban sector meets the demands of urban construction land and accelerated the process of urbanization and industrialization. Since the implementation of reform and opening up, China has made great achievements in economic development.

From 1978 to 2016, China's GDP has increased from367.87 billion yuan to 74,006.08 billion yuan, with an average annual growth of more than 9% (calculated with comparable price). At the same time, urbanization has been speeding up at an unprecedented pace. The urbanization level has increased from 17.92% in 1978 to 57.35% in 2016, with an average annual growth rate of 1% (Fig. 1.3). Relevant data show that there is a significant negative relationship between the rapid urbanization and the change of cultivated land area in China, and the cultivated land occupied by construction is one of the main reasons for the decrease of cultivated land (Table 1.1). From 1990 to 2000, 45.96% farmlands loss were attributed to construction land expansion. The ratio rose to 55.44% amid 2000–2010. The continuous decrease of cultivated land has led to significant changes in the regional pattern of agricultural production in China, and the contradiction among economic growth, food production and land use for people's livelihood has intensified (Liu et al. 2014). The conflict between land demand and supply shows spatial heterogeneity. In Huang-Huai-Hai region, Southeastern coastal region, the middle reach of the Yangtze River and Sichuan Basin, farmlands loss was mainly attributed to construction land expansion; while in the ecologically fragile areas of central and western China, farmland loss was mainly caused by the implementation of the policy of "returning farmland to forests".

China's urban–rural development has stepped into a new stage of transformation and upgrading. The non-agricultural process of rural elements will be further accelerated, and the pressure of urban development and industrial expansion to occupy cultivated land is still huge. In 2016, China's urbanization rate was 57.35%, there is still huge space for an economic transition from a traditional agriculture-dominated pattern to a modern urban-dominated pattern. It is obvious that over a period in the future, China's urban expansion will continue to accelerate, and the demands for construction land will keep increasing. The conflicts among farmlands protection, economic development and social security will be intensify. Currently, there are 83 metropolises with a population of more than 500 thousand in China, among which 73 are distributed in the 52 farmland aggregation areas. This means that there are spatial overlaps between the regions with the strongest economic growth and the areas where farmlands are concentrated and in urgent need of protection (Hu 2013), and the conflict between construction land expansion and farmland protection is especially obvious (Liu et al. 2011, 2014). Therefore, China has been trapped in a dilemma where economic development demands a great deal of

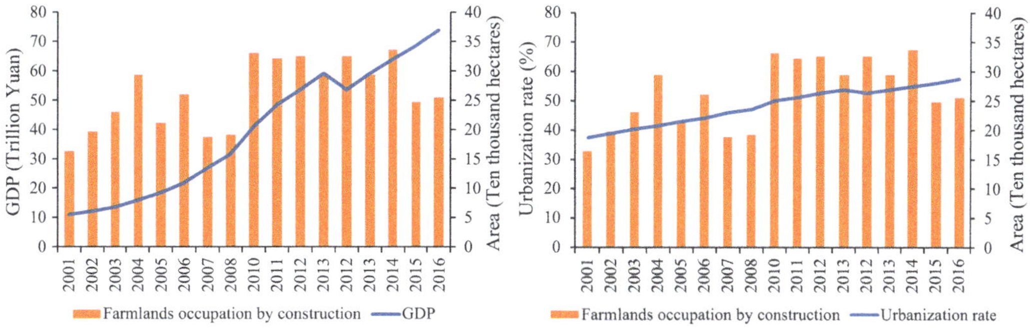

Fig. 1.3 GDP, urbanization rate and area of farmlands occupied by construction in China from 2001 to 2016

construction land, while the area of farmland needs to be maintained at a baseline of 120 million hectares to sustain food security.

There are two land ownership systems in China, i.e., state ownership of urban land and collective ownership of rural land. For a long time, due to the imperfection of land acquisition system and rural land property rights, it is difficult to effectively guarantee the rights of rural collective land circulation and transfer, which has become an important reason for the excessive occupation of agricultural land and ecological space by construction lands. To stick to the baseline of 120 million-hectare farmlands and guarantee food security, China has implemented a series of land use policies such as "balancing the occupation and compensation of cultivated land" and "linking the increase and decrease of urban–rural construction land". However, due to the lack of total factor farmland protection system design, including quantity, quality, ecology and space, the policies are not able to achieve desirable outcomes, and the problems of occupying high-quality cultivated land but supplementing poor quality cultivated land and paying attention to the area but ignoring the quality are widespread. Another land-use problem is the extensive utilization of rural homestead exemplified by the "hollow village", which is a result of imperfection in the withdrawal mechanism. To meet the demands of economic development, food production and ecological services, local governments have developed multiple forms of rural land consolidation and community construction. Although these projects were designed to advance sustainable rural development, the lack of technical support and fund guarantee make some rural communities confronted with problems such as obsolete infrastructures and undeveloped industries. Therefore, the rural land consolidation focusing on hollowing village is facing a series of system, mechanism, theory and technology problems that need to be solved urgently.

1.1.3 Imbalance of Regional Urban–Rural Transformation

China's opening to the outside world, system reform and institutional innovation have greatly promoted the regional economic growth and social development, but it also causes the imbalance of regional development. Since the 1980s, China's regional development strategy has changed from taking the leading development in the eastern region to seeking a relatively balanced development among regions, successively implementing the strategies of Great Western Development Strategy, Revitalization of Old Industrial Base of the Northeast, Rising Strategy of Central China. However, these ambitious strategies did not narrow regional gaps as planned. With heterogeneous natural and socio-economic conditions, the eastern China, central China and western China behaves differently in terms of industrial structure, employment structure, demographic structure and urban–rural spatial structure. Furthermore, with the rapid advancement of urbanization, the metropolises

Table 1.1 Reasons for farmlands loss in China from 2001 to 2016

Year	Total volume of farmalnds loss (10,000 ha)	Occupation by construction land		Destroyed by disasters		Returning farmland to forests		Adjustment of agricultural structure	
		Areas (10,000 ha)	Proportion (%)	Areas (10,000 ha)	Proportion (%)	Areas (10,000 ha)	Proportion (%)	Areas (10,000 ha)	Proportion (%)
2001	89.3	16.4	18.3	3.1	3.4	59.1	66.1	10.8	12.1
2002	202.7	19.7	9.7	5.6	2.8	142.6	70.3	34.9	17.2
2003	288.1	22.9	7.9	5.1	1.8	223.7	77.7	36.4	12.6
2004	147.8	29.3	19.8	6.3	4.3	73.3	49.6	38.9	26.3
2005	98.5	21.2	21.5	5.4	5.4	39.0	39.6	32.9	33.4
2006	102.7	25.9	25.2	3.6	3.5	33.9	33.1	39.3	38.3
2007	34.1	18.8	55.3	1.8	5.3	2.5	7.5	10.9	32.0
2008	27.8	19.2	68.9	2.5	8.9	0.8	2.7	5.4	19.4
2010	43.1	33.1	76.7	4.6	10.7	0.9	2.0	4.6	10.6
2011	40.9	32.2	78.7	2.2	5.5	0.9	2.3	5.5	13.5
2012	40.4	32.5	80.5	1.6	4.0	1.1	2.7	5.2	12.9
2013	35.7	29.4	82.5	0.7	1.9	0.8	2.2	4.8	13.5
2014	39.0	33.6	86.1	0.2	0.6	0.3	0.7	4.9	12.6
2015	30.3	24.8	81.9	0.4	1.2	2.5	8.2	2.6	8.7
2016	34.7	25.5	73.5	0.6	1.8	5.2	15.1	3.3	9.5

Data source China Land and Resources Statistical Yearbook (2002–2017).

and megalopolises are playing more important roles in economic development and social governance, and the regional gaps is more and more manifested as the "big urban–rural" difference between urban agglomerations and non-urban agglomerations. As a result of regional disparity, the urban–rural gaps and urban–rural transformation gaps also differed in different areas.

The urban–rural gaps are reflected in the differences of disposable income between urban and rural residents. In 2016, the top five provinces with the highest average disposable incomes of rural residents were Shanghai, Zhejiang, Beijing, Tianjin, and Jiangsu, and those with the lowest were Gansu, Guizhou, Qinghai, Yunnan, and Tibet. The top five provinces with the highest average disposable incomes of urban residents were Shanghai, Beijing, Zhejiang, Jiangsu, and Guangdong, and those with the lowest were Gansu, Heilongjiang, Jilin, Guizhou and Qinghai. Obviously, eastern China has made greater progress in urban–rural development and structural adjustment than western China. In 2016, the income ratio of urban residents to rural residents in eastern, central, western and northeastern China was 2.56, 2.45, 2.88, and 2.37, respectively. It is clear that western China had the largest urban–rural gaps.

Driven by export-oriented economy and rising township enterprises, eastern China achieved a higher level of industrialization and urbanization and transformed more rapidly in terms of rural employment and industrial structures than central and western China. Thus, an urban–rural spatial network in which cities, towns and rural settlements are treated as nodes and are linked by infrastructure elements (e.g., roads and pipelines) has formed in eastern China, demonstrating an integrated trend of urban–rural development. As important ecological functional and grain production areas, economic growth and urban development have proceeded slowly in central and western China since the restriction of the national functional orientation. Although the central government has adopted tendentious policies to support the development of central and western China since the beginning of the twenty-first century, it has been difficult for these undeveloped regions to obtain the international and domestic environment as eastern China in the reform and opening up period. Provincial capitals and prefecture-level cities have developed more rapidly than county-level cities in central and western China, and small towns lack facilities and capital to provide public services for citizens. Influenced by the economic crisis and international trade, although an increasing number of migrants in central and western China have chosen to work within the province, most migrants still flow into eastern China in pursuit of better job opportunities and higher salaries. In addition, labor-intensive industries are still concentrated in eastern China. So far, urban–rural links and the free flow of elements have not been established in the relatively undeveloped central and western China.

1.1.4 Increasingly Rural Hollowing

Hollowing village, featured by population emigration, depressed industry, abandoned lands and deficient infrastructures, is an unsustainable state of rural evolution that results from rural-to-urban migration and recombination of economic elements. Generally, hollowing of rural areas is often accompanied with high-speed non-agriculturalization of agricultural production factors, weakening of farmers' social subjects, abandonment of rural construction lands, and contamination of rural environment (Fig. 1.4). Influenced by dualistic urban–rural system and economic development model, rural areas are seriously lagging behind in the aspects of industrial transformation, land use transition and update of infrastructures.

First, the rural subject is getting older and weaker. From 1996 to 2016, the volume of rural resident population decreased from 851 million to 590 million, with an annual outmigration of 13 million, and the proportion of rural resident population in the total population dropped from 69.52% to 42.65%. As a result, the demographic structure of rural population changed significantly because the rural exodus was dominated by young labor force. In 2015, the population

(a) Non-agriculturization of farmland
(b) Weakening of rural resident
(c) Abandonment of rural construction land
(d) Contamination of rural environment

Fig. 1.4 Syndromes of "rural decline" in China

aged 60 and above reached 222 million, accounting for 16.2% of the total population, and the proportion in rural and urban areas were 18.5% and 14.3%, respectively (Liu 2018). The advancement of endogenous ability is an important precondition for rural sustainable development. Motivation and ability of rural households determined the endogenous ability of rural areas (Li et al. 2009). In general, rural development depends on the young and middle-aged labor with higher education attainment and skills. With the aging trend in rural areas, the destiny of future rural development attracts much more attention from the public.

Second, rural construction land is becoming increasingly abandoned. Impacted by the urban–rural dual structure and land governance system, "one household with several houses" becomes a common phenomenon in rural areas, resulting in the hollowing of rural space and the waste of rural land. Especially in the process of urban–rural transformation, a large number of rural populations migrate to urban areas, which directly leads to the vacant houses. Meanwhile, the homestead is generally "built new but not demolished", and the newly-built housing expands to rural periphery, which leads to the expansion of rural construction lands and the increase of idle and abandoned original homestead (Liu et al. 2010). According to the estimation of Institute of Geographic Sciences and Natural Resources Research, Chinese Academy of Science, the potential of the comprehensive renovation of hollowing village reached 7.6 million hectares (Liu et al. 2011). Under the background of urban–rural dual system, migrant workers can't afford to settle down in cities, and can't utilize rural lands efficiently as well.

Third, hollowing village is also exemplified by the deficiency of public facilities and infrastructures. Rural population emigration and abandonment of rural construction lands increase the difficulty of optimizing the allocation of public resources and the coast of infrastructure. The large-scale merger and withdrawal of elementary and middle school in rural areas provides a best example for this phenomenon. As documented in related statistics and reports, the number of rural elementary school decreased from 440.3 thousand in 2000 to 106.4 thousand in 2016, and the rural middle school decreased from 39.3 thousand in 2000 to 16.2 thousand in 2016. In 2001, only 38.2% teenagers finished their primary education in cities, while in 2016, 75.01% teenagers finished their primary education in cities. The excessive school merging and concentration not only leads to the problems of too long distance, dropout and increasing economic burden, but also further aggravates the migration of population due to the loss of educational function in a large number of villages.

Besides, rural poverty is a serious problem in China, and becomes a special "rural disease". Currently, the coexistence of hollowing village and rural poverty makes it hard to achieve rural sustainability. Until the end of 2015, there were 55.75 million poverty-stricken population and 128 thousand poor villages in rural China, most of which concentrated in the 14 concentrated areas with special difficulties. To implement targeted poverty alleviation strategy, more attention should be paid to the systematic research of hollowing village, rural poverty and rural aging to establish scientific support system for decision making.

1.1.5 Prominent Urban–Rural Environmental Problems

With the extensive growth of economy, over-exploration of resources and lack of efficient governance, the environmental pollution has worsened in urban and rural areas. At present, there is serious contamination of air, water and soil in China (She 2015). In recent years, the fog contamination incidents occurred frequently in central and eastern China, which seriously affected air quality, atmospheric visibility and public health and had received great attention from the government and society. With the industrial transformation and the improvement of environmental protection threshold in developed areas, some high-polluting industries began to transfer from eastern China to central and western China, from specific location to the whole catchment, from urban areas to rural areas. Rural environment is seriously destroyed by industrial waste, household garbage, agricultural non-point source pollution. In the past decades, the ecological environment in rural areas deserves to be worried as a result of insufficient investment and more debts in environmental governance (Huang and Liu 2010). According to the First Nationwide Survey of Pollution Source in 2010, approximately 50% of China's pollution emissions were from rural areas, in which the emission of COD, TN, and TP accounted for 43%, 57% and 67% respectively. It is estimated that approximately 9 billion tons of sanitary sewage, 280 million tons of domestic garbage and 260 million tons of human excrement are produced in rural areas each year, most of which are not properly disposed of. Some investigations revealed that metal mining and smelting, waste of breeding industries, chemical fertilizers, pesticides and electronic waste were the main sources of farmlands pollution (Wang et al. 2012). Some villages have been confronted with severe air pollution, water pollution and soil pollution issues. As shown in the Bulletin of the National Survey of Soil Pollution and Bulletin on the Quality Grades of Cultivated Land Nationwide, 19.4% of the sampling sites had soil pollutants that exceeded the threshold, and 40% of cultivated lands had been degraded. Soil pollution was more severe in southern China than in northern China, especially in the Yangtze River Delta, Pearl River Delta and Northeast Old Industrial Base. Environmental contamination is a potential threat to human health. In 2011, the number of villages affected by cancer had reached 351, and more in the east than in central and west (Gong and Zhang 2013).

1.2 Opportunities and Challenges for Urban–Rural Transformation in China

1.2.1 New Stage of Urbanization and Urban–Rural Relationship

In the past 40 years of reform and opening up, rapid urbanization and industrialization have induced the rapid transformation of industrial structure and the dramatic changes of urban and rural pattern, featuring significant increases in urban population, considerable growth of non-agricultural industries in rural areas, and increase of non-agricultural employment (Liu 2007; Long et al. 2011). Starting from the urban–rural system, the development of urban and rural areas is closely linked based on the circulation and agglomeration of labor, materials, capital and information (Li and Liu 2013). The support and contribution of agriculture and rural sectors to industrialization and urbanization show an "influx" effect, while the feedback of industry and urban sectors to agriculture and rural sectors basically belongs to the "income drop" effect (Hong 2007), and even has not formed substantial feedback (Liu 2014). Under the guidance of the strategy of balancing urban–rural development, this situation has changed to a considerable extent, but more effort is still needed to balance urban interests with rural interests.

China has entered a new stage of rapid overall urbanization since the beginning of the twenty-first century, with an annual growth rate of more than 1%. Featuring a surging urban population and dramatic urban sprawl, the urbanization in China is unprecedented in human history (Roger and Yao 1999). The rapid urbanization has had a profound effect on rural development, bringing about the transformation of rural population, land and industry (Liu 2007). With more than four decades of development since the reform and opening up, China has basically possessed the abilities of "promoting agricultural development through industry, driving rural development by urban areas", and the urban–rural relationship has been transformed from a dualistic structure to an integrated structure. Since 2003, when the per capita GDP exceeded 1090 dollars, China has entered a new transition stage characterized by the transformation from a traditional agricultural society to a modern industrial society and from planned economy to a market economy (Long, 2006, 2012).

With rapid urbanization and industrialization, cities have assumed a larger role in fostering economic growth and supporting human development. In 2016, the total volume of urban population in China increased to 793 million, with an urbanization rate of 57.35%, approaching the average level worldwide. The number of metropolises with a population of more than 5 million in China has reached 16. The urbanization process that has accompanied China's economic development is unique in the world. The European and American model proves that urbanization is an important approach to addressing rural issues, promoting regional coordinated development, increasing domestic demands, and advancing industrial updating. However, influenced by the dual urban–rural system and urban-biased policy, the regional imbalance of China's urban development is prominent, and the scale and grade differences among urban areas affect the flow of migrant workers, resulting in excessive concentration of migrant workers in cities (Gu and Liu 2012). The unconventional population expansion in urban areas has exceeded the resource and environmental bearing capacity, leading to a series of "urban diseases" such as traffic congestion, environmental deterioration, and polarization of wealth. Compared with the rapid expansion of large cities, small and medium-sized cities and towns have received less investment and developed slowly. Especially after the financial reform of "township finance and county management", the investment in the construction of small towns and rural areas has been greatly reduced, resulting in a serious shortage of infrastructure supply and public service supporting capacity in villages and towns. The declining function of villages and towns exacerbates their population loss and brain drain. The metropolis-dominated urbanization model could no longer adapt to the urban–rural integrated development policy and rural revitalization strategy.

Currently, China's urbanization development has reached a new starting point and confronts both challenges and opportunities. To sustain high-quality urbanization, we must take measures to deal with the problems such as how to solve the settlement of agricultural transfer population who has transferred to urban areas, how to improve the utilization efficiency of urban construction lands, how to optimize the layout and form of urbanization. In December 2013, China's Central Rural Work Conference confirmed the details of the urbanization schedule and proposed a goal of three "One Hundred Million" by 2020, namely, to settle one hundred million rural migrant workers in cities, refurbish shanty towns and "urban villages" for one million urban residents, and transform one million rural citizens into urban citizens in central and western China. All of these provide necessary conditions for promoting "people-oriented urbanization", improving the quality of urbanization, and optimizing the spatial pattern of urbanization. With the promulgation of the National New-Type Urbanization Plan (2014–2020) and the introduction of the top designs for national urbanization, it is undoubtedly conducive to promoting the allocation of urban resources to rural areas and the transformation of agricultural development mode. By strengthening the positive interaction between urbanization and new rural construction and promoting the parallel development strategy of new-type urbanization and beautiful countryside, small towns will play more important roles in attracting rural migration, supporting industrial development and serving rural residents, which will accelerate rural transformation and sustainable development in China.

1.2.2 Critical Stage of Agricultural and Rural Development

In recent years, new countryside construction and the improvement of rural living conditions have been the priorities of China's social development agenda. With more investment in rural areas, rural living conditions have improved significantly, and rural residents enjoy better education, medical care and public transportation services. The new countryside construction movement has achieved great success in socioeconomic development. Since the initiation of the "strategy of constructing a new socialist countryside", proposed by the 5th Plenary Session of the 16th Central Committee of the CPC in 2005, speeding up rural development and construction has become an important starting point for addressing the "three rural issues" that are embedded in China's modernization process. In 2016, the central government invested 245.5 billion dollars in supporting rural development, with a rising proportion in the overall national budget from 7.2% in 2005 to 8.93%. More financial capital was invested in the field of directly improving rural production and living conditions, including grain production, water security, road construction, methane supply and social welfare (Fig. 1.5). The investment of fixed assets by rural households also rose from 12.8 billion dollars in

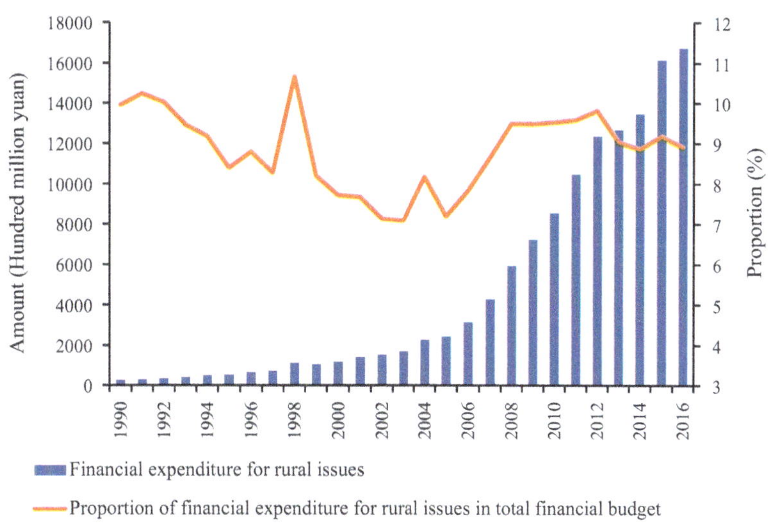

Fig. 1.5 China's financial expenditure for rural issues and its proportion in total financial budget

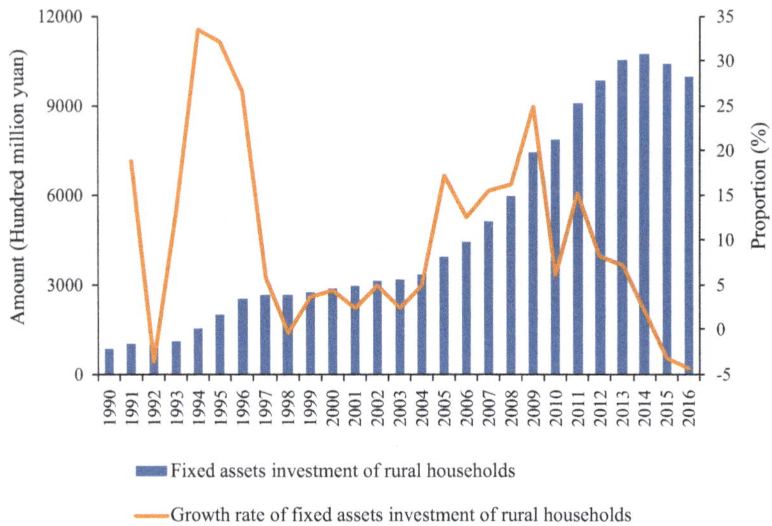

Fig. 1.6 Fixed assets investment of rural households and its growth rate in China

1990 to 145.4 billion dollars in 2016, with an annual growth rate of 9.8% (Fig. 1.6). The per capita disposable income of rural residents rose by about 10% annually from 2000 to 2016. Rural households enjoyed better medical insurance, social insurance and medical care. The efficiency of agricultural production improved significantly, accompanied by dramatic changes in rural living conditions. The success of rural development was exemplified by the typical models of new countryside construction driven by urbanization and industrialization, sustainable rural development guaranteed by the rural collective economy, and rural transformation stimulated by industrial restructuring.

With more rural–urban migration, changing demands of social consumption, and improvements of urban–rural infrastructures, agricultural and rural development in China was confronted with new opportunities and challenges. The new chances and challenges were as follows: (1) Agricultural production gained new momentum. Rural population emigration changed the human–earth relationship and increase per capita farmland, which led to a new trend of large-scale and specialized agricultural production. Although the proportion of rural land transfer reached 30% in China, this has not yet fostered new agricultural operators. Corporate farms struggled with inefficient management, expensive land leases, and low revenues from agricultural production. Collective corporations emphasized applying for financial subsidies more than business management. In addition, the traditional subsistence agriculture faced the challenges of the high demand for food quality and security. (2) Rural outmigration and rural aging coexist with the return of rural migrants. With the differentiation of rural population, a large number of farmers go out to work and became part-time farmers or full-time nonfarm employees. On the one hand, the outmigration of young and middle-aged laborers makes the social issues of left-behind children and left-behind elderly in rural areas increasingly prominent. On the other hand, the new technology, information, ideas and capital brought by returning migrant workers disturb traditional lifestyles, neighborhood relationships and land use in rural areas. (3) The trend of agricultural and rural multi-function is increasingly obvious. Currently, the non-productive function of agriculture and rural areas has been paid more and more attention, and leisure agriculture and rural tourism have become new forms of business in many rural areas. Rapid urbanization and industrialization not only promote the transformation of urban–rural development from isolation to integration, but also lead to a series of rural problems, such as rural hollowing, agricultural part-time, farmland conversion and left-behind population. In this complicated context, it is urgent to seek new ways of innovative development, seize the opportunity of urban–rural transformation, solve the practical problems of agricultural and rural development, thus promoting the relationship between urban and rural areas in China to gradually move towards urban–rural coordination, integration and equivalence.

1.2.3 Improving Perfection of Urban–Rural Policy System

In recent years, China has put more emphasis on propelling urban–rural integrated development and institutional innovation to give more priority to farmers, agriculture and rural departments rather than urban departments and industry, which gradually break the urban–rural dual structure. From 2004 to 2016, a sequence of policies, exemplified by the No. 1 documents from the national central government, have been introduced to support agricultural and rural development in China (Table 1.2). In 2005, the 5th Plenary Session of the 16th Central Committee of the CPC passed the resolution that took the "Construction of a New Socialist Countryside" as a major strategy to advance modernization in China. In 2008, the 3rd Plenary Session of the 17th Central Committee of the CPC passed another resolution that encouraged the establishment of a long-term mechanism of "promoting agricultural development with industrial development and leading rural areas with urban areas" and formed a new pattern of urban–rural integrated development. In 2010, the No. 1 document stipulated that balancing urban and rural development was the fundamental precondition of building an overall prosperous society in China. In 2012, the 18th Central Committee of the CPC put forward the strategic objectives of "promoting the construction of ecological civilization" and "urban–rural integrated development". In 2013, the 3rd Plenary Session of the 18th Central Committee of the CPC decided to comprehensively deepen reform, and proposed to coordinate the development of large, medium and small cities and small towns and integrate the development of industry and towns; balance urbanization with the new countryside construction; advance the equal exchange of urban and rural production factors and the balanced configuration of public resources; and optimize systems and mechanisms for the sound development of urbanization. Thus, it provided an effective way to solve the practical problems in urban and rural development. In 2014, the central government of China enacted the National New-type Urbanization Plan (2014–2020), which took the people as the core of the urbanization, orderly promoting the citizenization of agricultural transfer population, the equal exchange of urban and rural elements and the balanced allocation of public resources.

The implementation of these policies and measures reflected the transformation of China's urban–rural strategies from "supporting urban development with rural development" to "leading rural development with urban development" and "urban–rural integrated development". In particular, the reform of household registration system, land system, education and healthcare system provide guarantees for urban–rural integrated development. In July 2014, the State Council of China issued "advices about further promoting reform of the household registration system", proposing to establish a uniform household registration system in urban and rural areas and abolish the distinction between agricultural and non-agricultural household registration and the resulting blue print *hukou*. Meanwhile, differentiated settlement policies should be made according to the urban conditions, and fully liberalized the restrictions on the settlement of towns and small cities. At the same time, the reform of the land system in rural areas was also accelerated. In 2015, 33 typical counties were selected to implement pilot reform of rural land acquisition, marketization management of collective operating construction land and homestead system. Both the socioeconomic structure

Table 1.2 No. 1 documents from the central government and their themes from 2004 to 2016

Year	Document name	Themes
2004	Advice about several policies to increase farmers' incomes	Increasing farmers' incomes
2005	Advice about several policies to strengthen support for rural areas and improve agricultural production capability	Agricultural production capability; tax exemption in agricultural sector
2006	Advice about promoting new socialist countryside construction	Constructing a new socialist countryside
2007	Advice about promoting new socialist countryside construction and developing modern agriculture	Development of modern agriculture and construction of a new socialist countryside
2008	Advice about promoting construction of agricultural infrastructure, advancing agricultural development and increasing farmers' incomes	Agricultural infrastructure and agricultural modernization
2009	Advice about promoting stable development of agriculture and increasing farmers' incomes in 2009	Agricultural development, constant improvement of farmers' incomes
2010	Advice about advancing urban–rural integrated development and strengthening the foundation for rural and agricultural development	Urban–rural integrated development; strengthen foundation
2011	Advice about accelerating reform and development of water utility	Construction of water utility
2012	Advice about advancing agricultural technology innovation and guaranteeing agricultural product supply	Agricultural technology
2013	Advice about promoting the development of modern agriculture and stimulating rural vitality	Agricultural modernization; rural reform
2014	Advice about deepening rural reform and advancing agricultural modernization	Rural reform
2015	Advice about reinforcing reform and innovation and promoting agricultural modernization	Agricultural modernization
2016	Advice about implementing new ideas on agricultural modernization and realizing the objective of an overall prosperous society	Agricultural modernization

and national strategy reflected the transformational trend of urban–rural development in China. However, under the guideline of building a moderately prosperous society in an all-round way, the balance of urban and rural development in the period of 13th five-year plan still confronts huge challenges of socioeconomic restructuring and system reform. In the context of heterogeneous demographic structures, human–earth relationships and regional development patterns, it is imperative to further promote the fundamental innovation of the system of rural land, household registration, tax and fee, social security and public goods, and accelerate the construction of an integrated system of urban and rural development.

1.3 Perspectives and Themes of Urban–Rural Transformation Geography

1.3.1 Multidisciplinary Research on Urban–Rural Transformation

1. **Aspects of economic development based on benefit analysis**

Economic growth and structural adjustment are the economic basis and transformation preconditions for urban–rural development. The urban–rural dual theory indicates that there are different industrial sectors in urban and rural areas, the flow of labor promotes the transformation of urban–rural structure from dualistic to unitary. The urban–rural transformation mainly depends on the decisive role of market mechanism in the optimal allocation of resources. Guided by the goal of maximizing factor income, the urban–rural integrated market promotes the orderly flow and optimal combination of resources, capital and technology between urban and rural areas and among different sectors. Urban–rural relationship is the hot topic of economic and geographic research, which focuses on national strategy, urban–rural system and institutional restriction (Lu and Chen 2004). Since the beginning of the twenty-first century, economic growth in China has been confronted with many new situations and challenges, such as the transformation of economic growth momentum and the "middle-income trap". The traditional investment-driven development model depending on intensive investment in energy, resources and lands cannot adapt to the new situation. Therefore, it is urgent to promote the transformation of economic structure and economic development mode. With the rapid transfer of rural population to urban areas, Cai (2013) believed that China's economic growth had reached the Lewis turning point and needed to realize the synchronization of urbanization and non-agriculturization. The low-cost dividend that farmers provide cheap labor to support industrialization and urbanization is gradually disappearing, and it is necessary to reform the dualistic urban–rural management system and seek new drivers for economic growth.

2. **Aspects of social development based on demographic features**

The emergence of core families and heterogeneous characteristics of farmers are the main reasons for social restructuring and the reconstruction of the governance system in rural areas (Shi and Yang 2012). With more rural migrants flowing to urban areas and engaged in nonfarm employment, the stratification of rural society has intensified in China. Population migration, economic discourse improvement, and expanded social networks in rural areas have changed the traditional collective economy, neighborhood relationships, informal institutions and social values (Wen and Yang 2012). The outmigration of rural young and middle-aged laborers makes the rural population tent to be old and weak, and the deep poverty in central and western China further increases the difficulty of rural social governance. Under the urban–rural dual system, rural migrant workers differ from urban citizens in employment, income and access to social security. The separation of urban and rural areas has increasingly become an important obstacle for rural migrants becoming part of urban society and realizing urban–rural integrated development. Currently, rural migrant workers are dominated by young people who were born in the 1990s. It is urgent to deepen the social system reform of urban–rural separation and break the spatial order of "urban–rural double drift" and its latent social trap. With the implementation of the new-type urbanization strategy, rural areas need to pay more attention to the reform of rural democracy and township institutions, establish a sound rural governance system and restructure rural collective organizations. These adjustments are favorable to advancing urban–rural integrated development, promoting the modernization of rural governance and addressing "rural diseases".

3. **Aspects of regional development based on spatial heterogeneity**

As the common research object of geography, regional economy, urban planning and other disciplines, space emphasizes regional differences and their interactions, but different subjects focus on different things. Socioeconomic changes have triggered the transformation of spatial patterns between urban and rural areas, especially the transformation of urban–rural elements, which promotes the spatial pattern of "concentration in small scale and dispersion in large scale" in urban and rural areas, and the difference between plain area and hilly area is obvious (Liu and Yang 2015). The artificially high urbanization level and low quality of urbanization are represented by the semi-urbanization of populations and spaces (Lu and Chen 2015). Due to the existence of time and distance costs, the connection between urban and rural areas has the general characteristics of attenuation with distance, that is, from downtown to rural area. The collaborative conversion efficiency and correlation between urban and rural areas in core areas of urbanization are strong, while those in semi-urbanized and remote rural areas are weak. The direction, intensity and speed of this spatial transformation show obvious heterogeneity among regions. In addition to regional differences and interactions, geographers have fully considered the regional differences and comprehensive characteristics of development, brought urban and rural production and life into the regional ecosystem, and attached importance to the impact of economic growth and urbanization on regional water and soil resources, food production, ecological functions and rural development space. Scientists engaged in urban planning emphasizes the integration of urban–rural economy and society and the optimization of urban and rural space by spatial planning technology.

1.3.2 Characteristics of Urban–Rural Transformation from the Perspective of Geography

Region is the spatial organization form of interaction between specific geographical environment and human activities, and the evolution and spatial differentiation of regional system is one of the core issues of geography research. The study of urban–rural development in geography mainly focuses on rural transformation, urban–rural relationship and its mechanism, urban–rural development mode and so on. Urban areas relate to rural areas in terms of multiple factors and multiple sectors. On the other hand, there are land allocation conflicts between urban and rural departments. With the rapid urbanization and urban–rural socioeconomic development, the urban and rural elements, structure, function and spatial form have changed significantly.

Responding to the new-type urbanization strategy, geographers have conducted much empirical research in rural areas with a research paradigm of "process-pattern-mechanism-effects", which is exemplified by rural hollowing and hollowing village renovation, new countryside construction models, and comprehensive research on agriculture and rural development, etc. (Liu et al. 2011). Rural transformation is a comprehensive reflection of rural internal factors and external driving forces. Influenced by urbanization and industrialization, rural development is confronted with many challenges in terms of population, land, space, grain production and cultural traditions. Especially, driven by the land finance and distinct land property systems between urban and rural areas, land urbanization does not coordinate with industrialization and demographic urbanization. It is urgent to strengthen the economic and social transformation and reconstruction of rural areas and the optimization and remodeling of urban and rural regional space, thus facilitating coordination among population, land and industry. Geography has served the national economic construction and regional coordinated development for a long time, and has obvious disciplinary advantages and long-term accumulation in theory and practice. In the context of the new-type urbanization and urban–rural integrated development, geography should provide more knowledge to promote urban–rural transformation given its advantages in exploring comprehensive processes and spatial heterogeneity.

1. **Regional synthesis**

Urban and rural areas are territorially connected, like two sides of the same coin. Accordingly, rural construction should keep pace with urban development, like two wheels of the same car. Urban–rural integration and the new-type urbanization do not aim to eliminate rural departments or transfer obsolete and polluting industries to rural areas. Urban–rural integration does not pursue a centralized urban regime but rather seeks a new reciprocal industrial-agricultural and urban–rural relationship of promoting agricultural development through industrial development and leading rural development with urban development (Liu et al. 2016). There are interactions and identities between urban and rural areas in economic structure, demographic structure, social structure and spatial configuration. However, in practice, due to the obvious differences in regional landscape, scale efficiency and cultural characteristics between urban and rural areas, urban areas assume spatial functions different from those of rural areas. Based on the comprehensive viewpoint of geography, urban–rural development needs to fully consider the internal connection and functional complementarity of urban and rural areas in population, economy, society, culture, space and other aspects, integrate urban and rural areas into the same regional system and unified urban–rural market mechanism and governance system, and emphasizes the coordinated development of urban and rural regional systems. The urban–rural regional system is an important part of the human-earth areal system, the evolution of urban–rural economy and society has significant impacts on the quantity and quality of resources and environment. Therefore, geography takes the high ground in urban–rural development research by exploring the element-structure–function dynamics of urban–rural systems and evaluating these systems from a multifaceted perspective.

2. **Spatial heterogeneity**

Regional natural geography, resource endowment and other conditions and their composition and matching determine the resources and environmental carrying capacity, the suitability of natural environment and the adaptability of economic development, and also restrict the urban and rural economic development and regional transformation. Regionality is an important geographical feature formed by the interaction between natural geographical environment and human social activities. Based on specific natural conditions, ecological environments, socioeconomic factors and infrastructure elements, China is divided into various functional zones. For example, urbanization zones are dominated by construction land, major grain production zones are dominated by farmlands, and ecological zones are dominated by ecological lands (Fig. 1.7). Owing to China's vast territory and huge regional disparities, its urban–rural development confronts various challenges and limitations at the local scale. The pace, magnitude, strength and path of urban–rural transformation development in China vary at the regional scale due to varying stages of economic growth, rural development and urbanization. Meanwhile, the flow of elements, regional trade, and industrial transfer also have important impacts on regional demographic structures, industrial configurations, land-use patterns and urban–rural relationships. Therefore, to systematically analyze and solve these problems, we need to deeply explore the regional mechanism of urban–rural transformation, promote the investigation of different types of regions, finally put forward the sustainable ways and countermeasures to accelerate urban–rural transformation based on geographical perspective of regional differences and interregional connections.

3. **Scaling effect**

Scale generally refers to the spatiotemporal measurement of an object. The spatial organization of urban–rural regional system includes small towns, small and medium-sized cities, large and mega cities, urban agglomerations and other urban space, and also involves county economy, prefecture economy, provincial economy, national economy and other economic types at the level of administrative divisions. Factor flow, industrial transfer and administrative management between urban and rural areas are significantly affected by the scaling effect. In particular, with the enlargement and upgrading of regional scale, the urban–rural transformation follows a gradual changing rule of "element-structure–function-regional type", and the exploration of different scale issues needs to pay

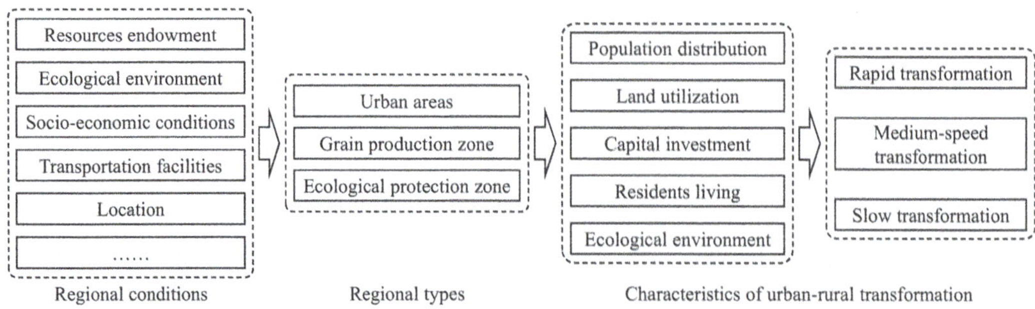

Fig. 1.7 Characteristics of urban–rural transformation

attention to spatial element organization, urban–rural mutual feedback, urban–rural linkage function and urban–rural regional type in turn (Fig. 1.8). The evaluation and optimization of urban–rural transformation should focus on the spatial heterogeneity and characteristics of urban–rural transformation. To coordinate the roles of the central government and local government in promoting urban–rural integrated development, we need to determine the specific problems and challenges at multiple spatial scales, such as the provincial scale, city scale and town scale. To advance the sustainable development of urban and rural areas, we also need to address the practical issues of "strong cities, weak towns" and "emphasizing urban areas while neglecting rural areas".

4. **Dynamic evolution**

The analysis of urban–rural evolution is the key to understanding the separation between urban and rural areas and revealing the mechanisms and patterns of urban–rural transformation. Rapid urbanization and industrialization in a certain region accelerate the flow and spatial agglomeration of urban and rural factors, bringing about changes in urban and rural structures, patterns and functions. To transform the urban–rural relationship from separation to integration, we need not only the traction forces of economic growth, population flow, and industrial updating, but also the driving force of agricultural industrialization, management organization and rural multi-function. The strategies and supporting

Fig. 1.8 Spatial scales and hierarchy of urban–rural transformation

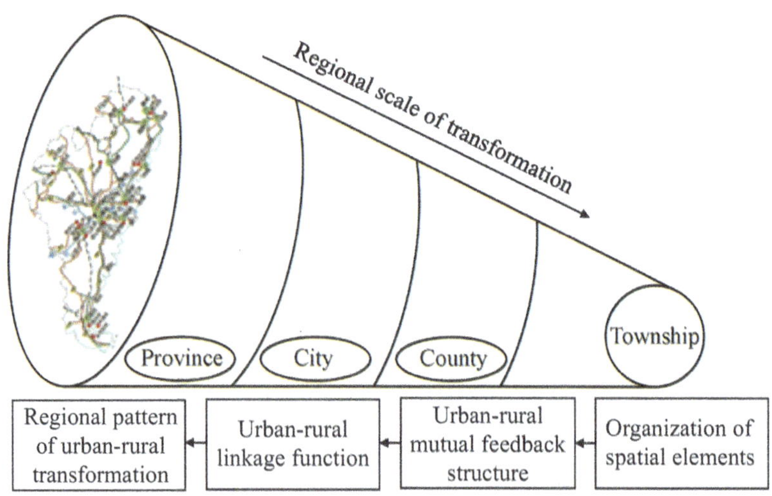

policies for urban–rural coordinated development are also important in promoting the transformation of the urban–rural system. Therefore, the dynamic analysis of urban–rural transformation should follow the regional law of urban–rural transformation; scientifically grasp the features, drivers and limitations of economic growth; systematically diagnose the driving force and binding force of urban–rural transformation in different periods; and identify the scientific path and typical mode of urban–rural transformation in different regions.

5. **Practice orientation**

In the context of the new-type urbanization and urban–rural integrated development, urban–rural transformation has become a complicated and systematic project for promoting socioeconomic transformation, mechanism transformation and strategy transformation. As a science that stresses practice and reveals laws, geography especially emphasizes the methods of interviews, questionnaires and other fieldwork, and provide scientific support for the spatial optimization of urban–rural areas, coordination of human-natural relationships and integration of regional development through making urban–rural planning, proposing land consolidation schemes and putting forward policies and suggestions. With the development of GIS and data mining technology, future research should pay more attention to data analysis and scenario simulation to identify typical models of urban–rural transformation and monitor the process of urban–rural transformation.

Featuring problem-driven, solution-oriented research, geography has a fine tradition of serving national strategy and social practice by linking knowledge with practice. In the early 1980s, geographers had made great contributions to agricultural resource surveys, comprehensive agricultural zoning, land use and rural geography in China. Since the mid-1990s, in view of the new mission of geographers and the increasingly prominent issues concerning agriculture, countryside and farmers, Chunjun Wu stressed at the 2007 annual meeting of the Chinese geographical society that "geography has comprehensive advantages and responsibility to implement the five overall plans and scientific development concept in the new era, and geography should serve to solve the issues concerning agriculture, countryside and farmers". To advance the new-type urbanization, which is people-oriented, it is necessary to understand economic, social and ecological problems from a practical perspective; reform mechanisms to coordinate populations, land, industry and capital in urban and rural areas; deeply explore the comprehensive integration and application of GIS, data mining and spatial simulation technology in decision-making of land consolidation, village-town construction and industry-city integration. Thus, it will help to explore the rural geographic engineering and regional mode of urban–rural transformation in different regions.

1.3.3 Main Research Topic of Urban–Rural Transformation Geography

Agricultural and rural development are traditional research domains of geography. As early as the 1950s, Kezhen Zhu, a famous Chinese geographer, stated that "geography should support national economic development, of which agriculture takes priority." In recent years, sustainable agriculture, rural urbanization, new countryside construction and hollowing village and consolidation have been the dominant research domains of rural geography (Liu et al. 2011; Long et al. 2014). With the development of industrialization and urbanization, China's urban and rural economy, society, population and culture have changed significantly, and the urban–rural relationship has been transformed from separation and opposition to coordination and integration, which is the basic trend of China's urban–rural relationship and modernization process. Urban–rural transformation has changed the stable structure of urban and rural areas. The reform of the household registration system and land system accelerated rural outmigration, nonfarm employment and nonfarm utilization of land resources.

The urban–rural dual management system and economy-oriented strategy led to an agglomeration of production factors in urban areas and an imbalanced economic structure. They are the main reasons for the uncoordinated and unbalanced development of urban and rural areas. At present, the research on urban–rural transformation needs to be deepened in several aspects. First, how can the rural declining trend be reversed? With the rapid non-agricultural development of rural production factors, rural development is faced with increasingly serious problems, such as the weakening and aging of rural population, hollowing village, obsolete infrastructure and environmental pollution. Second, how can the quality of urbanization be improved? Influenced by the urban–rural dual system, China's industrialization and urbanization seriously depend on abundant and cheap rural labor and land resources. Land finance and unfair land acquisition have led to an artificial urbanization level and an imbalance between demographic urbanization and land urbanization. Third, how can urban citizenship be granted to rural migrant workers? It is difficult for rural migrant workers to settle in urban areas as a result of the imbalance between urban and rural transformation. Approximately 300 million rural migrant workers drift to cities from the countryside every year.

Urban expansion and rural decline have resulted in increasingly severe "urban disease" and "rural disease". The traditional research paradigm, which separates urban and rural issues, cannot meet the practical needs of balancing urban and rural development and the institutional innovation of urban–rural integrated development (Liu et al. 2014). Woods (2009) claimed that it was necessary to strengthen communication between urban geographers and rural geographers to analyze the process of urban–rural development from the perspective of urban–rural integration. Urban and rural transformation is a systematic change of urban and rural areas, not only in the field of economic development, but also in the field of social life and ecological environment. Based on the research paradigm of "process-pattern-mechanism-regulation", geographers should pay more attention to exploring the evolutionary process, mechanism, effect and regulation of urban–rural systems, and provide scientific support for promoting balanced regional development, coordinated urban–rural development, sustainable socioeconomic development, and a harmonious human–earth relationship.

The main line of the research on urban–rural transformation geography should focus on the spatiotemporal heterogeneity of urban–rural regional system and solving the problems of urban–rural transformation and promoting urban–rural integrated development, take industrialization, urbanization and agricultural modernization as the driving forces, thus realizing the rural modernization, new-type urbanization, and urban–rural equivalence. By systematically screening urban and rural development problems, simulating the dynamic process, revealing the dynamic mechanism, combing the regional model, urban–rural transformation geography aims at exploring the scientific way to accelerate urban–rural transformation and realizing urban–rural integration, deeply changing the urban–rural development strategy, institutional mechanism and governance system, so as to provide scientific basis for realizing the strategic objectives of urban–rural structure optimization, function coordination and urban–rural integrated development.

Urban–rural transformation geography has rich scientific connotation and broad professional vision, which is based on urban–rural regional system and aims to explore the structures, processes, patterns, and effects and security of urban–rural transformation. According to the disciplinary boundaries, urban–rural transformation geography is a comprehensive integration of physical geography, human geography, technological geography and engineering geography, and is rooted in geographic science, urban–rural planning science, management science and engineering science (Fig. 1.9).

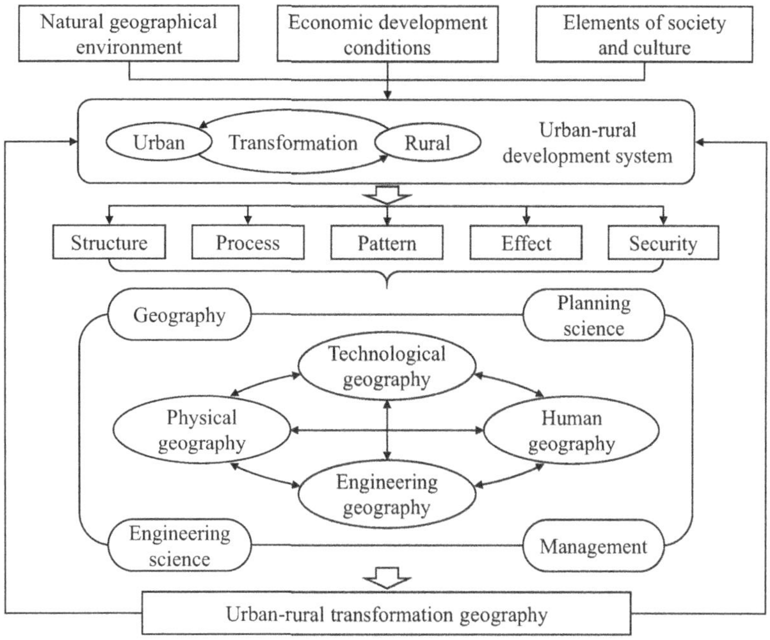

Fig. 1.9 Theoretical system of urban–rural transformation geography

References

Cai F. Bonus of labor resources and sustainable economic development in China. Gansu Soc Sci. 2013;1:1–4.

Gong S, Zhang T. Temporal-spatial distribution changes of cancer villages in China. China Popul Resour Environ. 2013;23(9):156–64.

Gu C, Li A. A framework for the integration planning between urban and rural areas in China: from "city exploiting rural area" to solving "three issues in rural area." Econ Geogr. 2013;33(1):138–41.

Gu S, Liu J. Urbanization transition in China: from a factor-driven to an innovation-driven approach. Population Res. 2012;36(6):3–12.

Hong Y. Research on the path of industry and city reversal feeding agriculture and country. Econ Res. 2007;8:13–20.

Hu C. Quality evaluation of farmlands in China and approaches to improving its quality. China agricultural news network, 2013-08-01.

Huang J, Liu Y. Environmental pollution in rural China and its driving forces. Chin J Manag. 2010;7(11): 1725–9.

Li Y, Liu Y. Investigation of the resource & environment issues in the urban-rural transition in China. Econ Geogr. 2013;33(01):61–5.

Li X, Luo Q, Fan X. A study on the formation and evolution of specialized rural villages. China Soft Sci. 2009;2:71–80.

Liu Y. Rural transformation development and new countryside construction in eastern coastal area of China. Acta Geogr Sin. 2007;62(6):563–70.

Liu Y. Research on the urban-rural integration and rural revitalization in the new era in China. Acta Geogr Sin. 2018;73(4):637–50.

Liu Y, Yang R. The spatial pattern measure of urban-rural transformation in the Bohai Rim region in China. Acta Geogr Sin. 2015;70(2):74–82.

Liu Y, Wang J, Long H. Analysis of arable land loss and its impact on rural sustainability in Southern Jiangsu Province of Chin. J Environ Manage. 2010;91:646–53.

Liu Y, Long H, Chen Y, et al. Report on rural development in China: hollow village and its consolidation strategy. Beijing: Science Press; 2011.

Liu Y, Long H, Chen Y, et al. Research report on rural development in China: Hollowing village and its renovation strategy. Beijing: Science Press; 2011.

Liu Y, Lu S, Chen Y. Spatio-temporal change of urban-rural equalized development patterns in China and its driving factors. J Rural Stud. 2013;32:320–30.

Liu Y, Long H, Wang J. Agricultural modernization and farmers in China. Beijing: Science Press; 2014.

Liu Y, Yan B, Wang Y. Urban-rural development problems and transformation countermeasures in the New Period in China. Econ Geogr. 2016;36(7):1–8.

Long H. Rural housing land transition in China: theory and verification. Acta Geogr Sin. 2006;61(10):1093–100.

Long H. Land use transition and rural transformation development. Prog Geogr. 2012;31(2):131–8.

Long H, Liu Y, Wu X, et al. Spatio-temporal dynamic patterns of farmland and rural settlements in Su-Xi-Chang region: implications for building a new countryside in coastal China. Land Use Policy. 2009;26 (2):322–33.

Long H, Li T, Zou J. Analysis of dynamic mechanism of rural transformation development in typical regions of China. Econ Geogr. 2011;31(12):2080–5.

Long H, Li Y, Liu Y, et al. Accelerated restructuring in rural China fueled by 'increasing vs. decreasing balance' land-use policy for dealing with hollowed villages. Land Use Policy. 2012;29(1):11–22.

Long H, Liu Y, Zhang X, et al. Recent progress in agricultural geography and rural development research. Acta Geogr Sin. 2014;69(8):1145–58.

Lu M, Chen Z. Urbanization, urban-biased economic policies and urban-rural inequality. Econ Res. 2004;6:50–8.

Lu D, Chen M. Several viewpoints on the background of compiling the "national new urbanization planning (2014–2020)." Acta Geogr Sin. 2015;70(2):179–85.

Roger C, Yao S. Urbanization and sustainable metropolitan development in China: Patterns problems and prospects. GeoJ. 1999;49:269–77.

She Z. The rural areas experiences conflicts under the urbanization process and its development issue. China Environ Manag. 2015;7(3):61–6.

Shi Z, Yang Y. Social status compensation through transnational practices: A comparative study of cultural remittances between two emigrant groups from South China. Soc Sci Res. 2012;3:157–81.

Wang J, Lin C, Chen Y, et al. Cultivated land pollution at township level in China: situation, factors and measures. China Land Sci. 2012;26(2):25–30.

Wen T, Yang S. Rural governance and rural development under the context of changing socio-economic structure in China. Theor Invest. 2012;6:76–80.

Woods M. Rural geography: blurring boundaries and making connections. Prog Hum Geogr. 2009;33(6):849–58.

2 Research Progress and Practical Enlightenment of Urban–Rural Transformation

Abstract

Regarding urban–rural relationship, scholars have deeply studied the elements, relevance and transformation of urban and rural areas from the perspectives such as urban geography, rural geography and urban–rural development. These studies have highlighted the dynamic evolutionary process of the structures and patterns of urban–rural relationship, depicted the processes and patterns of urban–rural separation, urban–rural integration and urban–rural coordination, and put forward the viewpoint of comprehensive geography, that is, the coupling of urban–rural transformation patterns and processes. This chapter is mainly a literature review, interspersed with the process analysis, theoretical evaluation and practical aspects of rural development and urban–rural transformation. To further explore and innovate the geographical theory, method and technology of urban–rural transformation, this chapter also puts forward the problems, strategic needs and frontier fields of urban–rural transformation.

2.1 Research Progress on Urban–Rural Transformation

2.1.1 Research Progress on the Non-agriculturalization of Urban–Rural Elements

With the development of industrialization and urbanization, China has transformed from a traditional agricultural society to a modern industrial society; thus, the non-agriculturalization of production factors has become an inevitable trend. Production factors essentially refers to the sum of all resources invested in social and economic activities, including land, capital, labor, technology, and entrepreneurial resources. The non-agriculturalization of rural production factors refers to the flow of land, capital, labor and other production factors from the agricultural sector to the nonagricultural sector, that is, the process of reallocating and repositioning agricultural production factors to nonagricultural industries, which is a kind of directional element flow (Qu et al. 2007). Currently, rural labor force

Y. Liu, *Urban-Rural Transformation Geography*, Sustainable Development Goals Series,
https://doi.org/10.1007/978-981-16-4835-9_2

transfer and agricultural land transfer are the two most important forms of non-agriculturalization of rural elements (Chen 2010). With the progress of science and technology and social development, technology and information have gradually become new urban–rural elements. Social, cultural, environmental, intellectual and other forms of capital have become important variables in determining urban–rural transformation, and the bidirectional flow between urban and rural factors and their relations has become the key link in urban–rural coordination.

1. **Research on the non-agriculturalization of land**

The non-agriculturalization of land refers to the process of transforming agricultural land into urban and rural construction land for such purposes as residence, transportation, industry and services (Zhang and Jia 2001). It is usually due to urban expansion, transportation needs, the construction of parks and other facilities, rural collective construction and rural individual housing, occupying which promotes an increase in construction land and a decrease in cultivated land area. In the process of rapid urbanization, the continuous increase in construction land and the continuous decrease in agricultural land has been an irreversible trend for a relatively long time (Liu and Lu 2007). The occupation of agricultural land by urban construction is an important manifestation of nonagricultural land use. Since the reform and opening up, urban industrial and mining land have expanded rapidly in China's Huang-Huai-Hai region, the southeastern coastal areas, the middle reaches of the Yangtze River, the Sichuan Basin and other densely populated, economically developed areas (Liu et al. 2014a), especially in the Bohai Rim, Yangtze River Delta, and Pearl River Delta (Shi et al. 2000; Yang 2001; Li et al. 2003; Liu and Yang 2015). Studies on some typical urban areas have shown that farmland and orchard land around cities have decreased significantly, and urban land along coasts and main highways has increased rapidly (Gu et al. 1993). These changes show the trend of the decreasing intensity of nonagricultural transformation from urban centers to rural areas (Li and Zhang 2013).

Urbanization, agricultural labor force changes and government ecological protection policies are the most important factors affecting land-use changes in China (Li 2009). In the process of industrialization and urbanization, population growth, the influx of foreign capital, the development of the real estate industry and the improvement of living conditions have promoted urban space expansion, which is the main reason for cultivated land non-agriculturalization (Tan et al. 2004; Shi et al. 2000). That is, the process of land conversion is deeply affected by multiple mechanisms, such as nature, the market, social values and political rights. Additionally, there have been great differences in the urban spatial expansion mode and expansion direction in different periods (Gu 1999; Liu et al. 2000; Li et al. 2003). The impact intensity of industrialization and urbanization determines the speed and regional differences in the land non-agriculturalization scale.

Studies have shown that the expansion of rural construction land is also an important form of land non-agriculturalization. The coupling analysis of cultivated land and homesteads from the country perspective shows that the area change of cultivated land and rural homesteads in China presents an obvious negative correlation, which means that rural homestead expansion is an important reason for cultivated land non-agriculturalization (Liu 2011; Long 2012). Based on village research in typical plain agricultural areas, the increase in the conversion of village construction land to homestead and threshing land has three stages: gradual expansion, mutation expansion and filling supplementary expansion. The great contrast between the internal conditions and the external environment of the village is the external driving force for the outward expansion of village land. Meanwhile, the increase in farmers' income and the lack of land planning and management also contribute to the expansion of village areas and the occupation scale (Wang et al. 2010).

China has implemented two landownership systems: state-owned land and collective-owned land. Due to unclear land property rights, lax planning and management, insufficient understanding of cultivated land value, unreasonable distribution of urban–rural land income, and unsound price mechanisms, arable land has long been occupied by nonagricultural activities such as industrial construction and urban expansion (Cai and Yu 2004; Qu et al. 2004; Long 2012). Therefore, how to effectively protect arable land and ecological land and coordinate the relationship among different needs, such as food production, economic construction and ecological land protection, has become an important topic in promoting the sustainable use of land. In addition to its economic value, agricultural land has functions such as ecological services, social security, and food security, and the simple market regulation mechanism lacks consideration of the externalities of agricultural land, making it difficult to achieve the optimal allocation of land resources. To prevent the rapid non-agriculturalization of agricultural land, there is an urgent need to establish an economic mechanism that improves the comparative efficiency of agricultural production, improves government regulation and policy control mechanisms, strengthens the concept of sustainable development of human–earth coordination, effectively protects the quantity and quality of cultivated land, and achieves the goal of efficient and intensive land use (Cai 2001; Zhang and Jia 2001; Liu et al. 2011). At the level of national development strategy, we will take the quantity, quality, ecology, space and time of farmland non-agriculturalization into account, innovate and form a compensation system for farmland protection value with Chinese characteristics, and improve the balance system of cultivated land occupation and compensation, which is an effective way to scientifically coordinate and rationally use cultivated land (Liu and Qiao 2014).

2. **Research on labor transfer and non-agriculturalization**

In the process of urbanization, the transfer of the rural labor force to cities and nonagricultural industries is a common feature of economic development worldwide. Compared with Western Europe, United States, Japan and other developed countries that have completed the transfer of the agricultural surplus labor force, most developing countries still have a rural labor force in the stage of rapid transfer (Li and Wang 2004). Since the reform and opening up of China, the rural household responsibility system has been implemented, which has greatly improved the efficiency of agricultural production; in this process, the rapid development of township enterprises has absorbed a large amount of agricultural surplus labor. With the deepening reform of China's household registration system and the development of the country's economy, the surplus rural labor force has been rapidly transferred to nonagricultural industries and urban areas, mainly including transfer to local township enterprises, transfer to provincial towns and cities, and inter-provincial and transregional transfer.

The transfer and employment of the surplus rural labor force is deeply affected by the urban–rural income gap, expected income, personal characteristics, employment scale agglomeration effect and urban pulling force (Michael and Todaro 1969; Wang 2005; Cheng et al. 2006). The "push–pull theory" of population migration holds the view that people can improve their living conditions through mobile employment, and favorable or unfavorable economic and social factors become the pulling or pushing force. This theory actually emphasizes the comprehensive effect of social, economic and institutional factors. There are different degrees of resistance and friction in the transfer of the agricultural surplus labor force, including individual factors such as transfer cost and psychological cost, as well as the interindustry supply and demand structure and institutional, policy and conceptual social factors (Jing and Chen 2004).

How to overcome the resistance and friction of the transfer of the rural surplus labor force to cities has become an important topic of urban–rural transformation research. There are great differences in the transfer mode of the rural surplus labor force in different regions. Most of the rural labor force that transfers to the city is engaged in dangerous, difficult and dirty work in industries such as construction, manufacturing, mining, and catering services (Hu 2007). To realize healthy urbanization, we must eliminate the institutional restrictions that affect the rational transfer of farmers from the aspects of employment, systems and urban development, and the key point is to change discriminatory policies on the transformation of the rural labor force (Wu 2002; Wang 2005). At the same time, we also need to accelerate the construction of small towns in rural areas by increasing government financial support (Zhou 2001).

3. **Research on the regional effect of the non-agriculturalization of elements**

Land has many functions, such as the means of production, living space, ecological services, and social security. Land non-agriculturalization has an important impact on regional agricultural production, farmer employment, economic development, etc. In terms of agricultural production, the non-agriculturalization of land has caused the proportion of landless farmers to increase, the degree of agricultural land-use agglomeration to decrease, and farmers' concurrent operation, which has held a great impact on the stability of food production and agricultural transformation. In rural society, these changes are mainly reflected in farmers losing land, such as farmers' loss of income, institutional arrangements, and rights and interests. The impact of land non-agriculturalization on rural areas is closely related to the stage of regional development. For developing countries, urban expansion and farmers' employment transfer provide opportunities for rural prosperity, while excessive occupation of cultivated land leads to inefficient land use, ecological environmental damage, and damage farmers' rights and interests, thus threatening national food security.

The impact of land non-agriculturalization on grain production and its relationship with economic growth have become focal topics of academic research (Chen and Qu 2006). Qu and Wu (2004) pointed out that there is an inverted "U"-shaped relationship between economic growth and cultivated land quantity. When the economy develops to a certain extent, the loss rate of cultivated land gradually decreases. With the advancement of the economic development stage, the contribution of construction land expansion to economic growth gradually decreases, and the intensive growth mode will alleviate the expansion of construction land to a certain extent (Jiang et al. 2009). Also, some scholars pointed out that the occupation of cultivated land by construction is only one of the factors affecting GDP, and the contribution to economic growth comes mainly from the increase in land, labor and capital investment in economic growth caused by cultivated land occupation. During the transformation of economic growth, the phenomenon of "decoupling" between economic growth and cultivated land occupation will appear. An empirical study has shown that the increase in the proportion of the tertiary industry slows the speed of land non-agriculturalization (Liu et al. 2014b). As in developed countries, the contradiction between economic development and land resource protection is reduced, and the process of land non-agriculturalization is related more to environmental impact and landscape ecological maintenance.

The non-agriculturalization of the labor force is also an inevitable trend of urban and rural economic and social development, and it is the process of the reallocation of labor factors. According to the theory of human capital, the flow of labor from areas with poor wages to areas with higher wages will inevitably promote the optimal allocation of social resources, improve the income of migrants and accelerate the economic growth of the whole country. The marginal productivity of the rural labor force is obviously improved by the shift from agriculture to nonagricultural industries. In the long run, the transfer of the rural labor force based on the continuous improvement of rural labor quality plays an important role in promoting economic growth and urban–rural transformation.

In the literature, the regional effect studies of element non-agriculturalization mainly analyze the impact of labor transfer on rural sustainable development in terms of structural adjustment, labor surplus and rural social capital. Due to the differences in the temporal and spatial scales, the research ideas of the impact of labor transfer on rural comprehensive development differ, which results in inconsistent conclusions (Chen 2013). On the one hand, labor transfer increases income, stimulates rural consumption and increases rural employment opportunities. Foreign empirical studies have shown that the positive effect of most rural population migration is greater than the negative effect, and rural population migration has promoted the development of the rural economy in Britain (Hodge and Monk 2004). Relevant domestic studies have also confirmed that rural population transfer alleviates the contradiction between rural people and land and improves rural development capacity (Lu and Yang 2008; Liu 2007). On the other hand, the transfer of the labor force leads to the lack of the main body of the rural labor force, the destruction of rural social networks, and the aggravation of rural hollowing. Under normal circumstances, the rural capable labor force is first transferred out, and the loss of the rural young and middle-aged labor force causes the lack of rural development subjects. Although migrant remittances can improve the income level of farmers, they are used mainly for the rural construction of new houses and the marriage of sons, which has a limited effect on the updating of production systems and economic growth in rural areas and will expand the income gap between backward and developed areas (Lipton 1980). The hollowing of rural areas has broken the traditional social division of the labor system in rural areas, and the rural governance system is facing difficulties that will damage to rural sustainable development (Ma et al. 2004; Liu et al. 2009).

2.1.2 Research Progress on Urban–Rural Integration

Most development theories assume that developing countries must undergo the process of transition from underdeveloped to developed; that is, the process of economic growth and modernization in developing countries must transfer the surplus labor force from the backward agricultural sector to the urban industrial sector to adjust rural resources, labor force and capital. The comprehensiveness and differences between the urban and rural regional systems determine the flow of population, products, capital and other elements between urban and rural areas, and there are material, energy and information connections as well as close economic, social, population and service links (Table 2.1). It is precisely because of these connections that urban and rural regional systems are connected. Preston (1975) divided urban–rural interaction into human movement, commodity movement, capital movement, social transaction, administration and service supply. He pointed out that informal social contact between urban and rural areas and the dissemination of cash, technology and management experience were also important. The interaction between urban and rural areas is the two-way flow of people, goods, technology, money, information and ideas between urban and rural areas. These flows are characteristics not only of the development process but also of the countryside and the city themselves. They form a variety of flow fields, such as flows of people, logistics, information, concepts, and capital. All kinds of flows move in a certain direction in a certain environment. McGee (1989) analyzed the connotations and modes of urban–rural interaction from the perspective of labor mobility. He considered that the interaction between urban and rural areas was manifested as the seasonal migration of the labor force, the flow of the labor force from the rural sector to the urban sector, the flow of the well-educated or those forced to leave the land from the agricultural sector to the urban sector, the flow from the urban sector to the agricultural sector, and the free flow between the rural sector and urban informal sector. Currently, nearly 300 million migrant workers travel to cities or return to their hometowns, and urban industrial and commercial capital and enterprises move to the countryside to start separate businesses, which is also the dynamic allocation of urban and rural labor elements. Although the

Table 2.1 Academic views of scholars outside China on urban–rural linkages and elements

Linkage type	Element composition
Material linkage	Roads, railways, communications and other infrastructure access, river basins, water networks, and ecosystem and other natural links
Economic linkage	Market form, flow of raw materials and intermediate products, capital flow, forward and backward vertical connection and income transfer of industrial chain, interdepartmental connection of enterprises, interregional production factors, commodity flow and free trade
Social linkage	Kinship, religious behavior, social group, scientific plan, visiting contact
Population linkage	Population flow and migration, commuting, urbanization population transfer, enterprise and entrepreneur alliance
Service linkage	Financial, network, education, training, medical, vocational, business, and technical service contacts
Technical linkage	Interindustry technical connection, agricultural production technology, communication technology, internet information technology
Organizational linkage	The relationship of power entrustment, government budget and investment, government management and informal decision making

Note Modified from Ma and Hu (1993)

nonagriculturalization of rural population is still the dominant factor, the movement of urban industrial and commercial enterprises and capital to the countryside is becoming a new driving force for the interaction between industry and agriculture and the integration of urban and rural areas in the new environment of "double innocations".

The systematic theory of urban–rural connection began in the 1980s. For example, on the basis of previous studies, Unwin (2017) initially constructed the theoretical analytical framework of "urban–rural connection, flow and interaction". He classified the various elements in urban–rural flow hierarchically and established the corresponding relations of urban–rural connection, urban–rural flow and urban–rural interaction. With globalization, technological innovation and management mode modernization, the scope of urban and rural element flow has been expanding. In the late 1990s, Douglass (1998a) further abstracted the theoretical analytical framework of "urban–rural connection, flow and interaction" as the "location development network mode" and proposed five "flows" of population, production, commodity, capital and information, and different spatial connection modes between them.

The cognition of urban–rural connections has received close attention in academic circles. Potter and Unwin (2017) believed that urban and rural areas are in a symbiotic relationship, and communication between the two is inevitable. The traditional interaction between urban and rural areas is the interaction between economy and population. With economic globalization and the reorganization of global political forces, the elements and forms of interaction between urban and rural areas will fundamentally change. According to Tacoli (1998), in addition to spatial connections between elements, such as the flow of population, products, funds, information and waste, there are also links between urban and rural sectors, such as the development of rural activities in urban areas and the operation of production and services with urban characteristics in rural areas (Table 2.2). Population migration between urban and rural areas is an important manifestation of urban–rural connections. In the process of urbanization, the population mainly migrates from rural to urban areas; however, migration from urban to rural areas is also increasing. In view of the problems of rapid urban growth and urban management, the government often creates policies to control population migration. In China, the household registration system and population policy are often used to limit the rural-to-urban flow and spatial agglomeration of the rural population. The relationship between urban and rural areas is deeply affected by the macroenvironment; for example,

Table 2.2 Urban–rural linkages and interactions

Urban areas		Rural areas
Agricultural product trading and transportation center	←------→	Agricultural production
Agricultural productive services (pesticides, fertilizers, maintenance, information, technology, etc.)	←------→	Agricultural intensification (renewal of rural facilities and agricultural technology)
Nonagricultural consumption market (agricultural product processing, private services, public services)	←------→	Income and demand for nonagricultural products and services
Agricultural product processing industry	←------→	Economic crops and agricultural diversification
Nonagricultural employment	←------→	All of the above

structural adjustment and economic reform, urban and rural populations, and employment are all affected (Potts and Mutambirwa 1998). Compared with rural functions and roles, most cities are the result of urban–rural interaction rather than one-way causality (Douglass 1998b). Cities and towns are market centers for the trade and flow of surrounding agricultural and rural commodities, on the one hand, urban areas need the supply of agricultural products from rural areas; on the other hand, rural need the products and services from urban areas (Table 2.2).

The gradual transition between urban and rural spaces and the objective existence of the role of small and medium-sized towns between urban and rural areas also reflect some limitations of urban–rural duality. Tacoli et al. (1998), UN-Habitat experts on urban and rural issues, noted that many development theories and practices focus on urban or rural development issues and ignore the relationship between them. It seems inevitable that urban and rural areas will be distinguished in order to describe the relationship between them. However, it is arbitrary to divide urban and rural areas into two parts that are closely related to regional space and industrial sectors. The ambiguity of urban boundaries, short-term and seasonal population migrations, separation of urban and rural employment, and nonagricultural employment of the rural population pose challenges to the urban–rural duality. According to the basic idea of "nature-space-human system design", the representative functions of cities are culture, modernity, entertainment and diversity, while the representative functions of rural areas are nature, space, interest and tradition (Takuro 1990). In view of the dualistic division of urban and rural areas in development theory and planning, Douglass (1998b) proposed a regional network mode to combine the relationship and interdependence between urban and rural areas. With the increasingly close relationship between urban and rural areas and the rapid development of small towns, the concepts of urban areas, urban outskirts and urbanization of rural areas have been proposed to describe the geographical areas between urban and rural areas and explain the dynamic relationship between urban and rural areas (Long and Zhang 2012). The transformation of urban and rural development should strengthen the relationship and interaction between urban and rural areas rather than separate and independent research on urban and rural areas (McGee 2008).

At present, in addition to industrialization, the driving force of urbanization in developing countries is also affected by a series of macro-situations, such as globalization, informatization and marketization. This is a new situation that has never previously been experienced in the urbanization process of developed countries and will have a certain impact on the flow and connection of urban and rural elements in developing countries (Fig. 2.1). Taking globalization as an example, the traditional urban–rural relationship is represented by the communication and exchange of population, commodities, currency and information between the urban and rural economies. In the context of globalization, the urban and rural economies of a country are both cooperative and independent and are connected with the world economy through foreign trade and in other interactive ways (Knox and McCarthy 2005).

2.1.3 Research Progress on Rural Transformation and Hollowing

1. **Research on rural transformation**

China is a traditional agricultural country with a large population. Agriculture, rural areas and farmers have long been dominant in national economic and social development. Since the 1990s, China's rapid industrialization and urbanization have intensified the urban–rural population flow and the reorganization and interaction of economic and social development factors. Therefore, rural participants respond and adjust to these effects and changes and then

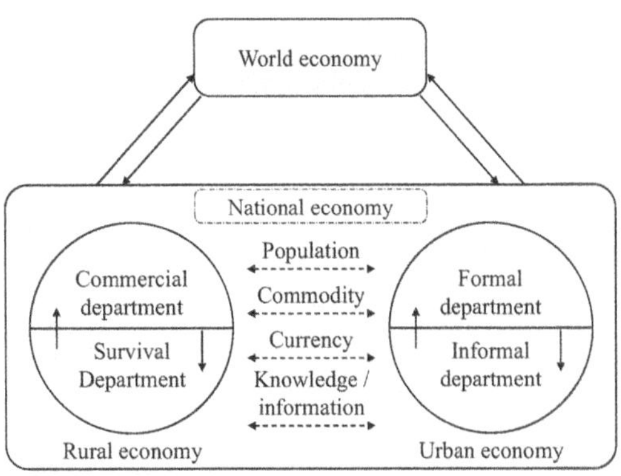

Fig. 2.1 The interaction between urban and rural economies and departments in less developed countries against the background of globalization

promote the reconstruction of economic and social morphology and regional spatial patterns in rural areas (Long 2012). Rural transformation are the regional responses of rural areas to economic growth and urbanization. They are manifested mainly in the development of agricultural industrialization, the gradual increase in farmers' incomes, the enhancement of access to social services, and changes in rural spatial structure and function, which constitute an important part of rural transformation.

The existing research focuses on rural settlement changes, rural development patterns, rurality, rural spatial systems, rural development types, spatial organization, etc. (Liu 2007; Chen and Zhang 2008; Long et al. 2009a; Zhang et al. 2009; Li et al. 2011). The mainstream view is that rural transformation is reflection of the comprehensive process of rural regional system transformation and has the characteristics of multidimensional, multi-objective and nonlinear factors. Long et al. (2011) constructed an evaluation index system to measure rural transformation in three dimensions: rural development degree, rural transformation degree and urban–rural coordination degree. The results showed that regional differences in China's rural transformation are obvious: from 2000 to 2008, the coordination between urban and rural areas in most areas decreased, the development of rural areas lagged behind that of urban areas, and rural transformation in some areas was divorced from the foundation of rural economic and social development.

Rural transformation is affected and restricted by terrain conditions, resource endowment, human capital, economic base, industry non-agriculturalization, urbanization policies and other factors (Zhang et al. 2010). The driving mechanism can be divided into three types: endogenous, exogenous and internal/external combination. At present, it is generally believed that rural transformation is result of the comprehensive effect of local and external forces (Terluin 2003). Zhang and Liu (2008) took the lead in studying the dynamic mechanism of China's rural development. They believed that the comprehensive capacity of rural development depends on rural self-capacity and the foreign aid driving force of industrialization and urbanization. Rural development must rely on the favorable external conditions of industrialization and urbanization, cultivate and strengthen rural self-development ability, accelerate the development of rural nonagricultural industries and modern agriculture, and pay attention to the efficient utilization of resources and rural environmental protection. In the accelerated stage of urbanization in China, rural capital, labor, land and other elements continue to flow to urban areas. This unidirectional and selective factor flow is usually not conducive to rural development. At the same time, the status of the agricultural industry in the national economy continues to decline, which makes the status of rural areas among urban and rural areas decline. The research shows that rural transformation in different regions is driven to different degrees economically and socially. For example, industrialization and urbanization in developed areas of eastern China have the most obvious driving effect on rural transformation (Wang et al. 2016).

In the period of late industrialization and post-industrialization, the surplus agricultural production capacity, the improvement of resource protection awareness, and the improvement of the rural landscape consumption demand as well as the transfer of urban capital and residents to rural areas accelerated the functional transformation of rural areas, forming a market-driven type of rural regional multifunctional evolution (Holmes 2008). Rural areas have evolved from the leading function of the agricultural economy to multifunctional spaces of agricultural production, culture, society and environment. To meet social needs, more attention is paid to organic agricultural production, rural environment and landscape, rural employment and services, showing that rural areas have entered a stage of benign development (Hall et al. 2004). In highly urbanized Western European countries, the development of rural areas has gradually integrated many functions, such as consumption, ecological services, agricultural production, and economic facilities, with agricultural economics and agricultural employees accounting for only a small proportion.

2. **Research progress on rural hollowing**

Rural hollowing is the continuous outflow of the rural population, industrial recession, weakening of rural functions and other adverse regional evolutionary processes. Due to the complexity and comprehensiveness of rural development, there are great differences in development trends. In some rural areas, the relationship between humans and land has changed rapidly, enabling them to achieve economic growth, structural transformation, affluence and environmental optimization. However, in most rural areas, such as plains areas, hilly areas, loess areas and underdeveloped areas, with the continuous outflow of the population, problems such as the aging and weakening of the population, land depletion, and industrial development lag have appeared, showing the rural hollowing phenomenon (Cheng et al. 2001; Xue 2001; Li and Li 2008; Liu et al. 2009; Long et al. 2009b; Qiao et al. 2011).

Cheng et al. (2001) believed that rural settlement hollowing is a process of farmers moving towards the areas surrounding the original settlements, resulting in the vacancy and collapse of the original settlements and gradually forming a "hollowing village". Xue (2001) believed that the hollowing village is the manifestation of the material form of the village in the complex social and economic evolution of China. It is the phenomenon of differentiating the spatial form of the village, with extensive development and internal decline caused by the contradiction between the rapid development of village construction and the backward planning and management system under the condition of urbanization lagging behind nonagricultural development. Liu and Liu (2010) summarized and put forward the scientific definition of rural hollowing based on existing researches, that is, in the process of urban and rural development and transformation, rural hollowing is caused by the nonagricultural transformation of the rural population, and the homestead is generally built without demolishing the old but expanding outward, resulting in the expansion of the scale of village construction land and the aggravation of idle and abandoned land, which is a kind of bad evolutionary process of "external expansion and internal hollowing". When the rural system continues to decline and production and living functions are seriously lost, the hollowing village comes into being. Rural areas are deeply affected by the flow of factors of production in different regions and stages and differences in the external environment. The process of rural hollowing is characterized by an obvious life cycle (Liu et al. 2009) and usually goes through five stages: emergence, growth, prosperity, stability and recession/transformation (Fig. 2.2).

Rural settlement hollowing, which appears to be the continuous vacancy and abandonment of rural construction land, actually reflects the dynamic evolution of the rural human-earth relationship in terms of economic and social changes. Space demand, number of families, economic income and land policy constitute the potential conditions for rural settlement hollowing, while low construction cost, low moving cost and low land cost constitute the driving forces of hollowing (Cheng et al. 2001). The objective factors of imbalance between centripetal and centrifugal force in villages, the subjective environmental factors such as economic development and backwardness, and the link between housing demand and planning lag together constitute the driving forces of rural hollowing (Wang 2005). Its essence is the combined effect of natural, economic, social, cultural and institutional factors in a rural area, leading to a process and representation of changes in the spatial pattern of land use within the village (Long et al. 2009b; Wang et al. 2010). Rural hollowing causes serious waste and inefficient use of land resources, scattered residential layout and rural landscape damage, which are not conducive to the construction of a supporting infrastructure. In addition, the aging population and lack of industrial developmental power seriously hinder the transformation and sustainable development of the rural economy and society.

In view of the increasing hollowing of rural areas and the increase in rural residential land and other problems, Chen et al. (2012) proposed a comprehensive renovation mode for hollowing villages, such as leading urbanization, central village integration and intensive villages, based on a survey of typical villages and farmers' questionnaires. Liu et al. (2011) put forward the regional differentiation mode, urban–rural integration mode, comprehensive mode of "one renovation, three returns" and coordinated decision-making mode of rural land consolidation, combined rural hollowing renovation with the new countryside construction, and promoted the "triple integration" of space, organization and

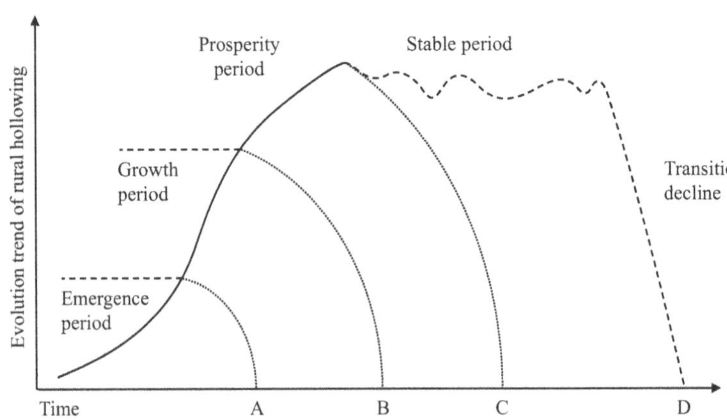

Fig. 2.2 The evolutionary cycle of rural hollowing (Liu et al. 2009)

industry of the rural regional system through strong governmental regulation and institutional reform (Fig. 2.3). Long (2013) proposed to reconstruct rural production, living and ecological space through agricultural land consolidation, "hollowing village" renovation and industrial and mining land consolidation. Since rural hollowing is a regional manifestation of rural transformation in the process of industrialization and urbanization, research on this topic should be placed in the overall regional framework, focus on scientific planning and regional promotion, optimize urban and rural land-use structure and improve land-use efficiency, and effectively promote regional urban–rural interaction, integration and coordinated development (Liu 2007).

2.1.4 Research on the Measurement and Evaluation of Urban–Rural Transformation

The transformation of urban and rural development is essentially the process and result of material exchange and factor interaction between urban and rural regional systems. The measurement of urban–rural transformation is the basis for identifying transitional stages and types and has become an important frontier issue in the study of the urban–rural relationship. In recent years, domestic and foreign studies have focused on the comprehensive evaluation of urban–rural relevance, coordination and integration, and measurements based on population, industry and land. In particular, multidimensional systematic diagnosis provides a scientific basis for revealing the situation, stage, pattern and mechanism of urban and rural development.

1. **Quantitative evaluation of the urban–rural relationship and interaction**

Since the founding of PRC, the urban–rural contact method has presented obvious stage characteristics, showing unbalanced orientation, nonmarket regulation, a state of non-openness, and unfair consequences (Hu and Ma 1993). Lu et al (2004) thought that there were various related infrastructure links between urban and rural areas, such as transportation and communication, as well as material, human, capital, technology, information and other element links formed by various facilities as carriers, which produce spatial and functional linkages in urban–rural areas. Spatial linkages refer to the network structure and infrastructure connection of the urban system, and functional linkages refer to the economic and social links between urban and rural areas in regional development. Zeng et al. (2002) used natural foundations, urban systems and infrastructure to reflect spatial connections, and economic and social connections to reflect functional connections. Then they conducted a static evaluation of the urban–rural correlation degree of 31 provinces in 2000, revealing the factor circulation between the two open systems of urban and rural areas and the actual development level of urban–rural relationship. The two-way flow of social and economic elements such as capital, labor force, material and information

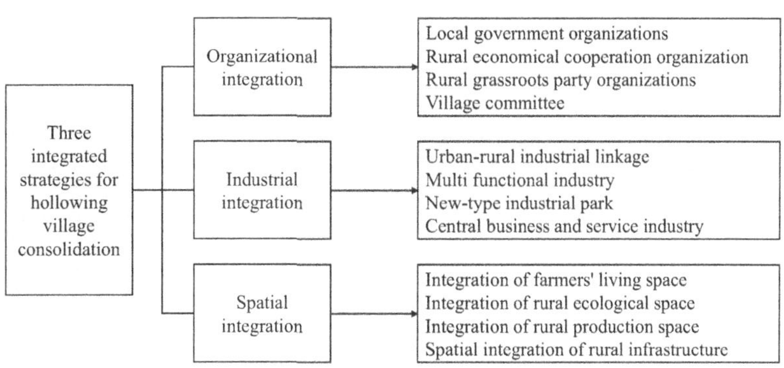

Fig. 2.3 The "three integration" regulation path for rural hollowing and new countryside construction

in urban and rural spaces forms the interactive development between urban and rural areas (Duan et al. 2006).

Luo and Li (2005) described urban–rural relationship by measuring the quality and quantity of the spatial medium that promotes the formation of the "flow" of various elements between urban and rural areas as well as the actual velocity and direction of factor "flow". Based on the urban–rural relationship theory, Chen (2005) performed a comprehensive analysis on the theoretical and practical level of the urban spatial restructuring of Hengyang city from the aspects of constructing new industrial relations, optimizing the spatial relationship and hierarchical scale structure of regional cities and towns, optimizing the urban–rural channel system, straightening out the urban–rural "flow" and exploring a new urban–rural spatial structure mode. Based on the theory of urban–rural relationship development and evolution, Chen (2010) focused on the urban morphology of Changchun city and proposed urban optimization measures for urban and rural development. Cheng et al (2010) used the spatial correlation analysis method to analyze and study the spatial relationship between urban and rural areas in the main grain-producing areas of central Jilin Province.

2. **Quantitative evaluation of urban–rural integration**

The integration of urban and rural areas is the advanced stage of the development of urban–rural relations and the goal of the evolution of urban–rural relations in the region. There is no consensus on the concept and connotation of urban–rural integration. Li and She (2000) believed that urban–rural integration refers to the rational flow, mutual cooperation, complementary advantages of resources, technology, capital, labor force and other production factors between urban and rural areas, realizing the optimal allocation of urban and rural resources and sustainable coordination of urban and rural economic social development through using urban development to drive rural development and using rural progress to promote urban progress. The connotations are reflected mainly in the following aspects.

The first is urban and rural spatial integration, that is, urban and rural regional entities form a continuous, unified, networked, multimode and permeable "regional complex"; the second is the complementary function and structure, integration does not mean the disappearance of the center-periphery economy but means only that they are interdependent; the third is the urban and rural infrastructure sharing, fast transportation and information networks make the production and life interactions of urban and rural residents convenient and close; the fourth is the highly smooth flow of urban and rural people, capital, technology, information and other elements; the fifth is the coordinated urban and rural ecological environments and the improved living standards of urban and rural residents (Huang et al. 2009). Wang et al. (2005a) held that the connotations, level and structure of urban and rural areas include four aspects. First, urban–rural integration is the scientific development concept in the period of China's comprehensive economic and social transformation. Second, urban–rural integration is the development strategy of China's overall promotion of modernization. Third, urban–rural integration is the development process of modern urban and rural convergence. Fourth, urban–rural integration is the modernization of the country and region, a set of development goals that are mutually undertaken in the same development stage. In a broader sense, urban–rural integration includes institutional integration, industrial structural integration, agricultural commercialization and farmers' citizenization as well as spatial integration, population integration, economic integration, market integration, ecological integration, social integration and system integration (Jiang 2004). There are six characteristics of urban–rural integration, i.e., long-term, holistic, regional, mutual mobility, bidirectional and universality (Deng 2001).

Many scholars have carried out research and in-depth analysis on the comprehensive measurement, dynamic mechanism, constraint factors and countermeasures of urban–rural integrated development and have achieved rich research results. Liu et al. (2012) evaluated the overall change in coordinated development between urban and rural areas in 31 provinces of China from 2000 to 2008 from three aspects of urban–rural economic ties, urban–rural economic growth and urban–rural economic differences and explored the causes. According to the principles of integration, scientificity, hierarchy and operability, the evaluation mode of the urban–rural correlation level was established, and a comprehensive evaluation of the urban–rural correlation degree in typical provinces was carried out. Zhan et al. (2003) evaluated the correlation between urban and rural areas in Shandong Province. Xia et al. (2003) thought that the coordination of urban and rural development should be reflected in the collaborative process of urban–rural differences and complementary development; only in this way can we achieve urban–rural integration. Zeng and Zhu (2002) believed that urban–rural differences remain within a socially tolerable limit, and the coordination of urban and rural development is the result of an unbalanced development process and tendency to reach equilibrium. Chen and Wu (1995) analyzed the role and significance of urban–rural coordination in the process of urbanization. From the perspective of "rural spatial change", Miao (1999) analyzed the influence of rural industrialization on the formation and development of new rural economic space, new rural social space and new rural geographical space, and studied the regional imbalance between rural industrialization and urban–rural transformation.

Zeng et al. (2002) believed that the essential attribute of urban–rural integration is the free flow of resources, economy, society and other elements between urban and rural areas, and the flow and speed depend on the degree of connection between urban and rural areas. Zeng (2001) believed that complex network fossils in urban and rural areas were an important way to realize the urban–rural integration. On the basis of constructing an evaluation index system of urban–rural coordination, Wang et al. (2009) comprehensively evaluated the static and dynamic coordination degree of urban–rural interaction in Shandong Province and Shanxi Province. From the perspective of the elements flow from the edge to the center, this paper explains the mechanism of urban–rural transformation: various elements flow from the edge (rural) to the city (center), which promotes the development of urbanization and industrialization and drives the urban–rural transformation (Zhang 1998). Xiu et al. (2004) proposed that the indicators of economic development level, rural non-agriculturalization level, social equity and welfare, transportation and daily contact constitute the evaluation indicators of urban–rural integration to comprehensively evaluate the process of urban–rural integration in northeast China. Ma et al. (2006) analyzed the evolution of the connotation of the urban–rural relationship from the perspectives of the urban–rural natural ecological relationship, economic relationship and social and cultural relationship on the basis of summarizing the theoretical progress of urban–rural relationship research in China. In addition, he analyzed the dynamic mechanism, mode, channel and regional impact of urban–rural interactions and elaborated the importance of system construction to the development of urban–rural relationships.

3. **Quantitative evaluation of urban–rural structural transformation**

The transformation of urban and rural structures includes population urbanization, industrial non-agriculturalization, land non-agriculturalization, employment non-agriculturalization and other aspects. Previous studies have described and measured these issues from different perspectives.

From the perspective of urban spatial expansion, the characteristics of urban and rural spatial form transformation are revealed (Wang and Duan 2010; Xie 2009). Liu and Chen (2012) used buffer zone and quadrant quantile analytical methods to research the regularity of urban space dispersion and concentration and to reveal the spatial pattern characteristics of urban growth. Taking Daqing city as an example, Wang et al.

(2012) analyzed the spatial characteristics of urban spatial expansion speed and intensity, urban spatial form compactness, a fractal dimension index, and urban growth rationality. Chen et al. (2010) measured the level of urbanization from two aspects, population urbanization and land urbanization, and revealed the transformation of urban and rural development from the coordinated development status of the two. Dai et al. (2010) comprehensively analyzed the internal motivation of urban spatial expansion from the perspective of economics and planning theory. Zhou and He (2006) comprehensively analyzed urban land-use structures, an expansion intensity index, a population elasticity index, an economic elasticity index, and a compactness change index and systematically analyzed urban land expansion from a time series and spatial structure evolutionary perspective. In addition, scholars have used GIS technology to analyze and study the spatial development of land urbanization (Lu et al. 2010).

From the perspective of the coupling relationship of population, industry and land nonagriculturalization, Li et al. (2012) decomposed urbanization into population urbanization, economic urbanization, land urbanization and social urbanization and believed that the subdimensions should be coordinated with each other. Taking the Pearl River Delta as an example, Zhang et al. (2003) analyzed the coupling relationship among land-use change, industrialization and urbanization and concluded that the coupling coefficient of industrial nonagricultural, urbanization and land-use change can be used to evaluate the types of land-use change. In the process of urbanization and industrialization, demand and industrial growth have always affected the transfer of land urbanization. There is a certain connection among population change, urbanization and industrial structure evolution (Chen et al. 2009; Li et al. 2009; Hu et al. 2009; Chen 2002). Based on the STIRPAT mode, Han et al. (2013) conducted an empirical analysis on the impact of industrial structure on land use by using the feasible generalized least square method (FGLS) and revealed the impact of industrial structure on intensive urban land use. Based on the coupling relationship between industry and land-use structure, this paper studies the interaction between land use and economic development (Ma et al. 2012).

From the perspective of urban and rural regional system evolution, the urban–rural regional system is formed by the coupling and interaction of rural and urban regional systems, including economy, society, population, resources, and environment. Liu et al. (2015), based on the four dimensions of population transformation, industrial transformation, land transformation and social transformation, found that urban and rural development and transformation in the Bohai Rim region showed obvious regional differentiation and stage characteristics.

2.1.5 Research on the Mechanism and Mode of Urban–Rural Transformation

1. **Influencing factors and mechanism of urban–rural transformation**

The transformation of urban–rural development is a process from separation and opposition to integration. Therefore, researches on the dynamic mechanism of promoting urban–rural coordinated and integrated development can be used for reference in urban–rural transformation. Domestic and foreign scholars have studied the dynamic mechanisms of urban and rural development from different perspectives. Canadian scholar McGee (1989) pointed out that the change in the urban–rural relationship in developing countries is not only affected by the external pull of urban radiation and diffusion but also strongly driven by the internal force of rural "non-agriculturalization", both of which promote the increasingly blurred boundary between urban and rural areas. Douglas (1998b) believed that rural areas can promote a virtuous circle of urban–rural ties and promote urban–rural integrated development through a series of "flows" and the external radiation and diffusion of cities. Li and She (2000) believed that the dynamic mechanism of urban–rural integration includes the centripetal force and centrifugal force of central cities, the development of township enterprises and rural industrialization, the development of small towns and rural urbanization, and the industrialization and modernization of agriculture. Zhang (2000) believed that the top-down diffusion mechanism, the bottom-up agglomeration mechanism, the driving mechanism of foreign investment injection, and the natural ecological dynamics mechanism jointly promote interaction between urban and rural areas. The radiation effect of the urban areas, the development of rural urbanization, and the joint role of external forces such as science and technology and administration promote the development of urban–rural linkages (Jiang and Zeng 2004). With the deepening of urbanization, urbanization has a direct impact on the development of towns and even remote villages (Antrop 2004). Hu (2009) divided the dynamic system of urban–rural integrated development into the internal and external dynamic factors and environmental factors.

Top-down and bottom-up diffusion and the combination of the two are important perspectives for analyzing the integration mechanism of rural development and urban–rural development (Terluin 2003). The former emphasizes the role of urban development in promoting rural evolution and regulating the administrative power of the government, with central cities and local governments playing a decisive role. While, the latter emphasizes the role of small-town development and rural industrialization in promoting the transfer of rural surplus labor force and rural economic and social development. The combined type starts from the interaction between urban and rural areas to explore the dynamic mechanism of urban and rural development (Murdoch 2000). Liu et al. (1997) believed that the dynamic mechanism of urban–rural transformation in Shanghai suburbs has three aspects: the mechanism of urban economic diffusion, the mechanism of factor agglomeration, and the external force mechanism of opening up. The mechanism of top-down urban economic diffusion is manifested as the construction of development zones, the radiation and diffusion of urban areas, and the suburbanization of urban industry. The mechanism of bottom-up factor agglomeration is represented by market establishment and adjustment mechanisms such as the labor market, land market, capital market and commodity market.

Different regions have different natural bases, location conditions and economic development stages, so there are great differences in the dynamic mechanism and path of urban–rural integrated development. McGee (1989) believed that Asian developing countries do not repeat the urban-based development mode of Western countries but realize the region-based urban–rural coordinated development mode (i.e., the "desakota" development mode) by mixing agricultural and non-agricultural development in urban–rural areas. However, domestic scholars believe that the "desakota" development mode is applicable only to areas with a large number of cities, strong radiation and developed transportation and thus has certain limitations for the central and western regions of China (Li and She 2000). In the coastal areas with rapid economic development in eastern China, urban–rural integrated development has been promoted by the agglomeration economy of cities (towns), the coordination of industrialization and rural modernization, the linkage of urban and rural factor markets, and the integration of urban and rural infrastructure (Kang 2010). Similarly, macro-policy, investment, market and radiation mechanisms play an important role in urban and rural development and the formation of urban areas (Ning et al. 1998).

In addition to economic and social factors, policy and institutional equity are important ways to promote urban–rural transformation and realize urban–rural integrated development. In particular, such policies should equalize urban and rural public services, provide equal opportunities for acquiring and sharing wealth between urban and rural areas, attach importance to education and training to solve farmers' problems, develop modern agriculture with urban and rural resource interactions, and strengthen independent rural development and characteristic rural space construction (Niu 2009; Gu and Li 2013). It is necessary to innovate the urban and rural element flow system and the urban and rural infrastructure investment system, improve the social security system and rural education system, develop the urban and rural construction land market, establish a coordinated mechanism of urban–rural interaction and industrial and agricultural interaction, and ensure an orderly urban–rural market system (Wang 2004; Luo and Li 2005). Against the background of globalization and regionalization, the urban–rural transformation development should also be based on regional characteristics and local embeddedness, comprehensively consider the transformation of urban–rural production-life-ecology, break the restrictions of the urban–rural dichotomy concept on urban–rural connection to promote urban–rural transformation (McGee 2008), and formulate corresponding policies for different types of urban–rural areas (Long et al 2011).

From a systematic point of view, the dynamic mechanism of urban–rural transformation is affected by a variety of factors and mechanisms. It needs not only the promotion of economic growth, urbanization and social demand but also the role of development strategy, policy and market mechanisms. However, there are great differences in the dynamic mechanisms of urban–rural transformation due to the differences in economic, social and natural conditions among regions. Therefore, to analyze the dynamic mechanisms of urban–rural transformation, we should not only summarize and explain the universal characteristics but also analyze the local characteristics of regional development. The transformation of urban and rural development involves social, economic, ecological, cultural and other aspects. The flow and reorganization of production factors between urban and rural areas is still the basic driving force of urban–rural transformation. Through the market mechanism and government macro-control, we can optimize the resources allocation, and identify the division of urban and rural industries and regional functions. In the new era, the exploration of rural land property rights and values and ecological function values will become the new power to promote urban–rural transformation.

2. **Typical mode of urban–rural transformation**

The development mode is the theoretical summary of the economic structure and economic operational mode with regional characteristics in the process of economic and social development in different regions. Its formation and development are affected by many factors, such as natural resources, economic resources, production conditions, historical traditions and government behavior. It is an effective way to coordinate urban and rural development by establishing the transformation mode according to local conditions. Existing researches on the regional development mode is mostly from rural and regional perspectives. From the perspective of rural areas, according to the mode of economic growth, the modes of urban–rural transformation can be divided into the southern Jiangsu mode, Wenzhou mode, Pearl River Delta mode and other modes. From the perspective of location, there are the suburban agricultural transformation mode, industrial rural development mode, traditional agricultural transformation mode and so on. The modes of urban–rural transformation from the regional perspective include the economic growth mode of developed areas, the development mode of ecologically vulnerable areas, and the development mode of purely agricultural areas. There are few studies on mode analysis based on the perspective of collaborative transformation between urban and rural areas. Based on existing researches and the literature review, this paper divides the urban–rural transformation mode into the industry-urban-driven development mode, rural comprehensive development mode, urban–rural coordinated development mode, urban–rural integrated development mode, network development mode, and urban–rural equivalent development mode (Table 2.3).

(1) Industry-urban-driven development mode. First proposed by Arthur Lewis and supplemented by Lanis and Fei, this is a classic mode for explaining the economic development process with the characteristics of developing countries and is also a typical spatial allocation model of resources,

Table 2.3 Different development modes of urban–rural transformation

Development mode	Characteristics	Applicable areas
Industry-urban-driven mode	In the 1950s, Lewis put forward that emphasizing the leading role of the city has an obvious urban industrial tendency, which is a kind of urban industry-oriented mode with the characteristics of top-down development	Developed cities and backward rural areas
Rural comprehensive development mode	After the 1980s, Todaro and others put forward the following suggestions, which are derived from amendments of the "urban bias" mode, which is oriented towards meeting people's basic needs, and emphasizes bottom-up development; the fundamental way is to strengthen the comprehensive development and construction of rural areas and narrow the income gap between urban and rural areas	Suitable for regions with good agricultural basic conditions, more developed agriculture, or major agricultural industrial belts
Urban–rural coordinated development mode	Based on rural development, we should strengthen the radiation of regional central cities to rural areas and emphasize the importance of rural non-agricultural industry development with township enterprises as the main body and rural small towns	Suitable for regions with obvious location advantages and a good rural development foundation
Urban–rural integrated development mode	Kotadesasi concept was originated by McGee in 1987 and then simplified to desakota, which refers to a new spatial form formed by the interaction and influence of two regional urban and rural systems; it breaks down the relatively closed space between urban and rural areas and emphasizes the interaction between urban and rural areas	Suitable for the suburbs of developed cities such as large cities or the areas covered by urban agglomerations
Network development mode	Cities, towns and townships form a symbiotic network system composed of homogeneous rural areas (domain), towns (nodes) and transportation channels (axis); the formation of the mode depends on three aspects: the improvement of node potential and expansion tension, the improvement of the capacity of undertaking, digesting and feedback in the domain; and the construction of a channel network	Regional urban and rural development under the conditions of opening up
Urban–rural equivalence mode	This mode emphasizes the regional systematicity of the urban–rural relationship; establishes a new relationship of equality, coordination and integration between urban and rural areas; promotes the coordinated development of new urbanization and new rural construction; and realizes the equivalence of urban and rural areas	Suitable for areas with strong centrality and rapid rural transformation. The core is to promote the integration of urban and rural development

Data sources Hu and Ma (1993), Shi (1998), Fan and Li (2005), Zhao (2009), Liu et al. (2013, 2015)

involving the transformation of industry, agriculture, and urban and rural areas. From the perspective of urban–rural development mechanisms, the industry-urban-driven development mode takes the modern industry located in the city as the main means of national economic development and absorbs the labor resources of agricultural departments in rural areas into urban industrial departments. Under the condition of constant wages, capital is formed by technological progress, which promotes the increase in profit sharing in the national income. The increased profits are continuously used for reinvestment, thus prompting the industrial sector to produce a sustained labor demand for the agricultural sector until the development and utilization of surplus labor resources in rural areas are completed. Although the dual economic model of Fei and Lanis emphasized that the improvement of agricultural production efficiency promotes the generation and transfer of the surplus labor force, the urban–rural relationship reflected by this mode still has an obvious unbalanced urban industrial development bias. The mode emphasizes the top-down-driven path of central city and industrial zone construction.

(2) Rural comprehensive development mode. In view of the neglect of agriculture and rural development and the one-sided emphasis on industrial and urban development, many scholars put forward rural comprehensive development mode from the rural perspective and Todaro is the representative. He believed that the reason for the relatively backward agriculture in developing countries is the excessive emphasis on the investment in urban industrial sectors to achieve rapid industrialization in the 1950s and 1960s. Therefore, Todaro stressed strengthening the comprehensive development and construction of rural areas, including paying more attention to agricultural development, actively developing rural nonagricultural industries, improving rural infrastructure construction; narrowing the gap between urban and rural employment opportunities, reducing the imbalance between urban and rural incomes; reforming the education system, adjusting the education structure, and developing practical secondary and vocational education to provide talents for agriculture and rural development. The rural comprehensive development mode ignores the characteristics of modern industrial development and urban agglomeration, which makes it difficult to coordinate urban and rural development from the perspective of urban–rural linkages and institutional change and weakens the market-oriented principle of factor allocation.

(3) Urban–rural coordinated development mode, also known as the path of urban and rural development proposed by Fei Xiaotong on the basis of the continuous investigation and summary of vast rural areas. This mode emphasizes strengthening the radiation of regional central cities to rural areas on the basis of rural development to promote urban–rural coordinated development. The development of rural nonagricultural industries and the construction of rural small towns in the form of township enterprises are of great significance. The development of township enterprises has broken down the closed state of urban–rural isolation to a great extent and promoted the free flow and combination of urban and rural elements in the urban–rural systems along with the continuous development of township enterprises in rural areas. Through the construction of market organizations, small towns in rural areas can establish an organic connection between urban and rural areas. This mode is based on the rapid development of rural industrialization and small towns against the background of the urban–rural division in the 1980s and 1990s. It emphasizes the important role of rural small towns and township industries in the coordinated development of urban and rural areas.

(4) Urban–rural integrated development mode. Professor McKee put forward a new pattern in the study of urban–rural transformation in Asia. This new is due mainly to the interaction between urban and rural economic and social regional systems in the process of factor flow and urban–rural linkage, namely, the desakota region (or "expanded metropolitan area", "gray area", etc.) and emphasizes the impact of urban–rural linkages and factor flow on regional development. In contrast to the highly centralized urbanization path based on cities in the West, the desakota mode is regionally based and is a relatively decentralized urbanization method implemented by developing

countries according to their own characteristics. It focuses not on the concentration of rural resources and production factors in large cities but rather on the guidance of urban factors for neighboring rural areas. A new type of spatial economy and residential settlement with the characteristics of both urban and rural areas has emerged in the rural areas around metropolises and along transportation corridors between large and medium-sized cities.

This kind of development mode is suitable for the suburbs or urban agglomerations of large cities. These areas have the characteristics of high population density and diversified economic activities, that is, not only operating small-scale farming and agriculture but also rapidly developing various nonagricultural industries. In addition, the land-use patterns are highly mixed, and the cultivated land, industrial district, real estate management and so on exist in the region simultaneously. The population is greatly mobile, and a large number of residents' work in cities owing to seasonal mobility, good infrastructure conditions, and convenient transportation. Because of the backward economy and low population density in the central and western regions of China, this mode has limitations in the central and western regions.

(5) Network development mode. This mode is a spatial process of forming and maintaining the symbiosis and coexistence of the network system of cities, towns and townships in a region. It is also a spatial process of the coordinated development of the regional urban–rural economy. The network consists of three parts: node, axis and domain. Node refers to towns, axis refers to media and channels that produce various connections, and domain refers to homogeneous rural areas. The potential and expansion of urban nodes at all levels are the driving force to promote the formation and movement of various flows. This mode has strong adaptability in more developed areas. Through the free flow of urban and rural economic elements and the effective allocation of resources, it promotes the effective combination of urban and rural functions. The connection function of the axis can realize the circulation of logistics, people flow and information flow and promote the formation of economic links among them. The capacity to undertake, digest and provide feedback in the domain is the foundation of network development.

(6) Urban–rural equivalence mode. This mode emphasizes the integrity of the urban–rural regional system and establishes a new urban–rural relationship of equality, coordination and integration. It originated in Bavaria, Germany. The core idea of this mode is that urban and rural areas are different but equivalent. Equivalence is fully reflected in the equivalence of infrastructure and living conditions (Liu et al. 2013, 2015). In the process of rapid industrialization and urbanization, it is necessary to promote the economic and social development in rural areas to enhance the endogenous power of rural development. Based on the interaction between urban and rural areas and the flow of rural elements, institutional platforms should be innovated and village and town construction platforms should be built. This mode is suitable for areas with strong urban centrality and rapid rural transformation. However, backward areas in the central and western regions have inconvenient transportation and scattered rural distribution, which makes it difficult to promote urban–rural equivalence.

The rise of the urban–rural transformation mode depends on the conditions of regional development and the stages of economic growth, and is affected by the national development strategy, location conditions, resource endowment and so on. Therefore, each model has strong applicability conditions. Although the choices of urban–rural resource allocation and urban–rural contact model are diverse, all development modes are based on the complex relationship between urban and rural areas.

Urban–rural transformation is the dynamic evolutionary process of urban and rural areas and the collaborative process of urban–rural relationship. Thus, the recognition and judgment of the national urban–rural macro-relationship and the identification and diagnosis of regional development characteristics are the basis for the selection of the urban–rural transformation mode.

2.1.6 The Impact of Urban–Rural Transformation on Resources and Environment

Urban–rural transformation leads to changes in resource utilization and environmental patterns. With the development of rapid industrialization and urbanization, the impact of urban–rural transformation on resources and the environment has been increasingly studied, mainly in terms of land-use change, industrial agglomeration and structural adjustment, urbanization and industrialization processes and other resource and environmental effects. In recent years, with the changes and transformation of the regional urban–rural relationship in China, scholars have attached great importance to the impact on rural resource consumption and environmental change. Relevant research has focused mainly on urbanization, industrialization and agricultural modernization.

1. **Impact of urbanization on resources and environment**

Urbanization is a major issue in contemporary China's socioeconomic development, involving the rational development and utilization of natural resources, strong environmental foundation, moderate transformation of rural populations and high-quality improvement of human living spaces (Yao et al. 2008). Urban land-use change is an important content of regional environmental evolution (Cai 2001). From 1978 to 2018, China's urbanization rate is increasing by 1% annually. The urbanization in China is basically a land resource-dependent development path. Rapid urbanization development and unreasonable land-use patterns not only lead to huge waste and destruction of land resources but also cause serious ecological and environmental problems (Yang et al. 2006). Wang (2009a) took Zhaotong in Yunnan Province as an example to discuss the resource and environmental effects of rural urbanization and pointed out that in the disordered expansion of urbanization, a large amount of cultivated land, forestland and water areas are occupied by construction activities, and the suburban land was idle, which caused land waste and rapid loss of cultivated land and increased the vulnerability of the ecological environment.

Due to the long-term restrictions such as urban–rural dual structure and household registration system, the level of urbanization in China is relatively lagging. Since the mid- and late 1990s, more than 11 million rural people have poured into urban areas every year, which has become an important means of rapid urbanization (Lu et al. 2006). However, due to the high cost of living and the institutional constraints of agricultural-to-nonagricultural transfer, the phenomenon of "amphibious" land occupation among rural population is widespread, and the phenomenon of rural hollowing has increasingly become serious. Taking Shigong village in Henan Province as an example, Li and Li (2008) found that a large number of cultivated lands were occupied and land resources were idle and wasted. The hollowing out of rural areas broke the spatial residential pattern of relative concentration and close families. The relationship of "four generations under one roof", close relatives and harmonious neighborhoods gradually weakened, which affected the reconstruction of various social and economic relations and resulted in the deterioration of living environment in rural areas. Based on the empirical research of more than 100 typical villages in Shandong Province, Liu et al. (2009) pointed out that rural hollowing essentially came from the urban–rural transformation promoted by rapid urbanization. In this process, the non-agriculturalization of rural population led to the phenomenon of "people leaving their houses empty" and "building new house but not demolishing the old one". The residential buildings have gradually expanded to the periphery of the village, which leads to the expansion of the village land scale and the aggravation of idle and abandoned land. It is a harmful evolutionary process of expanding outer space while leaving the inside vacant. In terms of the spatial distribution, Huaihe River, middle and lower reaches of the Yangtze River and North China Plain are the main areas of cultivated land occupied by rural residential expansion in China. Long et al. (2009a) analyzed the change of rural residential land in southern Jiangsu Province based on remote sensing image data, and found that 92% of rural residential land was converted from cultivated land in the process of rural hollowing expansion.

2. **Influence of industrialization on resources and environment**

Industrialization has an impact on resources and the environment through the industrial level and scale, industrial layout, economic growth mode and other factors. The resources and environmental problems associated with industrialization have a profound impact on the production and life of rural areas. Due to the huge consumption of resources, the use of fossil fuels and the emission of pollutants, such as harmful gases, smoke, wastewater, and solid waste, industrialization not only promotes economic development, but also leads to the deterioration and pollution of regional resources and environment. The main feature of China's rural industrialization, with township enterprises as the main body, is the on-site development, utilization and transformation of rural resources (including natural resources, labor resources and funds, etc.), which shows the characteristics of decentralized layout, small scale, and industrial convergence. The development pattern of "swarm to do something" directly brings serious waste of resources and environmental pollution and ecological destruction. Yu et al. (2008) used the method of a material metabolism analysis of regional eco-economic system, taking Tongzhou city in Jiangsu Province in the stage of rapid industrialization as an example and food, energy, water, nonfood substances and solid, liquid and gas three-phase wastes as objects, discussed the impact of industrialization on the natural environment and concluded that the rapid development of industrialization had accelerated the evolution of material metabolism in the county eco-economic system of Tongzhou city, and the metabolic rate was far higher than the natural productivity and self-purification rate. The rapid accumulation of hazardous wastes leads to the deterioration of resources and environment. Based on the investigation of water pollution in 54 small towns in Jiangsu Province, Wang (2009b) found that on the one hand, the rapid development of industrialization and township enterprises had caused a large amount of sewage and wastewater discharge, while the construction of sewage treatment facilities lagged behind, resulting in the low standard rate of industrial wastewater discharge and increasing water pollution; on the other hand, the recycling utilization rate of rural industrial water was low, and the waste of water resources was serious. Zhong (2010) analyzed the hazards of industrial solid waste, mainly including land occupation and soil pollution, water pollution and tailing in solid waste, also, the fly ash and particles in garbage entering the atmosphere would reduce visibility and pollute the atmosphere. Liu (2010) analyzed the causes of acid rain, considered that the social cause of acid rain was the large amount of SO_2 emissions in the process of industrial development and automobile use, and further discussed the resource and environmental problems such as soil acidification and forest degradation caused by acid rain.

3. **Influences of agricultural modernization on rural resources and environment**

Agriculture is a complex eco-economic system interwoven with natural reproduction and economic reproduction. In the transformation of urban and rural development, modern industrial products and advanced management technology have been widely used in agricultural production in the face of urban consumption and agricultural structural adjustment, but the basic strategies are based on the industrial agricultural mode with high energy and chemical input. The excessive use of industrial auxiliary raw materials has profoundly changed the traditional agricultural production mode and has a huge impact on resources and environment (Wang 1999), resulting in a series of problems such as soil fertility damage, soil consolidation, groundwater pollution, and visual landscape damage (Wen et al. 2008). For example, in the process of Japan's early agricultural modernization, high investment and large-scale use of chemical fertilizers and pesticides led to heavy resource and environmental costs, including soil pollution, pesticide and chemical fertilizer residue pollution, agricultural and livestock product processing effluent pollution, water quality and water pollution, and air pollution (Yi and Yan 2006).

Besides, many studies focus on the micro-mechanism of the impact of intensive agricultural production mode on rural resources and environment. Taking the application of chemical fertilizer as an example, Matson et al. (1998) analyzedthe use of excessive nitrogen fertilizer in agricultural production, which causes soil nitrate to spread through leaching, not only causing water pollution in agricultural areas but also increasing the nitrate concentration in downstream surface water, resulting in widespread water pollution. Many areas in China have also faced the problem of serious nitrate pollution of groundwater (Wen et al. 2008). Liu et al. (2010) analyzed the spatial variability of nitrate nitrogen content in the groundwater of planting areas in Shandong Province. The results showed that the nitrate nitrogen content of groundwater differed in different regions and decreased with increasing groundwater depth, thus showing obvious trend effects and variability. The spatial distribution law showed a gradient directional effect, gradually increasing from southwest to northeast, and the areas with high nitrate nitrogen content were distributed mainly in the Weifang, Qingdao and Yantai planting areas. The frequent application of various kinds of instant fertilizers can easily cause the coupling of rainfall and fertilization period, and lead to farmland nitrogen and phosphorus entering water body through runoff and leaching, resulting in water pollution and eutrophication. Yang et al. (2010) found that the continuous increase of pollutants into the lake caused by human activities led to an increase in the concentration of nutrients (TN and TP) in Taihu Lake and was the main reason for the ecological degradation of the lake and a decline in water environmental quality.

Compared with the Pearl River Delta and Yangtze River Delta, the natural conditions in the Bohai Rim are generally poor, water resources are scarce, and the ecological environment is relatively fragile (Mao and Yu 2001). Industrialization and urbanization have led to a large decrease in high-quality cultivated land around urban areas and along traffic corridors, the aggravation of the point source pollution such as urban industry, and the spread of the non-point source pollution in agricultural modernization, which has brought serious water and soil pollution. The increasing environmental pollution problem has become the resistance of economic development (Zhou 2006). Water and soil resources are poorly matched in plains areas, and the problem of water shortage is prominent. Urbanization and industrialization have led to an unprecedented increase in water resource demand, and the load of resources and environment in some areas is overload. In addition, the ecological environment in mountainous and hilly areas and the Bashang Plateau area is fragile, the unreasonable development of land resources has led to serious land degradation. In a word, the eco-environmental problems of urban–rural transformation in this region have become increasingly prominent, and some areas are aggravating, which should be paid enough attention to.

2.2 Practice and Exploration of Urban–Rural Transformation in China

Since the founding of the People's Republic of China, China implemented the strategies of agriculture supporting industrial development and rural areas supporting urban development, especially the most stringent household registration system and unified purchasing and marketing of grain based on urban–rural dual structure. The development mode and system in this special era boosted the evolution, formation and solidification of urban–rural dual structure and promoted the formation and development of urban–rural isolation. The premise of urban expansion is depriving rural areas of resource allocation rights and development opportunities, resulting in urban and rural areas becoming relatively independent regional systems. Among them, the most active labor force and population flow between urban and rural areas are strictly restricted by the household registration system. Therefore urban–rural development has evolved into an isolated or even antagonistic state under these institutional constraints and barriers, which has become an important obstacle to accelerate urban–rural integrated development in the new period.

2.2.1 Characteristics of Urban–Rural Transformation from 1949 to 1977

At the beginning of the founding of the PRC, China implemented the strategy of giving priority to the development of heavy industry, and established a national industrial system dominated by heavy industry during the period of 1st five-year plan. After the establishment of people's communes in 1958, the government set up financial and agricultural banks in the communes to fully extract the agricultural surplus and formed a dual structure of urban–rural division to ensure inward capital accumulation. Agriculture became the source of capital accumulation required by heavy industry. Against this political and economic background, the regulations of the PRC on Household Registration were passed at the beginning of 1958. Article 10 of the regulations stipulated that the conditions for peasants to move to urban areas were that they must have employment certificates from urban labor departments, admission certificates from schools, or urban planning admittance to legally restrict farmers from entering cities and restrict the free flow of urban and rural populations. These marked the formal formation of the dual urban–rural management system with the household registration system as the core, a series of complementary policy systems were established. The urban population enjoys a wide range of social services and security provided by the state and work units, including housing, medical care, children's education and pensions. Fuel, rice, oil and salt are supplied to urban residents with household registration and ticket certificates. The social services and security of the rural population are provided mainly by families and collectives, and the degree of security is low and unstable.

During this period, China's urbanization was dominated by the government from top to bottom, and the role of the market was very weak (Chen et al. 2004). The state invested a large amount of funds in the central cities to expand the distribution of industrial construction. Large and medium-sized state-owned enterprises and military enterprises serve the country's economic construction and strategic policies, which has nothing to do with local rural industrial chains and natural resources. The development of nonagricultural industries in rural areas was not permitted, and were transformed into a support system for industrial production and urban residents' welfare by means of unified purchase and marketing and lowering the prices of agricultural products. As a whole, urbanization was restrained and strictly controlled, and the development of urban areas was completely subject to national industrialization strategy. Compared with the rapid industrialization in the same period, the main body of urbanization lacked the right of independent choice under the urban–rural division, and farmers accounted for the vast majority of the population were confined to inefficient land, so urbanization developed slowly. There was no normal market connection between urban and rural areas, urban civilization did not radiate to rural areas in large areas, industrial and agricultural development was seriously imbalanced, and urban and rural development were isolated from each other.

2.2.2 Characteristics of Urban–Rural Transformation from 1978 to 1984

Affected by the elements flow between rural and urban areas over the long term, rural areas were in a poor and backward state, and it was difficult to sustain the original urban–rural segmentation system and economic system. At the end of 1978, the Third Plenary Session of the Eleventh Central Committee of the Communist Party of China adjusted the national economic and social development policy on the basis of reflection on economic and social development in the 30 years since the founding of CPC. Although the rural reform did not fundamentally change the urban–rural dual system, it stimulated rural productivity through the reform of rural grassroots production relations. These institutional reforms included the expansion of local autonomy, the household contract responsibility system, the double-track system of agricultural product prices, financial contracting and the development of township enterprises. Specific aspects included the cancellation of the people's communes, the establishment of the rural household contract responsibility system, and the implementation of unified and decentralized two-tier management. Although rural land was still owned by the collective, the right to use the land was gradually transferred from village collective to the farmers. Farmers gradually obtained decision-making power over production and products, and rural and farmers' autonomy was expanded. The household contract responsibility system quickly overcame the "big pot of rice" and "equalitarianism" systems, solved the problem of reverse incentive in the collective production process, mobilized farmers' enthusiasm for work, and solved the problem of farmers' food supply.

In terms of development strategy, the heavy industry development strategy of high accumulation, low efficiency and overtightening of people's consumption had changed, and the economic development strategy of giving priority to agriculture and consumer goods industry was established. After strategic adjustment, labor-intensive agriculture and light industry gained more development opportunities, and township enterprises began to emerge on a large scale and grow rapidly, reducing the proportion of capital-intensive heavy industry in regional economy. The number of township enterprises increased from 1.34 million to 12.22 million between 1983 and 1985, and the number of employments reached 38.66 million, exceeding the total increase of rural labor force in the same period. On the one hand, the success of rural reform brought about the liberation of the surplus rural labor force. Under the situation of urban reform lagging and population flow continuing, township enterprises became the main opportunity for rural migrant labor force. On the other hand, before the reform and opening up, the production of consumer goods was insufficient, which brought the sales market and profits to the township enterprises of light industry. More importantly, the central government implemented a relatively decentralized financial contracting system for local governments, which promoted the combination of local administration, collective rights, family capital and local social relations in grassroots society. It also closely linked the tax revenue of industrial and commercial enterprises with the fiscal revenue of local governments, and became a way for local governments to transfer extra budgetary funds, which greatly stimulated local governments to develop township enterprises.

In addition, the central government reformed the purchase and marketing system for agricultural products, replacing previous national completely unified purchase and marketing system with a combination of national brand purchasing, negotiated purchasing and market sales. In this stage, the single price was changed into multiple prices. Therefore, the more products farmers sold, the higher the average price, which formed a positive price incentive mechanism. Additionally, the government raised the prices of more than 10 kinds of agricultural products in 1979, including grain, cotton and oil. From 1978 to 1984, the national total price index of agricultural products increased by 59.7%, with an average annual growth rate of 7.1% (Table 2.4). During this period, rural areas occupied the dominant position for the first time in the changes of urban–rural relations. The rural household contract responsibility system fundamentally solved the problem of agricultural product shortages in China. At the same time, the rapid development of private enterprises and township enterprises, as well as increasingly significant differences in regional economic development, objectively formed the demand for population flow among regions and between urban and rural areas, and promoted the government to relax population migration control and adjust the relationship between industry and agriculture.

2.2.3 Characteristics of Urban–Rural Transformation from 1985 to 1996

Since 1984, great changes have taken place in China's economic and social systems. The most important is the implementation of the "dual-track system" in which the planned economy and market economy coexist. The dual-track system has characteristics of gradualness, path dependence and marginality and provides competitive conditions with the original planning structure remaining basically unchanged. The price double-track system, which is divided into planned and unplanned prices, opened the dual-track system of the economy, as is shown in different prices for the same product, different prices of state-owned and non-state-owned enterprises, and the liberalization of grain wholesale prices, which represents the transition from planned prices to market prices. The introduction of financial contracting has affected the relationship between the central and local governments, expanded the financial power of local governments, and greatly mobilized the enthusiasm of local governments for developing local economies. In the 1990s, the state gradually established the dominance of the market economy. At the same time, the focus of urban and rural system reform shifted from rural to urban areas. The reform of state-owned enterprises promoted market competitiveness, formed an attraction for rural factors, and developed a "push–pull" trend for rural factors between urban and rural areas.

After the reform in the previous period, the obstacles of management system have been eliminated to a certain extent in rural areas, and the production factor market in rural areas gradually opened up. In the mid- and late 1980s, township enterprises achieved super-rapid

Table 2.4 Economic and social development in rural China since 1978

Period	Purchasing price index of agricultural products (%)	Growth rate of output value of agriculture, forestry, animal husbandry and fishery (%)	Growth rate of per capita net income of farmers (%)	Increase of employees in township industry (10,000)	Increase of township industrial enterprises (10,000)	Rural resident population growth (10,000)	Average annual growth percentage of urbanization
1978–1984	7.1	6.9	15.1	1921	401	1326	0.84
1985–1995	12.5	4.0	4.0	3428	225	5190	0.53
1996–2003	−2.1	5.4	4.6	−3.9	−75	−8234	1.26
Since 2004	8.5	5.2	8.8	−700	−320	−11,483	1.35

growth, and became an important driving force for rural social and economic development and national economic growth in the following 10 years (Table 2.5). In 1987, the total output value of township enterprises exceeded that of agriculture for the first time, and the industrial added-value of township enterprises accounted for 50% of the national industrial added-value. Township enterprises played an important role in absorbing the employment population, increasing income and breaking the single-ownership structure. With the rise of township enterprises, the intermediate structure of township enterprises and rural industry has emerged in China's national economic structure except for the former dual structure of rural and urban, agriculture and industry. Some scholars believe that the development of township enterprises has transformed China's traditional urban–rural dual structure into a "ternary structure", and the traditional urban industrial mode presents the characteristics of urban–rural dual industrialization.

In response to the reform of urban and rural economies and systems, the State Council approved and transmitted the "Notice on the Issue of Farmers Entering Towns and Settlements" issued by the Ministry of Public Security in 1984. The document stipulated that the public security organs should allow those who have the abilities to operate, have fixed residences or work in township enterprises for a long time to stay in permanent residences, have a nonagricultural household registration, and purchase food at negotiated prices. Subsequently, the commodity grain distribution system was reformed, the barrier that farmers had to "bring their own rations" to enter the city was reduced. With the loosening of the household registration system, a small number of farmers began to enter the city, but this type of migration still did not become the mainstream. To curb the large-scale population flow, the general office of the State Council issued the "Urgent Notice on Strictly Controlling Migrant Workers' Going Out" in March 1989, requiring the people's governments at all levels to take effective measures to strictly control local migrant workers' exodus. Entering the 1990s, with the expansion of urban scale and the large-scale construction of industrial parks at all levels, urban employment opportunities increased. At the same time, the improvement of agricultural production efficiency in rural areas and the slow absorption of employment by township enterprises made it difficult for rural land, capital and labor to stay in the countryside, resulting in a large amount of cultivated land being occupied

Table 2.5 Development of township enterprises in China

Year	Total number of enterprises (10,000)	Employees (10,000)	Added-value of township enterprises (100 million yuan)	Added-value of township industry (100 million yuan)	Added-value of tertiary industry (100 million yuan)	Total profit (100 million yuan)	Funds for agricultural construction (100 million yuan)
1978	152	2827	208	160	21	96	30
1983	135	3235	408	302	52	137	25
1988	1888	9845	1742	1306	228	650	104
1991	1909	9314	2972	2227	427	815	155
1996	2336	13,508	17,659	12,628	3250	4351	268
2000	2085	12,820	27,156	18,812	5929	6482	167
2005	2250	14,272	50,534	35,661	11,068	12,519	187
2010	2742	15,893	112,827	77,693	26,323	19,624	401
2012	2844	16,186	–	–	–	32,425	–

and labor outflow. Economic development zones and urban expansion need a lot of land. Under the government's administrative approach and dual-track production, the urban government obtained cheap rural land from farmers and village collectives through land expropriation, and then transferred it to urban commercial land and residential land with high land income through the secondary land market. In addition, the capitalization tendency of urban reform was prominent. The land became capital flowing into the city through the transfer of the right to use it. Some powerful township enterprises also invested capital in urban industry and commerce. Although township enterprises still played a dominant role in the rural economy, rural elements began to flow from rural to urban areas.

2.2.4 Characteristics of Urban–Rural Transformation from 1997 to 2012

In the late 1990s, China accelerated the reform of state-owned enterprises and urban land systems, deepened the opening to the outside world, and the urban economy grew rapidly. First, state-owned enterprises gradually expanded their autonomy through rectification and reform. In addition, they reformed their internal distribution, implemented the economic responsibility system, and allowed enterprises to merge and go bankrupt. Since the mid-1990s, the vast majority of enterprises had carried out joint-stock reform. Second, China was fully open to the outside world and had established many joint ventures and wholly foreign-owned enterprises through the large-scale introduction of capital, technology and advanced management methods. Driven by the reform, the export-oriented economic growth promoted the rapid development of China's manufacturing industry, which drove the development of related services industry and construction industry and formed the market demand for labor. Third, the system of paid use of urban land has been gradually established, and the transfer of agreement became bidding for and auctions of land-use rights. This system boosted the capital accumulation of urban expansion and provided financial support for the construction of infrastructure and public services.

To reverse the situation of the proportion of the central government's fiscal revenue declining year by year, the implementation of the tax sharing system in 1994 had a significant impact on the industrialization mode and fiscal revenue of local governments, forming a situation of interaction and competition among the central government, local government, industrial and commercial capital, rural collectives and farmers. The proportion of land-related taxes in local taxes increased, which encouraged the local governments to develop land. As the management and use of land development are relatively free and large-scale, the local government had turned from developing enterprises to land development. The development zone was the engine of local economic growth, and the industrialization model based on it became an important driving force of land development and urbanization. Through the provision of low-cost industrial land and other preferential policies, the development zone became a space carrier for attracting investment and real investment in the construction of factories, and has continuously attracted new flows of people, logistics and capital, becoming the destination of urban–rural element flow and agglomeration. For example, during the period of 10th five-year plan, the GDP of national development zones increased by 34.5% annually, contributing more than 6.0% to GDP. Driven by changes of urban and rural systems and economic structure, the status of cities in urban and rural economic and social development continuously improved, which accelerated the transfer of rural populations to urban areas in the 1990s. From 1996 to 2003, the urbanization rate increased from 30.5% to 40.5%, with an average annual growth rate of 1.25%, much faster than the average annual growth of 0.62% from 1978 to 1995. In the mid- and late 1990s, the state refined management policies for migrant workers to work, business and housing in urban areas. Since the second half of 2000, the government had clarified and canceled various unreasonable restrictions on farmers' employment in cities, creating conditions for many farmers to enter cities to work. At this stage, 16 million rural people transferred to urban areas for employment every year.

With the rapid growth of urban economy, township enterprises also faced the pressure of property rights reform, supply–demand relationship changes, and market competition. The phenomenon of "swarm to do something" led to the deterioration of rural environment and inefficient utilization of resources. Township enterprises entered a slow development stage, and their abilities to absorb rural population transfer employment gradually declined. Township enterprises were closely related to local governments, and the contract system on revenue and expenditure for local governments encouraged them to support township enterprises and provide land, credit and other support for the development of township enterprises. However, township enterprises also undertook social functions such as public facilities construction and social security. The ownership of unclear property rights eventually led to the decline of township enterprises. At the same time, China's grain production rose sharply from 1995 to 1998, but the domestic grain price index showed a downward trend from 1998 to 2003. It is difficult for farmers to sell grain and the problem of increasing grain production without increasing income appears. The grain output dropped from 510 million tons to 430 million tons, which seriously threatened national food security. During this period, the growth rate of farmers' income slowed, and the income gap between urban and rural residents widened daily.

In 2004, China's per capita GDP exceeded 10,000 yuan. The rapid development of nonagricultural industries, especially the service industry, greatly increased national financial revenue, laying the economic foundation for the implementation of the policy of "taking the city to lead the township and industry supplementing agriculture". After 2003, China entered the stage of balancing urban and rural development. It became a consensus among all walks of life that "taking the city to lead the township and industry supplementing agriculture" was urgent. The central government and local governments strengthened investment in rural areas as well as financial transfer payments to underdeveloped areas, such as special funds for poverty alleviation, renovation of dilapidated houses, and wage increases. The 16th National Congress of CPC put forward the strategy of balancing urban–rural economic and social development for the first time. In 2003, the pilot project of agricultural tax and fee reform was launched countrywide. In 2004, the state reissued the central "No. 1 Document" concerning rural development after 18 years, requiring the implementation of the policy of "one exemption and three subsidies" for agricultural production. In 2005, the 15th Plenary Session of the 16th CPC Central Committee put forward new socialist countryside construction in the new period, and put forward the construction requirements of "production development, affluent life, civilized rural style, clean village appearance and democratic management". Since 2006, the state had decided not to tax agriculture alone, cancelled the agricultural tax which had been implemented for more than 2600 years, and implement preferential policies such as direct grain subsidy, improved seed subsidy and agricultural machinery subsidy to stimulate farmers' enthusiasm for growing grain. In order to promote agricultural and rural modernization as a whole, the central "No. 1 Document" for 12 consecutive years from 2004 to 2016 focused on the "three rural" issues, and provided systematic support for rural development from economic, social, cultural, environmental, agricultural and other aspects. However, a large number of practices show that, due to the economic strength and leading role of the city, the overall planning of urban–rural development had the characteristics of "unifying rural areas through urban areas", and urbanization depended on cheap labor and soil and water resources in rural areas, which exacerbated certain rural problems.

2.2.5 Characteristics of Urban–Rural Transformation Since 2013

Deeply affected by the international financial crisis in 2008, China's economic growth faced the new challenges of slowing growth rate and declining external demand, which were manifested in the rapid decline in the volume of foreign trade, overcapacity in iron and steel, coal and other industries, transformation of economic development and growth mode, as well as structure optimization and mechanism transformation. Therefore, it has formed a strong demand for the adjustment of urban–rural industrial structure and the reform of urban–rural management system.

In addition to the new normal of growth slowdown, structural optimization and power transformation of the economy, China's rapid urbanization since the 9th five-year plan experienced a decline in the growth rate, a decrease in urban employment, and an aggravation of rural hollowing. Meanwhile, the return of migrant workers from some cities and counties in western China and traditional agricultural areas had brought pressure on rural land use, industrial development and public service supply. In this process, due to the obvious advantages of large cities and megacities in terms of employment supply, policy bias and social resource allocation, the population growth of megacities such as Beijing, Shanghai, Guangzhou, and Shenzhen was particularly obvious, and the provincial capitals and prefecture-level cities became the main areas of population inflow, while a large number of counties and cities and the associated villages and towns showed a trend of continuous population outflow, and the imbalance of population flow among regions intensified.

The Third Plenary Session of the 18th CPC Central Committee proposed to speed up the construction of the system and mechanisms of urban–rural integrated development, which had become a programmatic document for the new government to coordinate the implementation of urban and rural development strategies. At the end of 2013, the central working conference on urbanization clarified the important role of urbanization in solving the issues concerning agriculture, rural areas and farmers, building a well-off society, and promoting the development of modernization. Also, six major-proposals were put forward, including promoting the citizenization of the agricultural transfer population, improving the efficiency of urban construction land, establishing a diversified sustainable fund guarantee mechanism, optimizing the urban layout and form, improving the level of urban construction and strengthening the management of urbanization. In 2014, the State Council's opinions on the reform of the household registration system and proposals of three pilot land system reforms accelerated the reform process of the household registration system and land system to adapt to urban–rural integrated development. Obviously, China's urban–rural development has entered the stage of deepening the overall planning. China needed to adapt to the new requirements of economic and social development, break the traditional mode of cheap elements and investment driven by innovation and entrepreneurship, and further reform the urban–rural dual systems, mainly including household registration, land management, and public service.

2.3 Rural Development Stages and Major Rural Policies in China

Since reform and opening up, with the acceleration of industrialization and urbanization, China's agricultural and rural modernization and urban–rural development experienced a sustained progress. Overall, the evolution of rural China presents an S-shaped curve (Fig. 2.4). Rural development in China could be divided into three stages according to its phase objectives, i.e., the society with adequate food and clothing (1978–2005), building a well-off society (2005–2020) and achieving prosperity (2020–2035) (Liu et al. 2018; Liu 2020). Accordingly, the rural system shows three modes from the perspective of regional function, including single agricultural

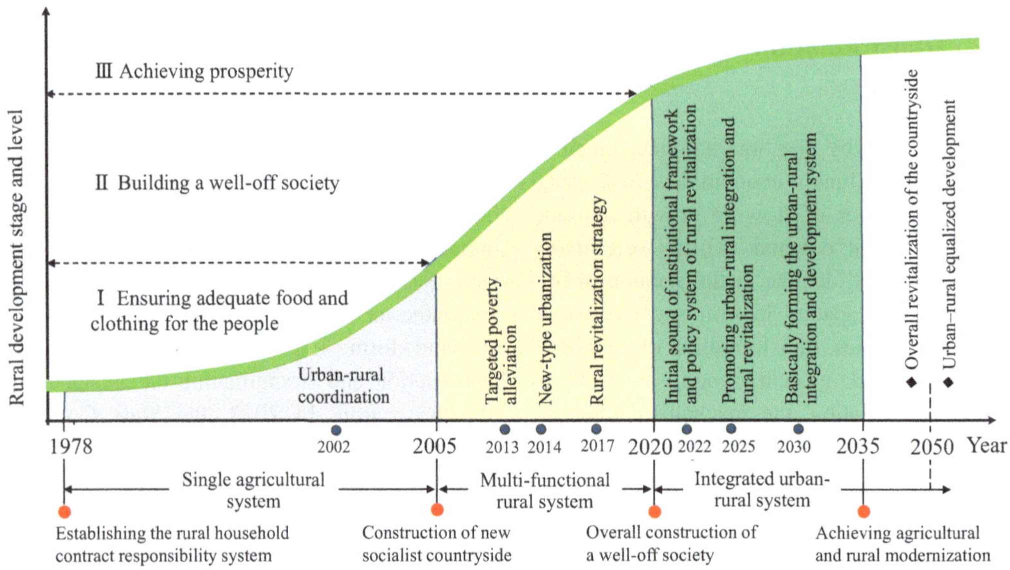

Fig. 2.4 Rural development stage and level of China from 1978–2050

system (1978–2005), multi-functional rural system (2005–2020) and integrated urban–rural system (after 2020).

China has built a dual system of urban–rural demarcation during the period of planning system. Then in 1978, the establishment of the rural household contract responsibility system realized the separation of ownership and contractual management rights, which mobilized the farmers' enthusiasm and facilitated the rapid increase of food production and agricultural modernization (Lin, 1992; Zhou et al. 2020). Yet, the subsequent urban reform led to the rapid development of cities and the widening gap between urban and rural areas. Due to the long-term influence of the urban-biased strategies, China experienced massive urban expansion and rapid economic growth at the beginning of the twenty-first century. And the three rural issues concerned agriculture, rural areas and peasants became increasingly prominent. Rapid urbanization led to an expanded city scale and extended industrial park construction, resulting in rural decline, a widening of the urban–rural gap, and severe rural problems (Liu and Li 2017). Rural functions emphasized the strengthening of social stability maintenance and the security guarantee of peasants' livelihoods. This emphasis introduced a long-term lack of traditional agricultural restructuring and a general imperfection of modern agricultural functions. Rural development lagged far behind urban development due to these unbalanced policies. Commentors and researchers have been concerned about these serious problems in rural China and have appealed to the government to promote rural development since the 1990s.

Responding to these public voices, the central government introduced a series of policies aiming to promote rural development. In 2002, balancing urban and rural development was proposed for urban–rural coordination. Through agriculture restructuring and rural construction, this strategy aimed to eliminate the influence of the urban–rural dual structure system, coordinate the urban–rural relationship by taking the combination of urban and rural development mode. Due to the increasingly severe rural problems, China successively launched several major development strategies, such as "five-coordinated strategies" (urban–rural development, regional development, socio-economic development, harmonious human-earth development, and domestic development) (2003), and "construction of new socialist countryside" (2005), etc. Consequently, important implementations were provided for achieving the development of urban–rural coordination and integration.

By 2012, the development of the urban–rural relationship in China had been further improved, and a gradual transformation from opposition to coordination and integration had taken place. Then, solving the "three rural problems" still remained an important content for rural development. In the context of a new era and a new normal, urban–rural transformation provides a guarantee for the further improvement of the urban–rural relationship, as manifested in the factor transfer, strategy change, and mechanism conversion of the urban–rural regional system. In the process of urban–rural transformation, important measures including targeted poverty alleviation (2013) and new-type urbanization construction (2014) were implemented. In particular, rural areas are experiencing a gradual development and improvement in terms of infrastructure, industry, and eco-environment. However, rural impoverishment still poses a significant obstacle to the rural development of China and, without solving this problem, either coordinated urban–rural development or a prosperous and well-off life for peasants are almost impossible to realize.

In 2017, China proposed the implementation of the rural revitalization strategy and pointed out that the path of integrated urban–rural development is the first-choice for the realization of rural revitalization; it represents a significant conceptual change that both rebalances and re-shapes the urban–rural relationship, and aims to create a new-type of urban–rural relationship characterized by mutual promotion of industry-agriculture, urban–rural mutual complementation, comprehensive integration, and common prosperity. This strategy aims to solve five main categories of rural problems, including the rapid loss of agricultural production elements (e.g., agricultural labor, cultivated land, etc.), excessively fast aging and weakening of agricultural social subjects, the increasing vacancy of rural houses and waste of construction lands, the severe pollution of rural ecological environment, and the deep impoverishment of rural poverty-stricken areas (Zheng and Liu 2018; Liu 2018). The strategy requires adhering to the priority development of agriculture and rural areas according to the overall demands of industrial prosperity, ecological livability, rural civilization, effective governance, and prosperous life (Fig. 2.5). In 2018, three stages were set to achieve rural revitalization, i.e., basically forming its institutional framework and policy system during 2018–2020; making decisive progress in rural revitalization and realizing the modernization of agricultural and rural areas by 2035; and fully revitalizing rural areas with realization of strong agriculture, beautiful countryside and rich farmers by 2050. Especially, the institutional framework and policy system of rural revitalization will be initially improved by 2022. Figure 2.1 also shows the future direction of rural development, i.e., promoting urban–rural integration and rural revitalization by 2025, basically forming the urban–rural integration and development system by 2030 and finally realizing urban–rural equalized development by 2050. In addition, rural revitalization should be promoted tightly with the dual circulation in the context of a changing world and the high-quality development. The world faces a profound change that has not been seen in a century and China enters a new stage of development during the 14th Five-Year Plan Period. More efforts should be made to promote the high-quality development and accelerate the establishment of a dual circulation development pattern in which domestic economic cycle plays a leading role while international economic cycle remains its extension and supplement.

2.4 Practice and Experience of Urban–Rural Transformation in Foreign Countries

2.4.1 Industrial Development and Regional Support

Industrialization is the basis for a country to coordinate urban and rural development. The higher the level of industrialization is, the stronger the overall economic strength of

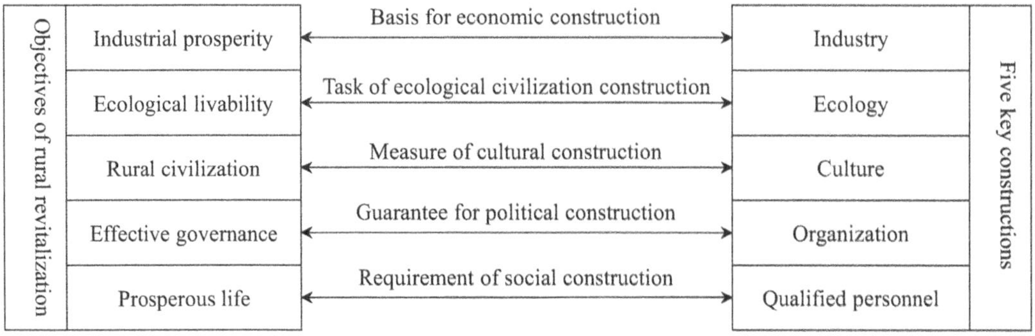

Fig. 2.5 China's rural revitalization and its objectives

the country, and then increase the national financial revenue and increase investment in backward rural areas. In addition, the improvement of industrialization level will help to improve production efficiency, enabling industries to transform from low-efficiency to high-efficiency production. Industrialization also accelerates the capital accumulation, creating new job opportunities, so as to meet the needs of the transfer of the agricultural surplus labor force. However, different countries perform differently in terms of realizing urban–rural coordinated development through industrialization. Before the middle of the eighteenth century, agriculture was the basic industry of all national economies; industry was in the stage of manual workshops and family workshops, and scattered in rural areas. From the second half of the eighteenth century to the middle of the nineteenth century, the developed countries took the lead in industrial revolution, fundamentally changing the socioeconomic structure. Industrial enterprises and service industries began to concentrate in cities and towns and areas close to coal resources and convenient transportation. With the promotion of industrialization and urbanization, the proportion of urban and rural populations had changed significantly. The developed countries, such as the United Kingdom and the United States, allocate production factors freely based on the market mechanism. Therefore, industrial development has the inherent demand of pursuing profits and seeking profits in agricultural production. Britain was the first country to carry out industrialization, and it led other countries in scientific and technological progress. Its urban industry created many jobs and constantly pull the agricultural surplus labor into cities. When industries absorbed agricultural labor in the United States, the mechanization and scale of agriculture develop rapidly based on the rich agricultural resources, ensuring the production income of agricultural producers. In the late stage of industrialization, the suburbanization of population and industry became an important driving force of industrial transformation in rural areas, promoting urban–rural coordinated development. In general, the United Kingdom and the United States have strong abilities to drive industrialization and rich agricultural resources; thus, the gap between workers and farmers had not widened in the process of industrialization (Feng 2011).

The industrialization in South Korea and Japan was relatively late and followed the mode led by the government. This mode sacrificed agricultural production to realize the aim of catching-up and surpassing other countries in terms of industrialization. In the midterm stage of industrial development dominated by capital-intensive industries, South Korea and Japan began to implement the policy of "back-feeding agriculture". Through the implementation of policies and measures to drive the countryside with cities and promote agriculture with industry, they reconstructed rural industry and society. After World War II, South Korea adopted the economic development strategy of leaping industrialized countries. Since the initiation of the 6th five-year plan, South Korea had leaped from a developing country to a developed country. The industrialization development strategy also changed from the import-substitution type to the export-oriented type, and it adopted the goose-shaped development mode closely linked with the economy of the United States and Japan, and established the industrial strategy of giving priority to the development of labor-intensive industries. In addition, the development mode of the central park city and industrial park was implemented, and the national economic center was distributed in Seoul, Busan, southeastern coastal cities, the west coast and other areas. Moreover, many industrial zones were established countrywide and became the main means of absorbing surplus rural labor. In the mid-1960s, the government guided the merger and reorganization of township enterprise parks in terms of policies, funds and industries to promote scale economy. Therefore, industrialization and the construction of towns and parks became the basis of urban–rural development. In this process, South Korea faced problems of rural population aging and the rural agricultural industry system being on the verge of collapse in sparsely populated areas. In 1970, South Korea began the "New-Village Movement", focusing on improving farmers' basic production and living conditions and developing rural public utilities, and promoting the development of rural production and income based on government support, farmers' autonomy and project development.

In view of the problems of rural population shortage and rural decline against the background of rapid urbanization, the Japanese government issued a series of agricultural and rural support bills beginning in the 1960s. In 1967, the "basic principles of structural policy" were formulated. In 1968, a comprehensive funding system was established. In 1969, the "regional consolidation law for agricultural revitalization" was passed. In 1970, the "farmland law" and the "farmers association law" were amended to create the pension system for farmers. In 1971, the "law on promoting the introduction of industry in rural areas" was formulated to encourage the transfer of urban industries to rural areas and provide nonagricultural employment opportunities for farmers. In 1984, the "agricultural law" and the "land improvement law" were amended to promote the construction of the rural regional environment. In 1987, the "village construction law" was formulated to regulate the land-use order of rural villages and their surrounding areas, and coordinated the development of different regions through the formulation of specific regional bills, such as the "island revitalization law", "mountain revitalization law", and "special measures for sparse region". In the middle and late stages of industrialization, the Japanese government made great efforts to promote the industrial transfer from metropolises to local cities and rural areas. Many satellite factories or branch factories of large enterprises, as well as many small-scale contract factories and smaller family factories belonging to large enterprise system emerged in rural areas. The development of urban–rural industries in Japan and South Korea showed that rapid industrialization could absorb the surplus rural labor force.

2.4.2 Planning System and Spatial Control

Spatial planning and spatial control are important means for the government to manage territorial space. Healthy urbanization and urban–rural coordinated development depend on a sound planning system and legal guarantee system. British urban planning began with the "public health law" at the end of the nineteenth century, which aimed to solve the serious social problems caused by the industrial revolution, urbanization and industrialization. In 1932, the British government promulgated the first "Urban–Rural Planning Act" to deal with urban sprawl and expansion in disorder, and promoted the integration of urban and rural areas from the perspective of spatial planning. The "Urban–Rural Planning Law" of 1947 clearly put forward for the first time that it was necessary to curb the expansion of cities to rural areas. In 1940s, the "Greater London Plan" regulated regional productivity and population through the construction of green belts, new towns and transportation corridors, and the development of new towns played an important role in promoting urban–rural integrated development. In addition to urban planning, Britain created planning guidance for urban villages and towns. The "Urban–Rural Planning Act" of 1947 focused on controlling the disorderly spread of large cities in rural areas from the perspective of urban land use. In 2004, the British government issued the "No. 7 Planning Policy Document: Sustainable Development of Rural Areas", which was used to guide the development and construction of rural areas and became the main basis of British rural planning and management. Local planning departments in UK pay attention to the protection of small towns when formulating urban and rural development policies, emphasizing the protection of natural environment, local culture and history, and areas with recreational value. If there were new development and construction needs in small towns, the planning should first consider the location of new development projects in local central areas where jobs and housing, service industry and other public facilities are concentrated. The spatial location of new services and public facilities needs to be specified and marked in the local planning document, which was called "embedded development" in British urban and rural planning (Yu 2013).

Although the United States is rich in land resources, urban–rural planning still attaches importance to the protection and construction of rural areas, and a large number of green villages and open spaces are used as strategic security reserve land. The land use of rural residents in United States is controlled by "Zoning Plan", which stipulates the width of roads and the proportion of pedestrian roads, and requires that rural houses should be properly arranged around rural roads. Due to the private ownership of rural land in United States, it is necessary to fully consider the opinions of indigenous people and guide public participation in the construction and management of land use zoning planning. In the construction of rural residential areas, great importance is attached to infrastructure and human settlements. Each residential area has sewage treatment facilities and medical and health facilities regulated by the government, and rural roads can be transformed according to the wishes of residents (Zhang et al. 2014).

Japan has formed a multilevel land planning system, including national comprehensive development planning, regional planning, urban planning, and village planning, aiming to control urban and rural development spaces through land planning. Since the 1960s, Japan has carried out five national comprehensive land development plans to guide social and economic construction. The "First Comprehensive Development Plan" selected a number of sites in Tokyo, Osaka, Nagoya and other surrounding areas to develop industrialization and urbanization, and closely linked the cities and the surrounding countryside to form local circles. The "Third Comprehensive Development Plan" focused on solving the problems of over-dense and over-sparse population and backward development of marginal areas, and restraining the concentration of population and industry to big cities. The "Fourth Comprehensive Development Plan" highlighted the network development of points (towns), lines (infrastructure) and areas (rural areas) and better dealt with urban and rural development problems in the process of rapid economic growth in modern society. The goal of the "Fifth Comprehensive Development Plan" still aimed at forming a multipolar pattern of land development (Wu et al. 2006). Village and town development planning is the most basic development planning in Japan and is used to guide rural development and space utilization. Meanwhile, the government actively promotes the construction achievements of grass-roots towns and villages, develops characteristic products of village towns in sparsely populated areas, and carries out the "one village, one product" campaign to promote rural construction.

2.4.3 Balanced Allocation of Public Resources

In the process of urbanization, Germany pays attention to the balanced development of large and medium-sized cities and small towns. In terms of industrial policies and infrastructure construction, it is inclined to small and midsized cities and towns, forming a unique mode of urban–rural overall planning, reasonable distribution and balanced development. Township enterprises have become the support for the development of rural industries in Germany. In areas with good locations and rich tourism resources, tourism, catering, entertainment and other service industries have developed. By the end of 2010, there were 2065 large, medium and small cities in Germany, of which 82 administrative districts with a population of more than 100,000 lived in 25.3 million people, accounting for 30% of the total population of Germany; the rest of the population was mostly distributed in small towns with a population of 2000–10,000. Based on the idea of balanced development, the German government divided the cities and towns into first-, second-, third- and fourth-level regional centers and sets standards for public infrastructure and municipal facilities. The construction of public infrastructure was arranged through regional planning and county planning, and the federal, state and local township levels share the cost of urban construction and development. Moreover, almost all rural areas were equipped with the corresponding sewage and solid waste treatment facilities, and implement the same charge management system as cities for sewage treatment and garbage disposal.

To coordinate the development of urban and rural areas, Britain has planned and guided urban villages and towns to develop a diversified economy. To reverse the decline of traditional centers and promote urban renaissance, the British suburbs attach importance to urban village planning, and attract the return of the middle class by building suburban villages. Urban villages have the characteristics of appropriate scale, compact form and reasonable density. They try to create an environment in which different classes live together, and provide convenient service facilities and beautiful rural environment. To solve the problem of rural employment, small towns with beautiful environment and excellent location conditions are selected as the focus of driving rural economic development in British planning. There are clear standards for rural planning and construction, including post offices, libraries, community offices, churches, rural primary and secondary schools, bus stops, etc.

From 1950 to 1960s, the industrialization and urbanization in South Korea accelerated, and a large number of rural young people swarmed into large cities. Traditional culture, ethics and social order in rural areas were impacted, the income gap between urban and rural residents widened, and agriculture in some areas was on the verge of collapse. In 1970, South Korea began the "New Village Movement", focusing on improving farmers' basic production and living conditions and developing rural public utilities. Through government support, farmers' autonomy and project development, rural production development and income increase were promoted. The first stage focused on rural reconstruction of houses, roads and public facilities; the second stage was to improve the farmland infrastructure and production and circulation facilities from the level of individual villages; the third stage was to promote regional cooperation, urban–rural cooperation and large-scale development of the agricultural product processing industry at the group level. In general, these reflected the working idea of orderly promotion from easy to difficult (Mao 2007). The implementation process attached importance to the enthusiasm of farmers and gave play to the leading role of the community, so that people could develop production and get rid of poverty through independent and cooperative development (Ge and Hua 2010).

2.4.4 Guarantee Mechanism for Urban–Rural Equal Rights and Interests

1. **Equal household registration system**

Population mobility in developed countries is not restricted by household registration system, and urban–rural residents enjoy equal rights in employment, medical treatment and schooling. The current laws in United States prohibit identity discrimination, so there are almost no artificial barriers to population movement, and migration registration is based on the tax location of individuals. To prevent gender or racial discrimination by employers, many state laws stipulate that recruitment companies do not allow identity restrictions, such as gender and race, in recruitment advertisements and registration forms. The free population flow allows the United States to successfully realize the transfer of the rural labor force in a relatively short period (Zhang 2010). Japan's constitutional stipulates that its citizens enjoy the freedom of migration. As long as individuals are qualified and can afford to pay tuition fees, they can freely choose a school regardless of region. or they successfully apply for employment, they can freely seek employment regardless of location. Also, if they can afford to buy a house or pay the rent, they can freely choose to reside anywhere in the country, become a local resident and enjoy all the local benefits.

2. **Balanced social public services**

In 1935, the United States passed the first social security code with the aim of social assistance, the "Social Security Act", and the social security system continued to improve thereafter. In the construction of rural medical security system, most American farmers participate in commercial insurance, and the government is responsible for medical insurance for the elderly and the poor. In the budget of the federal government and state governments, the proportion of rural social security expenditure in GDP is clearly specified, mainly involving food assistance and nutritional subsidies required by specific groups such as low-income farmers, women and children. In addition to the food and nutrition service of the federal government's Department of Agriculture, which is responsible for the implementation of federal and state aid programs, local charities are also involved in providing them free of charge. In 1946, Japan passed the "Promulgation and Implementation of the Life Protection Law". In 1959, it passed the "National Health Insurance Law" and the "National Annuity Law", and implemented a regional security system targeting farmers. The social security services enjoyed by farmers and urban workers are roughly the same, including social insurance system, national assistance system, social welfare system and public health system. In terms of specific implementation, the central government entrusts and cooperates with the local autonomous bodies to put these systems into practice. Japan has basically established a rural social security system based on medical insurance and endowment insurance in rural areas, which reduces the worries of farmers when they enter cities and has enabled a large number of farmers to successfully become urban citizens after relocating to urban areas. Germany has established a unified and sound social security system to reduce the threshold of urbanization. There is no obvious difference between rural and urban areas in society. Farmers and urban residents enjoy equal rights in election, education, employment, migration and social security. There is little difference between urban and rural areas. South Korea has implemented the integration of urban and rural social security systems in balancing urban and rural development. There is little difference between rural residents and urban residents in basic social security, such as pension, health and medical security, social insurance, social welfare, and public relief. Social insurance is the most important social security system in South Korea. The government includes farmers into the social insurance system, such as expanding the coverage of national pension to all kinds of farmers and giving free land subsidies to the elderly aged 65 and above to enable them to swap idle farmland.

3. **Education and training of migrant workers and their descendants**

The United States has issued "Human Development and Training Act", "Employment Opportunity Act", "Employment Training Cooperation Act" and other bills, which require the whole society to attach importance to and support the vocational training of migrant workers. Also, the United States set up the migrant children education plan in 1966 and then gradually adopted measures such as migrant children transfer record system. In addition, the school selection system was innovated to solve the education problem of migrant workers' children through the education voucher plan. According to the education funds allocated by the government, a certain number of securities will be issued to the parents of the students, which will be used as the expenses to help the parents of the students choose schools for their children, which is conducive to the realization of children from poor families to study in private schools.

Japan has implemented the urban–rural integration financial system of compulsory education, which promoted the balanced development of urban–rural compulsory education. By the 1980s, high school education had been popularized in Japan. 40% of the rural school-age population were able to enter universities for further study, and directly entered nonagricultural industry after graduation from various schools at all levels. The Japanese government protects the employment of the migrant workers through labor laws, such as the "Emergency Unemployment Countermeasures Law" and the "Temporary Measures Law for Coal Mine Leavers" in 1949, and the "Minimum Wage Law" in 1959, which required employers to pay more than the minimum wage for workers who apply the minimum wage provisions. At the same time, to guarantee the payment of wages, the government advance the labor insurance benefits of workers injured at work, which belong to the nature of industrial injury insurance, and the employer was obliged to repay those costs. In addition, the government has set up vocational training institutions throughout the country to provide skills training for farmers who want to seek nonagricultural work. Vocational training and pre-job training provide a variety of learning opportunities for rural job seekers and can help the agricultural surplus labor force adapt to the needs of nonagricultural jobs.

2.4.5 Agricultural and Rural Development Supporting Act

1. Promoting land system reform

The land systems and reform measures of different countries are different, but the common purpose is to improve agricultural production efficiency and farmers' income level. The United States is a typical representative of private ownership of land in developed countries, and land has a clear property right boundary. Although the U.S. federal, state and county governments retain the right to control, manage and benefit from agricultural land, it does not hinder the autonomous management of agricultural land. When the federal government wants to use state or local government land or private land, it needs to exchange or purchase the land. At the same time, the landowners have the rights of land transfer, lease, mortgage and inheritance. In 1951 and 1952, the Japanese government formulated and promulgated the "Land Acquisition Act" and the "Agricultural Land Act" successively, which established the permanent status of farmers' ownership and guaranteed farmers' property rights. At the same time, to protect farmers' land rights and interests, the government comprehensively controlled the transfer of agricultural land. To change the small-scale operation of agriculture and its disadvantages, Japan formulated the "Basic Act of Agriculture" in 1961. The primary policy objective of this act was to expand the scale of agriculture, improve the income and living standards of agricultural workers, and enable them to reach basic equilibrium with other industrial workers. In 1962 and 1970, the Japanese government twice amended the "Agricultural Land Act", abolishing the upper limit of land tenure and the restriction on land rent. The "Agricultural vibration Act" formulated in 1975 allowed farmers to freely sign or terminate short-term land lease contracts within 10 years after collective negotiation according to the terms of the agreement reached by both parties, and encouraged the transfer of land-use rights. These institutional reforms promoted land flow mainly in the form of land sales and land leases, prompted lower-level farmers to relinquish land ownership and land management, and promoted the transfer of the rural labor force to urban areas, which was conducive to the scale development of agriculture.

2. Introducing agricultural support policies

Since the 1930s, the U.S. government has constantly improved its legislation to protect agriculture. In 1933, the U.S. government promulgated the "Agricultural Adjustment Act", which became the beginning of government intervention in agriculture. From the "Agricultural Adjustment Act" of 1933 to the "Agricultural Reform, Food and Employment Act" of 2012, the U.S. government promulgated and implemented 17 farm acts in 80 years (Chen and Zhang 2013). Although the focus of the farm acts differed in different periods, the core content was always intended to promote agricultural production and increase farmers' income by intervening in and subsidizing farms to maintain the competitiveness of American agriculture. The agricultural subsidy policies in the United States are composed of three main types: direct payment, target price and target income subsidy, and disaster insurance premium subsidy and assistance (Peng et al. 2012). Thus, the objectives of agricultural subsidy policy tend to be diversified, from relying on the "yellow box" policy to relying mainly on the "green box" policy. While improving farmers' income and the international competitiveness of agricultural products, such policies also pay attention to environmental protection and promote rural development (Liang and Sun 2011). In terms of agricultural investment in infrastructure, large-scale irrigation facilities in the United States are funded and built by the federal and state governments, whereas small and medium-sized irrigation facilities are funded by farmers individually or jointly, with certain subsidies from the government. According to the characteristics and requirements of agricultural and economic development, the U.S. government has formulated the appropriate agricultural policies at different stages of agricultural modernization to promote the development of mechanization, agricultural technology and agricultural information. In response to the new social and economic environment, such as the development of bioenergy and WTO rules, the new agricultural act of 2012 canceled the measures of direct payment and counter-cyclical payment, and established the agricultural income risk protection plan and an extensive insurance plan (Cai et al. 2013).

The Japanese government attaches great importance to the formulation of agricultural laws and policies. The "Agricultural Land Act", the "Basic Strengthening Promotion Act for Agricultural Management", the "Regional Preparation of Agricultural Revitalization Act" and the "Land Improvement Act" are the main bodies of agricultural land laws and policies in Japan. Japanese government subsidies for agricultural production mainly include water conservancy construction subsidy systems, farmland consolidation subsidy systems, machinery and equipment subsidy systems, infrastructure subsidy systems, agricultural loan interest subsidy systems, disaster subsidies, and agricultural insurance subsidies (Zhou and He 2005; Dong 2012). To meet the requirements of the WTO agricultural agreement, the Japanese government promulgated the "Basic Act on Food, Agriculture and Rural Areas" in 1999, adjusting the way of agricultural subsidies, shifting from "yellow box" subsidies to "green box" subsidies, and from price-support to income-support. Since 2009, Japan has implemented a series of new direct subsidy pilot programs for farmers' income, including the direct subsidy policy for rice farmers in 2010 and the direct subsidy policy for dry-land farmers in 2011. In addition, Japan has established a national agricultural research and experimental network, established a perfect extension system, and developed agricultural education at all levels.

Land consolidation is often used by the government as an important means to support rural development. In Germany, the large-scale comprehensive land consolidation project has effectively stimulated rural investment and consumption, and promoted the steady and rapid growth of regional economy (Liu 2018). Before the 1960s, land consolidation in Germany was mainly reflected in single planning and measures of agricultural land consolidation, improvement of agricultural production facilities, land development and reclamation, and residential construction. After that, village renewal plan was implemented nationwide in Germany. In 1976, the revised "Land Consolidation Act" of Germany first incorporated village renewal into the law, providing legal guarantee for intensive land use, revitalizing land capital and rural development. Through open and transparent procedures, government financial assistance, villagers' active participation and planning with various options, the existing problems of village construction were corrected, and the development of villages was guided (Ye and Bi 2010).

3. Providing vocational training for farmers

The aging of farmers and the training of young farmers are the concerns of the European Union (EU). The latest round of the EU Common Agricultural Policy (CAP) reform plan proposed that 2% of the direct payment should be dedicated to guide young farmers under the age of 40 to engage in agriculture (The Ministry of Agriculture visits EU agricultural policy investigation group 2012). The formulation of relevant policies will help young farmers into agriculture, and provide all-round help to young farmers through educational institutions and associations. For example, the agricultural chamber of commerce not only provides information, agricultural enterprise registration, technical training and other services to young farmers, but also helps young farmers carry out career development planning. In addition to strengthening basic education, the Japanese government also attaches great importance to rural vocational and technical education. With the participation of the government and private enterprises, a hierarchical and focused rural vocational and technical education system has been formed, providing various learning opportunities for rural job seekers to adapt to the working environment and acquire labor skills. To strengthen the enthusiasm of farmers and the management of agricultural entities, the Japanese government put forward the certification system of agricultural operators in the law on strengthening the foundation of agricultural operation in 1980, reaching 249,000 in 2010, which strengthened the foundation of agricultural operation and effectively promoted the development of agricultural modernization (Zong et al. 2011).

4. Attaching importance to rural culture and environmental protection

Before the 1940s, Britain lacked laws to protect the environment, rural culture and rural landscape. Therefore, the rural natural landscape and wild animals and plants were threatened over a long period by industrialization and urbanization. In addition, under the long-term policy of the "concentrating efforts only on industry to the neglect of agriculture", British agriculture gradually declined and relied heavily on the international market. Subsequently, the British government introduced the environmental protection act and agricultural production act to support agricultural and rural development. In 1943, the government set up the Nature Reserve Committee and the Ministry of Urban and Rural Development. In 1949, the National Park Commission was established to provide employment, protect rural facilities and promote rural tourism. In 1986, the government introduced wildlife and rural act, which went beyond departmental interests. In terms of agricultural production, Britain promulgated the first agricultural law after World War II and established the agricultural subsidy policy. In 1957, the new agricultural law promised to establish and improve the price protection mechanism of agricultural products, and required the government to provide the corresponding budget to support the consolidation and development of agricultural land; in the 1960s, it focused on promoting the scale and marketization of agriculture. In the 1970s, in response to the phenomenon of "counter-urbanization", the British rural development policy mainly focused on how to alleviate the contradiction between the demand for leisure and entertainment and the protection of rural natural scenery, and the principle of agricultural development and rural protection is to help rural economic development, maintain and strengthen the ecological and natural environment.

5. The important role of farmers' associations

The construction of various rural associations has helped to enhance the self-organization ability of farmers, while various economic organizations have strengthened the development ability of rural industries. To promote rural development, Japan has established an excellent agricultural management organization association and formed a semi-official farmers' organization, namely, the Agricultural Association, which is mainly responsible for production guidance, organization and circulation, credit services and mutual assistance (Wu et al. 2006), and is divided into three levels, i.e., village farmers association, county agricultural association and national agricultural association. In July 1947, Japan formulated a special law, "the Combination Act of Agricultural Associations", to establish the agricultural association, which provides farmers with services such as agricultural guidance, agricultural means of production supply, agricultural product sales, credit services, agricultural insurance and information. At present, more than 99% of the farmers in Japan have joined the agricultural association, and more than 70% of the means of production and living are obtained from the farmers' association. The development of the association has boosted the rural economy, improved farmers' living conditions and enhanced their trading status. In addition, many local agricultural associations have replaced the government in providing public services, which has played a positive role in developing agricultural labor productivity and improving the quality of life of farmers (Guo 2007).

References

Antrop M. Landscape change and the urbanization process in Europe. Landsc Urban Plan. 2004;67(1–4):9–26.

Cai Y. The mechanisms of cropland conservation in Chinese rural transformation. Sci Geogr Sinica. 2001;21(1):1–6.

Cai Y, Yu F. Sticking points of fundamental policies for the farmland issue in China. China Land Sci. 2004;18(3):13–7.

Cai H, Han Y, Ni H. Main changes and characteristics of the U.S. agricultural act of 2012. World Agric. 2013;2:48–50.

Chen M. Urbanization of farmers and farmland conversion. Urban Issues. 2002;4:62–5.

Chen Y. Evaluation of the impact of non-agricultural production factors on the development of rural system and its mechanism: a case study of Shandong Province. Graduate School of Chinese Academy of Sciences. 2010.

Chen Y. Literature review and policy implications of effect of rural labor transfer on rural development. Guangdong Agri Sci. 2013;40(23):185–8.

Chen B, Hao S, Yang X. The dynamic mechanism of the rapid development of urbanization in China. Acta Geogr Sin. 2004;59(6):1068–75.

Chen J, Qu F. Land conversion and food security: theoretical framework and empirical analysis. J Nanjing Agric Univ. 2006;29(2):103–7.

Chen X, Zhang X. Rural transformation in south Jiangsu with 'The New South Jiangsu Model.' Probl Agric Econ. 2008;8:21–2.

Chen Z, Zhang H. The Enlightenment of American farm act on China's agricultural policy legalization construction. Econ Res Guid. 2013;12:113–4.

Chen M, Lu D, Cha L. Urbanization and economic development in China: an international comparison based on quadrant map approach. Geogr Res. 2009;28(2):464–74.

Chen X. Spatial structure reorganization of hengyang city based on interaction between urban and rural area theory and practice. Sci Geogr Sin. 2005;25(3):288-93.

Cheng Y, Deng J, Fang Y. Spatial analysis of urban-rural association to the major grain-producing areas of central Jilin Province. Geogr Res. 2010;29(4):727–36.

Cheng L, Feng W, Jiang L. The analysis of rural settlement hollowizing system of the southeast of Taiyuan Basin. Acta Geogr Sin. 2001;56(4):437–46.

Chen Y, Sun H, Liu Y. Reconstruction models of hollowed villages in key agricultural regions of China. Acta Geogr Sin. 2010;65(6):727–35.

Chen F, Zhang H, Wu Q, et al. A study on coordinate development between population urbanization and land urbanization in China. Human Geogr. 2010;25(5):53–8.

Chen W, Wu C. The role and development of urbanization in urban-rural relationship in China. Econ Geogr. 1995;15(3):25-9.

Cheng M, Shi Q, Xu J. From Malthus to Solow: an explanation for the motivation and obstacles effecting farmer labor emigration in China. Econ Res. 2006;4:68–78.

Dai J, Gao X, Du S. Expansion of urban space and land use control in the process of urbanization. Geogr Res. 2010;29(10):1822–32.

Deng L. On urban-rural integration. Urban Res. 2001;2:10–2.

Dong L. Japan's agricultural financial subsidy policy and its reference to China. World Agric. 2012;12:34–7.

Douglass MA. Regional network strategy for reciprocal rural-urban linkage. An agenda for policy research with reference to Indonesia. Third World Planning Rev. 1998b;20(1):1–33.

Douglass MA. Rural-urban linkages and poverty alleviation: toward a policy framework. International Workshop on Rural-Urban Linkages, Curitiba, Brazil. 1998a.

Duan J, Wen Y, Lu Q. Review and evaluation of researches on urban-rural interaction development in the last fifteen years. Prog Geogr. 2006;25(4):118-28.

Fan H, Li H. Discussion on the interactive development model between urban and rural areas. China Soft Sci. 2005;3:155–9.

Feng S. Research on the comparison of foreign countries' patterns of integrated development between urban and rural areas. Soft Sci. 2011;5:111–5.

Ge D, Hua C. New direction of rural planning: an urban-rural integration perspective. J Zhejiang Univ (Human Soc Sci). 2010;40(3):148–55.

Gu C. Study on phenomena and mechanism of land use/cover change in Beijing. J Nat Resour. 1999;14 (4):307–12.

Gu C, Chen T, Ding J, et al. The study of the urban fringes in Chinese megalopolises. Acta Geogr Sin. 1993;60(4):317–28.

Gu C, Li A. A framework for the integration planning between urban and rural areas in China: from "city exploiting rural area" to solving "three issues in rural area'. Econ Geogr. 2013;33(1):138–41.

Guo J. Background and lessons of the overall development of urban and rural areas in Japan. Agric Outlook. 2007;3(2):27–30.

Hall C, McVittie A, Moran D. What does the public want from agriculture and the countryside? A review of evidence and methods. J Rural Stud. 2004;20(2):211–25.

Han F, Wang Z, Yang H. The effect of industrial structure on intensive urban land use. Resour Sci. 2013;35 (2):388–95.

Hodge I, Monk S. The economic diversity of rural England: Stylized fallacies and uncertain evidence. J Rural Stud. 2004;20(3):263–72.

Hu F. Research on rural labor transfer in China: a literature review. Zhejiang Soc Sci. 2007;1:207–12.

Hu J. Study on the dynamic mechanism of urban rural integration development in China. Rural Econ. 2009;12:30–3.

Hu B, Ma A. Theory of urban and rural connections and the urban and rural relations in China. Economist. 1993;4(4):98–109.

Hu W, Zhang A, Qu L. Interrelationships among non-agricultural population, jobs and land. China Popul Resour Environ. 2009;19(5):104–10.

Huang G, Li T, Wang W, et al. Analysis on the new countryside constructor based on the ideas of urban-rural integration. Hum Geogr. 2009;24(4):16–9.

Holmes J. Impulses towards a multifunctional transition in rural Australia: Interpreting regional dynamics in landscapes, lifestyles and livelihoods. Landsc Res. 2008;33(2):211–23.

Jiang Z. A probe into the common target of countryside and cities development from an integrative view. South China J Econ. 2004;10:5–9.

Jiang Y, Zeng J. Dynamic mechanism and realization way of urban-rural related development. Res Develop. 2004;2:87–90.

Jiang H, Xia Y, Qu F. Study on contribution of construction land expansion to economic growth and its regional difference. China Land Sci. 2009;23(8):4–8.

Jing P, Chen Y. A study of mechanism for transfer of rural labor in the process of China's industrialization and urbanization. Southeast Acad Res. 2004;4:37–44.

Kang S. Urban rural integration: the evolution characteristics and path model of Zhejiang. Probl Agric Econ. 2010;6:29–35.

Knox P L, McCarthy L. Urbanization: an introduction to urban geography. Upper Saddle River, NJ: Pearson Prentice Hall; 2005.

Li X. Research priorities for land use change in the accelerated phase of urbanization. China Population Resour Environ. 2009;19(5):1–5.

Li J, Li X. The Microscopic analysis on village-hollowing in medium income and hilly land region of Henan province. China Popul Resour Environ. 2008;18 (1):170–5.

Li X, Wang C. Summary of national and international theories on transferring surplus labor force in countryside. Econ Rev. 2004;4:60–3.

Li Y, Zhang Q. Human-environment interaction in China: evidence of land-use change in Beijing-Tianjin-Hebei metropolitan region. Hum Ecol Rev. 2013;20(1): 26–35.

Li X, Fang J, Pu S. The intensity and modes of urban land use growth in Shanghai. J Nat Resour. 2003;18 (4):412–22.

Li C, Huang X, Liu Y. Research on mutual progress between development of industrial structure and urbanization in northeast. Econ Geogr. 2009;29 (2):231–6.

Li Y, Liu Y, Long H. Study on the pattern and types of rural development in the Huang-Huai-Hai region. Geogr Res. 2011;30(9):1637–47.

Li X, Li X, Ou M. Evaluation of urbanization development coordination and analysis of regional disparities in Jiangsu Province. Hum Geogr. 2012;27(3):50–4.
Liang J, Sun P. Analysis of the evolution trend of agricultural subsidy policy in the United States. Acad Exch. 2011;12:111–6.
Lin JY. Rural reforms and agricultural growth in China. Am Econ Rev. 1992;82(1):34–51.
Lipton M. Migration from rural areas of poor countries: the impact on rural productivity and income distribution. World Dev. 1980;8(1):1–24.
Liu Y. Rural transformation development and new countryside construction in eastern coastal area of China. Acta Geogr Sin. 2007;62(06):563–70.
Liu X. Formation of acid rain and its harm to environment. J Chifeng College Nat Sci Edn. 2010;26(5):96–7.
Liu S. Spatial patterns and dynamic mechanisms of urban land use growth. Prog Geogr. 2011;21(1):43–50.
Liu Y. Research on the urban-rural integration and rural revitalization in the new era in China. Acta Geogr Sin. 2018;73(4):637–50.
Liu Y. The basic theory and methodology of rural revitalization planning in China. Acta Geogr Sin. 2020;75(6):1120–33.
Liu D, Chen S. Diffusion and aggregate process of urban pattern in Suzhou, Wuxi and Changzhou cities over the recent 30 years. Sci Geog Sinica. 2012;32(1):47–54.
Liu Y, Li Y. Revitalize the world's countryside. Nature. 2017;548(7667):275–7.
Liu Y, Liu Y. Progress and prospect on the study of rural hollowing in China. Geogr Res. 2010;29(1):35–42.
Liu Y, Lu Y. Optimal land use strategies to guarantee urban-rural harmonious development in eastern coastal region of China. J Chongqing Jianzhu Univ. 2007;29(3):4–7.
Liu J, Peng Z, Xu Q. Rural urban transformation and coordinated development in Shanghai suburbs. City Plann Rev. 1997;21(5):44–7.
Liu S, Wu C, Shen H. A GIS-based model of urban land use growth in Beijing. Acta Geogr Sin. 2000;67 (4):407–16.
Liu Y, Liu Y, Zhai R. Geographical research and optimizing practice of rural hollowing in China. Acta Geogr Sin. 2009;64(10):1193–202.
Liu X, Xu J, Jiang L, et al. Spatial variability and distribution pattern of groundwater nitrate pollution in farming regions of Shandong Province, China. J Agro-Environ Sci. 2010;29(6):1172–9.
Liu Y, Lu S, Chen Y. Spatio-temporal change of urban–rural equalized development patterns in China and its driving factors. J Rural Stud. 2013;32:320–30.
Liu Y, Hu Z, Li Y. Process and cause of urban-rural development transformation in the Bohai Rim Region, China. J Geogr Sci. 2014a;24(6):1147–60.
Liu Y, Chen C, Li Y. Differentiation regularity of urban-rural equalized development at prefecture-level city in China. J Geog Sci. 2015;25(9):1075–88.
Liu Y, Yang R. The spatial pattern measure of urban-rural development transformation in the Bohai Rim region in China. Acta Geogr Sin. 2015;70(2):248–56.
Liu J, Kuang W, Zhang Z, et al. Spatiotemporal characteristics, patterns and causes of land use changes in China since the late 1980s. Acta Geogr. Sin. 2014b;69(1):3–14.
Liu Y, Long H, Chen Y, et al. Research report on rural development in China: Hollowing village and its renovation strategy. Beijing: Science Press; 2011.
Liu Y, Qiao L. Innovating system and policy of arable land conservation under the new-type urbanization in China. Econ Geogr. 2014;34(4):1–6.
Li T, She X. Discussions on the dynamic regimes and the evolution of the urban rural integration: taking Baoji city as example. J Northwest Univ (Nat Sci Edn). 2000;30(3):256–60.
Liu C, Li E, Qin C. Spatial patterns and its evolution of economic coordinated development between urban and rural areas in China. Hum Geogr. 2012;27(2):97–102.
Long H. Land use transition and rural transformation development. Prog Geogr. 2012;31(2):131–8.
Long H, Zhang X. Progress in international rural geography research since the turn of the new millennium and some implications. Econ Geogr. 2012;32(8):1–7.
Long H, Li Y, Liu Y. Analysis of evolutive characteristics and their driving mechanism of hollowing villages in China. Acta Geogr Sin. 2009b;64(10):1203–13.
Long H, Liu Y, Wu X, et al. Spatio-temporal dynamic patterns of farmland and rural settlements in Su-Xi-Chang region: implications for building a new countryside in coastal China. Land Use Policy. 2009a;26(2):322–33.
Long H, Zou J, Pykett J, et al. Analysis of rural transformation development in China since the turn of the new millennium. Appl Geogr. 2011;31(3):1094–105.
Long H. Land consolidation and rural spatial restructuring [J]. Acta Geogr Sin. 2013;68(8):1019–28.
Lu Q, Yang C. Characteristics of rural labor emigration in minority areas of Honghe, Yunnan and its impact on new socialist countryside construction in the 21st century. China Popul Resour Environ. 2008;14(4):85–9.
Lu Q, Zeng L, Wang G, et al. Spatial evolution of Chongqing urban-rural interaction and its comprehensive evaluation. China Popul Resour Environ. 2004;14 (2):81–7.
Lu X, Zhu S, Zhang Z. Empirical analysis on the scale of rural population migration in China. Chinese Rural Econ. 2006;1:35–41.
Lu B, Chen Z, Hou C, et al. Study on the special development of land urbanization in south outskirts of Xi'an city based on GIS. Hum Geogr. 2010;25(1):56–9.
Luo Y, Li T. Analysis on the systematic function of urban-rural integration in China. Hum Geogr. 2005;20 (4):47–50.
Ma A, Hu B. Urban-rural linkages: theoretical interpretation and mode selection. China Rural Econ. 1993; (8):51–5.
Ma Z, Zhang W, Liang Z, et al. Labor migration as a new determinant of income growth in rural China. Popul Res. 2004;3:2–10.
Ma Y, Zhang X, Li F, et al. Progress of spatial research on urban–rural relation in China and its discussion in geography view. Geogr Geo-Inf Sci. 2006;22(3):78–84.

Ma A, Li C, Liu S. Coupling relationship between urban industrial structure and land use: A case study of Shenyang. J Shanxi Finance Econ Univ. 2012;34(4):29–33.

Mao Z. The new village movement in South Korea and its inspiration to the construction of a new socialist countryside in China. On Econ Prob. 2007;4:87–9.

Mao H, Yu D. Regional carrying capacity in Bohai rim. Acta Geogr Sin. 2001;56(3):363–71.

Matson PA, Naylor R, Ortiz-Monasterio I. Integration of environmental, agronomic, and economic aspects of fertilizer management. Science. 1998;280(5360):112–5.

McGee TG. Managing the rural–urban transformation in East Asia in the 21st century. Sustain Sci. 2008;3(1):155–67.

McGee TG. Urbanisasi or Kotadesasi? Evolving patterns of urbanization in Asia. Urbanization in Asia: spatial dimensions and policy issues;1989:93–108.

Miao C. New rural spaces: the impact of rural industrialization on rural urban transition in China. Sci Geog Sin. 1999;18(5):409–17.

Ministry of Agriculture Delegation to EU Agricultural Policy. New trends of EU agricultural policy adjustment. Rural Work Bull. 2012;19:45–7.

Murdoch J. Networks: a new paradigm of rural development? J Rural Stud. 2000;16(4):407–19.

Ning Y, Shi Q, Cha Z. Study on the formation mechanism and regional expansion planning of the Yangtze River Delta metropolitan area. Shanghai Urban Plann Rev. 1998;1:16–20.

Niu W. Essential points of strategic design of Chinese new-approach urbanization. Bull Chin Acad Sci. 2009;24(2):130–7.

Peng C, Pan S, Duan Z. The direction of US agricultural subsidy reform: trends, causes and impacts of Farm Bill 2012. Probl Agric Econ. 2012;11:104–9.

Potter R, Unwin T. The Geography of urban-rural interaction in developing countries: Essays for Alan B. Mountjoy. Routledge; 2017.

Potts D, Mutambirwa C. Basics are now a luxury": Perceptions of structural adjustment's impact on rural and urban areas in Zimbabwe. Environ Urban. 1998;10(1):55–76.

Preston DA. Rural-urban and inter-settlement interaction: theory and analytical structure. Area. 1975;7(3):171–4.

Qiao J, Liu J, Xie M. Characteristics and microcosmic mechanism of rural hollow villages in less developed rural areas: a case study of Three villages in Lankao County. Hum Geogr. 2011;26(6):98–102160.

Qu F, Feng S, Zhu P, et al. Institutional arrangements, price system and farmland conversion. Econ (Quarterly). 2004;4(1):229–48.

Qu F, Chen J, Chen H. Economic development and land conversion in China. Beijing: Commercial Press; 2007.

Qu F, Wu L. Hypothesis and validation on the kuznets curves of economic growth and farmland conversion. Resour Sci. 2004;26(5):61–7.

Shi Y. The, "Desakota" model and its effects on the model of China's urban economic organization. Urban Studies. 1998;5:10–4.

Shi P, Chen J, Pan Y. Land use change mechanism in Shenzhen City. Acta Geogr Sin. 2000;55(2):25–34.

Tacoli C. Bridging the divide: rural-urban interactions and livelihood strategies. IIED;1998.

Tan Y, Wu C, Mou Y. Macro-analysis of cultivated land non-agriculturalization in Zhejiang. Sci Geogr Sinica. 2004;24(1):14–9.

Takuro K. Land planning towards the 21st century. Beijing: Science Press; 1990.

Terluin IJ. Differences in economic development in rural regions of advanced countries: an overview and critical analysis of theories. J Rural Stud. 2003;19(3):327–44.

Todaro MP. A model of labor migration and urban unemployment in less developed countries. Am Econ Rev. 1969;59(1):138–48.

Unwin T. Urban-rural interaction in developing countries: a theoretical perspective. The geography of urban-rural interaction in developing countries. Routledge 2017;11–32.

Wang K. Environmental problems derived from agricultural modernization and their countermeasure. Res Agric Modernization. 1999;20(5):270–3.

Wang G. Urban Rural Integration: transformation from dual structure to unitary structure. J Southwest Univ Nationalities (Philos Soc Sci). 2004;25(9):54–8.

Wang C. Income difference, mobility and regional employment assembly: Based on an empirical analysis of the rural labor transference. China Rural Surv. 2005;1:10–7.

Wang L. Suggestions on water pollution control in small and medium-sized cities in Jiangsu Province. Sci Technol Rev. 2009b;27(22):18.

Wang Y. Research on ecological effect of countryside urbanization and the control measures in ecological fragile area of southeast region by taking Zhaotong in Yunnan as a case. J Environ Sci. 2009a;28(3):35–8.

Wang L, Duan X. The expansion of urbanization area in Yangtze River Delta. Sci Geogr Sin. 2010;30(5):702–9.

Wang J, Li C, Cao R, et al. Tomorrow China: Towards urban-rural integration. Beijing: China Economic Press; 2005.

Wang F, Sun H, Sun F. Spatial differences of urban-rural development coordination in Shandong Province. Sci Geograph Sin. 2009;29(3):323–8.

Wang J, Liu Y, Chen Y. Spatial expansion pattern and its dynamic mechanism of typical rural settlements in Huang-Huai-Hai plain. Geogr Res. 2010;29(10):1833–40.

Wang S, Wang R, Wang Y. Analyzing Daqing's urban spatial form evolution: Basing on the technology of RS and GIS. Econ Geogr. 2012;32(6):67–72.

Wang Y, Liu Y, Yan B, et al. Spatial patterns and influencing factors of urban-rural coordinated development in China. Sci Geograph Sin. 2016;36(1):20–8.

Wen C, Zhang D, Tie Y. Environmental effects of agricultural non-point source pollution and its coupling impact on new rural construction. Soc Sci Guizhou. 2008;4:91–6.

Wu J. The transfer of rural surplus labor force and the problem of "three rural areas" issues. Macroeconomics. 2002;6:6–9.
Wu D, Yu X, Cha L, et al. National land-use planning and development of urban and rural construction in Japan. Acta Geogr Sin. 2006;61(7):771–80.
Xia A, Xu X, Xue D. A summary of the studies on the coordinated development of urban and rural areas in China. Hum Geogr. 2003;18(5):56–60.
Xie J. The strategy of city space development in urbanization process of China. Hum Geogr. 2009;24(4):87–90.
Xiu C, Xu D, Zhu X. Evaluation of urban-rural integration course in northeast China. Sci Geograph Sin. 2004;24(3):320–32.
Xue L. Study on the inner-decaying village and the countermeasures with Jiangsu Province as the case. City Plann Rev. 2001;25(6):8–13.
Yang G. The process and driving forces of change in arable-land area in the Yangtze River Delta during the past 50 years. J Nat Resour. 2001;16(2):121–7.
Yang Q, Lei L, Cheng X. Influence of urbanization on the land resource security in China. J Anhui Agric Sci. 2006;34(12):2821–3.
Yang G, Ma R, Zhang L, et al. Lake status, major problems and protection strategy in China. J Lake Sci. 2010;22(6):799–810.
Yao S, Wang C, Zhang L, et al. The influencing factors of resources and environments in the process of urbanization of China. Prog Geogr. 2008;27(3):94–100.
Ye J, Bi Y. Coordinated development of urban and rural areas in Germany and its implication for China: Taking Bavaria State as an example. China Land Sci. 2010;24(5):76–80.
Yi B, Yan D. Environmental problems and solutions in the process of Japanese agricultural modernization. Japan Study Forum. 2006;2:18–23.
Yu L. Urban-rural development policies in UK and some lessons for Chinese small-town development. Urban Stud. 2013;20(11):27–31.
Yu S, Huang X, Tan D. Evolution of material metabolism in ecological-economic system of rapid industrialization county: a case study of Tongzhou City. Chin J Ecol. 2008;27(9):1620–4.
Zeng J. Modern urban and rural network development model. Beijing: Science Press; 2001.
Zeng L, Lei J, Lu Q. Construction of evaluating indicator system of urban-rural interaction and the comparative analysis of regional urban-rural correlative-degree in China. Geogr Res. 2002;21(6):763–71.
Zeng J, Zhu Y. Urban-rural correlated development and cultural integration. Human Geogr. 2002;17(4):6–11.
Zhang X. On discrimination of rural definitions. Acta Geogr Sin. 1998;53(4):365–71.
Zhang A. The dynamic mechanism of urban-rural interaction and the transfer of urban-rural eco-economic element. Ecol Econ. 2000;6:51–5.
Zhang Q. The experience and reference of institutional arrangement in the overall planning of the urban and rural in the United States and Japan. Asia-Pacific Econ Rev. 2010;2:93–6.
Zhang H, Jia S. Analysis on the regulation mechanism of land conversion. Econ Res J. 2001;12:50–4.
Zhang F, Liu Y. Dynamic mechanism and models of regional rural development in China. Acta Geogr Sin. 2008;63(2):115–22.
Zhang W, Wang C, Lv X, et al. Coupling relationship between land use change and industrialization/urbanization in the Zhujiang River Delta. Acta Geogr Sin. 2003;58(5):677–85.
Zhan J, Lu Q, Deng X. Construction of urban-rural interaction evaluation model and its application in Shandong Province. Geogr Res. 2003;22(4):495–502.
Zhang Y, Jia D, Tan J, et al. The spatial structure of valley economy development in the mountainous areas of Beijing. Acta Geogr Sin. 2009;64(10):1231–42.
Zhang F, Liu Y, Zhang S. Analysis of rural development state and driving force in eastern costal China. J Nat Resour. 2010;25(2):177–84.
Zhang P, Zhang Z, Sun H, et al. Research progress of the urban-rural integration from international vision and the development experience in typical countries. Urban Plann Int. 2014;29(1):42–9.
Zhao Q. Strategic transformation of urban-rural relation in China and a discussion on urban-rural integration planning. J Urban Plann. 2009;6:47–51.
Zheng X, Liu Y. Scientific connotation, formation mechanism and regulation strategy of "rural disease" in China in the new era. Hum Geogr. 2018;33(2):100–6.
Zhong D. Research on municipal solid waste treatment. China Sci Technol Rev. 2010;26:176.
Zhou T. Defects of Todaro's model and its opposite policy implications: Thoughts on the transfer of China's surplus labor force and expansion of employment capacity. Econ Res. 2001;3:75–82.
Zhou Q. Researching trends and prospect of agro-environment science in the world. J Agro-Environ Sci. 2006;25(1):1–6.
Zhou J, He Z. Adjustment and enlightenment of Japanese agricultural subsidy policy. Rural Econ. 2005;10:123–6.
Zhou G, He Y. Characteristics and influencing factors of urban land expansion in Changsha. Acta Geogr Sin. 2006;61(11):1171–80.
Zhou Y, Li X, Liu Y. Rural land system reforms in China: History, issues, measures and prospects. Land Use Policy. 2020;91:104330.
Zong Y, Wei Y, Shen J, et al. The process of agricultural modernization in Japan and its enlightenment on China's modern agricultural construction. Agric Econ. 2011;4:14–5.

3 Geographical Basis and Theoretical Analysis of Urban–Rural Transformation

Abstract

This chapter discusses the basic theories of urban–rural transformation geography, and puts forward four core concepts of urban–rural transformation to comprehensively understand the transformation of urban–rural development, i.e., "transformation tree", "transformation law", "transformation threshold" and "fusion body". On this basis, the study of urban–rural transformation geography starts from the regional and comprehensive characteristics of geography, focuses on the "process, pattern, mechanism, problem, regulation" of urban–rural transformation, and takes solving urban–rural development problems and promoting urban–rural coordination as the guidance and the elements allocation, structural optimization and function upgrading of urban–rural system as the core. Further, it reveals the characteristics, types and regional differences of urban–rural transformation, and puts forward the paths and countermeasures to deepen the reform of the integration strategy, system and mechanism and governance system of urban and rural development based on following the laws and principles of urban and rural economic and social transformation.

3.1 Theoretical Basis of Urban–Rural Transformation

3.1.1 Spatial Structure and Optimization Theory

Spatial structure is the spatial mapping of agglomeration form originated from the interaction of economic and social activities in a specific geographical space. Location theory creates the precedent of spatial structure analysis, which has successfully given birth to agricultural location theory, industrial location theory and central place theory. Distance and transportation expenses play an important role in industrial spatial structure and layout. Location theory comprehensively analyzes the role of labor force, spatial agglomeration, administrative power, and other factors. its core lies in optimization of spatial allocation of production factors and the maximization of income driven by market mechanism.

Focusing on cities and regions, the theory of spatial structure includes urban interior space theory, urban-region spatial structure theory and interregional structure theory. Scholars from different disciplines have put forward the growth pole theory, core-periphery structure model and

Y. Liu, *Urban-Rural Transformation Geography*, Sustainable Development Goals Series,
https://doi.org/10.1007/978-981-16-4835-9_3

point-axis theory, which reveal the spatial evolution law of the city and its surrounding areas. The growth pole theory was firstly put forward by French economist Perroux in 1950, then it was introduced into the geographical space by Boudeville. Supplemented and extended by Hischman and other scholars, it was further organically linked with the formation and evolution of cities and urban agglomerations, forming a set of systematic theories. They believe the formation, development, decline and disappearance of growth pole will lead to changes in regional and urban spatial structure. Friedman put forward the core-periphery theory, pointing out the dual structure between urban center and small towns and rural fringe. In the new economic geography, the spatial factors and increasing returns of firms are included in the general equilibrium competition model, and the spatial inequality model of social and economic core-periphery is proposed. The effect of agglomeration and diffusion caused by circular and cumulative causation and path dependence lead to spatial non-equilibrium evolution (Zhang 2003). Point-axis theory combines the growth pole theory and the central place theory, and becomes a theoretical model to guide the distribution of productive forces, land and regional development (Lu 2002).

The change of spatial structure is closely related to industrialization and urbanization. Thus, the flow and agglomeration of urban–rural factors play an important role in spatial structure, and promote the formation of a spatial structure system with cities as nodes and infrastructure as channels. Since the 1970s, with the transfer of urban population and industry to suburbs and small towns, counter-urbanization appeared in developed countries, promoting the spatial restructure of population, industry and capital in urban and rural areas. The urban interior space and traditional rural space have changed obviously in the aspects of economy, society and environment. Due to the outward migration of urban middle class, the central areas of some cities are declining. Therefore, industrial restructuring, location dynamics, globalization, gentrification and urban regeneration are employed to guide urban spatial transformation in developed countries (Cumbers 2014). China is still in the stage of rapid urbanization. Promoted by industrialization and urbanization, socioeconomic and population changes promote the evolution and reconstruction of urban and rural spatial structure. In rural areas, it is manifested in rural economic and social transformation and spatial reconstruction (Liu 2011). Due to the disharmony between large and medium-sized cities and small towns and between regions in China, urban and rural spatial structure and organizational relationship are still difficult to meet the practical needs of urban–rural integrated development. Learn from foreign ideas and planning means such as smart growth, inclusive development and new rural construction, improving the relationship between urban and rural point-axis system and perfecting urban and rural spatial organization has become an important way to optimize urban and rural spatial structure in China.

3.1.2 Dual Economic Structure Theory

In general, economy in developing countries is consist of modern industrial sector and traditional agricultural sector. Development economics constructs a dual economic structure model, which is used to explain the process of dual economic structure transformation promoting economic growth, and promotes the transformation of dual economy to unitary economy through the transfer of agricultural surplus labor and capital accumulation. Due to the limited land and surplus agricultural labor force in traditional agricultural sector, the marginal productivity of labor force can be as low as zero. On the other hand, with the continuous expansion of modern economic sectors, the labor force has gradually transferred to the emerging industrial sector, forming the dual economic development process. The growth of labor demand exceeds the supply, and the wage will continue to increase until the Lewis turning point appears (Fig. 3.1).

In a certain area, the expansion of modern departments depends on absorbing the surplus labor force of traditional departments. Economic

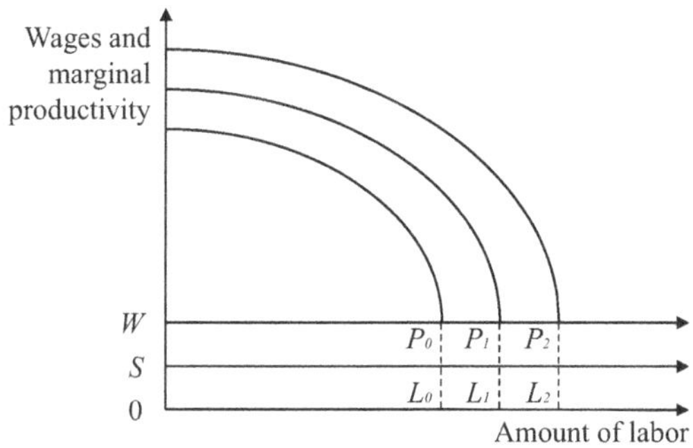

Fig. 3.1 The relationship between agricultural surplus labor force and wage and marginal productivity

development is a remarkable indicator, which means the transformation of labor force from traditional sector to the modern sector. This transfer can be divided into three stages. In the first stage, the outflow of the pure surplus labor force with zero marginal productivity does not affect the total output of traditional sectors; in second stage, the outflow of the labor force whose marginal productivity is greater than zero and less than the minimum average living cost will begin to affect the total output of traditional sectors; in the third stage, the surplus labor force of the traditional sector has been absorbed by the modern sector, thus the further development of modern sector must compete with the traditional sector for the labor force whose marginal productivity is greater than the minimum living cost. The supply–demand structure of labor force has undergone essential changes, and the phenomenon of surplus labor force disappears and is replaced by insufficient labor force. Promoted by market competition, the nature of traditional departments begins to change, the principle of community tends to disintegrate, and the capitalist management system is gradually established. Meanwhile, modern production technologies begin to rise in this sector, and the technical and operational characteristics of traditional sectors have gradually disappeared, forming a modern economic system. So far, the main task of industrialization has been completed, and the country developed from backward to developed level.

On this basis, Fei, Ranis and Jorgenson revise and improve the dual economic model, further emphasizing the importance of agricultural development and technological progress. Ranis-Fei model attaches importance to the development of agriculture and emphasizes that the contribution of agriculture to industry is not only that it provides the labor force needed by the industrial sector, but also provides agricultural surplus for the industrial sector and forms the accumulation of capital. Therefore, it becomes a classical model to analyze the dual economy under the framework of classicism. The dual economic theory points out that developing countries promote the non-agricultural transfer of agricultural surplus labor force, which has become an important theoretical support for analyzing the development of industry and agriculture and income distribution gap in developing countries. However, the hypothesis that surplus labor only exists in agricultural sector, the marginal productivity of surplus labor force is zero and barrier free transfer of population are not consistent with the reality. Although the neglect of the economic growth and urban–rural system has been criticized, the dual economic theory reveals the dual structure phenomenon of the coexistence of modern economy and small-scale peasant economy in developing countries. Only when the marginal labor productivity of agricultural sector and industrial sector tends to be relatively balanced, can the inefficient labor force be fully utilized, so that the dual structure turns to the unitary structure. In the process of urban–rural economic and social transformation, the transformation from dual economy to unitary economy means that urban and rural economic development can be promoted through the expansion of modern departments and the transformation of traditional departments, especially the introduction of modern factors in traditional agricultural production departments and agricultural products processing departments. Thus, rural development can be achieved by new technology and organizational integration.

3.1.3 Urban–Rural Regional System Theory

The interaction and connection between human system and natural system in the earth's surface constitute a giant system with large scale, vast space, long time, complex structure, numerous elements, and comprehensive functions, that is, the human-earth interaction builds the human-earth relationship system. Essentially, it is a dynamic structure of human-earth interaction in a specific region (Wu 1991). In the human-earth areal system, the change of each element may cause the transformation of other elements and the whole system. Therefore, mechanism exploring these processes is the research theme of human-earth areal system (Lu 2002). Industrialization and urbanization, the most obvious economic and social activities on the surface of the earth, have changed the traditional human-earth areal system, and affected relationship between urban and rural areas. Besides, cities and villages are two relatively independent subsystems. The coupling of elements and the relationship between them form urban–rural spatial structure and external resources and environment response. This means that urban–rural social, economic and ecological relations are also important contents of human-earth areal system.

Urban–rural regional system is a complex system formed by urban–rural connection within a certain geographical space, and is a spatial organization form with regional connotation and hierarchical system, including two subsystems of urban system and rural system (Fig. 3.2). Urban system is consisting of different levels of urban settlements, such as large city, medium-sized city, small city and suburban community; while rural system is composed of village-town settlements, such as village, central communities and town. Thus, the urban–rural interaction forms a unique urban–rural crisscross system (Zhang and Liu 2008). The urban–rural crisscross system originates from the dynamic, unconformity and relativity of urban and rural development, which makes it difficult to completely divide urban and rural areas (Zhang and Sheng 2002). Due to the different combination of industry, population, land, capital, information and other elements, there are many differences between urban and rural areas in industrial type, production mode, social culture and management system, forming different functions between urban and rural areas, including economic development, social security, food safety and ecological supply. Urban function refers to the role and labor division of political, economic, social and cultural activities within a certain region, which is determined by the composition and structure of various elements of the city and embodied through industrial development, political culture, transportation and tourism. Generally, urban function includes production, service, transportation, innovation, management and distribution. Rural function is the spatial combination of productive and non-productive functions, and refers to the role and labor division in promoting regional development, protecting ecological environment, ensuring food security and inheriting traditional culture. Therefore, there is a coordination and complementarity between urban and rural functions (Liu and Liu 2012).

In the process of industrialization and urbanization, the evolution of urban–rural structure and function and its impact on resources and environment are the most important characteristics of urban–rural regional system. The rural system exports population, food, raw materials to urban system, while the urban system provides technology, information and industrial products to the rural system. But due to the distance attenuation law and diffusion principle, this connection weakened with the increase of distance (Mao 1995). In the context of urban–rural transformation, the urban–rural regional system has evolved into a spatial structure system and related system of factor flow and agglomeration of population, land, industry and capital. They interact with each other, and gradually form a giant system with large scale, wide space, complex structure and comprehensive functions (Liu et al. 2014). The theory of urban–rural regional system is the theoretical basis for understanding the process, mechanism, structural characteristics, and development trend of urban–rural development, and seeking for its optimal regulatory approaches. It is also theoretical guidance for rationally allocating urban–rural resources, reconstructing rural space, promoting urban–rural transformation, and accelerating urban–rural integrated development. Based on

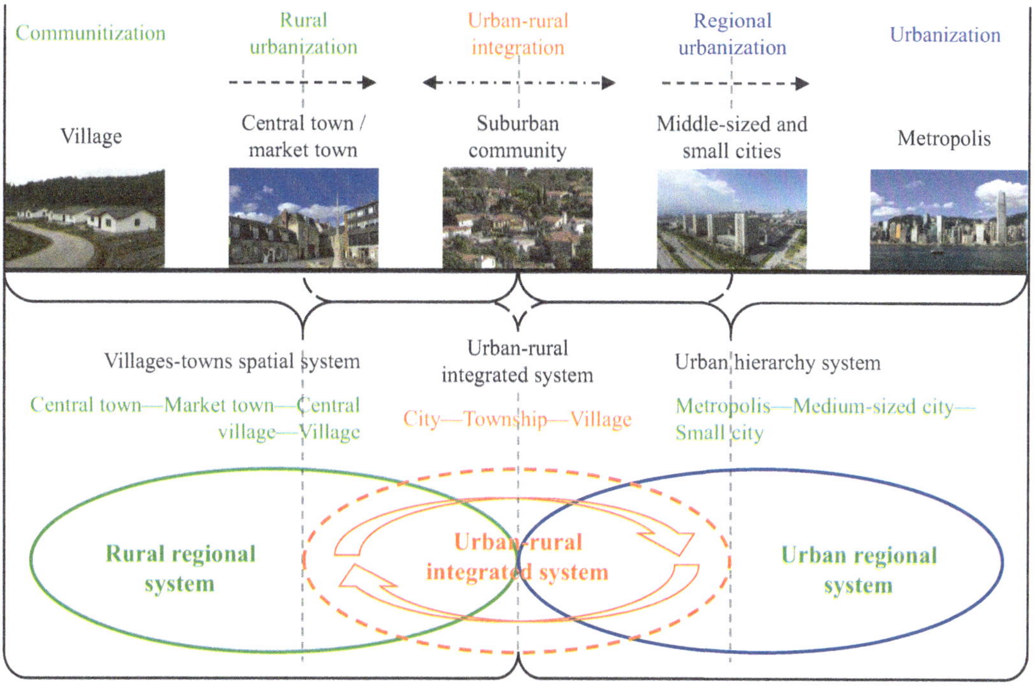

Fig. 3.2 Mechanism and hierarchy of urban–rural integration system

the concept of urban–rural regional system, researches on urban–rural development should clarify the relationship between system and element, structure and function, environment and behavior, thus promoting the rational allocation of urban–rural resources, equal exchange of urban and rural elements and optimization and reconstruction of urban–rural space. Eventually, the optimization of urban–rural relationship will be realized through the benign transformation of regional system.

3.1.4 Land Optimal Allocation Theory

In general, land resources allocation refers to the process of arranging the specific amount of land resources with selective development and utilization needed by human beings in the region through certain technical methods, thus achieving specific social, economic and ecological goals. On the one hand, land resource allocation ensures the reasonable distribution of land resources among various competitive uses, on the other hand, it effectively improve the efficiency of land use. With the development of industrialization and urbanization, the element, structure and function of urban–rural regional system have changed rapidly. In order to meet the needs of urban–rural transformation, the optimal allocation of land use should fully consider the principles of regional integrity, goal leading, dynamic allocation, structural coordination and spatial continuity.

1. **Principle of regional integrity**

Urban–rural transformation aims to realize the interaction of urban and rural elements, and gradually eliminate the dual structure between urban and rural areas. Land resources are not only the spatial carrier of urban–rural development, but also an important element in urban–rural transformation. The allocation of land resources should be in line with the regional goal of urban–rural integrated development. Therefore, the optimal allocation of land resources should not only consider the demand of urban economic and social development in the process of rapid urbanization, but also take the demand of rural

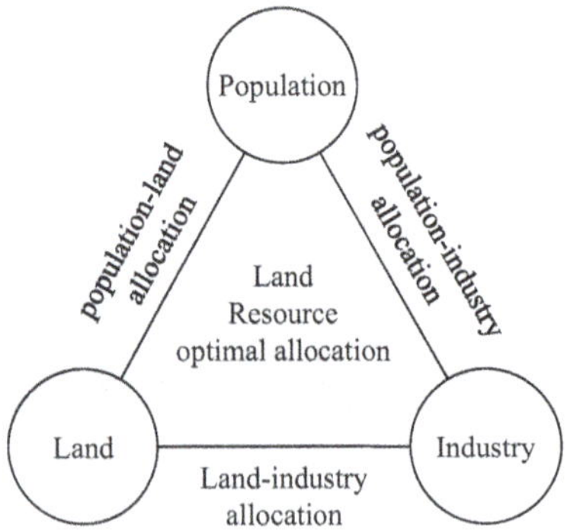

Fig. 3.3 Land allocation linkage mechanism adapting to urban–rural transformation

development as well as the optimization of rural land consolidation and management mode into consideration. Focusing on the population-land-industry linkage mechanism of land use in urban–rural transformation, the principles of population-land allocation, population-industry allocation and land-industry allocation should be analyzed systematically (Fig. 3.3).

2. **Principle of goal leading**

The regional differences and evolution stages of urban–rural transformation, as well as the diversity of driving mechanism, correspond to the differences of land use supply and demand and effect in urban–rural transformation. The principle of land resources allocation in different regional types of urban–rural transformation should take regional differences into consideration, characterized by the diversities in the dominant principles and objectives of the allocation of various regional types. For example, the area with rapid urbanization should take the regional industrial transformation and upgrading as the guidance to carry on the reasonable allocation of land resources. Rural area with serious hollowing should take the principle of spatial integration, industrial integration and organizational integration into consideration to reasonably allocate land resources.

3. **Principle of dynamic allocation**

Urban–rural transformation is a dynamic process. With the transformation of urban and rural areas, regional industrial structure, employment structure and land use structure are also in dynamic change, and the transformation and upgrading of industrial structure has a great impact on land use structure. Therefore, the optimal allocation of land resources adapting to the urban–rural transformation is a dynamic process. In land use system, the information entropy of land use types presents a feature of dynamic evolution, and gradually tends to a balanced steady state, and finally realizes the optimal allocation of regional land resources (Fig. 3.4).

4. **Principle of structural coordination**

The optimal allocation of land use provides a theoretical basis for the land use in regional urban–rural transformation. In this process, the optimal allocation of land use should be coordinated with land use planning, urban–rural development planning and industrial planning. According to the main strategies and planning objectives of regional industrial development, urban–rural transformation needs to follow the principle of optimal structure benefit of land use and sustainable land use, and must be coordinated with the requirements of optimal allocation of land use. Urban–rural development planning plans for the spatial structure of urban and rural development. The optimal allocation of land use not only makes the optimal allocation of the quantitative structure of land use linked with population-land-industry, but also requires high coordination with the regional characteristics and

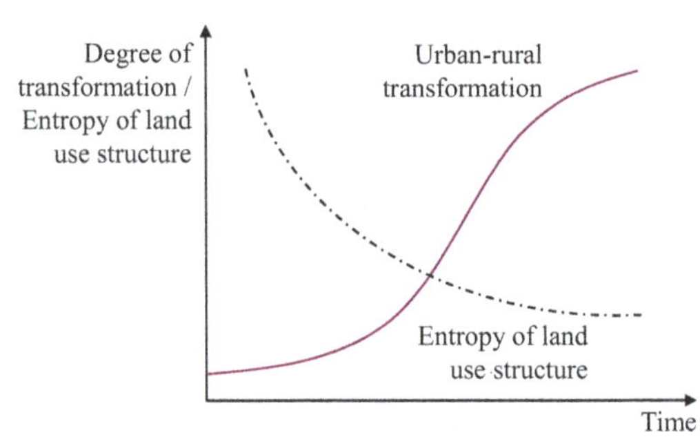

Fig. 3.4 Information entropy change between urban–rural transformation and land use structure

spatial structure of regional urban–rural transformation. Fully considering the industrial demand of the optimal allocation of land use and the high coordination of urban and rural spatial development provides strong support for land use planning, and ensure the sustainable and optimal use of land in urban–rural transformation.

5. **Principle of spatial continuity**

The optimal allocation of land use is an important way to ensure the reasonable and coordinated development of population-land-industry in urban–rural regional system according to the needs of urban–rural transformation. The quantitative and spatial structure of regional land use is formed in the process of long-term human economic and social development. In the evolution of land system, the optimization and promotion of land use system are based on the current situation. To adapt to the multi-objective and multi-scene optimal allocation of land use in urban–rural transformation, it is necessary to inherit the reasonable aspects of land use structure and spatial pattern. The main idea of the optimal allocation of land use is to reasonably adjust the structure of land use, and then optimize and upgrade the function of land system, achieving the adaptation and promotion between urban–rural transformation and land use transformation. Therefore, to optimize the allocation of urban–rural land use, it is necessary to fully consider the current characteristics of land use.

3.1.5 Urban–Rural Relationship Theory

Adam Smith, the originator of economics, pointed out in his famous book *An Inquiry into the Nature and Causes of the Wealth of Nations*, that the city and the village are a community of interests. On the one hand, the countryside supplies the city with means of living and raw materials for manufacturing, on the other hand, the city sends back some manufactured goods to the rural residents as a return. Essentially, city and village are the relative regional space in urban–rural regional system. There are many factors flow between urban and rural subsystem, such as labor, capital and technology, as well as many connections between industrial sectors. In the process of factor flow and structure change, dynamic changes and new relationship forms are produced. Focusing on the urban–rural relationship, many theories and social practices guiding urban–rural development are put forward. These theories and practices show staged and regional characteristics.

1. **The developmental stage of Marx and Engels**

Employing historical materialism, Marx and Engels analyzed the internal relationship between urban and rural areas. They believe that the social division of labor leads to the separation of urban and rural areas, which makes a large number of production factors flow to the city, and this flow reflects social progress. Due to the continuous concentration of production factors to the city, economic centers also begin to turn to cities. As a result, urban modern civilization has replaced rural traditional civilization and become the goal that people pursue. However, rural areas have become the pure suppliers of urban raw materials and resources, becoming an area where the development power is insufficient and the infrastructure is seriously lagging. Thus, the gap between urban and rural areas continues to widen, which eventually leads to the deterioration of urban–rural relationship. In *Manifesto of the Communist Party*, Marx pointed out that the village succumbs to the rule of the city under the unequal capitalist system, and the conflict between urban and rural areas is difficult to solve. Only by overthrowing the private capitalist system and establishing the public socialist system can the opposition between urban and rural areas be eliminated and the integration of urban and rural areas be realized.

2. **Unbalanced growth theory**

In 1950, French economist Perrour put forward the growth pole theory, and held the view that economic development is not balanced in geographical space, but is distributed in a point shape with different intensities and produce various effects on the whole region according to different ways of transmission. These dotted regional spaces are growth poles with growth and agglomeration. City is a growth pole. The connection between urban and rural areas is mainly realized by the diffusion of urban resources to rural areas through different channels. This process emphasizes the city as the center and drives rural development through the flow of elements from urban to rural areas. Furthermore, Myradal proposed the principle of cyclic accumulation causality in 1957, Hirshman put forward the polarization effect and trickle-down effect, and Friedman advanced the theory of spatial polarization development. These theories emphasize the imbalance of regional economic growth, and take the city as the development center, and the economic development further aggravates the regional gap.

3. **Urban bias theory**

Since the 1970s, the connection and communication between urban and rural areas have been strengthened continuously, and urban and rural areas have become an integral part of the whole region. Research on the interaction between urban and rural areas and the change of urban–rural economic and social structure became the mainstream in this period. Based on criticizing the city center theory, Lipton put forward the city bias theory. He believes that urban groups make use of their political power to make production factors and social resources flow to their own interest areas through the urban bias policy, which limits the development of rural areas, and this kind of unfair urban–rural communication objectively leads to the backwardness of rural areas and the widening gap between urban and rural areas (Lipton 1993). Corbridge thinks that urban–rural connection is the result of the interactions of some basic social structures and is dependent on these social processes. The root of urban bias lies in low food prices and a series of other price policies that are not conducive to rural areas. Correspondingly, the investment strategy that favors urban industry leads to the lack of technology in rural areas, and the common problems in rural areas, such as insufficient and low-quality infrastructure, especially in education and medical-care (Stöhr and Taylor 1981).

4. **Urban–rural connection theory**

Rondinelli believes that the hierarchical and scale structure of city is the key to the success of development policies. He focuses on the connection between rural areas and small towns, small cities and big cities, and holds the view that the social and economic basis of small and medium-sized cities can change the rural areas. If the top-level design of rural development goal is separated from the city, the bottom-up strategy will not be effective. On the contrary, Stohr and Taylor emphasize the bottom-up development. This kind of development takes the countryside as the center, establishes based on the maximum utilization of the natural, human and institutional resources, and takes meeting the basic needs of the residents as the primary goal. Also, this kind of development should face the problem of poverty directly and be initiated and implemented by the rural subjects (Ma 1994). In general, Rondinelli, Stohr and Taylor all focus on the impact of urban–rural coordination on regional development, especially how this relationship determines the distribution of relatively scarce resources in society (Luo and Wang 2005). The density of urban–rural connection determines the intensity of urban–rural interaction, and political and economic factors further affect the flow direction of urban–rural resources (Liu et al. 2014).

3.1.6 Urban–Rural Multi-body System Theory

In 1991, Chuanjun Wu, a famous geographer in China, put forward that the human-earth areal system is the core of geographic research (Wu 1991). He believed that the human-earth areal system is based on a certain region of the earth's surface and is a dynamic structure composed of the two subsystems of human society and geographical environment in a specific region. Bases on this theory, the urban–rural integrated system and rural regional system are the basic basis for a new understanding of modern urban–rural relationship and diagnosis of rural development problems. In general, urban and rural areas are complex regional systems with spatial mosaic, structural complementarity, functional coupling and interaction (Zhang and Liu 2008). The urban–rural relationship is the basic relationship between urban and rural dual socio-economic structure of a country or region. Rural area is an important foundation of economic and social development. In the process of rapid urbanization, rural development aims to continuously narrow the gap between urban and rural areas and gradually realize urban–rural integration (Du and Zhang 2006). Therefore, the pattern of regional urban–rural relationship has become the fundamental to understand the rural development, and also an important basis to promote urban–rural transformation.

According to the "point-axis" theory proposed by Lu (1988), regional economic development is always concentrated in a few areas with better conditions, and is distributed in a point shape. With the development of economy, the number of economic centers is increasing gradually, the point-axis system is formed by connecting the points through the development axes such as traffic lines, and the network is formed by the interweaving of multiple point-axis systems. The essence of urban–rural integrated development is to give full play to the diffusion effect on the basis of strengthening the polarization effect of urban–rural regional system, build an urban–rural destiny community, and form a three-dimensional space and grid structure of urban–rural development. The urban–rural transformation focuses on the rural regional multi-body system, which mainly includes urban–rural integrated system, rural complex, village-town organism, and housing-industry symbiosis, with hierarchical, regional and dynamic characteristics. To promote urban–rural transformation, it is important to achieve multi-level goals, which are composed of urban–rural infrastructure network, rural development area, village-town spatial field and rural development pole.

1. **Urban–rural integrated system and urban–rural integrated body**

The system is an organic whole with specific functions, which is composed of interactive elements according to a certain structure. In accordance with the pattern and form of urban–rural differentiation, urban–rural integrated system includes three levels of region, prefecture and county, which are connected and integrated through urban–rural infrastructure network. Urban agglomerations, megalopolis and new-urbanized areas belong to regional system, showing an urban–rural regional pattern with metropolitan areas as the main form; megalopolis, large cities and medium- and small cities belong to prefectural system, presenting a regional pattern with built-up area as the center and urban–rural equality; the county and its central town and rural community belong to county system, presenting a regional pattern characterized by big countryside and urban–rural integration. Cities and villages are an organism, only when both of them are sustainable can the support each other (Liu and Li 2017). Integration contains the meaning of harmony and penetration, and is a state in which different objects intersect, penetrate and integrate with each other. The urban–rural integrated system is an urban–rural crisscross system formed by the intersection, penetration and integration of urban and rural regional system (Fig. 3.5), and is composed of small and medium-sized cities, small towns, suburban communities and rural space. In China, "urban disease" and "rural disease" coexist, resulting in the imbalance between urban and rural areas and the increasing gap between urban and rural areas. Therefore, establishing a sustainable urban–rural integrated system and body is the basic premise of urban–rural transformation (Liu et al. 2018).

2. **Rural regional system and rural complex**

Rural regional system, also known as rural complex, is a rural spatial system with certain structure, function and interregional connection, which is composed of human, economy,

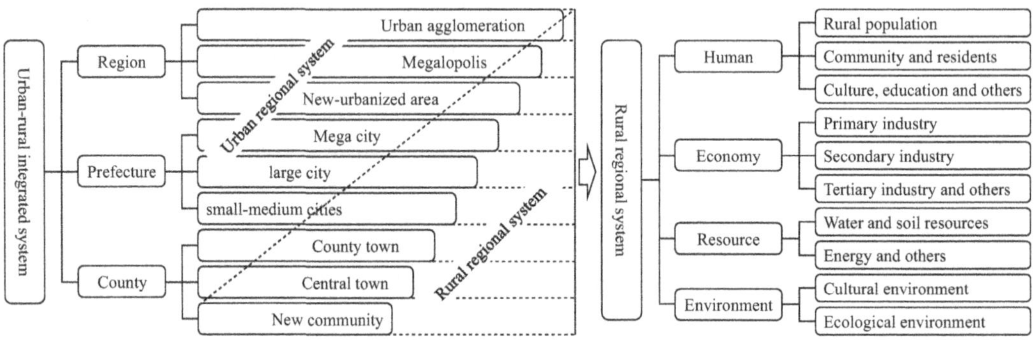

Fig. 3.5 The structure of rural–urban integrated system and rural regional system

resources and environment (Fig. 3.5). Compared with the "urban areas", the "rural areas" is a region including the vast areas outside the urban built-up area. "Village" is a kind of settlement, which is a collection of villages and towns within the county, and presents multi-level settlement space of different scale and level in rural areas, including county town, central town, market town, central village (community), administrative village and natural village, forming the village-town organism and its housing-industry symbiosis based on rural mountain, water, forest, field, lake and grass. In terms of function, it has developed different types of rural development areas, such as village-town community, agriculture-orientation area and industry-orientation area. Rurality is the essential feature of rural areas, which breed urban areas and support the urbanization. With China's economic and social transformation and the rapid development of industrialization and urbanization, the geographical space of the organization and development of agricultural production factors has changed significantly (Liu et al. 2018), and the problems of farmland conversion and village hollowing have become increasingly prominent due to a large number of rural populations leaving their hometown for the city. In this context, urban–rural transformation is to scientifically grasp the differences and dynamic characteristics of rural areas, optimize regional village-town pattern, solve the problems of rural regional system and promote urban–rural integrated development. Therefore, the overall strategic design must be carried out in different categories and in different areas, deepen the analysis of the process, spatial–temporal pattern and regional mode of rural transformation, and take the promotion of population employment, resource development, environmental governance, industrial upgrading, cultural heritage and institutional innovation as the main topics of urban–rural integrated development, laying the theoretical foundation for urban–rural transformation.

3. **Village-town construction pattern and village-town organism**

Villages and towns are important carriers of elements gathering and spatial organization in rural complex. Village-town construction pattern is the morphological characterization of spatial restructuring, organization reconstruction and industrial reshaping of rural regional system, and its core is to optimize the spatial layout, hierarchical relationship and governance system of county towns, key towns, central towns and central villages (communities) in rural areas (Liu 2013). In general, the village-town construction pattern is characterized by village combination, abolishing township to establish town, industrial park construction and spatial agglomeration. Through clarifying the status of villages and towns, adjusting the spatial structure and strengthening the central function, the spatial integration and functional fit between towns and villages can be realized, thus forming a clean and idyllic living space, intensive and park-oriented industrial space, civilized and beautiful ecological space, and the regionalization and diversification of cultural space (Liu et al. 2014). In view of the current widespread phenomenon of "weak towns and hollowing villages", urban–rural transformation needs overall planning, adjusting measures to local conditions and make key breakthroughs. Based on the integration of urban and rural areas, urban–rural transformation needs to focus on improving urban–rural infrastructure network, highlighting the characteristics of different rural areas, strengthen the advantages of central community and key villages and towns, accelerating the cultivation of village-town organism and housing-industry synergic agent integrating production, life and ecology, strengthening the space field of villages and towns and rural development pole (Fig. 3.6), promoting the formation of cities-towns-villages system feature by spatial coordination, reasonable grade and livable and efficient.

4. **Rural prosperity and housing-industry symbiosis**

Based on village-town organism, the housing-industry symbiosis is a village-town growth pole formed by the integration of residence and employment in rural areas. Housing-industry synergy and spatial polarization are the necessary endogenous driving forces for urban–rural transformation, and are the advanced forms of the integrated development of human, housing and industry in villages and towns. There are different types of villages and towns, and different modes of synergism. The village with development conditions should establish the concept of integration of primary, secondary and tertiary industries, cultivate leading industries and strengthen the rural and town economy based on village-town space, building livable and employable community. As to the villages with general conditions and large population, they should be in accordance with the requirements of the new-countryside construction (Liu 2011), take improving infrastructure and public service facilities as the premise as well as integrating land resources, developing characteristic industries and improving living conditions as the focus, constantly strengthen the organization construction, increase investment, accelerate the improvement of shortboard and cultivate the center in rural areas. While, as to the village with poor condition and small population, such as hollowing villages, they should adhere to the principle of central agglomeration and intensive development, respect the wishes of farmers, do a good job in overall planning, explore and promote the construction mode of village combination and housing-industry synergy, and develop new communities and small towns (Liu et al. 2018). The focus and foothold of urban–rural transformation is to seek the prosperity of rural industry and people as well as living and working in peace, constantly meeting people's growing needs for a better life.

3.1.7 Human–Earth Relationship and Human–Earth System

In 1981, Wu proposed that "the human–economy–nature system is an interdisciplinary research fields, and geographers focus on the

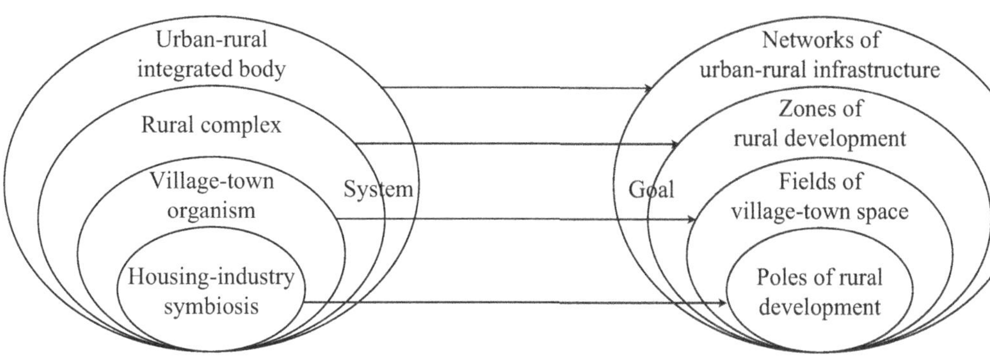

Fig. 3.6 The structure of rural–urban integrated system and rural regional system

regional system of human–earth relationship, which is the central research topic of geography". Also, he pointed that the methods to study the human–earth areal system mainly include classification, regionalization, evaluation, quantitative analysis and model establishment (Wu 1981). In 1991, Wu further explained the human–earth areal system and its key fields of geographical research. He believes that "the human–earth areal system is a complex and open giant system composed of the two subsystems of geographical environment and human activities. It is a human-earth relationship system based on a certain region of the earth's surface, and is a dynamic structure formed by the interaction between human and land in a specific region" (Wu 1991). The human–earth areal system is dynamic, and it will develop and change with the activities of human transformation and utilization of nature. Correspondingly, its content will become more and more abundant, which give new challenges and tasks for geographers (Wu 1981).

The surface of the earth is a spatial mosaic of different regions or landscapes. With the rapid development of global economy and the continuous progress of modern science and technology, the way and influence degree of human activities on the earth system are constantly changing and strengthening, and the era connotation, type structure and regional function of the human-earth areal system are constantly enriched and developed. Therefore, it is of great significance to the types, spatial–temporal differentiation and change laws of human–earth areal system at different levels. According to Wu's thought on human-earth areal system of its spatiotemporal differentiation, and focusing on the macro background of urban–rural transformation and territorial spatial governance in the new era, the human–earth areal system can be divided into urban regional system, urban–rural integrated system and rural regional system. Among them, the rural regional system can be subdivided into agricultural system, village system, rural system and urban system, forming the type structure of human-earth system with clear hierarchy and logic system (Liu 2020a) (Fig. 3.7).

These subsystems interact and depend on each other to constitute a specific relationship between the subject and the object or the relationship between human and earth or the relationship between biology and environment in a universal sense, namely, crop–soil relationship, population–settlement relationship, housing–industry relationship, and industrialization–urbanization relationship (Liu 2020a). The core of the agricultural system and village system is to shape the village settlement system. The core of village system and urban system is to construct urban hierarchy system, that is, the new village–town construction pattern, which focuses on spatial organization and structural optimization. The core of agricultural system and rural system is to establish rural production system, which focuses on the sustainability of rural regional system. The core of urban system and rural system is to build urban–rural integrated system, whose ultimate goal is urban–rural integration and even urban–rural equivalence.

3.2 Core Concepts of Urban–Rural Transformation

There are positive and negative transformation in urban and rural areas. Among them, positive transformation refers to that the urban–rural regional system develops towards the coordination and integration of urban and rural subsystems according to the specific transformation mode. This process includes population migration, industrial upgrading, spatial reconstruction, social change and environmental change between urban and rural areas. To fully understand urban–rural transformation and its scientific connotation, four core concepts of urban–rural transformation are proposed, including "transformation tree", "transformation law", "transformation threshold" and "fusion body".

3.2.1 "Transformation Tree" of Urban–Rural Development

The "transformation tree" of urban–rural development is a concrete exposition of the hierarchy

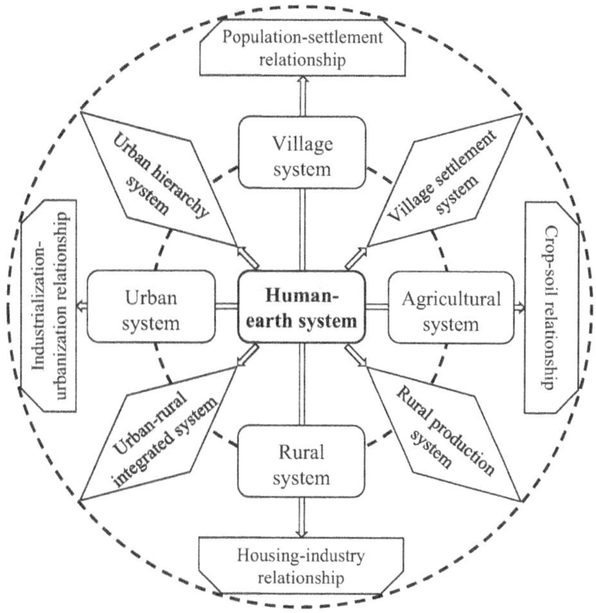

Fig. 3.7 Types and structure of rural human-earth system

and structure of urban–rural transformation from the factor level. From bottom to top, it mainly contains ecological space and public services, water-soil and industry, settlements and facilities, forming a complete tree system. The "transformation tree" reflects the organizational structure of the elements of different scales and levels in the process of urban–rural relationship changing from opposition to coordination.

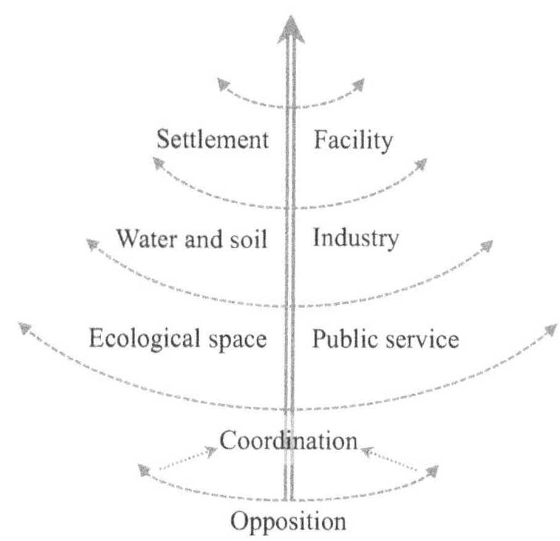

Fig. 3.8 "Transformation tree" of urban–rural development

Among the elements at all levels of the "transformation tree", the regional coverage of basic elements is wide, and its supporting role are strong, but its realization is difficult; while the coverage of the top elements is relatively small, and its supporting role is relatively weak, but its realization is less difficult (Fig. 3.8). Ecological space and public services are in the basic layer of the "transformation tree", which are the two groups of elements with the widest coverage. The former is the factor system which has the closest relationship with human activities and the most violent response. Due to the weak immunity of the system to the destructive disturbance in the process of industrialization, urbanization, and agricultural modernization, its self-regulation ability is limited, and the recovery is difficult. Ecological space is the platform of urban–rural elements flow and spatial reconstruction, and the spatial basis of urban–rural transformation.

Public service is the basic institutional environment to ensure the orderly development and transformation, which lead to factor flow and gives birth to the demand for optimal resources allocation. Therefore, it is urgent to promote the balanced allocation of urban and rural public resources by means of corresponding system guarantee and free and two-way flow of elements. Water-soil and industry are in the main layer of the "transformation tree", and are also the power engine of urban and rural development and transformation. Water-soil refers to the distribution and allocation of various natural resources such as water, soil, climate, and biology, which is the basic carrier for industrial development. Industry is not only the main body of factors utilization, but also the basis of economic development in urban and rural areas. Its development depends on the reasonable and orderly organization of elements. Correspondingly, the coupling allocation of various natural elements determines the resource-environmental bearing capacity and the potential productivity level, and will lead to the upgrading of industrial structure in the main urban and rural sectors.

Elements in the main layer are regional-orientation and problem-orientation, and play a supporting and leading role urban–rural transformation. Settlements and facilities are the top elements of the "transformation tree", and the foothold of the strategy for urban–rural transformation. The former refers to residential area, including rural communities, small towns, etc., which is the spatial carrier of transformation. The latter refers to the foundation and platform of public services and the hard environment to ensure development. Relying on the facilities construction and service guarantee, the industry can be established on the platform of existing ecological security pattern.

3.2.2 "Transformation Law" of Urban–Rural Development

The "transformation law" of urban–rural development is an evolutionary theory about urban–rural transformation. It measures the transformation from the perspective of process, involving urban–rural gap, urban–rural accessibility, industrial linkage, urban–rural interaction and so on. Since the process reflects the adjustment of pattern caused by the change of urban–rural elements, the theory of "transformation law" describes the process of urban–rural system from isolation and opposition to coordination and integration, and divides it into four periods (Fig. 3.9). Emergence period aims at the overall planning of urban and rural areas, rapid period aims at the coordination of urban and rural areas, integration period aims at the integration of urban and rural areas, stable period aims at the equalization of urban and rural areas. Different periods are divided according to the difference of actual growth rate.

The "transformation law" is characterized by the degree of urban–rural transformation, which is composed of economy, society, institution, and ecology. Each level includes two parts, development indicators and transformation indicators, which are used to measure the specific stages of urban–rural transformation. Therefore, a total of 11 indicators are selected, including gross domestic product (GDP), total retail sales of consumer goods, non-agricultural employment, income ratio of urban and rural residents, total collection of public libraries, number of beds in health institutions, and urbanization rate (Table 3.1).

3.2.3 "Transformation Threshold" of Urban–Rural Development

The "transformation threshold" of urban–rural development is the innovation order theory of urban–rural transformation, which describes the priority order and mechanism of urban–rural transformation based on the system level (Fig. 3.10). Strategic domain is the basis and platform of transformation threshold, and the basic consensus of promoting development and transformation of urban and rural areas. Starting from urban–rural relationship, strategic domain refers to the relevant urban and rural development strategies

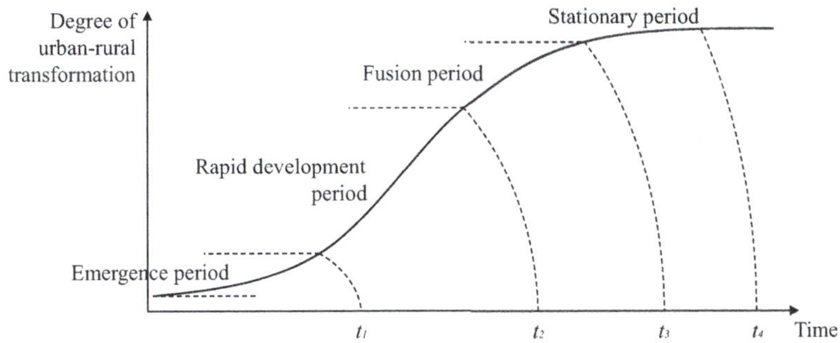

Fig. 3.9 "Transformation law" of urban–rural development

Table 3.1 Comprehensive evaluation index system of urban–rural transformation degree

	Level of transformation	Index	
Degree of urban–rural transformation	Economy	Development index	Gross domestic product (GDP)
			Total retail sales of consumer goods
		Transformation index	Non-agricultural employment
			Income ratio of urban and rural residents
	Society	Development index	Total collection of public libraries
			Number of beds in health institutions
		Transformation index	Urbanization rate
	Institution	Development index	Number of patent applications and granted
		Transformation index	Investment in fixed assets of non-state-owned economy
	Ecology	Development index	Green coverage rate of built-up area
		Transformation index	Standard rate of industrial wastewater discharge

Fig. 3.10 "Transformation threshold" of urban–rural development

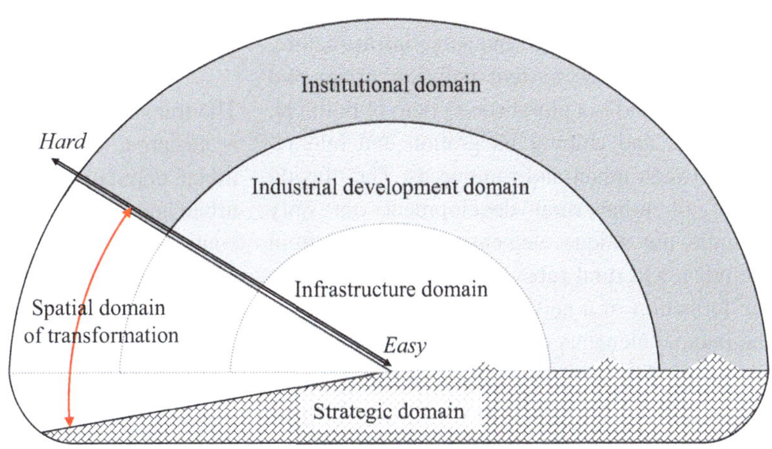

that have been defined and implemented by the national government. The infrastructure domain is in the core layer, and corresponds to the top elements of transformation tree, which is the guarantee of transformation and includes roads, greening and other infrastructure driven by investment. Its development and transformation are less difficult. Compared with the core threshold, the transformation of peripheral threshold is more difficult, and the innovation space and effect brought by transformation are also stronger. Industrial development domain is in the periphery of infrastructure domain, corresponding to the main layer of "transformation tree", and is the leading driving force of urban–rural development and transformation. Based on the premise of strategic domain and the guarantee of facility domain, accelerating industrial development and structural upgrading meets the needs of urban–rural transformation. The institutional domain is in the outermost layer of the transformation threshold, which is the most difficult to innovate, the most extensive transformation space and the strongest triggering effect. As the wind vane of regional development policies and stage goals, the transformation and development of institutional domain will drive the common changes of other domains.

3.2.4 "Fusion Body" of Urban–Rural Development

"Fusion body" of urban–rural development is a spatial pattern theory about urban–rural transformation. It carries the industry, infrastructure, and public service system between urban and rural areas, and is a pluralistic system of political, economic, and cultural integration and interaction between urban and rural areas. The "fusion body" of urban–rural development not only includes the unique element connection within the urban and rural subsystems, but also leads to the formation of a new correlation mode among the internal elements of the integration body due to the mobility of population and the demand for scarce resources (Fig. 3.11). The characteristics of "fusion body" are mainly reflected in the

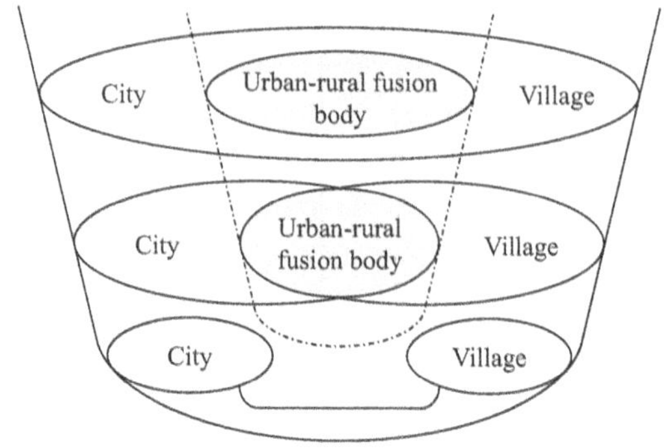

Fig. 3.11 "Fusion body" of urban–rural development

elements flow, function transformation, scale shift and so on, and are embodied in the realization of two-way flow between elements and the simultaneous occurrence of integration in multi-level (ecological, living, production and other functional levels), multi-scale (urban and rural areas, different levels of cities), and multi-dimensional (time series transformation). The "fusion body" depicts the aggregation force and centrifugal force of regional economic and social activities, reveals the change trend of the two forces and their influences on geographical structure and spatial form of regional economy.

3.3 Theoretical Exploration of Urban–Rural Transformation Geography

3.3.1 Changes of Urban and Rural Structure and Its Law

1. **The transformation of urban–rural population and industry**

The transformation of urban and rural population is the process of rural population changing into urban population, that is, the transformation of urban and rural population is characterized by the continuous growth of population urbanization rate. Based on summing up the development experience of all countries in the world, Northman, a famous American economist, has constructed a S-curve model to reflect the process of urbanization (Fig. 3.12). Taking the urbanization rate of 25%, 50% and 75% as the dividing line, the

Fig. 3.12 Theoretical curve of urban and rural population process

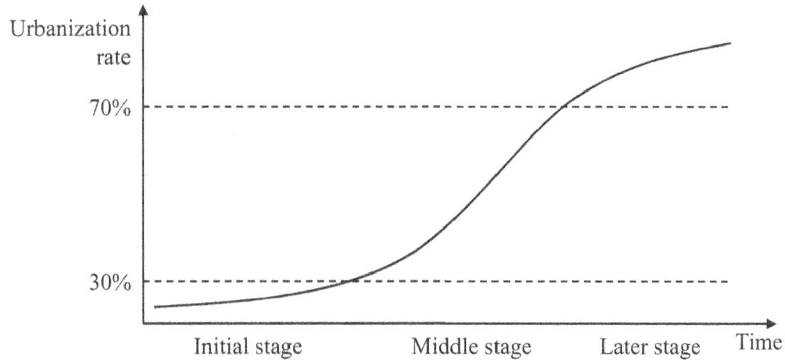

development and evolution of urbanization can be divided into four stages, and each stage has different economic connotation and performance. In the early stage of urbanization when the urbanization rate is lower than 25%, the material basis of urban development is weak, the scale is small and the development is slow. The stage of urbanization rate between 25% and 75% is the leap stage of urbanization development, that is, the growth stage of urban and rural tertiary industry; At the stage when the urbanization rate reaches more than 75%, all aspects of economic and social development tend to be mature, and the speed of urbanization obviously drops, which means that urbanization has entered the later stage. Economist Chenery divides the process of urbanization into non-urbanization stage ($U < 20\%$), industrialization preparation stage ($20\% \leq U < 32\%$), initial stage of industrialization ($32\% \leq U < 36.4\%$), middle stage of industrialization ($36.4\% \leq U < 49.9\%$), mature stage of industrialization ($49.9\% \leq U < 65.2\%$), late stage of industrialization ($65.2\% \leq U < 70\%$) and high urbanization stage ($U \geq 70\%$). From the process of urbanization in the world, most countries have experienced the transformation process from urban–rural differentiation, isolation and opposition to urban–rural integration and integrated development. Among them, the urbanization rate in the stage of urban–rural integration is generally 50–70%, and the industrialization rate is 40–50% (Chi 2010).

Industrial structure refers to the composition among and within the industrial sectors of the national economy, as well as the economic connections and quantitative contrast relations that are interdependent and mutually restricted. The changes of industrial structure show the emphasis of national economy and the distribution of production factors among different sectors. Urban–rural transformation is a process of upgrading the industrial structure, and is characterized by the gradual transformation from primary industry to secondary and tertiary industry. Generally, industrial transformation and upgrading is accompanied with the urban–rural employment transformation, which is mainly reflected in the continuous improvement of non-agricultural employment level in urban and rural areas. Among the various theories on the evolution of industrial structure, the most famous one is the theory of industrial gradient transfer put forward by British economist Clark. With the development of economy and the improvement of per capita national income, the proportion of labor in the three industries shows a trend of transferring from the primary industry to the secondary industry, and then from the secondary industry to the tertiary industry. Based on Clark's research, Kuznets, a famous American economist, focuses on the statistical analysis from the perspective of the proportion of primary, secondary, and tertiary industries in the national income, combines the evolution trend of national income and labor force among the three industries, and further demonstrates the evolution law of industrial structure. In terms of the change of the proportion of the employed population among the three industries, there is a trend of transferring from the primary industry to the secondary industry, and then to the tertiary industry. As to the change of the proportion of national income, there is a decline in the proportion of national income realized by primary industry, while the proportion of the national income realized by the secondary and tertiary industries shows an upward trend. The transformation and upgrading of regional industrial structure are one of the main manifestations of urban–rural transformation. The flow of urban and rural elements is directly related to the scale and structure of regional economy, and the evolution of regional urban–rural industrial structure directly drives the evolution of land use structure and spatial pattern. Urban–rural employment structure is the embodiment of industrial structure in the proportion of population employment, which is dependent on the change of industrial structure and reflects the stage of regional economic development.

2. **Interaction process of urban–rural population and industry**

The evolution of industry and population in urban and rural areas constitutes the basic economic and social form of urban–rural transformation (Liu 2020b). It is assumed that in the early stage of economic development, there are urban and rural areas and two sectors of industry and agriculture in a specific region. The return to scale of agricultural sector in rural areas remains unchanged, while urban industrial sector pursues economies of scale, and there is an obvious differentiation in productivity. The difference of labor productivity between urban and rural industrial sectors promotes the transfer of labor force from villages to cities, and the location advantages of cities promote the agglomeration of enterprises, which boosts the diversification of product market and consumer market. This has greatly met people's diverse needs, and result in the increase of the actual marginal utility of people's consumption, leading to the further agglomeration of rural population in urban areas. The agglomeration effect of population transferring to urban areas reduces the labor cost of enterprise searching and training, and promotes the agglomeration of urban non-agricultural industries, thus reducing the transaction cost of enterprises. Due to the higher productivity of the industrial sector, the factors of production are constantly transferred to the industrial sector, realizing the rapid development of the industrial economic sector. With the development of economy, the expansion of urban scale, and the change of people's demand structure, the service economy in urban and rural areas is developing rapidly, and the proportion of service economy and employment is increasing. When the elasticity of social demand for services is higher than that of industrial products, people will spend more on purchasing services, which will increase the proportion of urban–rural service economy. On the premise of lagging behind and diminishing returns to scale of living service industry, the growth rate of service industry is lower than that of labor input, and the decreasing returns of service industry offset the increase of returns in manufacturing sector, so the industrial expansion speed in urban areas will gradually be consistent with the growth rate of labor force (Tan and Zheng 2012).

Figure 3.13 shows the change curve of the share of urban population or industry, which shows that the proportion of urban industry and population changes with the improvement of economic level. The horizontal axis represents the level of economic development, and the vertical axis represents the share of urban population or industry. The evolution of urban–rural industry and population can be divided into three stages, where the first two stages are the agglomeration stage, and the third stage is the decentralized stage. The first stage is the leading period of urban industrial economy, when the mismatch between urban industry and population expands. The second stage is the leading period of urban service industry economy, when the proportion of urban industry and population increases, the gap between them has narrowed, and the balance between industry and population is improved. In the third stage, urban industry and population begin to transfer to rural areas, the proportion of industry and population in rural areas is rising, and the balance of population is increasing.

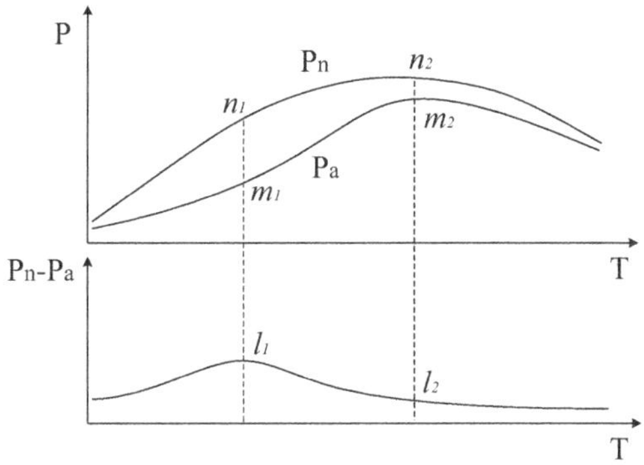

Fig. 3.13 The balancing process of urban–rural industry and population

Considering regional differences and regional policies, there are differences in the evolution of urban and rural industries and population. In the areas with superior location, technology, information and traffic conditions, urban economy develops rapidly. In addition, affected by a series of biased regional policies and investment preferences, the city scale effect is improved, which makes the production cost of enterprises decrease and the speed of industrial agglomeration is faster than that of other regions. The share curve of urban industry and population in these areas has increased faster. Under the interaction of urban–rural agglomeration and diffusion, urban–rural population transfer enters the third stage, and the curve forms a trend to upward. However, influenced by urban–rural dual system, the migrant population is restricted to enter the city, the cost of population mobility has increased, thus the share curve of population will decrease, and the imbalance between industry and population will increase. In some villages, township enterprise, rural tourism, and health care industry have developed rapidly since the geographical advantages and urban and rural policies, which make it possible for rural population to be urbanized locally and to transfer employment nearby. Therefore, characteristic and professional small towns become important places for population agglomeration, which slows down the transfer speed of rural population to urban areas, and the regional balance level of industry and population is relatively high.

At present, China is in the middle-late stage of industrialization and acceleration period of urbanization. Economic growth and population flow still show a trend of spatial agglomeration, which is reflected in the continuous increase of the proportion of various industries and the increase of the proportion of urban population. Affected by the changes in the interrelation between urban and rural areas, especially the agricultural industrialization, the rise of non-agricultural sector and concurrent business, the equilibrium between industry and population is characterized by fluctuation. From 1979–1984, the equilibrium between industry and population in China showed a downward trend, and reached the minimum in 1984; from 1984 to 1995, the equilibrium increased in fluctuation, while it declined briefly in 1995–1998, the equilibrium state further is on the rise as a whole after that (Fig. 3.14).

3. **Urban–rural transformation and land use transition**

Land use transition is a theoretical hypothesis based on the study of land use change. The theory assumes that there is a direct corresponding relationship among regional urbanization, industrial evolution and land use, and a certain stage of social development corresponds to a certain land use structure. Therefore, the urban–rural transformation at a specific stage should also have corresponding land use structure and spatial configuration. This kind of adaptive relationship is the goal of land use optimal allocation in urban–rural transformation. Generally, during the process of urban–rural transformation, the types of land with rapid changes in land use form are mainly concentrated in cultivated land, urban and rural construction land.

Industrial transformation and population transformation are the core contents of urban–rural transformation. According to the leading function of regional economy and the comparative advantage of regional industry, the urban and rural industrial structure in a region has gradually evolved into a certain industrial structure with the leading industry as the core. In

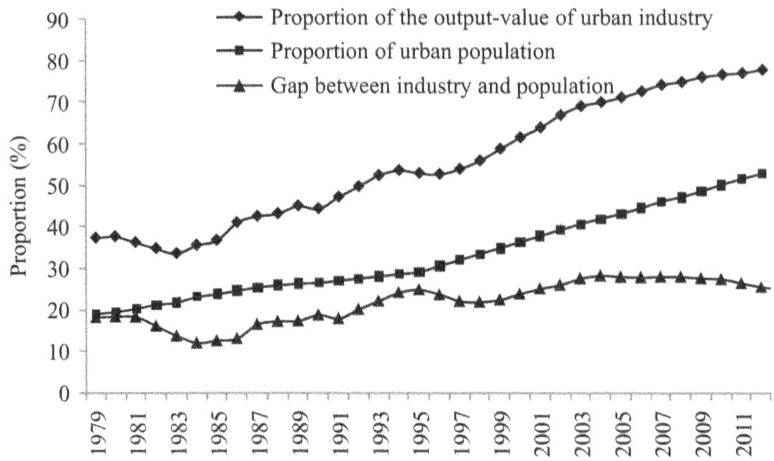

Fig. 3.14 Evolution trend of China's urban–rural industries and population

different stages and regions of urban–rural industrial transformation, there are significant differences in the demand and goal of land use and land resource development and utilization technology, and there are great differences in demand of land resource and utilization efficiency of land resources between urban and rural economic and industrial departments. Driven by the maximization of economic, ecological, and social benefits, the allocation of land resources in various industrial sectors has changed, and the change of regional urban and rural industries will lead to the change of the quantitative structure and spatial pattern of land use. The quantitative structure and spatial change of land use types also have a prominent impact on the evolution and change of the urban–rural economic development, industrial structure and layout.

The urban–rural population transformation is reflected in the process of population urbanization. Regional urbanization is the inevitable process and stage of economic and social development, and it is also the only way to regional modernization. With the urbanization stepping into an accelerated stage, the regional conflict between human and earth is becoming increasingly prominent, which brings about the rapid non-agricultural transformation of land resources and the continuous decline of regional per capita cultivated land, posing a certain threat to regional food security (Liu 2020a). The plain area with the highest quality and concentration of cultivated land resources is the main spatial carrier of urbanization expansion. Generally, scale expansion, industrial concentration and population growth in urban area play a significant role in regional land use structure and land use spatial structure (Fig. 3.15).

The changes of spatial demand caused by urban–rural transformation of population and industry directly and indirectly affected the process of farmland conversion. Therefore, the agglomeration of urban–rural population and the change of industrial structure are the main causes of land use transition (Hu et al. 2009). The development of regional industry increases the residents' income and promotes the level of urbanization, and the concentration of industry and population leads to the increasing demand of land for industrial development and urban infrastructure construction, as well as residential land and commercial land. The increasing income has enhanced the farmers' ability to pay for the cost of citizenization. This has given birth to the spatial demand for urbanization, resulting in the gap between supply and demand of urbanization space, which promotes the expansion of urban scale, especially the residential land. In terms of the payment level of the citizenization cost of farmers, when they can get more income, they will first consider returning to their hometown to build houses, which leads to the phenomenon that the old buildings are not replaced by new ones, and a large number of

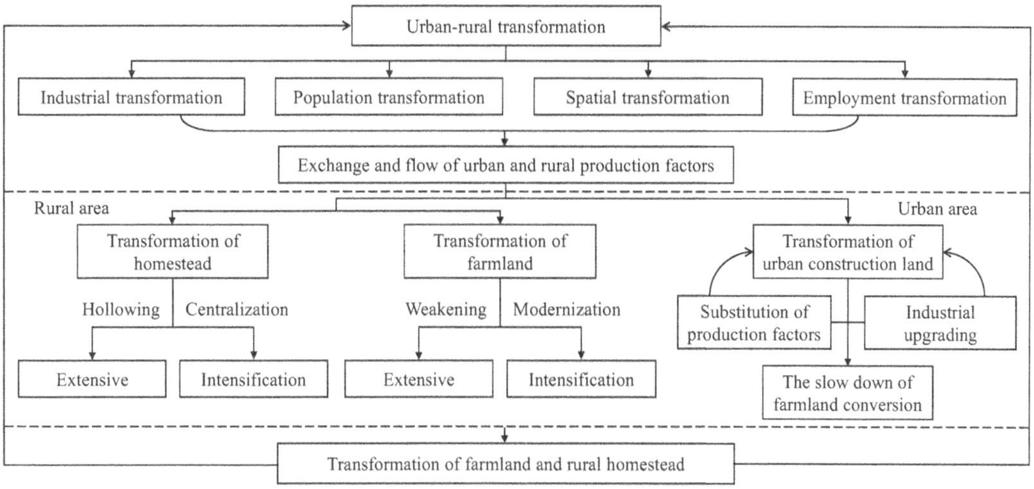

Fig. 3.15 Mechanism between urban–rural transformation and land use transition

rural homesteads are idle, resulting in extensive use of rural construction land. In the initial stage of urbanization, farmers are the main demanders of urbanization space in small towns. With the continuous improvement of farmers' income, its influence on the supply and demand of urbanization space in big cities continuously increases.

Urban–rural transformation is the result of the agglomeration of population and economic activities in urban space through the interaction of agglomeration and diffusion. Cities and towns are market nodes, and the market proximity effect, profits of manufacturers and cost of living of consumers jointly drive the concentration of population and employment. The scale and industrial cluster effects brought by externalities lead to the spatial concentration of population and economic activities. The transformation of rural economy, the decline of land security function and the decrease of farmers' dependence on land are the main driving forces for farmers to transfer to cities and towns and the transformation of non-agricultural employment and livelihood. With the adjustment of industrial structure and the change of population flow direction, the high land price in the central urban area hastens the change of urban industrial spatial pattern. On the one hand, the high value-added financial and other industries remain in the city center, and the phenomenon of "shifting from labor-intensive industry to service economy" in urban industrial development is widespread. On the other hand, the "big city disease" in the central urban area urges the middle class to transfer to the suburbs, promoting the development of suburbanization. Due to the industrial transfer and urban scale expansion, the non-agricultural conversion of farmland land is further intensified, which promotes the change of urban–rural land use structure and layout. The new-type urban–rural spatial system structure should be a healthy urbanization system with the orderly and reasonable, overall and coordinated development of big city, small and medium-sized city, central town, central village, central community. These settlements develop reasonably according to the industrial carrying capacity, market radiation force and central function, which promotes the urbanization pattern of urban–rural integration and the reasonably shift of industry and population spreading by big cities, thus ensuring the efficient utilization of urban–rural land.

Urban–rural transformation promotes the intensity and speed of urban and rural elements flow, especially the production factors from rural areas to urban agglomeration. The rapid non-agricultural transformation of rural production factors has a great impact on the development and evolution of the whole rural regional system. Cultivated land is the main labor object of farmers, while the non-agricultural labor force in rural areas is speeding up, the weakening of the main body of agricultural production has a certain impact on the intensive and economical use of cultivated land. In the backward mountainous areas, the marginal land with low land rent gradually withdrew from farming, which led to the widespread phenomenon of farmland abandonment. Agricultural land marginalization existed in some rural areas, and the form of cultivated land use changed. Rural homestead is the most important land use type of rural construction land. With the continuous non-agricultural transfer of rural population, the rural population is decreasing. As a result, many rural old houses are empty and idle, and the land use of rural residential land is generally extensive. However, the phenomenon of building new houses without demolishing the old ones is widespread, and the area of rural homestead increased instead of decreasing. Under the dual drive of new urbanization and new rural construction, it is of great significance to construct a reasonable city-town-village system and to speed up the construction of rural urbanization and centralization on the theoretical basis of "spatial integration, organizational integration and industrial integration". Meanwhile, it is necessary to promote the comprehensive renovation of "mountains, rivers, forests, fields, lakes, roads and villages" on the platform of major land consolidation projects, and reasonably and orderly promote the comprehensive renovation of hollow villages in different areas. Through the regulation of returning farmland, garden and forest, the form of rural homestead use will change from extensive form to intensive form with city, town, and village centralized construction as the center. Under the influence of rapid urbanization and industrialization, many rural populations gathered in cities and towns, boosting the transformation and upgrading of urban industry, increasing the demand for urban space, so the scale of urban construction land is increasing. On the other hand, the regulatory role of the price mechanism of land market has been strengthened, urban land use is becoming more intensive, the contribution of land input per unit output is gradually reduced, the substitution of production factors in capital, labor force and science and technology is strengthened, the non-agricultural land occupation of the increase of unit urbanization rate declines, the rate of continuous decrease of cultivated land slows down, the structure of cultivated land use changes with the development of economy and science and technology, thus the cultivated land use begins to transform.

3.3.2 Interactive Features Between Urban and Rural Areas

1. **Stage of the interaction between urban and rural areas**

Urban and rural development is affected by both internal and external factors, such as economic globalization, decentralization of central government, marketization, etc. Due to the differences in infrastructure supply, resource utilization and innovation system, there are differences in development capacity between urban and rural areas, which induces the interaction between urban and rural areas and promotes the transformation of urban and rural regional system. In this context, urban–rural elements, structure, and function gradually change from disorder to equilibrium, showing a balanced evolution under the unbalanced state. The field theory in theoretical physics holds that "when there is an object, there is a gravitational field coexisting with it, and the two are closely related." Urban and rural areas are composed of different levels of urban and rural settlements and natural geographical environment, which is directly manifested in the relationship between the city and the countryside. Therefore, the urban and rural regional system can be regarded as the field of interaction, that is, the urban–rural regional field, which contains state, potential, energy, and matter (Fig. 3.16).

Urban and rural areas, as relatively independent entity units, are field sources, and there is interaction force between urban and rural areas. This force has direction and strength. The former is called potential, the latter is called energy, and

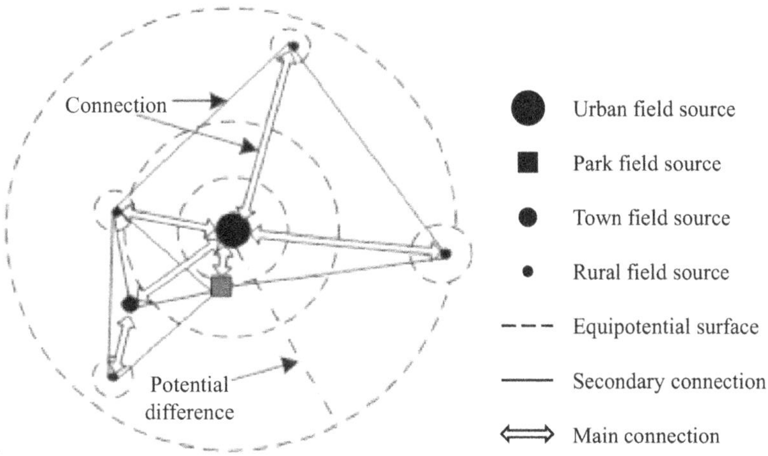

Fig. 3.16 Interactive field of urban and rural regional system

the potential energy is the dynamic mechanism of urban–rural transformation. When the potential energy is towards the urban area, material in rural areas is mainly transferred to urban areas, and the speed of transfer depends on the amount of energy; while when the potential energy gap between urban and rural areas is small and towards the countryside, material in urban areas is mainly transferred to rural areas. Material refers to the labor force, land, capital, technology, information, products and so on in urban–rural areas, and can flow and interact between urban and rural areas through the role of potential energy. State is the external expression of the structure, function and relationship of urban and rural regional field in a certain period, and reflects the potential energy and material conditions currently. In different stages, there are great differences in the intensity, direction and form of urban–rural interaction. Accordingly, the development of urban–rural interaction field can be divided into primary stage and advanced stage.

In primary stage, the urban and rural areas are initially integrated, and the function of urban center is relatively low. Although urban expansion will affect rural areas, the impact of cities on rural areas is mainly manifested in the non-agricultural factors of agricultural production (i.e., polarization effect) due to the organization and management ability, the stage of urban development and other factors. The urban diffusion effect is relatively weak, and the scope of influence is limited. During this stage, the rural areas with diffusion effect greater than polarization effect are only limited to some parts of the urban fringe, which will promote the agglomeration development of rural elements in the city and its surrounding areas.

In advanced stage, urban centrality is gradually enhanced with the development of urban and its surrounding regional elements, and the surrounding small and medium-sized cities and central towns begin to appear and develop continuously. As a result, there are not only the cross flow of production factors such as personnel and technology between cities at all levels, but also the elements flow between villages and central towns of small and medium-sized cities. The superposition of radiation field intensity and influence area of cities at all levels leads to profound changes in the curve of urban field intensity and total effect. On the one hand, with the increase of the centrality of cities at all levels, the ability of cities to drive rural development has been continuously improved, and the density of urban field intensity curve has increased, and the diffusion effect curve has evolved upward. On the other hand, the role of cities in rural development is not only reflected in the simple flow of elements, but also reflected in the value complementarity, functional coupling, and industrial interaction between urban and rural areas under the background of rapid urbanization. Some rural areas have evolved into new small and medium-sized towns relying on the element agglomeration, and further generate field strength to drive the rural development in the surrounding areas; and some rural areas develop modern agriculture and non-agricultural industries based on the inflow of scarce elements, and gradually construct an endogenous development mechanism to achieve sustainable development (Liu 2011).

As shown in Fig. 3.17, y represents not only the degree of urban–rural interactive development, but also the city's self-organization ability, and x represents the distance from the city center. In the primary stage of urban–rural interactive development, self-organization ability of urban area is weak, and the influence intensity and scope of city on rural area is relatively limited; while in the advanced stage, the centrality of cities at all levels has increased significantly, and the ability to drive rural development has been remarkably enhanced. Therefore, the urban–rural interaction has been promoted gradually and organically from point to surface, from part to whole. The joint effect of urban radiation field strength and rural endogenous force develops different rural areas, gradually enriches urban space at all levels, and finally realizes the coordinated development of urban and rural areas.

2. **Operation law of urban and rural gradient nodes**

The diversity of economic activities determines the complexity of economic space, and the element agglomeration in different spaces constitutes an urban–rural hierarchy in accordance with the law of "rank-scale". If several key nodes are selected in the urban and rural space with different field intensity (economic radiation capacity), it can be found that the interaction of each field strength leads to different potential of each node, corresponding to different quantity scale, factor structure and industrial characteristics. The distribution of each node is subject to Christaller's central place theory, and the highest-level nodes usually correspond to the metropolis, which is small in quantity, wide in range of service, and complete in range of goods. The intermediate nodes usually correspond to small and medium-sized cities and central towns, with a large quantity, a small range of services, and less types of goods and services. While the lowest-level nodes usually correspond to small towns and villages, with the largest number, the smallest range of services, and the least types of goods and services.

There are significant differences in the marginal returns of factors among different nodes, which leads to the flow of production factors between different nodes. The combination of flowing production factors and the original production factors of each node changes the factor allocation structure and efficiency of each node, increases the output and service capacity of each node, and improves the potential and development level of each node. Under the common influences of different production factors, industrial structure (and the production function determined by it) and supporting service capacity of each node, on the one hand, the production factors between nodes generally flow in the order of "low-level industry → intermediate industry → high-level industry" or "high-level industry → intermediate industry → low-level industry"; on the other hand, the effect of factor flow on each node will be different, which will change the relative potential of each node. Therefore, the intermediate node is the transfer platform of factor transfer and the important subject of rural development driven by radiation, and becomes the key node of urban–rural coordinated development (Fig. 3.18).

The intermediate nodes play the role of link, agglomeration, diffusion and support in the interactive development of urban and rural area. Specifically, as a link between city and village, the intermediate node is an important platform for "top–bottom" absorption of modern production factors such as advanced technology, information and management in the city, and "bottom-up" absorption of rural labor, capital, raw materials and other initial production factors, but also an exchange platform of industrial and agricultural products between urban and rural areas, which is conducive to the smooth flow of elements and products between urban and rural areas. Because the industrial gradient between

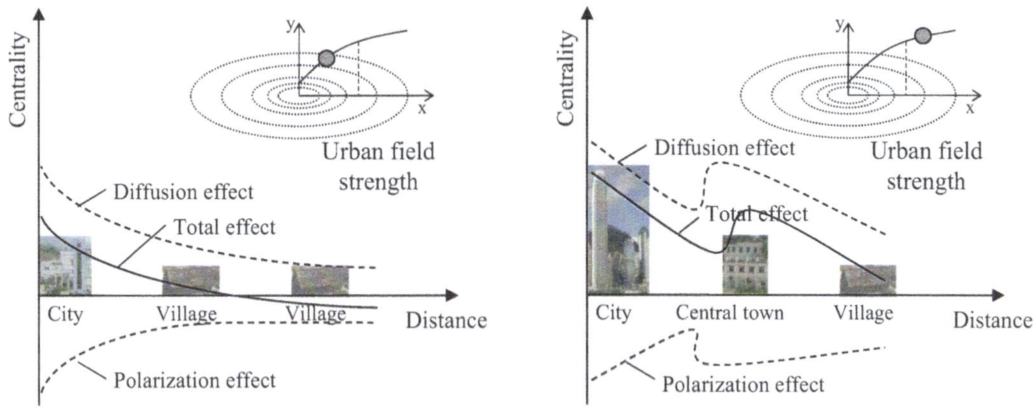

t_0: Primary stage of urban and rural interactive development

t_1: Senior stage of urban and rural interactive development

Fig. 3.17 The development stage of the interaction between urban and rural areas

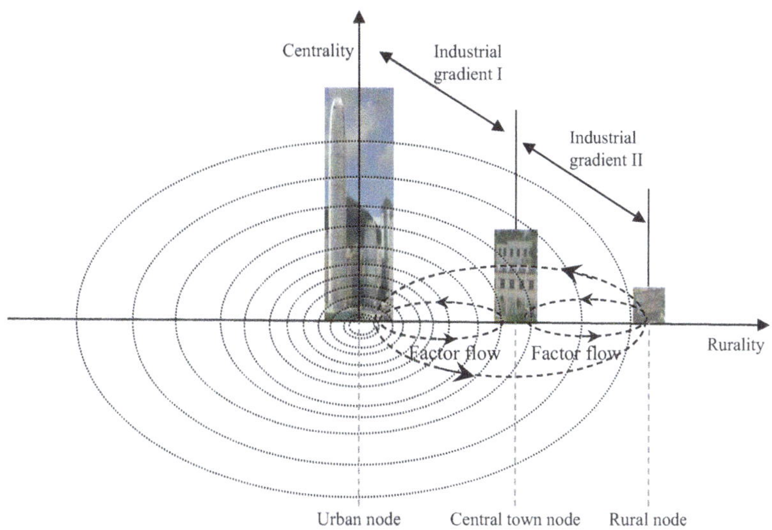

Fig. 3.18 Urban–rural interactive development and gradient node (Liu 2011)

the intermediate node and the high-level node and the low-level node is the least, the intermediate node becomes the primary gathering place of non-agriculturalization of rural elements and ruralization of urban elements, and is the main carrier of rural non-agricultural industry which has obvious agglomeration effect on township enterprises, culture and education, entertainment services and other industries. Meanwhile, as a regional economic growth pole, the intermediate node is conducive to promote the diffusion of modern elements to rural areas, the extension of infrastructure to rural areas, the popularization of public services in rural areas, and the extension of urban civilization to rural areas, so as to drive the development of surrounding villages. By promoting population concentration, land intensive use, industrial clusters and other ways, intermediate nodes can build strategic nodes of urban and rural development, the core carrier of industrial intensive agglomeration, and an important platform for equalization of public services, which plays a supporting role in the overall development of urban and rural areas. In a word, the agglomeration and diffusion abilities of intermediate nodes, such as small and medium-sized towns and small and medium-sized communities, determine the coordinated development level and development ability of urban and rural areas, which is an important fulcrum for balancing urban and rural development and promoting new rural construction.

3.3.3 Evolution of Urban–Rural Transformation

Under the influence of human activities and environmental changes, the land system has undergone profound changes, resulting in urban and rural systems. The evolution of urban and rural system depends on the mode and intensity of human activities, presenting a system evolution characteristic of point-line-area-network (Fig. 3.19). The interaction between urban system and rural system is the urban–rural relationship. Generally, the urban–rural relationship refers to the general relationship and interactive relationship formed by the interaction and mutual restriction between cities and villages, and is a concentrated reflection of the relationship between urban and rural areas under certain social and economic conditions. In a broad sense, the urban–rural relationship should also include the competition in land resources, water resources, mineral resources and development space between urban and rural areas. The factor flow, structure transformation and function change in urban and rural areas not only cause the change of urban–rural relationship, but also reflect the operation state of urban–rural development process through the change of urban–rural relationship.

In terms of the historical evolution, urban–rural relationship has generally experienced serval major stages from separation and opposition to integration, coordination and even equivalence, which is usually closely related to the level of regional economic and social development. If the law of urban–rural relationship changes with time is expressed as the "transition law" of urban–rural development, the transition law is a comprehensive cognition of urban–rural relationship from the perspective of process, involving urban–rural interaction, urban–rural gap, factor conflict and so on. Based on combing the stages of human development, the evolution of urban and rural regional system can be roughly divided into four stages, i.e., low-level coordination between urban and rural areas in agricultural society, conflict stage between urban and rural areas in the early stage of industrialization, urban–rural conflict-coordination stage in the middle stage of industrialization, urban–rural integration stage in the late industrialization, and urban–rural equivalence stage in post industrialization. Elements, structures, functions and relationships in different stages show corresponding relations and close ties. The goal of the "transition law" of urban–rural development evolves along the overall planning, coordination, integration and equivalence of urban and rural areas, and urban–rural relationship evolves along the path of separation dependence → conflict dependence → coordination dependence → symbiosis dependence (Fig. 3.20).

Stage I: Low-level coordination between urban and rural areas in agricultural society

In the traditional agricultural society, agricultural production is the basis of urban and rural economy and occupies a dominant position. Due to the low proportion of handicraft industry and commerce, cities are economically dependent on rural areas. Constrained by the agricultural economy and the production mode of small farmers, folk beliefs, customs and urban culture are rooted in agricultural civilization. Although the city is in a dominant position politically and militarily, urban and rural areas have achieved "the unity of urban and rural areas without distinction", and the urban–rural relationship shows a natural relationship of pastoral style (Cai 2003). In this stage, land element is the foundation and support of urban and rural economic and social development, and local society is the main social form.

Stage II: Conflict stage between urban and rural areas in the early stage of industrialization

In the early stage of industrialization, technological innovation promoted the separation of handicraft industry from agriculture and rural areas,

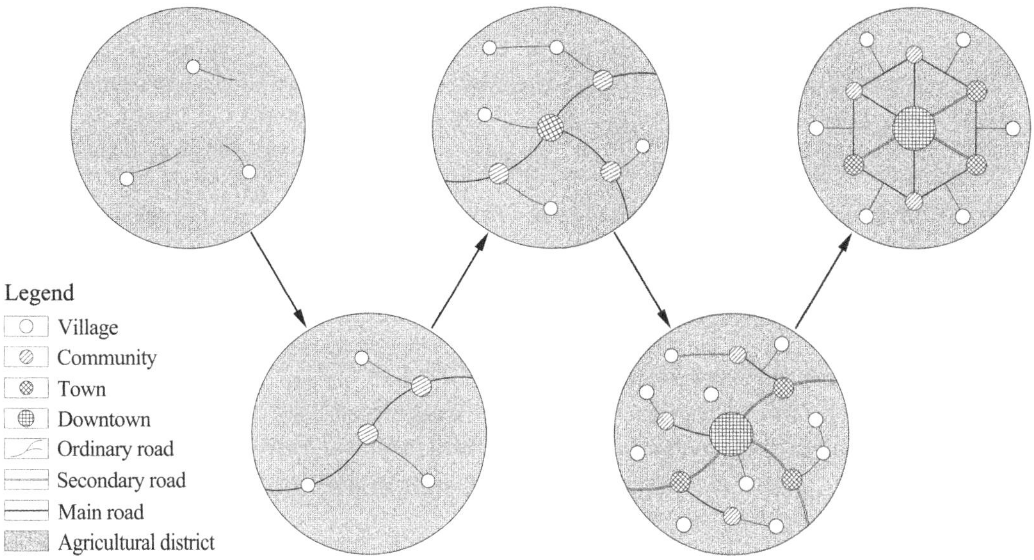

Fig. 3.19 The stage state of the transformation of urban–rural relationship

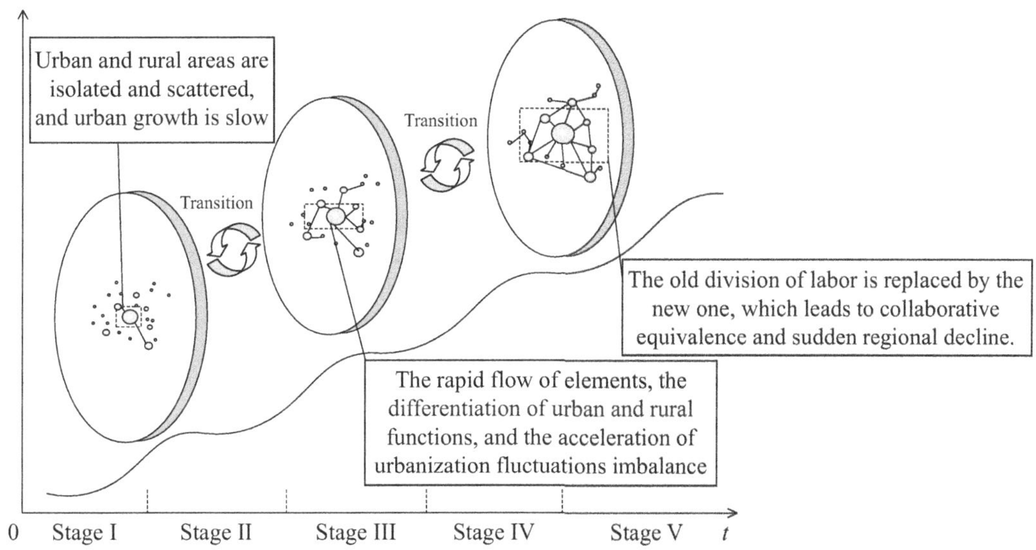

Fig. 3.20 Stage states and transition laws of the transformation of urban–rural relationship

and gave birth to the basic energy and mineral resources industry. Therefore, the position of industry in the economic system is gradually rising. As industry is mainly concentrated in cities, urban and rural areas begin to differentiate in economic functions, forming the regional division of labor of "city-industry" and "rural-agriculture". And due to the difference of economic benefits, rural elements transfer to the city, and there are great differences in industrial structure, organizational structure and lifestyle between urban and rural areas. With the development of industrialization and urbanization, the role of cities in economic, social, political and other aspects has become increasingly prominent, and has become the center of regional economic growth and non-agricultural products supply. During this period, local society and urban civilization exist at the same time, but the balance of rural agricultural economic and social system began to be tipped, while urban civilization and industrial economy gradually formed, and thus urban and rural areas began to separate and oppose.

Stage III: Urban–rural conflict-coordination stage in the middle stage of industrialization

Affected by economic growth and continuous urbanization of population, the influence of urban areas on rural economy and society is more and more significant. The level of rural development is directly affected by the level of urban development and urban–rural development strategy, and remote rural areas are also affected by product supply and population migration. The expansion of urban scale and the spatial agglomeration of population have changed the landscape structure in urban and rural areas, and urban production and demand occupy a dominant position in non-agricultural products. While the industrial structure and employment structure have changed in rural areas, many young and middle-aged people go out, resulting in serious population aging and land waste. The production and living functions of some villages are deteriorating and even declining. Due to the extensive economic growth and disorderly urban expansion, the interference of human activities on the physical and geographical environment has increased, and the human-earth relationship has gradually deteriorated. Meanwhile, economic growth and urban expansion overuse rural resources, occupying rural development space and destroying the ecological environment.

Based on the analysis of the urban–rural disparity, rural decline and ecological environmental problems, more and more attention has been paid to social and ecological development. Developed countries have put forward a series of strategies and policies to achieve the goals of poverty eradication, environmental protection, ecological agriculture and sustainable development. At the same time, social forces such as farmers' organizations and environmental protection organizations have become an important force to coordinate urban and rural economic, social and ecological relations. With the weakening of agricultural economic role, the rural function is generally in the weakening stage. Therefore, how to make full use of the advantages of rural areas in housing characteristics, neighborhood relations, ecological environment, pastoral scenery, clean air and other aspects to promote rural reconstruction has become an important issue of agricultural policy innovation and rural development. In this stage, the urban–rural relationship is in the transition period from opposition to integration.

Stage IV: Urban–rural integration stage in the late industrialization

Entering the middle and late stage of industrialization, service industry has become the leading industry of national economy and employment supply, and agricultural production has changed significantly, forming three main modes of large-scale mechanical production, intensive production and part-time production. The purpose of agricultural production is to make profits, the main body of agricultural management takes the form of enterprises. The state can increase the income of agricultural producers through agricultural production fund subsidies and policy support. Under the background of social demand transformation, rural neighborhood, ecological environment, rural scenery and clean air attract urban population inflow, and the increasingly prominent rural non-production function has become an important force for the balanced development of urban and rural areas. Industry and commerce have become the main employment channels of rural residents, and the lifestyle of rural residents has been gradually citizenized. Thus, the division of city-industry and rural-agriculture has been completely broken, and the urban–rural relationship shows a trend of integration and coordination.

Stage V: Urban–rural equivalence stage in post industrialization

Equivalence between urban and rural areas is the core goal of urban–rural transformation, and it is also an ideal state for the urban–rural regional system to maintain stability and sustainability. In the post-industrial society, the economic production capacity has been greatly improved and guaranteed, people's demand for quality life and self-leisure has been constantly improved, and the differentiation and individualization of consumer goods are obvious. In the stage of urban–rural equivalence, the coordination of urban–rural industrial and spatial systems has been realized, the declining rural area and backward area promote spatial reconstruction through agricultural scale operation, village-town construction and service industry development. At the same time, it also realizes the integration of urban and rural systems, the two-way and equal flow of urban and rural population, capital and other elements, and the balanced allocation of urban–rural public services and resources (Liu et al. 2013). Thus, urban and rural areas achieve different but equivalent, and the production and living conditions of residents are equivalent.

From "attaching importance to the city and neglecting the countryside" to "leading rural development with the cities", the function of rural regional system is becoming more and more diversified. On the basis of the traditional functions, including the birthplace of agricultural civilization, concentrated areas of agricultural production and farmers' life, place of raw materials for industrialization and urban consumption, as well as strategic highland for ensuring ecological and food security, the rural area is also an important hinterland for the healthy development of modern cities and a prosperous place for innovation, entrepreneurship and health culture. These functions are of great significance to promote rural development and realize urban–rural integration.

3.3.4 Mechanism of Urban–Rural Transformation

1. **Dynamic system of urban–rural transformation**

Urban–rural transformation is the change process of urban and rural elements composition and allocation efficiency caused by the adaptation of urban and rural areas to the changes of internal and external environment. City is the center of regional economic growth and factor spatial agglomeration, which can produce economies of scale and cluster effect. Thus, economic growth, population agglomeration, technological innovation and management innovation promote the transformation of urban and rural areas from rural function leading to urban function leading. However, this change does not necessarily bring about the benign evolution of urban and rural areas. For example, excessive population concentration without industrial support and "rural disease" without rural construction support make urban and rural development unsustainable. Under the guidance of fairness, policy and system adjustment is inevitable, especially in the aspect of public resources which non-market allocation is needed. Obviously, the gap of potential energy between urban and rural areas does not depend on the urban and rural regional system itself, but is determined by various factors that affect the development capacity of rural and urban areas. These factors include physical and geographical environment, economic and social factors and institutional environment. They have different impacts on urban and rural development, and then lead to the change of urban and rural potential energy. In general, the driving mechanism of urban–rural transformation contains supporting system, power system, potential energy system, state system and response system (Fig. 3.21).

The supporting system is an important basis of urban–rural transformation, which mainly includes natural resources, topography and other natural and geographical conditions, as well as location condition, traditional culture, economic foundation, urban and rural human capital, etc. When the external conditions and systems change, these elements determine the self-development ability of urban and rural areas. Power system is the direct driving force of urban–rural transformation, which is embodied in the modernization process of industrialization, urbanization and agricultural modernization. Behind them are the promotion of innovation, technology, social needs, globalization and regionalization. In addition, policies and institutions are also important factors to promote urban–rural transformation in a certain region. For example, since the 16th national congress of the Communist Party of China, a series of supporting and benefiting agricultural policies, such as the new rural construction and the cancellation of agricultural tax, will contribute to the change of rural industry and living environment, thus promoting urban–rural transformation.

Potential energy system is the expression of the direction and degree of the interaction between urban and rural areas, which directly determines the direction and intensity of urban and rural elements flow. Physical geographical environment, economic basis and power system are the main factors that determine the potential energy of urban and rural areas. For example, with the improvement of the level of industrialization in urban and rural areas, urban economy gains growth momentum and attracts rural population to gather, which improves the workers' income. Promoted by the increase of population size and consumption level, urban service industry gets development opportunities and further attracts rural labor transfer. In general, the potential energy of city to countryside in the early and middle stage of industrialization continues to increase, attracting the transfer of rural elements to cities. As an open system, urban–rural potential energy system is also directly affected by the inter-regional potential energy system, which is not only manifested in the inter-regional industrial division of labor formed by the inter-regional factor flow, raw material and product trade, but also manifested in the complex

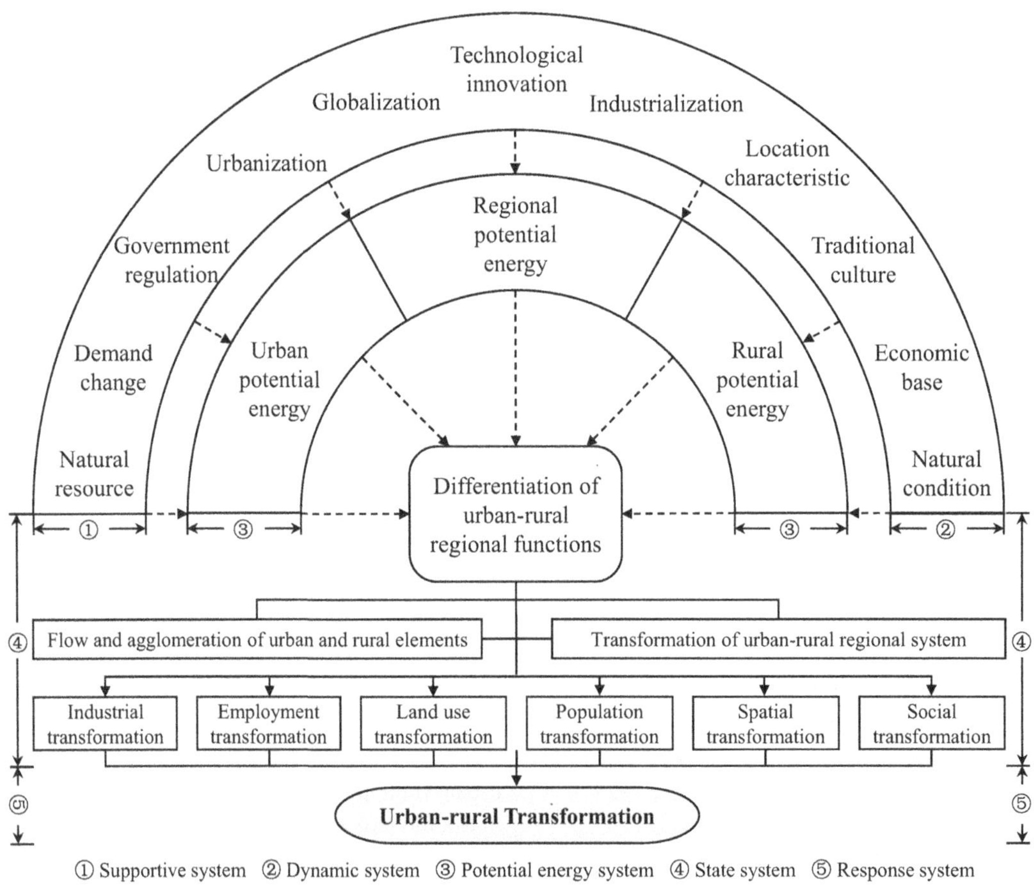

Fig. 3.21 Dynamic system of urban–rural transformation

potential energy system of city-city, city-village and village-village based on the government management function, financial and banking institutions, headquarters economy and medical system. State system refers to the flow of urban and rural elements and the transformation of regional form, and is an important external representation of urban–rural transformation, which is manifested in the transformation of industrial structure, employment structure, population structure, spatial structure and social structure. The response system is the development and transformation of urban and rural areas. Under the influence of the above driving factors, urban and rural areas have formed a response to external changes and internal foundation. The core of the system is to realize the optimal allocation of urban and rural elements based on market mechanism, and then promote the balanced allocation of urban and rural public resources through policy regulation.

2. **Influencing factors of urban–rural transformation**

Physical geographic environment, economic and social structure, as well as system and policy are the main factors leading to regional differences in urban–rural transformation (Fig. 3.22). Specifically, physical geographic environment provides basic conditions for economic growth, industrial agglomeration, human production and life. Economic system and its evolution are the basic driving force of urban and rural development. Economic growth promotes people to improve their living conditions, change consumption demand and mode, and provides conditions for improving infrastructure and public services, which leads to the development and change of urban and rural social system. The changes of industrial structure, employment structure and population structure inevitably lead to the change of urban and rural spatial structure, which pose new challenges to urban and rural social governance, environmental protection and sustainable land use. System and policy are the important influencing factors of urban–rural transformation, and are subject to the macro demand of national development strategy. The top-level design system of relevant systems and policies in a certain period constitutes the basic constraint on the flow and exchange of urban and rural elements. Meanwhile, there are obvious externalities in urban and rural infrastructure, public services, agricultural and rural multi-function, which usually require the government to carry out macro-control through administrative means and relevant policy innovation.

(1) Physical geographic environment

Natural conditions and natural resources constitute the natural background conditions for economic and social development, and their impact on the development of different regions is different due to their regional, limited and integrated characteristics. First, the physical geographic environment provides the basis for economic and social development. Water and soil resources, mineral resources and fossil energy not only provide material basis for the development of agriculture and raw material industry, but also provide conditions for the life of urban and rural residents. Besides, the regional differentiation of mineral resources and fossil energy determines the spatial pattern of raw material industry and energy industry to a certain extent. In the middle and late stage of industrialization, good ecological environment has become the support of economic activities and living and a new driving factor of local economic development. Second, the regional nature of physical geographic environment causes regional functional differentiation. The combination of natural resources and natural conditions forms the comparative advantages of different regions, which provides the basic material basis and development environment for regional participation in labor division and industrial division. The natural regional function formed by the physical geographical environment is the basis of the formation of regional differences and regional division of labor. Agriculture, energy and mining industries rely heavily on natural geographical environment, while modern industry and service industry rely less on natural environment. Third,

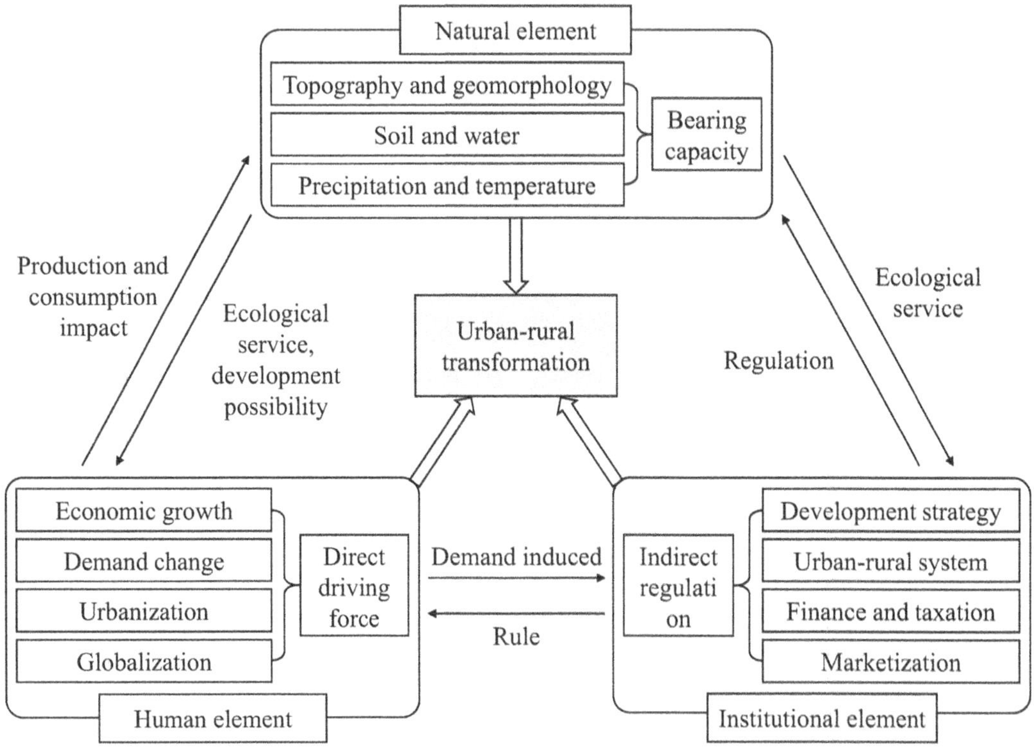

Fig. 3.22 Framework system of influencing factors of urban–rural transformation

the human-earth interaction restricts urban–rural transformation. The comprehensive carrying capacity and environmental carrying capacity of the physical geographical environment are limited. The regions with good natural regional function have better self-regulation ability, which can realize the sustainable utilization of resources and the absorption and diffusion of pollutants through renewal, circulation and other mechanisms. Overgrazing, over reclamation, over urbanization and industrialization inevitably weaken the ecological service function of natural regional system, and the results of human-earth interaction feedback to human society in the form of insufficient resources, destruction, and even disasters.

The carrying capacity of resources and environment is jointly determined by the conditions of resources, the types of economic activities and social needs. Different resource conditions restrict the change of urban–rural regional system, and affect the speed and intensity of urban–rural transformation. Generally, the impact of physical and geographical environment on the urban–rural transformation is mainly reflected in two aspects, i.e., natural resource supply and natural ecological function service. In terms of the former, the demand of urban population and industry for food and energy far exceeds the supply of the city itself, the change of population concentration and development mode in urban and rural areas has changed the relationship between decentralized urban–rural areas and physical geographical environment. As to the ecological function service, the emission of production and living materials in urban and rural development is increasing rapidly, which is beyond the absorption and decomposition capacity of natural environment. As a result, the demand capacity of urban expansion, urban and rural industrial development on the ecological function of the surrounding areas is unbalanced with the supply capacity of the surrounding areas, forming the environmental constraints of urban and rural development transformation.

(2) Industrialization and urbanization

Industrialization is the initial driving force of urban and rural elements flow and structural transformation since modern times, and causes urban and rural regional economic and social changes through resource allocation and population flow. First, the rapid development of industry provides many non-agricultural employment opportunities, attracting many rural labor forces to transfer to industry and towns, forming the initial driving force of population urbanization. Second, the upgrading of industrial structure and the change of production factors make the influence of location, transportation, market and other factors on industrial development increasingly prominent. Population and industry tend to gather in urban areas with good location conditions, and the coupling of industrial function and urban service further improves the driving effect of city-industry and promotes the improvement of urban function. Third, industrialization promotes the upgrading of industrial structure and the adjustment of production mode in urban and rural areas, changes the life style of urban and rural residents, and affects the relationship between economic activities and natural environment. Overall, industrialization promotes the transformation of urban and rural economic form from agriculture-led to industrial-led and then to service-oriented, which provides economic development power and transformation foundation for urban renewal and rural reconstruction.

Urbanization is an important manifestation of the change of urban and rural spatial structure. Its basic characteristics are population agglomeration, urban scale expansion and urban number increase. As a dynamic and multi-level social spatial process, urbanization involves the interaction of population, society, culture, economy, ecology, material and regulation, and includes rural and urban phenomena. In the case of giving priority to cities, the examination of rural economy and rural lifestyle also has a certain urban perspective (Friedman and Tang 2007). The transfer of rural surplus labor force to cities and towns not only causes the change of urban and rural population structure, but also promotes the economic accumulation of urban and rural areas through demand and specialized production (Yang 2001). Because of the increase of labor supply and local expenditure in urban areas, and the rise of urban wages is restrained and the production returns are increased, thus the urban productivity and social demand are improved. This will promote the improvement of urban accumulation capacity and the formation of new capital, and further attract the entry of rural labor force. Meanwhile, the information, technology and culture in urban areas spread to rural areas to accelerate the improvement of agricultural and rural production capacity. Agricultural industrialization is a process in which farmers specialize in a certain link in the chain of agricultural industry management, and then gradually marketize other value-added links. With the rural labor force flowing into the city, the expected cost of agricultural management increases, and farmers tend to reduce the scope and scale of other aspects of agricultural operation, and purchase relevant services from the market, thus promoting the expansion of agricultural value-added market. Agricultural industrialization improves agricultural labor productivity, which promotes the transfer of agricultural labor and land to cities and promotes the growth of agricultural output (Fig. 3.23).

As the core of regional economic growth, cities and towns are the core of effective integration and optimal allocation of urban and rural resources. Due to the differences in urban–rural elements composition, returns to scale and facilities, industrialization and urbanization affect the speed and direction of urban and rural development and transformation. For example, China's economic growth and city centered development strategy ignores the urban–rural linkage, social costs and resource and environmental costs, resulting in the excessive urbanization of land, large occupation of cultivated land, environmental degradation, rural hollowing and other problems. The excessive concentration of political, educational, medical and other resources in large urban areas has become an incentive for the spatial agglomeration of

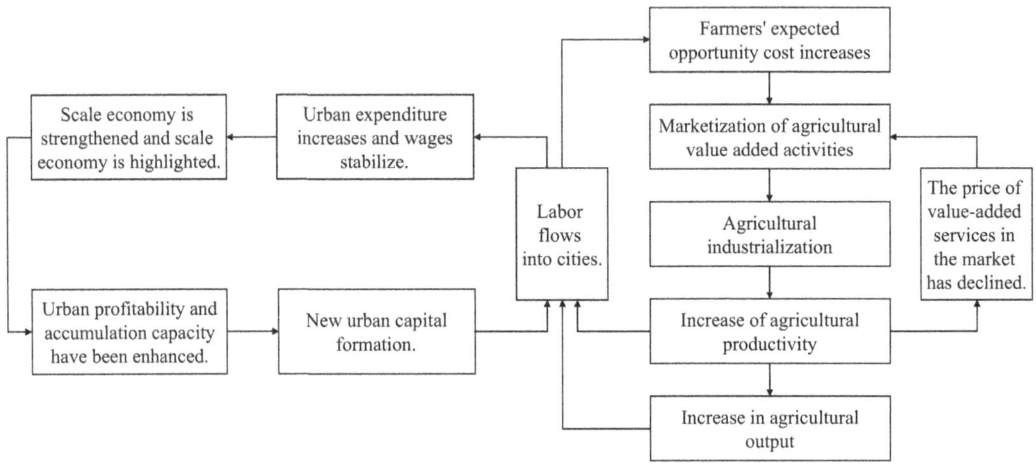

Fig. 3.23 The impact of urbanization on urban and rural economy

industry and population, which indirectly leads to the uncoordinated construction of large and medium-sized cities and small towns, and the separation of industrial and urban development. In the middle and late stage of industrialization and urbanization, resource and economic problems in urban and rural areas have gradually turned into social problems, especially the inclusiveness of urban infrastructure and social security system, which determines whether the agricultural transfer population can fully integrate into the urban society.

(3) Transportation and information technology

Distance and location are the important factors affecting the spatial differentiation of urban and rural areas. In classical location theory, the freight cost based on distance occupies an important position in agricultural production layout, industrial layout and commercial layout, and is also an important factor in the formation of spatial pattern. The scale of traditional rural settlement is restricted by the scale of cultivated land and the distance of cultivation, and the balance of living and working space is also an important factor. With the construction of transportation and information facilities, the time distance between cities and villages, and between cities and cities is constantly shortened, which reduces the distance and time cost of urban and rural economic and social links, provides the basis for the inter-regional factor flow and product circulation, and facilitates the entry of enterprises and the flow of labor force. Meanwhile, the scale and spatial form of urban area have changed greatly. Expressway, high-speed railway, national highway and other rapid transportation modes improve the accessibility of urban and rural traffic, and urban space expands along the highway and high-speed rail. The new mode of transportation has also accelerated the construction of industrial parks and new urban areas.

Entering the information age, information technology and facilities have changed the traditional way of contact, management and life style. Communication technology has surpassed the traditional regional space, and many funds and businesses can be circulated and operated through the network. The reduction of communication cost makes some non-technical and labor-intensive industries flow to the countryside, and realizes the spatial diffusion of urban industry, technology and information. In the new era, relying on transportation and information facilities, Internet plus logistics network, urban and rural areas are more closely related, and these factors will contribute to rural development. The rural e-commerce represented by "Taobao village" combines the Internet with rural economy, which has changed the traditional rural economic development mode and has significant differences in the impact on different rural areas (Guo 2015).

(4) Science, technology and innovation

Science and technology are important factors affecting economic and social development, and has a significant impact on the succession of industrial structure, the transformation of traditional industries and the differentiation of spatial economic structure. In general, the role of science and technology in the industrial structure and various industries is constantly improving, and the proportion of high-tech industry in the national economy is gradually increasing; the role of science and technology in transforming traditional industries and driving economic and social development is becoming more and more obvious; the spatial agglomeration of financial institutions, information departments, scientific research institutions and innovative enterprises forms the core driving force for the upgrading of regional industrial structure and economic development. Therefore, science and technology are important driving forces to promote urban–rural transformation. At the same time, the regional and urban–rural differences in science and technology innovation ability and application ability also lead to the economic and social gap between the region and the urban and rural areas.

Schumpeter holds the view that innovation is to rearrange and combine the original production factors, so as to form new products, new production methods, new markets, new raw material supply and new organizations. The role of science and technology is fundamentally reflected in the industrial innovation ability. Technological progress triggered by innovation not only promotes the development of modern industry, but also provides the possibility to transform traditional agricultural production mode and enhance agricultural production capacity. Under the premise of agricultural labor force and agricultural product surplus, the urbanization process driven by industrial development can be smoothly promoted.

Science, technology and innovation factors affect the speed and quality of urban–rural transformation. Innovation always takes place where some innovative elements gather. The differences of regional innovation capability promote the spatial flow of production factors and the spatial agglomeration of economic activities. The adjustment, optimization and upgrading of industrial structure and the emergence of new industries in these regions have promoted the changes of industrial structure and layout in larger regions. In turn, the flow of production factors and the agglomeration of economic activities provide the possibility for various types of innovation activities. The difference of innovation in urban and rural areas forms the fundamental driving force of industrial and population evolution, and promotes the flow and agglomeration of urban and rural elements. When innovation activities spread to rural areas, the industrial system in rural areas gets the impetus of transformation and upgrading.

(5) Economic globalization

Globalization, decentralization and marketization are the three important reasons for promoting regional economic growth and regional development differentiation (Wei 2001). With the deepening of China's reform and opening-up, China's economy is gradually integrated into the global industrial system. Foreign direct investment and foreign import and export have become important factors to promote regional economic development. To promote economic development, Chinese government has set up open cities, open regions and border cities since 1978. As a result, the pattern of opening to the outside world is expanding from coastal areas to border areas and inland areas. The "Belt and Road" initiative has set up a platform for the further development of the export-oriented economy in the central and western regions. Currently, national economic and technological development zones and national high-tech development zones are the forefront of attracting foreign investment and participating in the economic global division of labor. The export-oriented economic development promotes the changes of industrial structure and employment structure, which promotes urban and rural elements to gather to urban areas and development zones. Thus, foreign investment has also become an important driving force for urban space expansion.

The impact of globalization on urban and rural development is characterized by indirect and regional differences. Globalization promotes the level of regional economic development through foreign investment and industrial chain, and influences the transformation of urban and rural economic structure and employment structure through economic growth and environmental improvement. In addition, there are differences in the location preference of foreign investment. Although the investment environment and the level of opening-up in central and Western China have been continuously improved, the Eastern China is still the main area of foreign investment, which determines that the impact of globalization on urban–rural transformation has obvious regional differences. In the Eastern China, the development of export-oriented labor-intensive industry provides a lot of employment opportunities, which become an important reason for labor inflow.

(6) Demand structure

Demand is the basic attribute of human being, and it is the internal reason stimulating people's action. The evolution of human history shows that human development has formed self-demand, product demand and natural demand, and the evolution order of these needs has obvious stages. Maslow, an American psychologist, believes that human needs can be divided into five categories from low to high, i.e., food and clothing needs, security needs, social needs, respect needs and self-realization needs. Essentially, it reflects the transformation from survival needs, material needs to spiritual needs. Product demand can be divided into agricultural product demand, industrial product demand, service demand and ecological product demand, which is consistent with the succession of economic growth and industrial structure. Natural demand is composed of production function demand, ecological environment demand and spiritual quality demand. Among them, production function demand refers to the provision of basic material product production service; ecological environment demand is the environmental regulation service composed of safe atmosphere, water, soil, biology and ecosystem itself; spiritual quality demand is the service provided by ecological service system in knowledge, culture and landscape.

Demand structure promotes the change of industrial structure and regional function. The process of interaction between human activities and natural geographical environment has realized the transformation from low-level needs to higher-level material needs, and then to high-level spiritual needs. The change of self-demand structure also changes the functional requirements of natural geographical environment. To meet the basic livelihood, people are engaged in picking, food production and other activities, and urban–rural region forms agricultural oriented-development activities. In the stage of pursuing material needs, people exploit the natural environment on a large scale, and industry is the main development activity in urban and rural areas. In the stage of pursuing spiritual needs, people pay more attention to the cultural services and environmental services provided by the ecological environment and change the traditional extensive development, thus forming many protected areas. The change of demand structure causes the evolution of regional function, which is finally reflected in the form of land use.

With the development of economy and society, the succession of different demand levels also forms the competition for limited resources. When human beings seek material improvement, they constantly promote the redistribution of production resources in different industries and regions. Because of the scale effect of the city, the demand structure promotes the change of urban and rural industries, population and land form, and realizes the maximization of economic returns and the modernization of service industry. When the spiritual demand is enhanced, the economic and social development realizes the comprehensive transformation of social and ecological needs, and the formation of rural areas' cultural and ecological functions is driven by the market. In addition, to meet the needs of food security and ecological protection at the national level, the government will intervene in the flow direction of urban and rural regional elements through certain laws and regulations.

(7) Regional culture

The formation and development of culture are influenced and restricted by the natural environment of human existence. Regional culture is the product of the interaction between human beings and the natural geographical environment of a specific region, which has obvious regional characteristics, such as Qilu culture in Shandong, Wuyue culture in Yangtze River Delta and Lingnan culture in Pearl River Delta. As a result, the mode of thinking and value orientation have been formed, which affects the behavior of people and the production and management, living consumption, social responsibility and other links of social and economic development (Fig. 3.24). Regional culture has a variety of impacts on economy, such as industrial development and agglomeration. In general, industrial agglomeration is dependent on the environment of specific regions, including system, social history and culture, values, customs, implicit experience knowledge, and relationship network. This dependence shows the characteristics of rooting in the region (Yan 2007). In the process of implementing the opening-up strategy, Guangdong, Fujian and other places have obtained a large amount of investment from Hong Kong, Taiwan and overseas Chinese by virtue of the geographical, blood and interpersonal ties formed based on common culture. The export-oriented development model of "three-plus-one" trading-mix in Guangdong is closely related to regional culture.

Social capital refers to the social connection, trust and behavior norms generated by human personality activities, which can promote the investment of material capital and human capital. Generally, it is closely related to regional culture, and has obvious regional characteristics. People's participation in the network, reciprocal norms and trust formed based on both constitute the main content of social capital. The more developed the network, the higher the level of trust, the more conducive to enhance the cooperation between groups and members, thus bringing positive externalities. The influence of social capital on economic development plays a role through two types of social capital, i.e., structural (interpersonal network) and cognitive (trust, norms and values) (Li et al. 2016). At the macro level, social capital helps to make up for the deficiency of the formal system, carry out innovation activities and strengthen material investment, improve the governance ability of the government, and enhance the supply of public resources. At the micro level, families can obtain information through social networks, such that they can obtain more opportunities and resources to participate in activities, and increase family income by reducing opportunism and opportunity cost in transactions. Obviously, the relatively conservative and closed cultural atmosphere and the lack of social capital in backward areas are not conducive to regional economic growth and urbanization.

(8) Development strategy and industrial policy

Development strategy is the summation of the development direction, speed, quality, planning and tactics of a certain thing in a certain period. The change of urban and rural regional system, economic growth and urban–rural development strategy are closely related, and largely depends on the transformation of development strategy. National economic strategy is the basic premise of other development strategies, and is affected by domestic and foreign political environment, geographical environment, economic environment and so on. Before the reform and opening-up in 1978, China was seriously affected by the imperfect domestic industrial system, the deterioration of international security environment and the idea of catching up, forming a highly centralized planned economic system and the economic development strategy of giving priority to the development of heavy and chemical industrial system. To solve the problem of eating, "taking grain as the key link" was the basic strategy of agricultural development. And to realize the rapid development of heavy industry, China had implemented a series of supporting systems, which indirectly caused the long-term

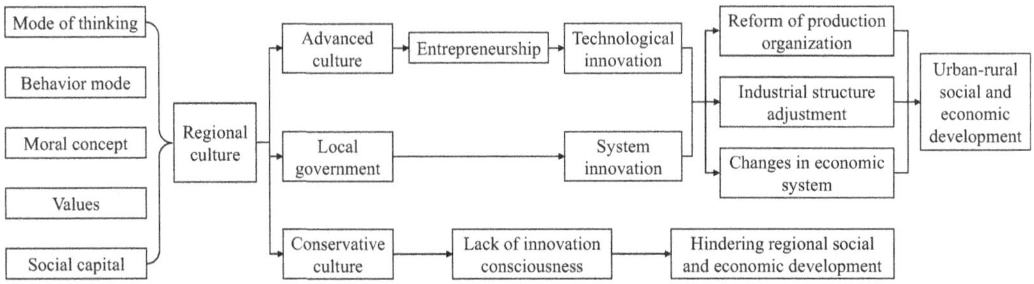

Fig. 3.24 The influence of regional culture on urban–rural economic development

deprivation of villages by cities and agriculture by industry, and formed the development idea of "urban affairs are handled by the government and rural affairs by farmers". Because the capital-intensive enterprises have little demand for labor and large demand for capital, the heavy industrial structure promotes the separation of urban and rural areas in terms of economy, population and society, thus forming the urban–rural dual economic structure.

Since the reform and opening-up, the Chinese government had established a development strategy centered on economic growth and the goal of economic reform with the establishment and improvement of market economic system as the core. In this context, economy with different types of ownership developed rapidly, external investment, foreign-funded enterprises and private economy have become the important driving forces of regional economic growth and urbanization. However, the development of township enterprises is slow by the end of the twentieth century, and rural production factors were rapidly transferred to cities. To solve the increasingly serious issues concerning agriculture, countryside and farmers, the Central Committee of the Communist Party of China (CPC) put forward the development strategy of balancing urban and rural areas at the 16th National Congress of the CPC, thus promoting "industry nurturing agriculture and cities driving rural areas". The change of strategy points out the direction for the adjustment of system and policy, and then influences the transfer of urban and rural elements and the development of urban and rural areas in many aspects.

Urban–rural development strategy is the directional opinion and strategy of urban and rural development, which directly affects the mode and relationship of urban and rural development (Chen and Lin 2013). Since the reform and opening-up, the changes of rural institutional environment and economic development promote the transformation of the strategy of urban–rural separation to the urban–rural coordinated development. With the gradual establishment of the market economic system and the loosening of the urban–rural household registration system, the relationship between urban and rural areas is becoming closer and closer. At the beginning of the twentieth century, issues concerning agriculture, countryside and farmers became a major issue in China's economic and social development. The government timely put forward the major strategy of balancing urban and rural economic and social development, and promoted the development of rural areas through a series of policy measures and investment means.

Government investment refers to the behavior that the government uses fiscal expenditure to invest in specific sectors and regions to promote economic and social development, thus realizing its functions and meeting the public needs of the society. Generally, government investment is closely related to development strategy and regional strategy. In developing countries and socialist countries, the ratio of government investment is large, which plays an important role in balancing social investment, adjusting investment structure, creating a good investment environment, arranging industrial layout and supporting national key projects. The difference of government's investment between urban and rural areas is constrained by urban and rural development strategies and institutional mechanisms. The city-biased investment behavior affects the rational allocation of resource elements and widens the income gap between urban and rural areas to a certain extent (Li and Hu 2015).

Industrial investment and urban–rural infrastructure investment are the main aspects of government investment to promote the urban–rural transformation. The government supports the development of existing industries through financial funds and finance, improves the industrial production capacity, technical level, and the efficiency of resource allocation. At the same time, it guides social capital to invest in related industries, change the industrial input–output pattern, and enhance the multiplier effect of investment. The driving force of investment on industry affects the speed and direction of urban and rural development through industrial scale expansion, industrial structure optimization, technological transformation and upgrading. Investment in infrastructure such as roads, water conservancy and electric power has formed an important support for urban expansion and improvement of urban and rural conditions. In general, the improvement of infrastructure directly promotes the construction of industrial parks and residential development, and supports the urban–rural economic and social development.

(9) Urban and rural policies and systems

The policy system is consistent with the economic development stage and national strategy of the country, and is also deeply influenced by foreign politics and economy. In the primary stage of economic development, China has focused on the implementation of the development strategy of urban–rural dual structure to adapt to the development policy of "taking grain as the key link", heavy industry first and national defense security strategy under the planned economy, especially in land, household registration, industry and facilities construction. This management mode of urban–rural separation is realized through a highly administrative planning system. The government controls the population, materials, production and their flow in urban and rural areas, forming a clear opposition and separation between urban and rural areas, industry and agriculture, citizens and farmers. After the reform and opening-up, measures such as the household contract responsibility system, abolishing the system of unified purchase and marketing of agricultural products, the development of township enterprises, market-oriented reform and rural tax and fee reform has been carried out in rural areas. Therefore, the system and mechanism environment of rural development has been continuously improved. In the case of the loosening of dual system, the development of township enterprises has obtained a loose institutional environment, and in the 1980s and 1990s, and it becomes an important force in the transfer of labor force and economic growth in rural areas. As a result, the southern Jiangsu model, Wenzhou model and Pearl River Delta Model appeared.

Land system and *hukou* system are the two most significant systems. In China, the land is divided into state-owned land and collective land, which correspond to two market systems and bear different industrial and social functions. The relevant laws stipulate that only state-owned land can be traded in the market, and local governments can expropriate rural collective land according to the needs of urban development. In 1958, the implementation of the "Regulations of the people's Republic of China on household registration" indicated that the urban–rural dual structure had been fixed in the form of law, and the population was divided into agricultural population and non-agricultural population. Non-agricultural population can go to school and work in the city, and get supplies such as grain and oil; while agricultural population can only go to school and work in the village, and only a small number of agricultural populations can be transformed into non-agricultural population through the national plan of recruitment or employment. Affected by the *hukou* system, farmers are strictly bound to the land. With the increase of floating population, temporary residence permit, blue print residence permit and residence permit have appeared successively since the end of 1980s.

China's *hukou* system reform has always focused on small and medium-sized cities, and the access to large and medium-sized cities has always been strictly controlled. However, there are huge differences between big cities and small towns in public services, employment opportunities and urban welfare. Generally, the *hukou* reform is focused on limiting population mobility, which is difficult to effectively coordinate the development of large cities and small towns. While strictly controlling the population scale of megacities, the national new-type urbanization planning (2014–2020) proposes that megalopolis can adopt the way of credit entry to realize step settlement. But it has not fundamentally solved the demand for *hukou* system from the coordinated transformation of urban–rural regional systems. Under the background of accelerating market-oriented reform and population mobility, the difference of urban–rural dual *hukou* system and industrial system, infrastructure and public services based on it become institutional obstacles to the healthy development of economy and society as well as the integration of urban and rural development. At present, the urban–rural coordinated development is still faced with the institutional constraints of *hukou*, land, social security and other aspects, and the inter-provincial population flow faces the institutional constraints of inter-regional coordination.

(10) Space control and regional management

Regulation is a measure between the market and the government, which can be used to coordinate the interests of different subjects such as governments, groups, enterprises and residents at all levels in the region. Also, it can be used to guide coordination and cooperation between different regions, thus making up for the deficiency of relying solely on market and government regulation. According to certain planning principles, methods and processes, space control delimits the regional space of economic growth, social development and ecological environment, defines the development standards and control guidance requirements of various types of areas, so as to give full play to the advantages of different types of space, form a regional pattern with reasonable division of labor and complementary advantages, and realize the rational allocation and efficient utilization of space resources and the coordination of the interests of all parties. Different types of planning, such as socio-economic development planning, urban–rural planning, land use planning and environmental planning, delimit regional space according to different principles and objectives, and are implemented and guaranteed through policies of investment, finance and taxation, population, environment and land.

Spatial control based on various planning is the spatial control of land development right (Lin and Xu 2014). China's unique urban and rural land ownership system has formed two levels of market, i.e., state-owned market and rural market. However, the rural market is not mature, and the market value of state-owned land is much higher than that of rural land. Under the land level development right, a game has been formed between the state and the local, the higher-level government and the lower-level government, the city and the village, the government and the collective. Therefore, the regulation of urban and rural regional space is of great significance to protect the rights and interests of urban and rural land development. For a long time, rapid urbanization has led to construction land occupying many rural production and ecological space, increasing the pressure on rural population resources and environment. Generally, it is difficult to establish an equal spatial control system between urban and rural areas in the economic-growth-oriented development strategy. As a result, the rights and interests of rural land development are transferred to cities and towns, which damages the rights and interests of farmers and is not conducive to urban–rural equity.

Administrative region is an effective means for the central government to manage all parts of the country, and is a major strategic issue concerning the overall situation of national political, economic and social development (Zhu et al.

2015). The adjustment of administrative division is a kind of macro adjustment behavior of the government to meet the needs of economic growth, regional relations and urbanization development, which affects the scope of regional management, financial control and interregional links. With the development of urbanization, the expansion of central city, the growth of population concentration, and the improvement of traffic and communication conditions, the original administrative region does not meet the requirements of urban and rural regional control. The administrative authority of different levels of administrative units in China is very different. According to the evolution process of urban and rural areas, the proper adjustment of administrative divisions is conducive to the coordinated development of urban and rural areas and expand urban space, thus providing conditions for cities to drive rural areas. The main forms of administrative division adjustment in China include the establishment of a new administrative region, the upgrading of counties into cities, the transformation of counties and cities into districts and the establishment of new urban space; in rural areas, it is mainly manifested as township to town, township merger, multi village merger, new community construction and so on.

If we break away from the law of urban and rural regional evolution and economic development, and simply promote the transformation of rural space to urban space through state administrative means, it will easily lead to the problem of "pseudo urbanization", that is, the population adjusted by administration is passively counted as urban residents, and they are passively changed the living and production conditions. Due to the lack of professional skills in the process of transformation from farmer identity to citizen identity, it is difficult for them to obtain corresponding jobs. If we carry forward the change in land ownership and rural land expropriation by adjustment of administrative division, some farmers may become a social group without land, employment and insurance due to the low compensation for land expropriation, which will infringe on the legitimate rights and interests of farmers and increase social instability.

3.3.5 Allocation Mechanism and Regulation of Urban and Rural Elements

1. **Market-oriented allocation of urban and rural production factors**

There are many element connections between urban and rural areas, and a variety of factor flows have been formed. In different stages, the direction and intensity of urban and rural elements transfer are different, especially in the accelerated stage of industrialization and urbanization, when rural elements transfer rapidly to cities and towns. Based on the law of diminishing marginal returns of factors, production factors such as population, labor force, capital and land are allocated among industrial departments and between urban and rural areas to maximize the benefits of decision-makers. According to the firm theory of economics, the change trend of production factors inputted into a certain industry or urban–rural regional space is analyzed (Fig. 3.25). *Q* refers to the quantity of factor input, *R* is the product output value or GDP. In the stage of OQ_1, the marginal return (MR) of factor input in the industry/region is greater than the average return (AR), which indicates that the input of production factors is insufficient, so we should continue to input production factors. With the continuous increase of production factor input, MR of factors begins to decline. When the quantity of Q_1 is reached, MR equals to AR. In the stage of Q_1Q_2, MR is greater than 0, and the total return (TR) continues to increase, which

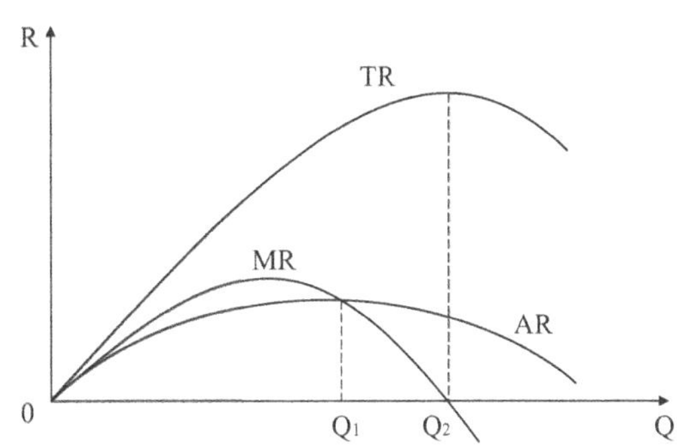

Fig. 3.25 The change trend of the profit of production factor

means the production factors can be continuously input. When the quality of factor input exceeds Q_2, MR of factor is negative, that is, the factor is over-invested and should be reduced.

Equilibrium refers to a state in which a thing is still, motionless and balanced, or a state of relative balance and competition achieved by the interaction and adjustment of several things. Although equilibrium assumes a relatively static state, potential equilibrium itself is constantly changing. The final state of factor transfer realizes the equilibrium of factor income, that is, the income of the same factor in different industries and regions is equal. Adam Smith put forward the idea of "laissez-faire" under the market mechanism, and hold the view that the capitalist market can achieve the goal of balance of supply and demand without external intervention. However, because the transfer of urban and rural elements is also constrained by its own characteristics, external institutional environment, urban and rural main body behavior, it is difficult for the elements transfer between departments and urban and rural areas to flow and exchange equally according to the market value. Affected by the urban–rural *hukou* system, employment system and social security system, it is difficult to realize the citizenization of China's rural population, which has transferred to cities and towns, thus resulting in the serious "urban–rural wandering". Land elements have non-productive public functions such as ecological function, social function and space bearing. Overemphasizing the economic function of land is easy to bring negative impact on ecological environment. In addition, the urban–rural regional system has multi-layer attributes such as economy, society and ecology. The public attribute and externality make it difficult to realize the transformation of urban–rural functions completely relying on the market mechanism. Therefore, it is necessary to restrict the excessive non-agricultural transformation of rural elements and the uncoordinated transfer of elements by means of legal constraints, planning guidance and investment guidance.

Urban–rural elements flow and spatial distribution are the basis of urban–rural regional transformation. Equilibrium is only a kind of relative balancing state, and urban and rural factor flow cannot achieve the complete equality of factor marginal income. In the process of industrialization and urbanization, urban–rural transformation is accompanied by the repeated transformation of balanced-unbalanced-balanced factors between urban and rural areas, realizing the balanced allocation of different industries and uses. Based on the perspective of transformation, urban–rural transformation, as a process of transition from a balanced state to another, implies a turning stage, and there are great differences in the quantity, structure and function mode of production factors in different stages before and after. Population urbanization and land urbanization are the core contents of urban and rural elements flow and transfer, which focus on further analysis of allocation and transformation of elements based on the population factors and land factors.

2. **Optimization and regulation system of urban and rural region**

The coordinated transformation of urban and rural areas is the process of urban–rural balance, which is the superposition of multiple processes, and is affected by the law of economic growth, the law of human social demand and the system mechanism. Restricted by human capital, age structure, industry type, government investment, policy system and other factors, there is always a certain economic potential gap between urban and rural areas. To promote the balance of urban and rural areas through the internalization of non-productive functions, it is necessary for government regulation and policy constraints to take effect. Corresponding to the natural order of urban–rural transformation, in addition to creating a market environment of equal exchange of production factors, we should promote the balanced urban–rural development from the perspective of public services and institutional innovation, thus narrowing the urban–rural gap.

To solve the problems of the widening gap between urban and rural areas and the uncoordinated transfer of factors, the government adjusts relevant policies to coordinate urban–rural development. Transformation threshold is the priority of government policy implementation and overall planning, and it is the expression of the difficulty degree of coordinating urban–rural development in all aspects. The transformation threshold includes strategic domain, infrastructure domain, industrial development domain and institutional domain. The strategic domain is the macro direction of the national development strategy in a certain period, and is always consistent with the main problems and objectives of the country. The infrastructure domain is the material basis of economic development, serving the needs of national strategy, and can be expressed by national investment. The industrial development domain is the basic content of economic growth, involving growth rate, economic structure and transformation mechanism. The institutional domain is the basic guarantee for the implementation of national strategy, government investment and industrial development environment, and solidifies the original relationship.

The urban–rural transformation domain presents the transformation order of strategic domain, infrastructure domain, industrial development domain and institutional domain, which is from easy to difficult. When the imbalance of urban and rural development reaches a certain degree, the balancing urban and rural development will first be realized in the strategic domain, that is, the government will change the mode of economic growth and the strategy of urban–rural development. Meanwhile, the government improves infrastructure and public services by increasing investment in rural areas, and then coordinates urban and rural development. Therefore, the infrastructure domain is relatively easy to realize. As the development of enterprises is guided by the maximization of production profits and optimized allocation of resources through market mechanism, it is difficult for the government to regulate the layout and types of industrial enterprises. Although small towns have advantages in terms of population transfer cost and urban–rural development, they are difficult to attract factors and enterprises due to insufficient service function and poor location conditions. As a result, it is difficult to optimize and coordinate urban–rural relations through industrial development.

The system involves urban–rural *hukou* system, land system, employment system, social security system, medical system, etc. China's dual system of urban–rural inequality is difficult to protect the rights and interests of farmers and rural areas in land, finance, free migration and so on, and does not meet the needs of urban–rural population mobility and contact. Based on Coase's theory of institutional economics, a new system will occur only when the expected net benefit of innovation is greater than the expected cost. Therefore, it is more difficult to change the urban–rural system and mechanism to promote urban–rural transformation. Only when the problems in urban and rural development cause huge social burden or loss of economic benefits can the reform be possible. In terms of time sequence, the government's regulation of urban–rural regional transformation generally follows the sequence of strategy, infrastructure, industrial development, and institutional system. With the transfer of rural population to urban area, it is essential to develop modern agriculture, agricultural products processing industry, township service industry and other industries through the support of industrial policies in rural areas, and it is also necessary to change the urban–rural dual system and management mode.

3.3.6 Theoretical Framework of Urban–Rural Integration

Urban–rural integration is the first way to realize China's rural revitalization strategy in the new era which is aimed at solving the main social contradictions and prominent problems of rural regional system and narrowing the gap between urban and rural areas. The object of urban–rural integration is a regional multi-body system, which mainly includes urban–rural integrated body rural complex, village-town organism, and housing-industry symbiosis.

Human-earth science is a new interdisciplinary subject which focuses on the coupling mechanism, evolution process and complex interaction effects of human-earth system (Liu 2020b). The human-earth areal system offers a theoretical basis for the research of urban–rural integration and rural revitalization. It is necessary to consider the urban–rural integration within the territorial rural system for the research of rural revitalization and development, given that integrated urban–rural development is a dynamic process with an ultimate goal of urban–rural equivalence. The urban–rural integration represents interactions between urban and rural systems; moreover, the urban–rural integration is manifested in the factor transfer, strategy change, and mechanism conversion of the urban–rural regional system, including the transformation of the population pattern, industrial structure, land use, and spatial form. Population and land constitute core factors of the human-earth relationship, and the industrial structure offers a basic framework for the regional system of human-earth relationship. Moreover, rights, reflecting the relationships between people and between people and land, are requested to be equal between urban and rural areas as China's urbanization has sacrificed farmers' rights and interests. In this context, a solution-oriented rural revitalization strategy should fully implement the "four-revitalization" development of rural areas (i.e., revitalization of population, land, industry and right), effectively stimulate rural vitality, capacity, motivation and competitiveness, and systematically promote the urban–rural integration, coordination and equivalence.

Cities and rural areas constitute an organic whole, and a sustainable development can only be achieved through reciprocal interactions between urban and rural department. The urban–rural integration was created and has evolved by the interaction mechanisms between urban and rural systems. It has four cores: urban–rural strategic position equivalence, equal rights and interests for urban and rural residents, urban–rural factor allocation balance, and an integrated urban–rural development process. The theoretical framework of China's urban–rural integration can be summarized as shown in Fig. 3.26.

The key for the success of urban–rural integration lies in the "people". Currently, the rapid non-agricultural conversion of agricultural production elements has induced the loss of rural labor. Those migrating to cities are mainly young and middle-aged people, while those left behind are mainly the "three types of left-behind groups" (i.e., the elderly, women, and children) who cannot support modern agriculture and new countryside construction. In the process of urbanization, peasants should be entitled to freely choose between cities and rural areas, and efforts should be made to rationally solve problems related to peasant workers and their return to hometowns. At present, no institutional environment exists in which peasants are encouraged to move their whole families into cities, especially when considering the most important factors restricting their settlement in cities, such as housing and children's education.

Land provides an important resource for the survival and development of mankind, and it constitutes the core of rural socio-economic development. However, rapid industrialization and urbanization have caused a massive loss of high-quality cultivated land. Moreover, the scattering and fragmentation of small-scale cultivated land and the lack of adequate farmland infrastructure have resulted in a low utilization rate of cultivated land. Existing rural problems related to people's well-being, ecology, industry, and facilities can all be attributed to the problem of land use. Rational land use provides a valid foundation for reinvigorating land resources, optimizing spatial pattern, improving the eco-environment, and supporting industrial development, which constitute the highlights of the urban–rural integration.

The focus of urban–rural transformation is industrial prosperity and rural industrial development is a motivational guarantee and inner mechanism of urban–rural integration. On the one hand, traditional rural industries must be protected while modern agriculture must be developed; on the other hand, efforts should be made to cultivate both emerging and strategic industries as well as to promote the integration of primary, secondary, and tertiary industries in

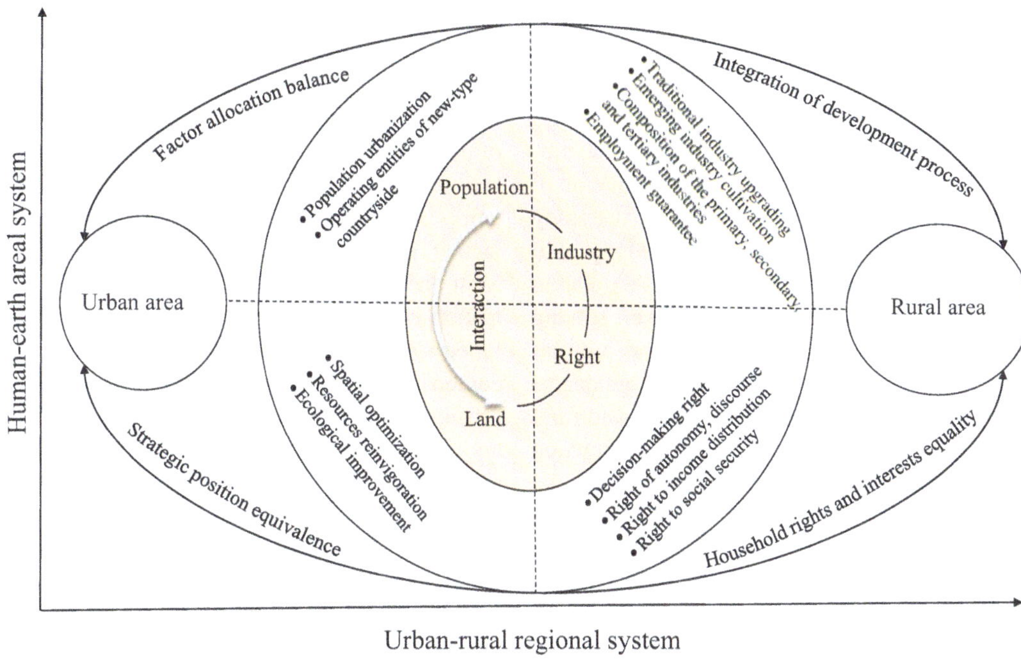

Fig. 3.26 Theoretical framework of the urban–rural integration

rural areas. For instance, unique tourism resources can be relied upon for the development of the rural tourism industry, the sightseeing industry, the leisure agriculture and service industry, and the rural new energy and new material industries (including e-commerce, informatization, and other industries). To develop rural industries, an overall consideration should be focused on the regional differences of rural areas, the bearing capacity and property rights of land, and the related rights enjoyed by peasants. The operating entities of the new-type countryside constitute a major force for rural revitalization, and by developing new techniques, new modes, and new commercial forms of rural industries, urban–rural factor flows can be promoted (such as labor flow, the flow of funds, technology flow, information flow, and product flow).

The right of peasants to make their own decisions is important in rural revitalization, and their rights of discourse, to autonomous development, to income distribution over collective land, and to social assistance and social security must all be safeguarded. In the process of coordinated urban–rural allocation and the equal exchange of urban–rural factors, the key lies in guaranteeing equal rights and interests to both urban and rural residents. During the as-sets-oriented innovation of land, the property rights of peasants must be safeguarded, such as their rights and interests over land assets both before and after land consolidation as well as their right to income. At present, many regions have explored the possibilities of implementing a joint-stock land consolidation mode, with the purpose of safeguarding peasants' rights to income and their right of participation after land consolidation.

3.4 Main Contents of Urban–Rural Transformation Geography

Urban and rural development involves many aspects, such as space, industry, employment, land, culture, system, resources and environment. The essence of urban–rural transformation is to realize the multi-factors coupling of population-land-industry under the support of various systems and policies of urban–rural development, and is spatial evolution process of urban function diffusion and urban–rural factor flow and agglomeration (Fig. 3.27).

Based on the comprehensive and regional perspectives of geography, it is necessary to strengthen the systematic research on the process, mechanism, pattern, effect and regulation of urban–rural transformation. Therefore, the study of urban–rural transformation geography will focus on the process-power-problem-regulation of urban–rural transformation, take solving urban–rural development problems and promoting urban–rural coordination as the guidance, and take the spatial allocation of urban and rural elements, structural optimization and function upgrading as the core. Further, it reveals the characteristics, types and regional differences of urban–rural transformation, and puts forward the paths and countermeasures to deepen the reform of the integration strategy, system and mechanism and governance system of urban and rural development, which follows the laws and principles of urban and rural economic and social transformation (Fig. 3.28).

3.4.1 Spatial Pattern of Urban–Rural Transformation

Under the condition of market economy, the flow of production factors is rooted in the relationship between urban and rural areas and regional systems. However, the flow and agglomeration of factors in different regions directly affect the form and characteristics of urban–rural transformation in a certain range. Thus, a regional evolution system with different regional urban–rural composite characteristics is formed. In terms of urban–rural development, urban and rural areas are the two subsystems of the region, which show different evolution patterns in urban–rural transformation. Affected by different agglomeration forces, urban and rural regional space has gradually evolved into urban and rural settlement spatial form with megacities and big cities as primary nodes, small and medium-sized cities and small towns as secondary nodes, and rural settlements as hinterland, which forms the spatial pattern of point-axis or network and urbanization

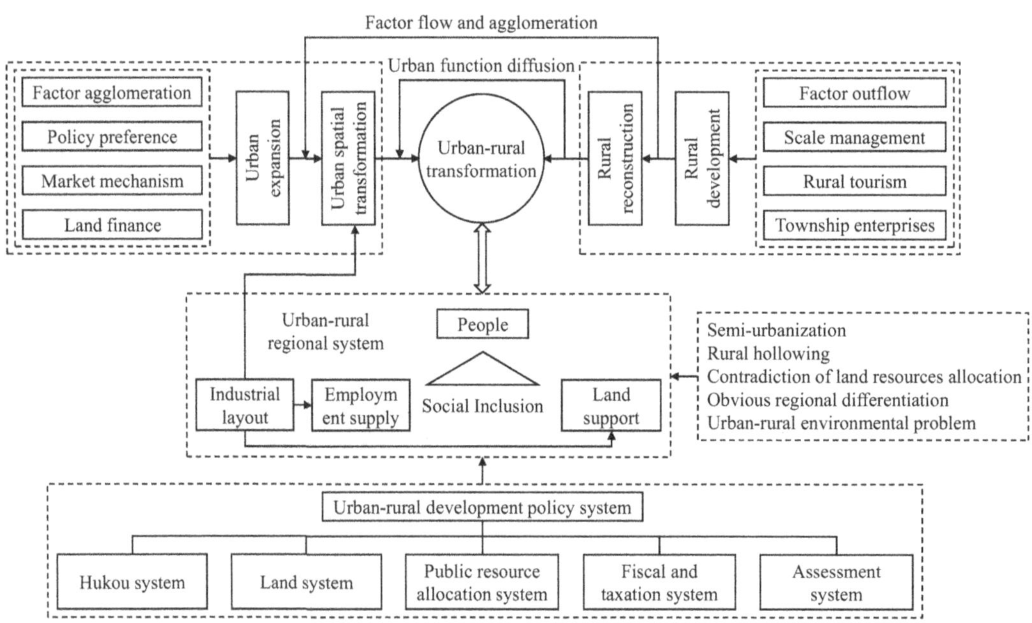

Fig. 3.27 The process of urban–rural transformation and its scientific mechanism

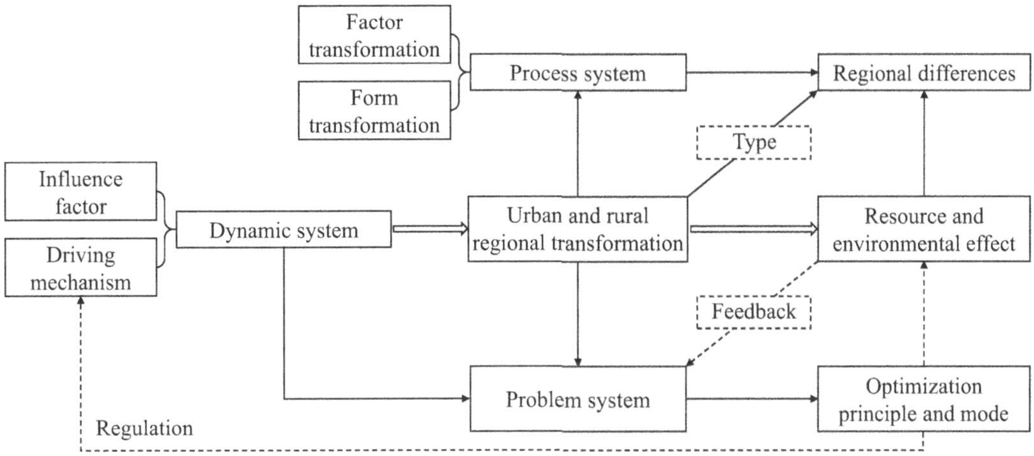

Fig. 3.28 Basic framework of the research on urban–rural transformation

area through the connection of transportation, water conservancy and other infrastructure. In the rapid urban–rural transformation stage, the impact of urban areas and industrial parks on rural development through labor demand and industrial investment is more and more obvious. From the urban center to the periphery, the response area of urban area, urban–rural transition zone and rural hinterland is formed, which is facing transformation development and spatial reconstruction in the aspects of population, industry, land and culture. Rural reconstruction and urban evolution are the regional responses of urban space and rural space in urban–rural regional system under the influence of various factors such as factor flow and industrial transfer, and are important components and spatial manifestation of urban–rural transformation. And this change and expression form is in the whole process of urban–rural transformation.

There are obvious differences in population concentration and evolution, economic structure and development level, urban structure and pattern dynamics, and regional ecological environment conditions in different regions. Therefore, the spatial form of urban–rural transformation and the composite function of development and protection have different performance. Based on the state of economic growth and urbanization, there are great differences in the relationship between urban and rural areas between economically developed areas and underdeveloped areas, rapid urbanization and lagging urbanization areas, so the main contradictions and key problems faced by urban and rural development are obviously different. Among them, economically developed areas have convenient transportation, perfect infrastructure, the cities and small towns are developing rapidly, the agglomeration conditions of production factors are superior and the ability is strong. Thus, the "trickle-down effect" of urban development on the surrounding rural areas is obvious, which not only provides important opportunities for rural population transfer and employment, but also stimulates and promotes the endogenous development of rural areas through the transfer of technology and funds to the countryside. Therefore, the research on the pattern of urban–rural transformation should integrate the urban and rural sub-system and urban–rural regional system, focusing on the research and evaluation of the pattern and dynamic evolution characteristics of urban–rural transformation.

3.4.2 Comprehensive Process of Urban–Rural Transformation

Under the joint action of many factors, there are obvious stage and dynamic characteristics in urban–rural transformation. The combination of this process and specific urban and rural areas forms a geographical process of multi-dimensional evolution of urban and rural areas. Geographical process shows the temporal and spatial evolution of geographical phenomena, which is helpful to understand the mechanism of things change (Fu 2014). With the changes of economic society, science and technology, globalization and other external environment changes, population, employment, industry, land and other elements in urban–rural areas and their structure, as well as the functional relationship between urban and rural areas are in dynamic changes. Through the analysis of the process of urban–rural transformation, we can identify the evolution characteristics of urban–rural relationship, and find out the problems of rural development and their causes. Based on the research of process, we also can analyze and grasp the mechanism of elements flow and agglomeration and function evolution of urban and rural regional system. Therefore, the analysis of process domain is an important basis for the study of dynamic domain and effect domain.

3.4.3 Driving Mechanism of Urban–Rural Transformation

The mechanism of the factors influencing the transformation of urban and rural development constitutes the dynamic domain of urban–rural transformation, and it is key to study the transformation mechanism of urban and rural development to analyze and sort it out. In general, market and government are the "two hands" of urban and rural economic development, which also play an important role in promoting the transformation of urban and rural areas. Due the economic growth and technological innovation, the market mechanism, which is expressed by agglomeration diffusion, endogenous extraneous and innovation, can increase the efficiency of urban and rural economic development, and promote the sustainable evolution of urban and rural industrial structure and employment structure; the change of economic level and social demand promotes the continuous change of industrial types and urban and rural regional functions.

With the transformation from planned economy to market economy, market mechanism has gradually become the main driving force of urban–rural economic and social transformation. The maximization of capital return and externality of economic activities under the action of market mechanism cannot consciously achieve the maximization of social equity and social welfare. The government plays an important role in maintaining social equity, ecological construction and public facilities supply in urban and rural areas. Therefore, it is urgent to clarify the way of action and division of labor of market mechanism and government regulation, rationally define the scope and jurisdiction of the government in coordinating urban and rural development, and scientifically study and judge the overall ideas and path selection of deepening the system reform, thus providing scientific suggestions for the countermeasures formulation of urban and rural transformation.

3.4.4 Comprehensive Effect of Urban–Rural Transformation

Based on the theory of human-earth areal system and human-earth coordination of sustainable development, the regional system of urban–rural relationship can be divided into three subsystems: natural environment, economic society and system policy. In the economic development and governance decision of urban and rural areas, the economic society and institutional policies of urban and rural areas are usually regarded as the representation of human initiative, which are used to depict the stages, situations and types of urban and rural development, focusing on the evolution of urban and rural regional system in structure, space and functional form. In the process of urban and rural elements transfer and rapid structural transformation, the urban and rural areas will have regional response to this process. According to the law of urban–rural spatial structure differentiation, the transition trend from urban (core) to rural (peripheral) has formed the urban area, urban–rural transition zone, and rural area, which includes urban transformation, urban–rural transformation and rural transformation. Meanwhile, there are obvious differences in the transformation process, mechanism and effect in different regions. The urban–rural transformation not only reflects the changes of urban and rural regional economic activities, but also strongly affects the natural resources and geographical environment supporting urban and rural development. In particular, the regional land use structure, water resources utilization pattern and ecological and environmental response process constitute the influence area of urban–rural transformation from the perspective of "human-earth system" (Liu 2020a). Generally, the influence domain of urban–rural transformation not only includes the continuous impact on the resources, ecology and environment of urban and rural regional system, but also includes the response between the elements of urban and rural subsystem.

3.4.5 Optimal Decision of Urban–Rural Integration

The overall goal of the research on urban–rural transformation is to explore the changing characteristics and coupling mechanism of each elements and the overall pattern of the system in the dynamic evolution of urban–rural relationship regional system, and then understand the long-term mechanism of structural adjustment, function optimization, overall balance and effective regulation and control of the urban–rural relationship regional system from the perspective of time evolution, spatial structure and overall effect. Based on the comprehensive research on the process, structure and mechanism of urban and rural system, this study reveals various problems in the process of urban–rural transformation, and discusses how to effectively regulate and solve these problems, forming the "regulatory domain" supporting the optimal decision-making of urban–rural regional system. The main body of the regulatory domain involves the government, enterprises, urban and rural residents, and social groups, which are the executors to optimize the economic and social development of urban and rural areas.

Building a moderately well-off society in all aspects is the main goal and strategic direction of China's development in the new period, and the key and difficult points are in the vast rural areas. Under the background of rapid industrialization, urbanization and economic globalization, it is necessary to systematically consider the transformation and modernization of agriculture and rural areas, the balance between economic development and people's livelihood security, the coordination between the optimization of human land relationship and the protection of ecological environment, and realize multi-objective coordinated development in the process of system dynamic evolution. In view of the spatial and temporal characteristics of the rapid transformation of urban and rural regional system in the new period, the core purpose of strengthening the study of urban–rural transformation geography is to summarize and put forward the ways and countermeasures to scientifically coordinate regional relations, urban–rural relations, and human-earth relations, so as to promote sustainable urban-rural integrated economy and social development.

References

Cai Y. On the relation of town-country and the problem of modern China urbanization. J Southwest Univ (Soc. Sci. Ed.). 2003;29(5):117–21.

Chen B, Lin Y. Development strategy, urbanization and the rural-urban income disparity in China. Soc Sci China. 2013;4:81–102.

Chi F. The second transformation. Beijing: Economic Press China; 2010.

Cumbers A. Advancing theory in urban research. Urban Stud. 2014;51(5):865–7.

Du Z, Zhang X. Trend and policy analysis of global rural development and the evolution of urban-rural relationship. World Surv Res. 2006;(7):7–8, 47.

Friedman J, Tang M. Four theses in the study of China's urbanization. Urban Res. 2007;(7):4–6.

Fu B. The integrated studies of geography: coupling of patterns and processes. Acta Geogr Sin. 2014;69(8): 1052–9.

Guo C. Research on rural e-commerce model based on survey of Taobao village. Reform Econ Syst. 2015; 5:110–5.

Hu W, Zhang A, Qu L. Interrelationships among non-agricultural population, jobs and land. China Population Resour Environ. 2009;19(5):104–10.

Li Y, Hu Z. Approaching urban-rural integrated development in China: the changing institutional roles. Sustainability. 2015;7(6):7031–48.

Li Y, Wang Y, Liu Y. Impact and mechanism of social capital in poverty alleviation of China. Bull Chin Acad Sci. 2016;31(3):302–8.

Lin J, Xu C. Land development rights, space control, and synergetic planning. City Plan Rev. 2014;38(1): 26–34.

Lipton M. Urban bias: consequences, class and causality. J Dev Stud. 1993;29(4):229–58.

Liu Y. New-type urbanization should cure "rural diseases". People's Daily, 2013–09–10.

Liu Y. Geography of new countryside construction in China. Beijing: Science Press; 2011.

Liu Y. Modern human-earth relationship and human-earth system science. Sci Geogr Sin. 2020a;40(8):1–14.

Liu Y. The basic theory and methodology of rural revitalization planning in China. Acta Geogr Sin. 2020b;75(6):1120–33.

Liu Y, Li Y. Revitalize the world's countryside. Nature. 2017;548(7667):275–7.

Liu Y, Liu Y, Guo L. Connotations of rural regional multifunction and its policy implications in China. Hum Geogr. 2011;26(6):103–6.

Liu Y, Long H, Zhang X, et al. Research progress and prospect in the disciplines of agricultural geography and rural development in China. Prog Geogr. 2011;30 (12):44–51.

Liu Y, Lu S, Chen Y. Spatio-temporal change of urban–rural equalized development patterns in China and its driving factors. J Rural Stud. 2013;32:320–30.

Liu Y, Liu Y. Progress and prospect in the study of rural region multifunctions. China Population Resour Environ. 2012;22(10):164–9.

Liu Y, Chen C, Li Y. The town-villages construction pattern under new-type urbanization in China. Areal Res Dev. 2014;33(6):5–10.

Liu Y, Zhang Z, Wang J. Regional differentiation and comprehensive regionalization scheme of modern agriculture in China. Acta Geogr Sin. 2018;73(2): 203–19.

Lu D. Location theory and the methods of regional studies. Beijing: Science Press; 1988.

Lu D. Theoretical studies of man-land system as the core of geographical science. Geogr Res. 2002;21(2):135–45.

Luo J, Wang D. The summary and enlightening on the theories of the relation of town-country. Res Dev. 2005;1:29–31.

Ma A. Regional economic development and urban rural linkages: a new theoretical framework for spatial economic transformation in developing Asia. Econ Geogr. 1994;14(1):11–5.

Mao H. Research on man-earth system and regional sustainable development. Beijing: China Science and Technology Press; 1995.

Stöhr W, Taylor DRF. Development from above or below? The dialectics of regional planning in developing countries. Chiehester: Wiley; 1981. p. 9–26.

Tan H, Zheng J. The mystery of coexistence of rapid economic growth and lag of service industry in China: based on the perspective of sector TFP. China Ind Econ. 2012;9:5–17.

Wei YD. Decentralization, marketization, and globalization: the triple processes underlying regional development in China. Asian Geogr. 2001;20(1–2):7–23.

Wu C. Special research fields and future tasks of geography. Econ Geogr. 1981;1(1):5–10.

Wu C. The core of study of geography: man-land relationship areal system. Econ Geogr. 1991;11(3): 1–6.

Yan B. An analysis of the dynamic mechanism of the growth of Zhejiang industrial clusters: a perspective of regional culture. Chin Rural Econ. 2007;(s1):87–92.

Yang K. Mechanism and conceptual model of economic growth driven by urbanization in China. Urban Probl. 2001;3:4–7.

Zhang W. Study angles of the new economic geography. Prog Geogr. 2003;22(1):95–103.

Zhang F, Liu Y. Dynamic mechanism and models of regional rural development in China. Acta Geogr Sin. 2008;63(2):115–22.

Zhang X, Sheng M. The reorientation of rural geography in China. Hum Geogr. 2002;17(1):81–4.

Zhu J, Chen T, Wang K, et al. Spatial pattern evolution and driving force analysis of administrative division in China since the reform and opening-up. Geogr Res. 2015;34(2):53–64.

4 Research Methods of Urban–Rural Transformation Geography

Abstract

This chapter analyzes the main research methods of urban–rural transformation geography. The geographical research thinking paradigm of urban–rural transformation mainly includes empiricism, positivism, behaviorism, and structuralism, and the research methods can be divided into qualitative description and quantitative analysis. The evaluation process of urban–rural transformation mainly includes determining evaluation subjects, constructing index system and collecting basic data. At the theoretical level, the comprehensive evaluation index system of urban–rural transformation involves three levels: urban–rural social and economic transformation, urban–rural factor flow and structural change as well as the relationship between urban and rural development. The comprehensive zoning method of urban–rural transformation includes regionalization classification method and transect research method. The dynamic mechanism of urban–rural transformation can be analyzed through the regression analysis, structural equation model and Geodetector. The simulation and forecast of urban–rural transformation can be realized through the linear programming, multi-objective optimization model, system dynamics method and gray prediction method.

4.1 Methodology of Urban–Rural Transformation Geography

4.1.1 Geographical Thinking of Urban–Rural Transformation

From the perspective of local uniqueness, regional difference and human-earth relationship, geography uses field investigation, field observation, remote sensing, map, geographic information system and spatial econometric analysis to explain population, resources, environment, development and other issues, and to support relevant decision-making. The main features of geography research are the comprehensive analysis of the natural environment on different scales and the relationship between humans and environment, which constitutes the unique thinking way in geography. The main thinking paradigm includes empiricism, positivism, behaviorism, and structuralism. Urban–rural transformation is the inherent requirement for transforming the economic development mode, coordinating the development of urban–rural population-land-industry, and promoting the coordinated development of "three modernizations". It is a complex system engineering project with the characteristics of multidisciplinary integration, total factor integration, and long-term tasks. The application

Y. Liu, *Urban-Rural Transformation Geography*, Sustainable Development Goals Series,
https://doi.org/10.1007/978-981-16-4835-9_4

of geographical thinking to the study of urban–rural transformation forms the geographical thinking and analytical logic of the study of urban–rural transformation.

1. **Empiricism analysis**

Empirical analysis, also known as normative analysis, is a traditional thinking way in geography. It mainly defines, classifies and measures disordered things based on perceptual experience, so as to form a coherent system, and then induces and generalizes to obtain a theory or model, which is finally used to explain phenomena or problems. Therefore, it can be said that the empiricism emphasizes observation, accumulation of knowledge, induction and summary. This method can be used to discover problems and typical cases of urban–rural transformation, summarize the process of urban–rural transformation and the law of regional differences, and put forward the successful model of urban–rural integrated development.

2. **Positivism analysis**

In the 1960s, the quantitative revolution brought about the thinking of positivism in geography, which emphasizes the use of statistical analysis and mathematical tools to explore the formation of the spatial structure of economic activities. The application of this thinking mode and method promotes geographic research from experience to empirical research, from qualitative analysis to quantitative research, and from perceptual thinking to rational judgment. Positivism analysis can be applied to the quantitative research on urban–rural transformation, mainly including the evaluation of the difference degree and dynamic degree of urban–rural transformation, the optimal allocation of urban–rural space, and the evaluation and measurement of the resource and environmental effects of urban–rural transformation.

3. **Behaviorism analysis**

Behaviorism analysis, also known as humanism analysis, takes people the starting point and attaches importance to the rich meaning of human behavior and the important role of the social value system. However, due to the potential and indirect effects of these factors, they are invisible, and a direct analysis method is not proposed. Humanism emphasizes the humanistic orientation, and puts the research subject in the subjective experience world. The humanistic thinking way shows a strong guiding significance for the major social issues in the urban–rural integrated development of China. It is suitable for the study of the interactions between government, enterprises, residents, and the object environment. It is mainly used in the study of land use and transformation, public resources allocation, and the willingness of urban–rural residents to transfer.

4. **Structuralism analysis**

Structuralism is an important thinking way in geographical research. It takes the research object as a whole, analyzes and studies problems with holistic thinking, respects the complexity, comprehensiveness, systematicness, dynamics and organic connection of things, so as to get close to things and easily reveal their essence, which also echoes and is consistent with the research thinking of the human–environment relationship system.

The thinking modes of empiricism, positivism, behaviorism, and structuralism run through the different system levels and spatial scales of the urban–rural transformation. Thus, it forms research thinking system for urban–rural transformation, which is suitable for different specific research methods at different levels and scales (Fig. 4.1).

As far as the research on the mode and mechanism of urban–rural transformation in China is concerned, its geographical research focuses on the zonal law of regional urban–rural transformation in the macro scale, implements the differentiated development strategy based on regional main function, establishes the interregional compensation mechanism and regional urban development and balancing urban–rural development policy, coordinates the interest relationship between different regional entities, and establishes a fair, reasonable and efficient macro platform for urban and rural transformation. In meso scale, the research on urban–rural transformation focuses on

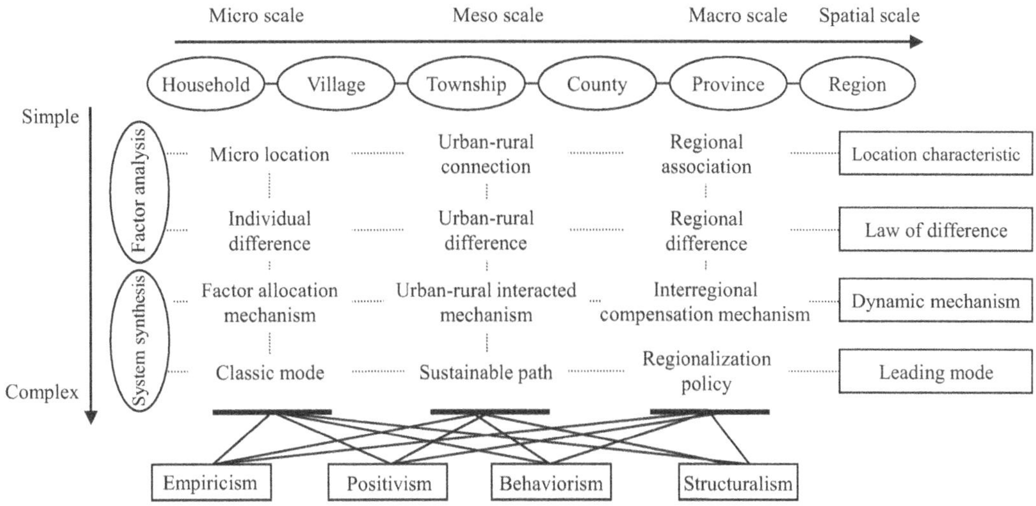

Fig. 4.1 Research thinking system of urban–rural transformation geography

balancing urban–rural development, establishing a long-term mechanism of urban–rural interactive development with the core of urban support for rural areas and industry feeding agriculture, and seeking sustainable way of urban and rural coordinated development. In micro scale, the research emphasizes the micro-location characteristics and individual differences of urban and rural regional evolution, and forms the spatial allocation of urban and rural elements under the guidance of economic maximization and social equity. In particular, it is necessary to establish a mechanism for urban–rural coordinated development in elements reallocation and infrastructure equalization, and form a typical model of urban–rural transformation and regional sustainable development.

4.1.2 Methodological System of Urban–Rural Transformation Geography

A perfect methodological system is an important guarantee for the study of urban–rural transformation geography. From the perspective of the overall way of thinking, the methods of urban–rural transformation geography can be divided into empiricism, positivism, behaviorism and structuralism. From the overall process of urban–rural transformation, it can be divided into urban–rural problem investigation, urban–rural transformation evaluation, urban–rural development planning, urban–rural transformation prediction and simulation, as well as urban–rural transformation decision support. From qualitative and quantitative perspectives, it can be classified into qualitative descriptions and quantitative analyses, the former can be mainly divided into mathematical statistical method, system analysis methods, and model simulations; the latter mainly includes observation method, interview method, comparative analysis method, historical analysis method, etc. From the perspective of data processing and analysis, it can be divided into data collection, data management, data processing, and decision support methods. Based on the above analysis, a geographic methodological system of urban–rural transformation has been constructed (Fig. 4.2).

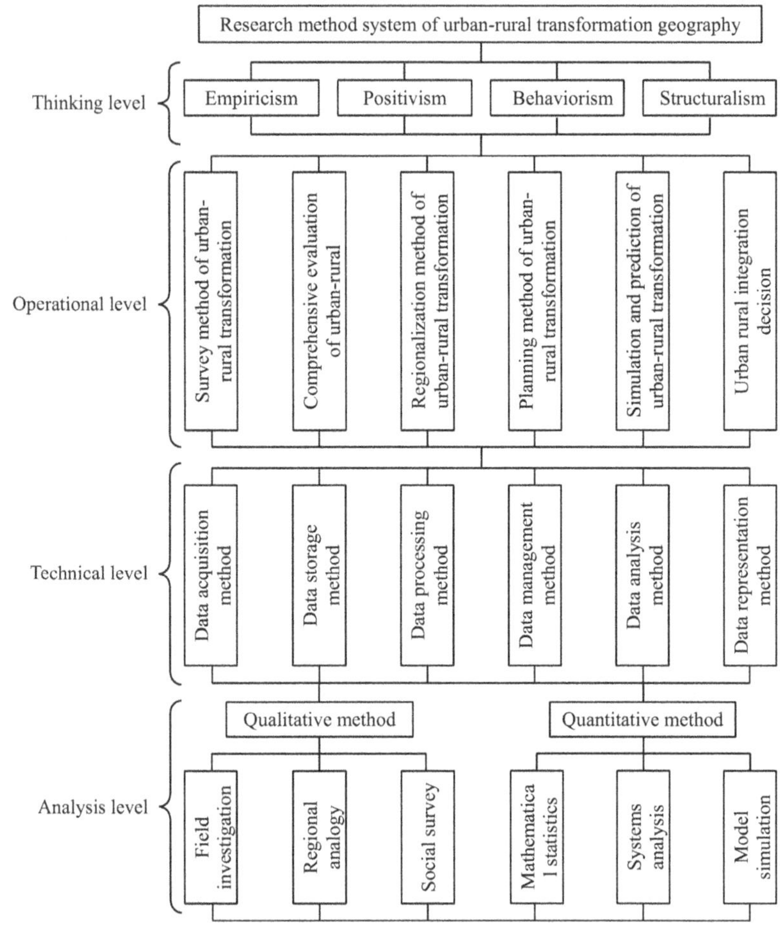

Fig. 4.2 Methodological system of urban–rural transformation geography

4.1.3 Quantitative Analysis Method of Urban–Rural Transformation

The research on urban–rural transformation has obvious characteristics of complexity, regionality, periodicity and randomness. Thus, it is an important direction to carry out standardized and quantitative research on urban–rural transformation. Quantitative research can more accurately describe the state of urban–rural development, reveal the interaction between urban and rural system and its subsystem, and provide reliable decision-making support to balance urban–rural development. Quantitative research methods include mathematical, system analysis and modeling methods. From the perspective of geography, the quantitative research method system of urban–rural transformation is summarized as shown in Fig. 4.3.

The mathematical methods of urban–rural transformation research mainly include mathematical-statistical analysis methods and fuzzy mathematical analysis methods. Mathematical statistical analysis is mainly used in the study of the non-agriculturalization of urban and rural elements and the mechanism between them, including probability analysis, sampling analysis, variance analysis, regression analysis and correlation analysis methods.

The system analysis methods of urban–rural transformation research mainly include structure–function analysis, control-feedback analysis, linear system analysis, nonlinear system analysis, grey system analysis and so on. Based on the theory of the urban–rural regional

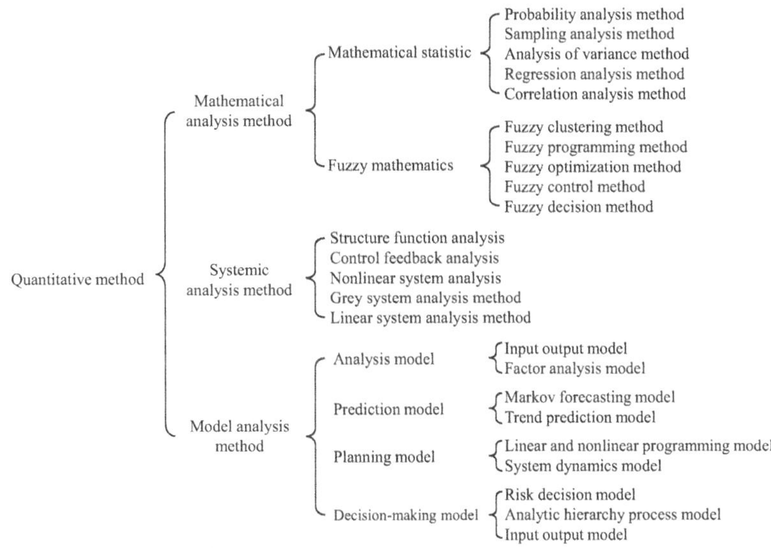

Fig. 4.3 Quantitative research method system of urban–rural transformation geography

relationship system, the transformation structure of urban and rural elements and the functional structure of urban and rural areas are constructed, and the system analysis method is used to quantitatively analyze the function mechanism of element structure and the system function utility. The model analysis method is based on mathematical methods and system analysis methods to simulate the structure, function, and evolution law of urban–rural regional systems, and then establish analytical, prediction, planning, and decision-making models.

4.2 The Comprehensive Evaluation System of Urban–Rural Transformation

4.2.1 Evaluation Process

Geography has the advantage of comprehensive analysis, and the evaluation of sustainable use, land intensification and comprehensive development level is relatively mature, while the quantitative identification and comprehensive evaluation of urban–rural transformation are less. Based on the existing evaluation index system and methods of sustainable development, the evaluation method system and general technical process of urban–rural transformation is formulated. The state evaluation of urban–rural transformation is mainly based on the internal relationship of urban–rural factor flow, structural transformation, and mechanism transformation. By selecting a certain evaluation index and adopting appropriate technical model method, it describes the current situation and function of urban–rural development, evaluates the output results, and forecasts the future development trend. As the urban–rural transformation and its mechanism involve social, economic, cultural and other dimensions, the evaluation of urban–rural transformation needs to form a scientific cognition, determine the evaluation subject, collect relevant information, and build an index system. In general, the comprehensive evaluation process of urban–rural transformation mainly includes the following aspects (Fig. 4.4).

Step 1: Under the guidance of theories such as system theory and sustainable development theory, the main content of urban–rural transformation evaluation is determined based on the analysis of the connotation of urban–rural transformation, and the change degree of urban–rural elements and the characteristics of urban–rural regional morphology are systematically analyzed.

Step 2: Basic data and map information of urban and rural development are collected to

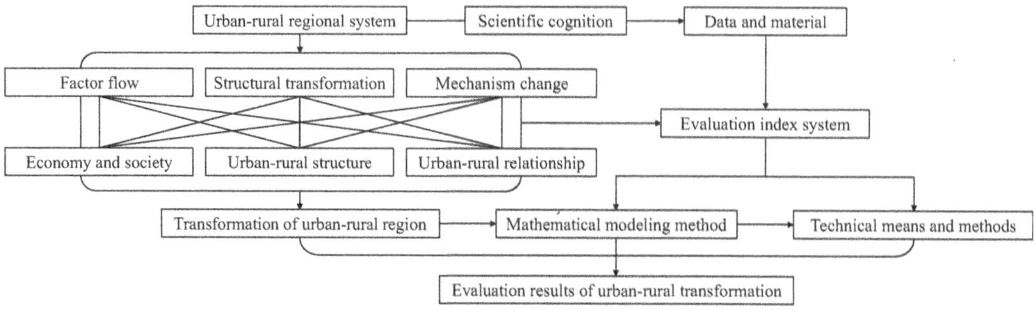

Fig. 4.4 Evaluation flow chart of urban–rural transformation

accurately describe the characteristics and spatiotemporal laws of urban–rural development. The basic data includes statistical data, image data, literature and so on.

Step 3: The core index system of urban–rural transformation is determined, including single index and comprehensive index. Then, the mathematical model of quantitative evaluation is constructed. In addition, combined with ArcGIS and spatial statistical methods, the monitoring, evaluation and prediction analysis of urban–rural transformation are realized, which provides auxiliary decision-making for the overall planning of urban–rural development and rural development.

4.2.2 Construction of Index System

1. **Urban–rural social and economic transformation**

The degree of urban–rural social and economic development is a comprehensive status of economy, society, ecology, environment and facilities, which can be summarized from three aspects, i.e., economy, society and facilities. The economic criterion is the urban–rural production system, the social criterion is the urban–rural social system, and the facility criterion is the supporting system for production and life. Economic transformation mainly includes four aspects, i.e., economic level, economic extroversion, economic benefit and economic growth mode, all of which represents the comprehensive state of economic system. Social transformation mainly consists of income, consumption, education, medical treatment, and social security. Facility transformation includes transportation facilities, communication facilities and living facilities (Table 4.1).

2. **Urban–rural factor flow of and structural change**

Urban–rural economic and social development promotes the changes of industrial structure, employment structure and consumption structure in urban and rural areas, which leads to the transformation of population and land use mode and forms the transformation of urban and rural regional structure and function. The urban–rural structural transformation is based on the structural changes of key elements. Analyzing the structural changes of urban–rural regional systems is an overall analysis of regional urban structure. On this basis, the indicator system for urban–rural structural transformation is constructed. The degree of urban–rural structural transformation includes four criteria: economic structure, employment structure, consumption structure, and spatial structure. The spatial structure is the spatial mapping of urban–rural economy and society, which can be characterized by urban–rural population pattern and settlement landscape pattern (Table 4.2).

3. **The relationship between urban and rural development**

The evaluation of urban–rural coordination involves in the comparison of the development level of urban and rural areas. The relationship

Table 4.1 Measurement of urban–rural social and economic transformation

Functional index	Structural index	Specific index	Description
Economic transformation (A)	Economic level (A_1)	Per capita GDP (Yuan)	GDP/Total population
		The productivity of primary industry (Yuan)	Output value of primary industry/Employment in primary industry
		The productivity of secondary industry (Yuan)	Output value of secondary industry/Employment in secondary industry
		The productivity of tertiary industry (Yuan)	Output value of tertiary industry/Employment in tertiary industry
	Economic extroversion (A_2)	Per capita exports (USD)	Total exports/Total population
		Per capita utilization of foreign capital (Yuan)	Total use of foreign capital/Total population
	Economic benefit (A_3)	Per capita fiscal revenue (Yuan)	Total fiscal revenue/Total population
		The profit rate of industrial output value (%)	Profits of industries above designated size/Total output value of industries above designated size
	Economic growth mode (A_4)	Energy intensity of GDP (t/10,000 yuan)	Total energy consumption/GDP
		CO_2 emissions of unit energy consumption (t/t)	Total CO_2 emissions/Total energy consumption
Social transformation (B)	Income level (B_1)	Per capita disposable income of urban household (10,000 yuan)	–
		The year-end growth rate of savings of urban–rural residents (%)	(Year-end balance of savings of urban–rural residents in current period/Year-end balance of savings of urban–rural residents in base period) − 1
		Per capita disposable income of rural household (10,000 yuan)	–
	Consumption level (B_2)	Per capita consumption of urban household (10,000 yuan)	Per capita consumption expenditure of urban household
		Per capita consumption of rural household (10,000 yuan)	Per capita consumption expenditure of rural household
		Per capita total retail sales of consumer goods (10,000 yuan/person)	Total domestic retail sales of consumer goods/Total population
	Educational conditions (B_3)	Average schooling years (Years)	Average years of education
		Per capita investment in education (Yuan)	Education expenditure/Total population
	Medical conditions (B_4)	Number of beds per thousand people (Number/1000 people)	Total number of beds/Total number of people
		Number of doctors per thousand people (Number/1000 people)	Total number of doctors/Total number of people
	Social security (B_5)	Incidence of civil cases (Cases/10,000 people)	Number of civil cases/Total population

(continued)

Table 4.1 (continued)

Functional index	Structural index	Specific index	Description
Facility transformation (C)	Transportation conditions (C_1)	Railway density (km/km^2)	Length of railways in operation/Area
		Road density (km/km^2)	Length of highways/Area
	Living facilities (C_2)	Proportion of villages benefiting from tap-water (%)	Number of villages with tap-water supply/Total number of villages
		Per capita electricity consumption (kW/person)	Total electricity consumption/Total population
	Communication facilities (C_3)	Internet broadband penetration rate (Households/10,000 people)	Number of broadband households/Total households
		Number of post offices per 10,000 people (Units/10,000 people)	Number of post offices/Total population
		Mobile phone penetration rate (Households/10,000 people)	Mobile phone users/Total population

between urban and rural developments can be analyzed in two ways. One is to construct the index systems of urban and rural development, calculate the development level of urban system and rural system respectively, and the coordination degree model is used to analyze the coordination level of the two systems. The other is to select indicators that can be directly compared to evaluate the urban–rural gap or coordination status, such as urban–rural income difference, industry difference and service difference (Wang et al. 2016). Considering that rural China generally lags behind urban China, the study selects the relative values of urban and rural indicators as the criteria. If the related urban–rural indicator values are closer to 1, the degree of urban–rural coordination is higher, and if the indicator values are farther from 1, the urban–rural coordination is lower. Therefore, the index system for urban–rural coordinated development is constructed, includes five criteria: policy difference, economic difference, social difference, public service difference, and infrastructure difference (Table 4.3).

4.2.3 Quantitative Evaluation Method

1. **Entropy method**

The methods for determining the weight of index involve subjective and objective weighting methods. The former evaluates the relative importance of each index based on the subjective knowledge of the evaluator, thereby determining its weight; the latter determines the weight of the index based on the amount of information provided by the original value of the index, and the weighting information comes from the objective environment. In this study, the entropy method, an objective weighting method, is used to determine the weight of the index, and the two equal weights are then weighted to obtain the final weight value of the indicator.

The entropy method calculates the index weight by measuring the uncertainty of system state. In natural science, entropy is a measure of the disordered state of the system. When it is applied to a social system, information entropy is

Table 4.2 Index system for urban–rural structural transformation

Functional index	Structural index	Specific index	Description
Economic structure (D)	Industrial structure (D_1)	Proportion of the output value of non-agricultural industry (%)	Output value of secondary industry and tertiary industry/GDP
		Hoffman coefficient	Net output value of consumer goods industry/Net output value of capital goods industry
		Proportion of agricultural processing industry in total agricultural output value (%)	Output value of agricultural processing industry/Total agricultural output value
		Proportion of forestry, animal husbandry, and fishery in total agricultural output value (%)	Output value of agriculture, animal husbandry, and fishery industry/Total agricultural output value
Employment structure (E)	Employment structure (E_1)	Proportion of employment in secondary and tertiary industries (%)	Employees in secondary and tertiary industry/Total employment
		Proportion of non-agricultural employment in rural areas (%)	Rural non-agricultural employment/Total rural population
Consumption structure (F)	Consumption structure (F_1)	Engel coefficient of urban household	Food consumption of urban household/Total consumption of urban household
		Engel coefficient of rural household	Food consumption of rural household/Total consumption of rural residents
Spatial structure (G)	Population distribution (G_1)	Urbanization rate (%)	Total urban population/Total population
	Settlement landscape (G_2)	Proportion of urban construction land in urban–rural construction land (%)	Area of urban construction land/Area of total urban–rural construction land
		Regional density of city (units/km^2)	Total number of cities/Regional area
		Density of small towns (units/km^2)	Total number of small towns/Regional area

a measure of the uncertain state of the system. It is generally believed that the higher the information entropy is, the more balanced the system structure is, and the smaller the difference is or the slower the change is; on the contrary, the structure is unbalanced, the larger the difference is or the faster the change is. Therefore, the smaller the information entropy, the greater the index weight.

Here, extreme value standardization is used to standardize the data. X_{ij} matrix is transformed into $X_{ij}^{'}$. The proportion of index j of the ith region in the index is calculated as follow:

$$p_{ij} = X_{ij}^{'}/\sum_{i=1}^{n} X_{ij}^{'}(i = 1, 2..., n,\ j = 1, 2..., m) \quad (4.1)$$

The entropy of index j is calculated as follow:

$$e_j = -k\sum_{i=1}^{n} (p_{ij}\ln(p_{ij})) \quad (4.2)$$

where $k > 0$, $k = 1/\ln(n)$, and $e_j \geq 0$.

The calculation of information entropy redundancy: For the index j, the greater the difference of index value is, the greater the impact on the scheme evaluation is, and the smaller the

Table 4.3 Index system of urban–rural relationship transformation

Functional index	Structural index	Specific index	Description
Policy difference (G)	Investment difference (G_1)	Ratio of urban and rural fixed asset investment (%)	Per capita fixed asset investment in urban areas/Per capita fixed asset investment in rural areas
	Financial investment (G_2)	Proportion of agriculture related expenditure (%)	Proportion of local financial expenditure on agriculture, forestry and water conservancy/Proportion of primary output value
Economic difference (H)	Industrial development (H_1)	Industrial duality	(Output value of secondary and tertiary industries/Employees in secondary and tertiary industries)/(Output value of primary industry/Employees in the primary industry)
Social difference (I)	Resident income (I_1)	Per capita urban–rural income gap	Per capita disposable income of urban household/Per capita disposable income of rural household
	Resident consumption (I_2)	Comparison of urban and rural consumption capacity	Per capita consumption expenditure of urban household/Per capita consumption expenditure of rural household
		Comparison of urban and rural consumption structure	Engel coefficient of urban household/Engel coefficient of rural household
Public service difference (J)	Human capital (J_1)	Comparison of urban–rural education levels	Average years of education in urban areas/Average years of education in rural areas
	Pension (J_2)	Comparison of urban–rural pension insurance rates	Proportion of urban endowment insurance population/Proportion of rural endowment insurance population
	Medical conditions (J_3)	Comparison of medical beds per capita in urban and rural areas	Per capita hospital beds in urban areas/Per capita hospital beds in rural areas
Infrastructure difference (K)	Pollutant treatment rate (K_1)	Urban and rural solid waste treatment rates	Municipal solid waste treatment rate/Rural solid waste treatment rate
		Urban and rural sewage treatment rates	Urban sewage treatment rate/Rural sewage treatment rate

entropy value is. Information entropy redundancy is defined as follow:

$$g_j = \frac{1 - e_j}{m - E_e} \tag{4.3}$$

where $E_e = \sum_{j=1}^{m} e_j$, $0 \le g_i \le 1$, and $\sum_{j=1}^{m} g_j = 1$.

Weight of index:

$$w_j = g_j / \sum_{j=1}^{m} g_j (1 \le j \le m) \tag{4.4}$$

Score of single index:

$$S_{ij} = w_j \times X_{ij}' \quad (i = 1, 2, \ldots n) \tag{4.5}$$

Comprehensive score of each region:

$$S_i = \sum_{j=1}^{m} S_{ij} \quad (i = 1, 2, \ldots n) \tag{4.6}$$

2. **AHP analysis method**

Analytic hierarchy process (AHP) is an objective and subjective comprehensive evaluation method. Based on the connotation of urban–rural transformation, we refer to many studies on the evaluation index system of urban and rural development, design a set of expert questionnaires after careful screening, and invite experts in the field of urban–rural transformation to rate the relative importance of each index. Then, the consistency of the results is tested, and the weight of each index is obtained. The specific steps are as follows:

(1) Establishing a hierarchical structural model. The hierarchical structure model is a multi-level structure model which makes the problem organized and hierarchical. All factors belonging to the same level remain independent of each other, and a given factor is affected by the factors at the upper level and exerts a dominant role in the factors at the next level. The hierarchical structural model is divided into target layer, factor layer and index layer, and consists of the first-level, second-level and specific index.

(2) Constructing a pairwise comparison matrix. The hierarchical structural model reveals the relationship between various factors, but the proportion of each criterion in the criterion layer is different in the target measurement. Therefore, it is necessary to compare and analyze the importance of the two indexes, and construct the judgment matrix A.

$$A = \begin{pmatrix} a_{11} & a_{21} & \cdots & a_{i1} \\ a_{12} & a_{21} & \cdots & a_{i2} \\ \vdots & \vdots & \ddots & \vdots \\ a_{1j} & a_{2j} & \cdots & a_{ij} \end{pmatrix}, where\ a_{ij} > 0,$$

$$a_{ij} = \frac{1}{a_{ji}},\ a_{ij} = \frac{1}{a_{ji}},\ a_{ii} = 1 \tag{4.7}$$

When constructing pairwise comparison judgment matrix, the "1–9 scale table" is used (Table 4.4), and the importance of indexes is compared and judged through consulting domestic experts in this field.

(3) Weight calculation under single criterion. There are many methods to calculate the weight. When the accuracy requirements are not high, the average value method can be used. The calculation process of the weight vector W_i and the maximum eigenvalue $\lambda_{\max}$ is as follows: (i) Finding the sum of each column of the pairwise comparison matrix. (ii) Dividing each element of the pairwise comparison matrix by the corresponding sum, the new matrix is called standard pairwise comparison matrix. (iii) Calculating the average value of each row of the pairwise comparison matrix, and the average value is the weight of this index.

(4) Consistency test of comparison matrix.

Step 1: Test method. Due to many practical factors, it is difficult for judgment matrix A to have strict consistency, so it is necessary to test the consistency of judgment matrix A. That is, when A is a consistent matrix, $\lambda_{\max} = n$, otherwise, $\lambda_{\max} > n$, and $\lambda_{\max}$ is much larger than n, the more serious the degree of inconsistency of judgment matrix A is. The detailed steps of the consistency test are as follows:

(i) Multiplying the eigenvector by the tested pairwise comparison matrix to obtain the weighted sum vector; (ii) Multiplying the reciprocal of the component of the corresponding feature vector by the component of each weighting and vector; (iii) Calculating the average value in (ii) and recording it as $\lambda_{\max}$; (iv) Calculating the consistency index C_I:

$$C_I = (\lambda_{\max} - n)/(n - 1) \tag{4.8}$$

(v) Checking the average consistency index R_I of the same order matrix (Table 4.5).

(vi) Calculating the consistency ratio C_R, and the calculation formula is $C_R = (C_I/R_I)$. When $C_R = 0$, matrix A has complete consistency; when $C_R < 0.1$, A has satisfactory consistency and is an acceptable judgment matrix; when $C_R > 0.1$, A has unsatisfactory consistency and should be readjusted or discarded. Through the purposeful allocation and adjustment of the weight of each index, the evaluation index system in the evaluation process becomes more useful and flexible.

Step 2: Testing results. Using AHP software *yaahp* 0.5.3 to check the weight and the consistency of indexes.

Table 4.4 Scale and its meaning

Scale	Meaning
1	Indicates that the two factors are of equal importance
3	Indicates that the former factor is slightly more important than the latter
5	Indicates that the former factor is obviously more important than the latter
7	Indicates that the former factor is more important than the latter
9	Indicates that the former factor is extremely more important than the latter
2, 4, 6, 8	Represent the intermediate values of the above adjacent judgments
Reciprocal	If the importance ratio of factor i to j is a, then the importance ratio of factor j to i is $a_{ji} = 1/a_{ij}$

Table 4.5 Average random consistency index

Order n	1	2	3	4	5	6	7	8	9
R_I	0	0	0.58	0.89	1.12	1.24	1.32	1.41	1.45

Step 3: Determining the weights of the index. Through consistency test and calculation, the weight of urban–rural transformation evaluation index is obtained.

4.3 Comprehensive Zoning Method of Urban–Rural Transformation

4.3.1 Regionalization Classification Method

Geographical regionalization has always been the core content of geography, which is the results of geographical rules of differentiation and formation. Geographical zoning has been applied to the study such as physical geographical regionalization and comprehensive natural regionalization, especially in comprehensive agricultural regionalization. According to the law of natural zonality, agricultural regionalization fully reflects the characteristics of vegetation, soil and climate in a region, and reveals the ability of regional economic and social development. Physical geography is the basis of economic and social development. Industrialization and urbanization are still mainly concentrated in the plain, coastal and riverside areas with superior environment. In the practical application of regionalization method, there are economic regionalization, main functional regionalization and so on except for the natural-oriented regionalization. Among them, the main functional regionalization is a functional region divided in accordance with the regional resource and environmental carrying capacity, the existing development status and the future development trend (Fan 2007). With the progress of science and technology and the improvement of productivity, the restraint of natural conditions and resources on economic growth and social development declines, and the effects of factors such as innovation, location, globalization and institution become increasingly significant. This makes the geographical regionalization of the socioeconomic system more complex and diverse.

Regionalization involves the principles of genealogy, relative consistency, regional integrity, etc. (Zheng et al. 2005). The principle of relative consistency means that regionalization needs to establish a relatively complete regional unit hierarchy system to objectively reflect the hierarchical affiliation among regional units corresponding to a certain spatial scale. In general, geographic region can be divided into type and regional divisions. The former allows mutual isolation and spatial repetition, while the latter emphasizes regional adjacency and spatial non-repetition (Zheng et al. 2008). The urban–rural regional system is a composite system of natural geographical system and socioeconomic system. In terms of the regional environment of urban and rural development in

China, there are obvious physical and geographical differences and socioeconomic differences between the east, the central, and the west. Therefore, the regionalization of urban–rural transformation needs to comprehensively consider the type and regional division. Based on the basic conditions of China's natural geography, economic development and ecological environment, it is necessary to carry out macro and overall control, and build a corresponding index system, which can cluster and overlay the multiple-dimensional analyses of urban–rural transformation and forming a regionalization of urban–rural transformation (Fig. 4.5).

4.3.2 Transect Research Method

Promoted by the International Geosphere-Biosphere Programme (IGBP), the transect research method has been widely used in global change research. The terrestrial transects in IGBP emphasize the use of gradient differentiation to explain global change. Each transect is composed of a series of research sites distributed in a large geographic area (1000 km long and 500 km wide), which contains existing gradients of basic control transects of ecosystem structure and function, such as climate, soil, and land-use gradients. Since transects are used as a series of coherent research points to determine how land use and related land cover occur as results of social and environmental factors, thus the transect method has been successfully applied in land-use research. Long and Li (2001) used transect research method to analyze and process the land-use data of the transect along the Yangtze River in the mid-1990s, analyze the socioeconomic development trend along the Yangtze River, and make statistical analysis on the land use and its influencing factors of the transect along the Yangtze River combined with the correlation analysis method, so as to quantitatively diagnose the contribution of land-use pattern formation.

As a linear research area that changes regularly along the gradient of the dominant driving factor or a linear regional type with obvious differences, transect is applied to the study of the evolutionary pattern of regional spatial economy and society, which helps to reveal the spatial difference law of urban–rural transformation and the regional differences of its formation mechanism. Domestic scholars have selected specific transects such as areas along the Yangtze River, Longhai-Lanzhou-Xinjiang Railway, northern border and National Highway 106, as well as counties within 100 km of the eastern coastal areas to analyze the problems such as agricultural regions and urbanization, and found that there are regular characteristics of economic and social development along the traffic line and along the river (Liu and Yang 2012; Lu et al. 2013). Combined with the methods of mathematical statistics, spatial econometrics and geographic detection, the transect analysis is also applied to the study of development mechanism (Liu and Li 2017).

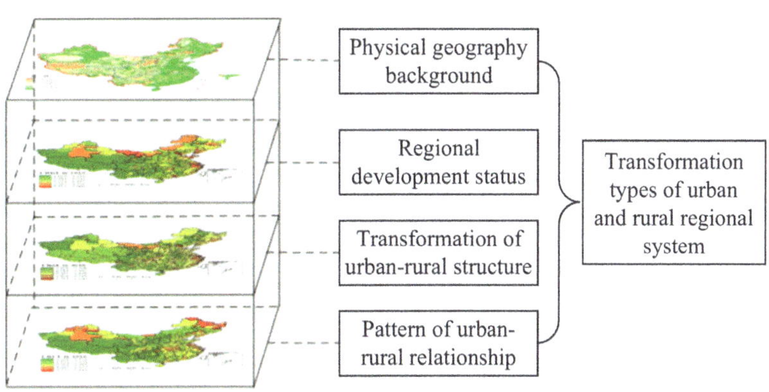

Fig. 4.5 The thinking of urban–rural transformation regionalization

4.4 Dynamic Mechanism of Urban–Rural Transformation

4.4.1 Regression Analysis Method

Correlation analysis and regression analysis are often used in analyzing the influencing factors and mechanism. Due to the spatial correlation of the variables, that is, the spatial distribution of the data is not independent, geographers usually use the maximum likelihood method to estimate the spatial effect model based on the traditional least square method, and combine with the spatial econometric model for comparative analysis. In spatial econometric model, the spatial effects of variables should be considered, and a series of model are set, estimated, tested and predicted to study how to deal with spatial autocorrelation and spatial heterogeneity in cross-sectional data and panel data regression model. Spatial connection variables are introduced into spatial econometric models to estimate and test spatial effects, and the most commonly used models are spatial lag model (SLM) and spatial error model (SEM) (Anselin 1988; Elhorst 2014).

SLM mainly explores whether there is diffusion phenomenon (spillover effect) of the variables in a region. This model is used when the spatial dependence among explained variables is critical to the model and leads to spatial correlation. Since SLM is similar to autoregressive model in the time series, the SLM is also called the spatial autoregressive model (SAR). The model expression is:

$$y = \rho Wy + X\beta + \varepsilon \tag{4.9}$$

where y is the dependent variable; X is the explanatory variable and W is the spatial weight matrix; ρ is the parameter of the spatial lag term Wy, which measures the degree of spatial interaction between observations; β is the parameter vector of X, reflecting the influence of the independent variable on the dependent variable; and ε is the interference term.

SEM describes the spatial disturbance correlation and the overall spatial correlation, and measures the error impact of the dependent variable on the observed values in the neighboring area. When there is a spatial interaction with the error term process, that is, there is a spatial covariance between errors from different regions, another form of spatial dependence will appear, i.e., SEM. Since the SEM is similar to the sequence correlation problem in time series, it is also called the spatial autocorrelation model (SAC). The mathematical expression of SEM is:

$$y = X\beta + \varepsilon,\ \varepsilon = \lambda W\varepsilon + \mu \tag{4.10}$$

where ε is the random error term vector; β is the regression residual vector; λ is the spatial error coefficient of the n*1 order cross-sectional dependent variable vector, which measures the spatial dependence of the sample observation value, that is, the direction and extent of the influence of the observation value y in the neighboring area on the observation value y in the area; and μ is the random error vector of the normal distribution.

4.4.2 Structural Equation Model

Structural equation modeling is a statistical analysis method which appeared in the 1960s, and is known as one of the three major advances in applied statistics in recent years. It is a method of establishing, estimating, and testing causality model, which contains both observable explicit variables and latent variables that cannot be directly observed. At present, there are mainly two kinds of estimation techniques to solve structural equation models. One is the covariance structural analysis method based on maximum likelihood estimation, which is called "hard modeling" and is represented by the LISREL method. The other is based on partial least squares (PLS) and is called "soft modeling". This is a multivariate prior model to test the relationship between observed variables and latent variables, among latent variables (Wu et al. 2010). The LISREL method requires that the samples involved in the structural equation calculation conform to the normal distribution, and the samples with non-normal distribution can be solved by the PLS method, thus obtaining a relatively robust evaluation.

The measurement model describing the relationship between latent variables ζ and η and observed variables x and y is as follows:

$$y = \Lambda y\eta + \varepsilon,\ x = \Lambda x\xi + \delta \quad (4.11)$$

where y is a vector composed of endogenous observation variables; x is a vector composed of exogenous observation variables; η is an endogenous latent variable; ξ is an exogenous latent variable, which has been standardized; and Λy is the factor load matrix of endogenous observation variables on endogenous latent variables, which represents the relationship between the endogenous latent variables and the endogenous observed variables; Λx is the factor loading matrix of the exogenous observed variables on the exogenous latent variables, which represents the relationship between the exogenous latent variables and the exogenous observed variables; and ε and δ are the residual matrices of the measurement model.

The structural model describes the causal relationship between latent variables, and its equation is as follows:

$$\eta = B\eta + \Gamma\xi + \xi \quad (4.12)$$

where B is the interaction effect coefficient between the endogenous latent variables; Γ is the effect coefficient of exogenous latent variables on endogenous latent variables, also known as the path coefficient of the influence of exogenous latent variables on the endogenous latent variables; and ξ is the residual vector of η.

4.4.3 Geodetector

Geodetector was initially used to study of the risk of endemic diseases and related geographic factors (Hu et al. 2011). When dealing with such problems, there are many assumptions in traditional statistical methods, such as homoscedasticity and normality. However, real cases rarely conform to these assumptions, which will affect the performance of the model. Geodetector is less restricted in assumptions, which gives it a unique advantage in model analysis. Geodetector includes four modules, i.e., risk detector, factor detector, ecological detector and interaction detector. Among them, the core idea of factor detector is that geographical objects always exist in specific spatial locations, and the environmental factors that affect their development and evolution are spatially different in space. If a certain environmental factor and the change of geographical objects have significant spatial consistency, it means that environmental factors have decisive significance for the occurrence and development of geographical objects (Wang et al. 2010).

Affected by regional economic, social and natural factors, there are obvious regional and different characteristics in urban–rural transformation. It is an important content of geography research to explore its formation mechanism based on the spatial theory of geography. The calculation model to explore the influencing factors of urban–rural transformation process is as follows:

$$P_{D,U} = 1 - \frac{1}{n\sigma_U^2}\sum_{i=1}^{m} n_{D,i}\sigma_{U_{D,i}}^2 \quad (4.13)$$

where $P_{D,U}$ is the detection power of detection factor D, $n_{D,i}$ is the number of samples in the second-level region, n is the number of samples in the whole region, σ_U^2 is the variance of the change values of regional urban–rural transformation, and $\sigma_{U_{D,i}}^2$ is the variance of the second-level region. If $\sigma_{U_{D,i}}^2 \neq 0$, the model is established, and the value range of $P_{D,U}$ is [0,1]. When $P_{D,U}$ = 0, it shows that the distribution of urban–rural transformation is random. The larger the value of $P_{D,U}$ is, the greater the influence of zoning factors on urban–rural transformation. Based on relevant literature, the study selects 10 indicators, including per capita net income of rural households, the proportion of secondary and tertiary industry output value in GDP, urban fixed-asset investment, distance from central city, population density, per capita food possession, per capita GDP, average distance from main transportation lines, annual average precipitation and average altitude, to explore and analyze the impact of each indicator on the spatial pattern of urban–rural transformation and its mechanism.

4.5 Simulation and Forecast of Urban–Rural Transformation

Urban–rural transformation involves many factors flow and structural transformation, and is the reflection of urban and rural economic, social, ecological and other systems. As a systematic project of urban–rural regional transformation, urban–rural transformation is a multi-objective, multilevel continuous fitting and decision-making process. Therefore, the construction of an optimization model for urban–rural transformation needs the use of multiple disciplines and methods, such as dynamic simulation, mathematical programming, system dynamics and engineering.

4.5.1 Linear Programming

Linear programming is an important branch of operations research, which has developed rapidly and been widely used. It is a mathematical method to assist people in scientific management. In economic activities, such as economic management, transportation, industrial and agricultural production, improving the economic performance is an indispensable requirement for people. In general, there are two ways to realize this goal. One is technical improvement, such as improving production process, using new equipment and new raw materials; the other is the improvement of production organization and plan, that is, reasonable arrangement of human and material resources. Linear programming research focuses on how to rationally arrange human and material resources to achieve the best economic effects under certain conditions. The problem of finding the maximum or minimum value of a linear objective function under linear constraints is called linear programming problem. As an important basic production factor, the optimal combination of land resources directly reflects the development of regional industries. Linear programming is a common method for the optimal allocation of land resources. In addition to solving certain linear problems, some nonlinear problems can be solved by logarithm method. The linear programming method can realize the optimal allocation of land use at the regional scale. However, it cannot consider all optimization objectives and constraints in the process of optimal land-use allocation. Some influencing factors that are difficult to quantify cannot be considered, and sometimes the optimal scheme is not consistent with the reality.

4.5.2 Multi-objective Optimization Model

Multi-objective analysis is an emerging discipline developed rapidly in recent 30 years. It mainly studies the simultaneous optimization of multiple objectives in a specific sense. In 1951, Koopmans proposed the multi-objective optimization problem from the analysis of production and distribution activities and put forward the concept of the Pareto optimal solution for the first time. In 1968, Johnsen systematically put forward the research report on multi-objective optimization models, which became the earliest monograph on multi-objective decision making. Since the 1970s, multi-objective decision analysis has been widely used to solve many problems in engineering technology, economic management and systems engineering. Canada is the first country to applied multi-objective decision analysis to land research. In 1984, a Canadian land appraisal team used the multi-objective analysis method to study land use appraisals. In urban–rural transformation, the multi-objective programming method plays an important role in the modeling of the optimal allocation of land resources, thus ensuring the sustainable use of land resources. The multi-objective programming method applies the theory of multi-objective decision-making to optimize the land use structure, with the characteristics of multi-objective and multi-scheme. Decision makers can choose the most suitable optimal allocation scheme according to different criteria to optimize the allocation of land resources and enhance the scientificity of decision-making. The basic model is expressed as follows:

$$\max Z = c_1x_1 + c_2x_2 + \cdots + c_nx_n \tag{4.14}$$

$$\sum_{i=1}^{m}\sum_{j=1}^{n} a_ix_i \leq (=, \geq) b_i \tag{4.15}$$

$$x_1, x_2, \cdots x_n \geq 0 \tag{4.16}$$

Equation (4.14) is the objective function of the model, and Eqs. (4.15) and (4.16) are constraints, where Eq. (4.16) is a non-negative constraint, and the variable x_j represents the planned area for each industry. As land-use structural optimization involves many fields, such as environment, economy and policy, multi-objective analysis method is widely used in land-use structural optimization.

4.5.3 System Dynamics Method

The system dynamics (SD) method is based on the feedback control theory and the computational simulation technology. It has a strong ability to simulate large-scale nonlinear dynamic multiple feedback system, which is different from linear programming and other methods. In the modeling of land resource allocation at regional and national scales, the SD method adopts the structure–function simulation method which combines qualitative and quantitative analysis, emphasizes the system structural analysis, and has the characteristics of less dependence on data, flexible operation and strong plasticity. It can not only predict the future, but also review the historical behavior process of the system, which is easy to reflect the processes that are difficult to express in mathematical forms, such as nonlinearity and delayed response. The demand of land use is affected by many factors, such as economy, society and policy, and the planning objectives and planning programs should be constantly adjusted accordingly. The SD model can solve this problem. SD model can reflect the relationship among the structure, function, and behavior of the land-use system from a macro-perspective to investigate the changes and trends of the system under different scenarios. It establishes an information feedback mechanism through the causal relationship between planning objectives and planning factors. The behavioral mode and results of the model mainly depend on the model structure rather than the parameter value. In addition, the model has characteristics of dynamic and simulation. According to the principle of system theory, an SD model reflecting the optimization degree of the land-use structure can be established by analyzing the feedback relationship between the structure of land use and the internal components of the system. However, the establishment of SD model needs a thorough study of the simulated system and a deep understanding of various feedback mechanisms within the system, and it is inconvenient to use this model when causality is not clear.

4.5.4 Gray Prediction Method

The linear programming model is static. However, land use against the background of urban–rural transformation is in a process of dynamic change. Therefore, it is necessary to predict the degree of urban–rural population transformation, employment transformation and industrial transformation in the optimization and adjustment of land use structure. Since the causality of some factors in the prediction of the degree of urban–rural transformation is not clear, gray system prediction provides a useful technical approach. Gray theory argues that although the behavior of a system is obscure and the data are complicated, it is nonetheless orderly and has overall functions. Gray numbers are generated to determine the law that organizes the apparent mess. Additionally, gray theory establishes a generative data model rather than a primitive data model.

Therefore, the data od gray prediction are the inverse processing results of the predicted values obtained by the GM (1,1) model of the generated data. During urban–rural transformation, the optimal allocation of land use based on the three linkage mechanisms of "population-land-industry" plays an important role in the process of multi scenario simulation and prediction.

Gray prediction method is an important and commonly used prediction method in geography. The method of gray sequence prediction is among them. Sequence prediction predicts the development and change of an index, and the prediction result is the specific value of the index at various moments in the future.

The basis of the sequence prediction is based on the GM (1,1) model of accumulating sequence. The specific calculation model is as follows:

Assuming that $x^{(0)}(1)$, $x^{(0)}(2)$, …, $x^{(0)}(m)$ are the original data of an index to be predicted. To eliminate fluctuation of the original data, they are accumulated once, namely:

$$\begin{aligned} &x^{(1)}(1) = x^{(0)}(1) \\ &x^{(1)}(2) = x^{(0)}(1) + x^{(0)}(2) \\ &x^{(1)}(3) = x^{(0)}(1) + x^{(0)}(2) + x^{(0)}(3) \\ &\cdots \\ &x^{(1)}(k) = \sum_{i=1}^{k} x^{(0)}(t) \\ &\cdots \\ &x^{(1)}(m) = \sum_{i=1}^{m} x^{(0)}(t) \end{aligned} \tag{4.17}$$

Compared with the original data, the randomness degree of the new sequence is greatly weakened and the stability degree is greatly increased. The changing trend of the new sequence can be approximately described by a differential equation:

$$\frac{dx^{(1)}}{dt} + ax^{1} = u \tag{4.18}$$

a and u can be obtained by the following least-squares fitting:

$$\begin{bmatrix} a \\ u \end{bmatrix} = \left(B^T B\right)^{-1} B^T Y_m \tag{4.19}$$

Y_m is the column vector:$Y_m = \left[x^{(0)}(2), x^{(0)}(3), \cdots, x^{(0)}(m)\right]^T$, B is a construction data matrix:

$$B = \begin{bmatrix} -\frac{1}{2}\left[x^{(1)}(1) + x^{(1)}(2)\right] & 1 \\ -\frac{1}{2}\left[x^{(1)}(2) + x^{(1)}(3)\right] & 1 \\ \vdots & \\ -\frac{1}{2}\left[x^{(1)}(m-1) + x^{(1)}(m)\right] & 1 \end{bmatrix} \tag{4.20}$$

The time response function corresponding to the differential formula (4.20) is:

$$x^{(1)}(t+1) = \left[x^{(0)}(1) - \frac{u}{a}\right] e^{-at} + \frac{u}{a} \tag{4.21}$$

According to formula (4.21), the original data restoration value can be obtained from the predicted value $x^{(1)}(t)$ of the sequence generated by one-time accumulation:

$$\hat{x}^{(0)}(t) = \hat{x}^{(1)}(t) - \hat{x}^{(1)}(t-1) \tag{4.22}$$

Then, it can be further sorted out:

$$\hat{x}^{(0)}(t) = Ae^{-wt} \tag{4.23}$$

where A and w are undetermined coefficients, t = 1, 2, …, m, and $\hat{x}^{(0)}(0) = 0$. The residual values between the restored value and the actual value of the original data are:

$$\begin{cases} \varepsilon^{(0)}(t) = x^{(0)}(t) - \hat{x}^{(0)}(t) \\ q(t) = \frac{\varepsilon^{(0)}(t)}{\hat{x}^{(0)}(t)} \times 100\% \end{cases} \tag{4.24}$$

The method to test the accuracy of the predicted value is as follows:

First, calculate:

$$x^{-(0)} = \frac{1}{m}\sum_{t=1}^{m} x^{(0)}(t) S_1^2 = \frac{1}{m}\sum_{t=1}^{m}\left[x^{(0)}(t) - x^{-(0)}\right]^2$$
$$\varepsilon^{-(0)} = \frac{1}{m-1}\sum_{t=2}^{m}\varepsilon^{(0)}(t) S_2^2 = \frac{1}{m}\sum_{t=21}^{m}\left[\varepsilon^{(0)}(t) - \varepsilon^{-(0)}\right]^2 \tag{4.25}$$

Second, calculate the variance ratio, $c = \frac{S_2}{S_1}$, and small error probability, $p\{\left|\varepsilon^{(0)}(t) - \varepsilon^{-(0)}\right| < 0.6745 S_1\}$.

References

Anselin L. Spatial economy: methods and models. Dordrecht: Kluwer Academic Publisher; 1988.

Elhorst JP. Spatial panel data models. Spatial econometrics. Berlin, Heidelberg: Springer; 2014. p. 37–93.

Fan J. The scientific basis of our country's main function zoning. Acta Geogr Sin. 2007;62(4):339–50.

Hu Y, Wang J, Li X, et al. Geographical detector-based risk assessment of the under-five mortality in the 2008 Wenchuan earthquake, China. PloS one. 2011;6(6): e21427

Liu Y, Li J. Geographical exploration and optimal decision-making of differential mechanism of rural poverty in China's counties. Acta Geogr Sin. 2017;71 (1):161–73.

Liu Y, Yang R. The spatial features and formation mechanism of urbanization in Chinese counties. Acta Geogr Sin. 2012;67(8):1011–20.

Long H, Li X. Analysis of land use pattern and influencing factors of transect along the Yangtze River. Acta Geogr Sin. 2001;56(4):417–25.

Wang Y, Liu Y, Yan B, et al. Spatial patterns and influencing factors of urban-rural coordinated development in China. Sci Geogr Sin. 2016;36(1):20–8.

Lu S, Guan X, Liu Y, et al. Recognition of agricultural area types and its evolution characteristics: take the typical transect area along the National Highway 106 as an example. Geogr Sci Progress. 2013;32(4):637–48.

Wang J, Li X, Christakos G, et al. Geographical detectors-based health risk assessment and its application in the neural tube defects study of the Heshun Region, China. Int J Geogr Inf Sci. 2010;24(1):107–27.

Wu W, Liu Z, Zhang W. Determinants of residential land price: Structure equation model analysis using land-leasing parcel data in Beijing. Acta Geogr Sin. 2010;65(6):676–84.

Zheng D, Ge Q, Zhang X, et al. Retrospect and prospect of China's regionalization work. Geogr Res. 2005;24 (3):330–44.

Zheng D, Ou Y, Zhou C. Understanding of and thinking over geographical regionalization methodology. Acta Geogr Sin. 2008;63(6):563–73.

5 Strategic Objectives and Regional Orientation of Urban–Rural Transformation in China

Abstract

In line with the new trends of economic growth, social development and population flow, principles such as people-oriented, intensive and efficient, overall consideration and regional coordination are important premises to promote the development of new-type urbanization and urban–rural integration. The key tasks of urban–rural transformation are to improve modern agricultural production system, strengthen the division and connection of urban–rural industries, accelerate the orderly reconstruction of urban–rural space and deepen the comprehensive reform of urban–rural systems, thus continuously promoting regional economic growth and urban–rural structure optimization, improving social and living conditions and solving the increasingly prominent problems of urban–rural development. In term of the spatial pattern, the agglomeration of the high-value areas of urban–rural socio-economic transformation from 1990 to 2015 is more obvious in the eastern coastal areas, while the vast western region is the main distribution area of low-value areas; the level of urban–rural structural transformation in the eastern and northeastern regions was higher than that in the central and western regions; the spatial pattern of urban–rural relationship transformation in 2015 was similar to that in 1990, but the urban–rural synergy in provinces such as Inner Mongolia and Jilin decreased significantly. According to the cluster analysis, urban–rural transformation in China during the period of 1990–2015 can be divided into seven regions and 19 sub-regions, and there are obvious differences in the characteristics, formation mechanism and development orientation of different regions.

5.1 Principle and Criterion of Urban–Rural Transformation in China

5.1.1 Principle of Urban–Rural Transformation

In 2003, the third Plenary Session of the 16th Central Committee of CPC proposed the policy of coordinated development in five aspects, including the coordination of urban and rural development, regional development, economic and social development, human and nature harmonious development, domestic development and opening up, among which coordinating urban and rural development was placed in the first place. In 2005, the Resolution of the fifth Plenary Session of the 16th Central Committee of CPC raised the policy of "building a new socialist countryside" as a major strategy to promote the modernization of China. In 2008, the Decision of the Third Plenary Session of the 17th Central Committee of CPC further emphasized the vision of

Y. Liu, *Urban-Rural Transformation Geography*, Sustainable Development Goals Series,
https://doi.org/10.1007/978-981-16-4835-9_5

"establishing a long-term mechanism of 'promoting agriculture by industry and leading rural areas by cities' to construct a new pattern of integrated urban and rural economic and social development". In 2010, the No. 1 Document of the CPC Central Committee put forward coordinating urban and rural development as a fundamental requirement of building a moderately prosperous society in all respects. In 2012, the 18th National Congress of the CPC put forward the strategic goals of "promoting the construction of ecological civilization" and "integrating urban and rural development", and proposed "improving institutions and mechanisms for promoting urban–rural integrated development" and "promoting the equal exchange of elements and the balanced allocation of public resources between urban and rural areas". In 2014, the National New-Type Urbanization Plan (2014–2020) emphasized people-centered urbanization, requiring coordinating urban and rural development, orderly promoting the citizenization of agricultural transfer population, and accelerating equal exchange of urban and rural elements and balanced allocation of public resources.

The integration of urban and rural development has become an important strategy and orientation for current rural construction and institutional system innovation, the core of which was to emphasize people-oriented, overall planning and all-round consideration, ensuring people's livelihood, and regional balance (Table 5.1). On the whole, China's urban and rural development strategy has undergone a transformation from the separation and independence of urban and rural areas to coordination and integration, which is not only related to the transformation of national development strategy, but also closely related to national economic development stage and industrial structure. In 2003, China's per capita GDP reached $3000, making it a moderately developing country as a whole. With the economic growth, the proportion of added value and employment of China's primary industry has been declining, and the production function and employment security function of rural agriculture have also been declining. Instead, new demands such as scale agriculture, characteristic agriculture, leisure agriculture, rural landscape and culture have been constantly increasing. Currently, in the context of the "new normal" of decreasing economic growth rate, production structure adjustment and transformation of development mechanism, the population urbanization rate tends to decrease, the proportion of internal population mobility in the central and western regions increases, and the urbanization pattern shows a new development trend in China.

Adapting to the new trend of economic growth, social development and population mobility, and innovating the principle of urban–rural transformation is important prerequisite for promoting the new-type urbanization and urban–rural integrated development. The first is to strengthen the principle of people-centered development, which means changing the traditional orientation of land-oriented urbanization, orderly promoting the citizenization of agricultural transfer population, steadily accelerating the equalization of urban and rural public services, and promoting the all-round development of people based on the improvement of population quality, so that the agricultural transfer population can finally share the achievement of modernization. The second is to strengthen the principle of intensive and efficient development, which means changing the extensive and low-cost development pattern of urbanization and industrialization, improving the utilization efficiency of water and soil resources and energy use, promoting the efficient utilization of urban development areas, industrial parks and rural construction land by expanding urban boundaries and scientifically demarcating red lines of cultivated land and ecological land, and strengthening the protection of ecological environment by land remediation and environmental remediation. The third is to strengthen the principle of overall planning and all-round consideration, which means strengthening the coordinated development of urbanization, industrialization,

Table 5.1 Principles and tasks of urban–rural transformation

Principle	Core	Main task
People-oriented	Transforming the land-oriented urbanization to people-oriented urbanization	Promoting the citizenization of agricultural transfer population in an orderly manner, steadily promoting the equalization of urban and rural public services, and promoting the all-round development of people based on the improvement of the population quality, so that the agricultural transfer population can finally share the modernization achievement
Intensive and efficient	Transforming extensive and low-cost urbanization and industrialization to efficient and comprehensive model of intensive economic growth	Delimiting the boundary of urban expansion, optimizing the internal structure of the city, improving the land efficiency of industrial parks and renovate the empty and waste land in rural areas; utilizing land, water, energy and other resources sparingly and intensively
Overall consideration	Emphasizing the coordination of "four modernizations" and promotes the coordinated development of economy, society and ecology between urban and rural areas	Enhancing the capacity of industrialization and urbanization to attract migrant workers and drive rural development; overall considering improving farmers' quality, increasing farmers' income and strengthening comprehensive agricultural production capacity; enhancing the functional value of rural social services, ecological security, rural culture, and food supply
Regional coordination	Coordinating the development of large, medium-sized and small cities, and balancing the development of eastern, western and internal regions	Improving the services and institutional systems for building small and medium-sized cities and towns; improving the function compensation system of the main function orientation; improving the interregional trading system; defining the objectives and tasks of urban–rural transformation in each region, and compiling the urban and rural development plans and implementation plans in light of local conditions

informatization and agricultural modernization, in particular, enhancing the leading ability of urbanization for rural development and the ability of rural self-development, and strengthening the coordinated development of the economy, society and ecology, and fully reflecting the values of rural areas in culture, ecology and food. The fourth is to strengthen the principle of coordinated development among regions, which means further improving the regional policy system, promoting economic and social development in the central and western regions, and gradually narrowing the disparity in employment, education, medical care, and social security between the east and the west; coordinated promoting the development of large, medium-sized and small cities and towns, improving the service system and employment supply capacity of small and medium-sized cities and towns, and compiling urban and rural development plans and implementation plans in light of local conditions.

5.1.2 Mechanism of Urban–Rural Transformation

1. **Boundary between market and government**

Urban–rural transformation involves multiple levels of economic, social, and ecological dimensions, which needs to strengthen the dual role of market allocation and government regulation, and rationally divide the boundary between market and government. Therefore, the focus of economic institutional reform is to properly treat the relationship between government and market. Economic growth and optimal allocation of resource elements are the core of economic and social development. The third Plenary Session of the 18th Central Committee of CPC clearly put forward that "establishing a unified, open, orderly competitive market system is the basis of ensuring the decisive role of the market in resources allocation". In the field of economic growth, it is necessary to gradually establish the decisive role of the market in urban and rural element allocation and sectoral industries based on the maximization of element returns. In the traditional monopolistic industries such as electricity, coal, mineral resources, water resources and other fields, it is necessary to gradually introduce the market mechanism, accelerate the equal exchange of urban and rural elements, and promote the rural areas to obtain benefits from resource development. However, in the social and ecological fields, due to the existence of market externalities, non-market value is difficult to be effectively reflected and market mechanism may fail, so the government needs to better play its regulatory role. Obviously, in the fields of public resource allocation, food safety and ecological security, the government plays an irreplaceable role, and making the non-market value of some regional functions valuable, such as grain production function, ecological service function and cultural inheritance function, is an important path to promote urban–rural coordinated development. Therefore, it is necessary to identify the boundaries of market and government regulation during the urban–rural transformation development process, and reflect the fundamental role of market in element allocation and the role of government in public service balance and pubic resource non-market value, so as to promote urban–rural economic, social, and ecological coordinated development by market allocation and government regulation.

2. **Urban and rural development policy bias**

In a specific historical period of China, influenced by the priority development strategy of heavy industry and the demand orientation of national defense construction, in order to take the lead in establishing a complete industrial system, China has implemented the development strategy of emphasizing industry over agriculture, valuing cities over rural areas, and formed a urban–rural dual system under the planned regulation system, in which urban and rural population, employment, society and public services were strictly separated. After the reform and opening up, the reform of the land system and the loosening of the household registration system have enabled the flow of population and elements of production between urban and rural areas. With the increasingly close relationship between urban and rural elements, under the social need of economic structure transformation, China's urban–rural dual system was becoming more and more difficult to meet the needs of urban–rural economic and social transformation development. Since the 2000s, although urban and rural coordinated development has become a national strategy, due to the urban–rural dual system, urban and rural public resources were still allocated in cities first, especially in metropolitan areas, with obvious urban bias. China's rapid urbanization relied heavily on rural migrants and rural land, while the supply of social services for migrant workers was seriously insufficient, which exposes the inequality in the implementation of urban and rural development policies. With the transfer of the rural population and the aging and weakening of rural subjects, rural areas cannot get effective support and guidance from the public resource allocation. To promote the integration of urban and rural development, it is urgent to change the strategy and policy orientation of giving priority to urban development, emphasize the economic and social coordinated development of urban and rural areas, construct an institutional platform for the free flow and equal exchange of urban and rural elements, build an interactive industrial system between urban and rural areas, and improve the rural social service functions and infrastructure, especially for the central villages and towns.

3. **Functional orientation transformation of rural areas**

Agricultural multi-function and rural multi-function have gradually sprung up in the West in recent years, and have become an important theoretical foundation and experience to guide the agriculture and rural transformation development. In the process of industrialization and urbanization, the continuous decrease of the proportion of agricultural economy rural population was an inevitable trend, and how to avoid the demise of traditional farming culture and rural civilization as well as rural decline has become a common proposition worldwide. When the Western Europe, the United States, Australia and other developed countries entered the middle and later stage of industrialization, they began to re-recognize the urban and rural space and lifestyle. Environmental protection, ecological sustainability, food quality and localization have become the trend of agricultural production. Agricultural development showed a multi-functional trend. The multi-function of agriculture emphasizes that in addition to producing food and fiber, agriculture can also produce the functions of renewable resources management, providing ecological services and pleasant environment, agricultural cultural protection and biodiversity, which has the characteristics of diversification, non-production, ecology, etc. On the basis of the multi-function of agriculture, the multi-functional development of the countryside has been enhanced. Rural multi-function emphasizes the attributes of rural agricultural production space, ecological space and settlement space, and the emergence of this multi-functional cognition is the result of the continuous evolution of rural areas driven by the change of human demand for rural production, consumption, ecology and other multi-functional needs in the process of social development (Holmes 2006). With the increasing demand for food safety, ecological environment and rural culture, it is an important force to explore and develop the multi-function of rural areas to promote the transformation development of rural areas in China. Based on the exploration of rural multi-function, it is necessary to carry out reform and innovation in urban and rural household registration system, land system and infrastructure services, promote the free flow of urban and rural elements between urban and rural areas, improve rural production and living conditions, and emphasize the quality of rural environment and the inheritance of rural culture.

5.1.3 Criterion of Urban–Rural Transformation in China

Urbanization is the potential of China's economic growth, and rural construction is an important path to stimulate domestic demand. Urban–rural transformation needs to coordinate the construction of the "four modernizations", urban–rural economic and social ecosystem, and realize the equal exchange of urban and rural elements and the balanced development of urban and rural areas through system innovation. Since large population and little land, low agricultural efficiency, weak rural foundation and large regional disparities are the basic national conditions that China has been facing for a long time in the process of modernization, the transformation of urban and rural development needs to pay full attention to the urban and rural elements transfer and regional development, especially promoting the economic and social development of rural areas through the transfer of rural population, land and other elements. Under the background of new-type urbanization, coordinating urban and rural development and innovating the institution and mechanism of urban–rural integration have become an important way to solve the problems of agriculture, rural areas and farmers. Although the proportion of urban population will reach 60% in 2020, there are five to six billion people in rural China, and promoting the healthy development of rural areas is an important proposition during the 13th Five-Year Plan period. Urban–rural transformation is to achieve the coordinated development of urban and rural areas and the coordinated evolution of economy, society and ecology during the urban and rural elements flow and departmental interaction. Therefore, from a systematic perspective, it is necessary to combine township-to-city and city-to-township to promote the development of urbanization and rural construction as a whole.

1. **Rural migrants live and work in peace and contentment**

Urbanization is not only the process of urban population increase and urban space expansion, but also the process of transforming rural migrants into urban residents. Since the 1990s, although small and medium-sized cities and small towns have been supported by policies, China has carried out the urbanization strategy of developing big cities, which was also a land resource-dependent road. The scale expansion of large cities and megacities and the rapid development of infrastructure construction have created conditions for the non-agricultural transfer of rural population. However, due to the soaring urban housing prices and the increasing survival threshold, it is difficult for the farmers to live and work in peace and contentment in the cities. It is difficult for a large number of migrant workers to become citizens by settling down, resulting in the speed of land urbanization much faster than population urbanization. The growth elasticity coefficient of urban land, which is the ratio of urban land growth rate to urban population growth rate, is usually used to measure the rationality of urban land use growth worldwide. The generally accepted reasonable value is 1.12, and the larger the value, the more unreasonable it is. The value in China during the period of 1999–2007 was close to 1.8.

From the perspective of urban–rural structure, there were 792.98 million urban permanent residents in 2016, with an increase of 21.86 million compared with 2015, and the rural permanent resident population was 589.73 million, with annual decrease of 13.37 million. The urban population accounted for 57.35% of the total population in 2016, but the urbanization rate of registered population was only 41.2%. The rapid transfer of rural population into the cities and the lag of urban social services have resulted in a virtual high urbanization rate, which makes it difficult for migrant workers to enjoy the same treatment as urban residents in education, medical care and social security. The National New-type Urbanization Plan (2014–2020) proposed that the urbanization rate of urban registered population should reach about 45% by 2020. In the process of promoting the household registration reform, some areas have simultaneously carried out the removal of counties and districts, and the removal of towns and townships to set up streets. By changing statistical caliber, a large number of farmers have been "urbanized", dissociated from the urban education, medical care, housing and other social security systems, facing the problems of "difficult employment", "difficult to see a doctor", "difficult to go to kindergarten" and "difficult to go to school". In addition, because the agricultural transfer population is difficult to live and work in the cities, most rural migrant labor force still occupies a large number of homestead and agricultural land in rural areas. Under the background of long-term existence of a large number of amphibious population and the weakening of rural subjects, it is impossible to realize the fundamental transformation of rural social structure and resource utilization mode. To solve the "hollow village" problem of abandoned homestead and idle land in rural areas, it is necessary to guarantee effective citizenization of rural migrants.

2. **Combination of urbanization and ruralization**

Urbanization is an important process of China's modernization, which not only related to the construction of the city itself, but also concerns the strategic issue of coordinating urban and rural development. In the next 5–10 years, China's urbanization will still be in a rapid development stage. The new-type urbanization requires the transformation of path and mode of traditional urbanization, promoting the intensification of land, the citizenization of farmers, the socialization of services, the cleanliness of the environment, and the equality of rights and interests, realizing the value of rural elements and the effective transfer and citizenization of population through the healthy and orderly urbanization, and solving the problems of the migration of rural migrant workers and the dual occupation of resources.

Ruralization is the new trend of urban–rural regional transformation development in the new period. First of all, as the first generation and part of the second-generation migrant workers return from the city to the countryside, the age structure, intellectual structure and governance system of rural labor force have changed significantly. These farmers with certain skills have become an important driving force for the development of scaled, specialized and multi-functional agriculture in rural areas. Secondly, traffic congestion, environmental pollution, rising housing costs and other "urban diseases" make some urban residents transfer to the surrounding small cities and rural areas, and the development of traffic conditions and information technology makes it possible for urban residents to transfer to rural areas. Thirdly, the beautiful scenery, local culture, quiet environment and leisure agriculture in rural areas attract more and more urban residents to rural areas for short-term leisure, tourism and health preservation. Meanwhile, with the aging of population in China, providing the elderly care services in rural areas become a new development direction. Therefore, with the advancement of urbanization, ruralization has become an important force and characteristics to promote the transformation and development of rural areas.

To strengthen the combination of urbanization and realization, it needs the benign interaction among the government, enterprises and farmers. Reforming the household registration system and improving the social security system is undoubtedly a breakthrough to break the dual structure between urban and rural areas, so that the qualified agricultural transfer population, especially the new generation of rural migrant workers, can gradually migrates to the cities for employment and settlement, and enjoy the same rights and interests as urban residents. Employment is the foundation of people's livelihood, and improving people's livelihood is the fundamental purpose of economic development. With the adjustment of industrial structure and the transformation of economic development mode under the background of globalization, both the transfer and employment of rural labor force and the re-employment of rural migrant workers who return to hometowns due to industrial transformation or old age are facing great challenges and pressures. The key to overcoming these problems is to speed up the innovation of government management system and enterprise system, establish multi-agent employment alliance, improve the market mechanism to promote the coordination of employment and entrepreneurship, and effectively promote the institutionalization of vocational training, the socialization of employment services and the integration of urban and rural social security. At the same time, to meet the needs of ruralization development, it requires to promote rural land demutualization, agricultural professionalization, service community, ecological civilization and resource institutionalization based on the above urban and rural institutional innovation.

3. **Highlighting regional comparative advantages and regional disparity**

China has a vast territory with great regional disparity. Not only does the foundation and level of regional economic development show obvious regional disparity, but also the process of urban development has stage and subject differences. Therefore, it is necessary to formulate strategies suitable for local development according to the characteristics of different regions, and promote the transformation of elements, structures and mechanisms of urban and rural regional systems by stages and priorities. From the perspective of system theory, scientific planning mainly refers to the government-led urban and rural planning, including new urban and rural economic development, social transformation and urban and rural spatial reconstruction, and overall arrangement in both spatial and temporal dimensions. In terms of spatial planning, it mainly includes two aspects: first, different policy preferences and investment priorities should be adopted among regions with different geographical and economic environments, for example, there should be differences in urban and rural development and spatial planning between mountain areas and plains, coastal and western areas, poor areas and developed areas, and regional adaptive urban–rural transformation models and supporting policies should be adopted. For example, the eastern coastal and metropolitan areas should realize the regional transformation of urban and rural development through the cultivation of urban agglomeration, and enhance the inclusiveness of urban society and the intensive allocation of elements, while the small and medium-sized cities in mountainous and hilly areas should strengthen the cultivation of nodes. The second is the decomposition of tasks at different levels within the same administrative division. Strategic planning and layout at all levels from the province to the village should be carried out as a whole and coordinated. For example, the implementation of the Five-Year Plan within the province should be hierarchically implemented. In terms of temporal planning, it mainly means that different development stages of the same region should have different planning priorities and development goals.

4. **Ensuring national grain and food security**

Grain security is a major strategic issue concerning China's economic development and social stability. For a long time, the Chinese government has attached great importance to grain security, put agriculture in the first place in the development of national economy and tried every means to promote grain production, thus better solving the problems of people's eating. However, agriculture is still a weak link in the national economy, and China's grain supply and demand have been in a tense balance for a long time. With the rapid development of industrialization and urbanization, the massive outflow of rural population leads to the abandonment of cultivated land, the weakening of agricultural subject and the shortage of rural laborers. Meanwhile, the non-agricultural construction takes up a lot of cultivated land, and the trend of non-grain utilization of cultivated land is increasing. From the perspective of medium and long-term development trend, these situations will be difficult to reverse due to the changes of factors such as population, cultivated land, water resources and international market, and bring great challenges to China's grain and food security.

In terms of the spatial dimension, the regional imbalance of cultivated land and grain production in China is extremely significant. Over the past 40 years, great changes have taken place in the grain output pattern of China. For a long time, the major grain production areas in China have been the Middle-Lower Yangtze Plain, North China Plain, and Northeast China, forming a pattern of "grain transportation from the south to the north" since the shortage of grain supply in the north. However, this situation has been gradually reversed since the reform and opening up. Due to the rapid urbanization and industrialization in the south, the large-scale expansion of urban areas has occupied a large amount of cultivated land, which leads to the continuous reduction of cultivated land area and grain production in the south. In 1993, the grain output of northern China exceeded that of southern China for the first time, gradually forming a pattern that the grain supply in the north was greater than that in the south (Fig. 5.1). These results show that the grain circulation pattern in China have changed fundamentally, from the traditional pattern of "grain transportation from the south to the north" to the present pattern of "grain transportation from north to south" (Liu et al. 2009).

5.2 Key Tasks of Urban–Rural Transformation

5.2.1 Improving Modern Agricultural Production System

1. **The core goal of modern agricultural development**

The construction of modern agriculture is not only the primary task and industrial foundation of the new socialist countryside construction in China, but also the key measure to promote agricultural development, increase farmers' income, guarantee national food security and rural sustainable development. The development of modern agriculture can help to enhance the comprehensive production capacity and international competitiveness of China's agriculture,

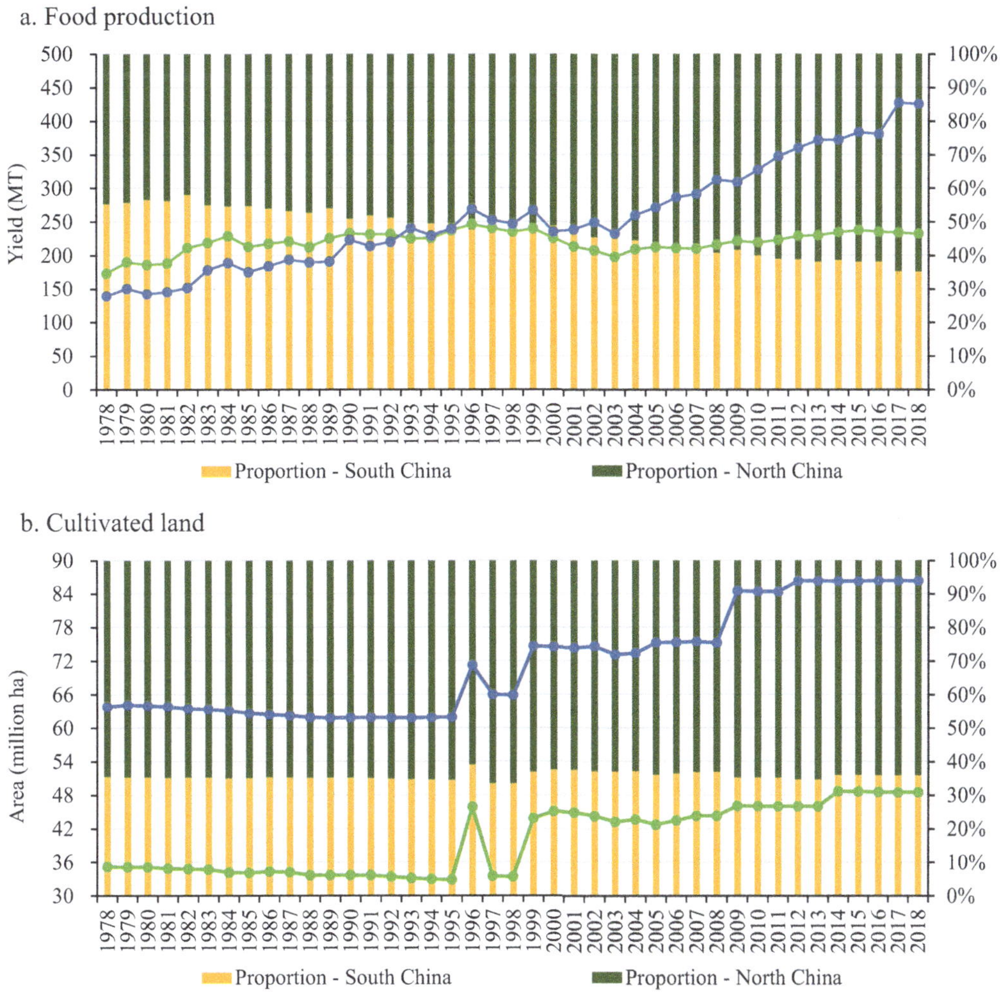

Fig. 5.1 Food production and cultivated land from 1978 to 2018 between the south and north China

consolidate the important position of agriculture in the national economy, promote the effective support of social and economic development for agriculture and rural development, and promote the sustainable development of agricultural production. It can also help to deepen the division of urban and rural functions and economic and technological exchanges, improve the comprehensive competitiveness of rural areas, and improve the relationship between urban and rural areas and the relationship between industrial and agricultural sectors. In addition, it is helpful to improve the efficiency of intensive utilization of agricultural resources, reduce the pressure of resources and environment of agricultural production, and promote the coordinated, stable and sustainable development of human land relationship in rural areas.

Promoting the construction of modern agriculture is based on the premise of ensuring national food security and the continuous improvement of comprehensive agricultural production capacity under the background of rapid and stable social and economic development and industrialization and urbanization. It requires to promote the improvement and popularization of traditional agricultural production technology, and at the same time, vigorously strengthen the diversified and integrated application and development of biotechnology, material technology, information technology and other emerging technologies in agricultural production, and further promote the reform of agricultural technology. On the basis of rural land reform and market-oriented, it also requires to promote scaled production and industrial management of agriculture, improve the efficiency of saving and intensive use of cultivated land, and promote the adjustment and upgrading of agricultural industrial structure, so as to achieve the long-term and efficient development of agriculture and the stable increase of farmers' income, and lay a solid industrial foundation and driving force for the construction of a new socialist countryside.

2. **Key measures for the construction of modern agricultural development**

At the specific operational level, it is necessary to establish and improve the development system of modern agriculture in China. On the one hand, taking ensuring national food security as an important strategic goal, the basic national policy of "effectively protecting cultivated land" and effectively improving the comprehensive production capacity of cultivated land resources should be implemented, and at the same time, the planting proportion of food crops in various regions should be rationally adjusted and the agricultural production structure in various regions should be rationally optimized, according to their local agricultural production functions and market demand; closely relying on the construction and development of national large-scale commercial grain production bases, the mechanization, scale, breed improvement and ecology of grain production should be comprehensively promoted, the natural environment advantages and economic benefits of regional grain planting should be highlighted, the financial transfer payment system between grain production and marketing areas should be gradually established, the reasonable regional division of national agricultural production should be realized, and the risk sharing and benefit sharing should be implemented. The priority should be given to solving the problems of grain production and rural development in major production areas, and promote the long-term and stable development of national food production. On the other hand, the investment and application of modern emerging technologies in agricultural breeding, pest control, facility agriculture, water-saving agriculture should be continually strengthened, so as to promote agricultural technology innovation, and improve the output and efficiency of agricultural products; village collectives and farmers' cooperative organizations should actively broaden their development ideas, according to their own geographical and resource advantages, vigorously develop characteristic agriculture, agricultural products processing industry and resource product processing industry, increase their market competitiveness and added value of products, actively cultivate and absorb their own surplus labor force for employment, so as to effectively promote the growth of farmers' income while enhancing the overall strength of rural economy with stability, consolidate the foundation of agricultural production and improve the comprehensive agricultural production capacity. At the same time, the application of information technology in agricultural production and management should be strengthened, agricultural information service and network platform should be set up, and a geographic information database of agricultural resources should be established, which integrates regional market information, agricultural production and technical information, and spatial information, to form a monitoring, forecasting and management system covering the whole country; timely and effective information services should be provided for agricultural production, market selection, scientific research and education, industrial structure adjustment and technology popularization, so as to promote the development of agricultural production and increase farmers' income.

5.2.2 Strengthening the Division and Connection of Urban–Rural Industries

1. **Improve the system of regional division of labor in urban and rural industrial development**

For a long time, China's rural economy has been dissociated from the urban industrial development system, and agricultural production mainly provided food and industrial primary raw materials for the development of urban industry. In the 1980s, under the background of insufficient social supply and strong demand for daily necessities, township enterprises initially participated in the division of labor and cooperation of urban industries. With the reform of state-owned enterprises and the reform of urban system, the scale of urban industry was growing, with large-scale agricultural land converting to construction land, a large number of rural labors flowing to urban employment and directly participating in the development of urban industry. However, under the traditional industrialization and urbanization development model, the unequal exchange of urban and rural elements, the lack of industrial connection mechanism, the loss of rural labor force and the continuous deprivation of development opportunities are prominent, which seriously restrict the sustainable development of rural areas and regional coordination and stability. At the same time, with the continuous outflow of population, the evolution of rural hollowing out and the lack of growth power have become the deficiency of urban and rural industrial development. Agricultural modernization lagged behind the process of industrialization and urbanization. Therefore, strengthening the division of industrial functions in urban and rural areas, constructing urban and rural industrial development system and highlighting the important value of rural areas are the objective requirements and important measures to achieve urban–rural coordination and regional sustainable development.

To promote the division of regional functions in urban and rural development, the key lies in strengthening the functional orientation of urban and rural development and the construction of industrial system. According to the urban and rural development pattern and resources and environmental conditions in different regions, and on the basis of establishing the urban and regional development strategy, the rural areas should be gradually brought to the spatial system of regional industrial development, some labor-intensive and agricultural raw material processing enterprises should be guided to gradually transfer to rural areas, especially to carry out production and operation activities based on key villages and towns in rural areas, and agriculture and rural non-agricultural industry development should be promoted to build a complete urban and rural industrial chain and spatial development system. Closely relying on the development of rural economy and industrial structure, the comprehensive development of rural economy, society and culture should be deepened.

2. **Establishing the interest connection and sharing mechanism of urban–rural economic development**

On the basis of increasing the government's investment in agriculture and rural areas and expanding the coverage of rural public finance, the interest linkage and sharing mechanism of urban and rural economic development should be constructed, and the institutionalized development of the strategy of "driving the countryside by city and promoting agriculture by industry" should be realized. At the same time, the role of small towns as a link between urban and rural areas should be vigorously strengthened, and their comprehensive functions of economic support, information transmission, service diffusion, product absorption, employment transfer, etc. should be emphasized, so as to promote the efficient, sustainable and stable implementation of the urban–rural overall development policy. Furthermore, the reform of rural tax system and market system should be promoted, the burden of farmers should be gradually reduced, the price of agricultural products and agricultural income level should be steadily increased, the agricultural finance and insurance system should be established and improved, and market mechanism should be adopted to stabilize the risk of agricultural production.

5.2.3 Accelerating the Orderly Reconstruction of Urban–Rural Space

1. **Optimizing urban and rural spatial reconstruction**

Urban and rural spatial reconstruction is a process of economic and social development that promotes the redistribution of production factors in urban and rural space, which can be affected by industrial structure upgrading, layout change, population flow, urban expansion and so on. In the accelerated stage of industrialization and urbanization, due to the agglomeration advantages and economies of scale of cities and towns, urban and rural space is characterized by rural element center agglomeration, industrial space agglomeration and urban scale expansion. The central position of cities and towns in economic, social, political and other aspects is constantly promoted, and a city-centered and hierarchical urban–rural point axis system is gradually established. With the upgrading of industrial structure and the adjustment of spatial layout, urban industry is transferred to the suburbs or small towns. The structural changes make the core and periphery of the city establish a new coordination and layout relationship. In the middle and late stages of urbanization, with the flow of population and industry, urban and rural space is usually reconstructed again based on the diffusion power.

The process of urbanization in China promotes the transfer of rural population to urban areas and realizes the reconstruction of urban and rural space. Urban spatial transformation focuses on the change of urban internal land use structure and function, as well as the spatial reallocation of industries, which is the spatial response of urban space to the changes of population, economy and society. Rural spatial reconstruction is the reconstruction of rural social and economic structure, the optimization and adjustment of rural production space, living space and ecological space and even fundamental change process in the process of rapid industrialization and urbanization, which are caused by the comprehensive effect of rural endogenous development demand and external driving force (Long 2013). With the development of industrialization and urbanization, urban spatial transformation and rural spatial reconstruction have been widely concerned in different fields. The reform of China's fiscal and taxation system and the establishment of urban state-owned land system make urban space reconstruction fully rely on the power of market mechanism and adjust land use and structure based on land rent income. However, rural spatial reconstruction is still mainly constrained by the rural land system, affected by the unreasonable income distribution of land acquisition and the separation of urban and rural land system, and the rural space reconstruction is facing the problems of damaged rights and interests, passivity and so on.

2. **Accelerating the construction of village and town pattern**

The construction of village and town system is based on the division of regional functions of urban and rural economy and industrial development, closely relying on the economic development goals and strategies of the town, to formulate the industrial development direction and functional orientation of each village according to local conditions, support the industrial cultivation of each village, strengthen the planning of village construction, enhance the purpose and service of rural infrastructure construction, and form a relatively complete village and town industry and infrastructure network, promote the rationalization and efficiency of the overall regional economic structure. In order to meet the needs of regional economic and social transformation and development, it is necessary to reconstruct the system of villages and towns and build a new pattern of villages and towns to coordinate urban and rural development.

The construction pattern of villages and towns refers to the spatial layout, hierarchical relationship and governance system of counties, key towns, central towns and central villages (communities) in rural areas, mainly including residential space, industrial space, ecological space, cultural space, etc. (Liu et al. 2014a). Facing the current situation of scattered, disorderly and empty land use in rural residential areas and the low efficiency of land use, it is urgent to promote the internal pattern of villages and spatial reconstruction between villages and towns to promote the construction of village and town pattern. Therefore, it is necessary to strengthen the planning of village and town system construction, scientifically arrange the industrial layout and infrastructure construction within the village, promote the comprehensive renovation of rural stock construction land with "hollow village" as the main body, and strengthen the key construction and service allocation of central villages and towns and central communities, so as to smoothly promote the agricultural and rural development and rural reform, and realize the efficient utilization of rural land resources and the overall optimization of village and town spatial pattern.

3. **Improving the equity level of urban and rural public services**

Restricted by the slow development of social economy and the lack of effective government investment for a long time, the development of rural infrastructure in China is generally lagging behind, with extremely significant contrast with urban development, which greatly restricts the development of rural economy and the improvement of farmers' life. It is the objective demand and inevitable choice to promote the construction of rural infrastructure, improve the rural production, living and ecological conditions, and establish a new style of rural life, based on the actual situation of various regions.

While increasing government financial input, rural production and agricultural positioning should be strengthened, with concentration of limited resources on roads, bridges, energy, water conservancy, electric power, communications and other key facilities, and avoidance of decentralized and inefficient use of funds and repeated construction, so as to provide strong support for agricultural production and the development of rural industries. On this basis, with improving the life of farmers as the fundamental goal, under the guidance of village development planning, the transformation of rural housing and living service facilities should be steadily promoted, and simple, clean, stable and harmonious new socialist countryside should be built, and the synchronous improvement of rural material and spiritual and cultural life should be promoted.

5.2.4 Deepening the Comprehensive Reform of Urban–Rural Systems

Institutional reform has always been an indispensable part of China's social and economic development. The institutional reform carried out in different stages of development provides institutional guarantee for economic growth, social harmony and ecological security. Urban institution, rural institution, regional institution and various special institutions all played important role in the urban and rural transformation development. Currently, urban–rural transformation urgently needs to build the institution and mechanism of urban–rural integration, which provides strong institutional support and development power for the stable and orderly development of agriculture, rural areas and social economy as well as the balanced development of urban and rural areas. When steadily building the regional division system of urban and rural industries, comprehensively deepening the reform of economic, political and social security system is the inevitable choice to promote the rapid development of rural economy, social harmony and stability and the steady improvement of farmers' life, promote the functional division and exchange cooperation between urban and rural areas, and promote the coordinated and orderly development of regional social economy.

The focus of deepening urban and rural institution reform is in the economic field, and the core of urban and rural economic reform is to effectively promote land institution reform. The land property rights relationship is the basis of rural production relations and income distribution of urban and rural land assets. It requires to earnestly implement the spirit of the resolution of the Third Plenary Session of the 18th Central Committee of CPC, deepen the reform of rural land reform, deepen the reform of three rights division of agricultural land, which means the separation of ownership, contracting right and utilization right, and promote the pilot reform of collective economic construction land entering the market, homestead voluntarily withdrawing with payment, and land expropriation income distribution experience, promote the scaled, industrialization and characteristic management of cultivated land resources. The agricultural development and the accumulation of capital and technology can be effectively promoted by increasing the productivity of cultivated land and the competitiveness of agricultural products due to rural land institutional reform. The proportion of land property income of farmers in rural areas can be increased, and the rural areas can share the achievement of urbanization equally, by reforming the land acquisition system and the distribution mechanism of land acquisition value-added income.

On the basis of building a mechanism for linking and sharing interests in urban and rural economic development and deepening urban–rural economic reform, the urban–rural collaborative governance system should be improved. With the organic law of the villagers committee as the core, it should improve the villagers' self-government system, steadily expand the scope of rural grass-roots democratic autonomy, gradually promote the institutionalization, standardization and routinization of villagers' autonomy, strengthen the effective connection and benign interaction between township management and villagers' autonomy, improve the restriction and supervision mechanism, and promote the smooth operation of rural democratic management. In particular, under the context of the continuous transfer of rural population to the cities, it should take the opportunity of rural community construction to build new hierarchical organizations such as community management committee, explore the coordination mechanism of occupation, utilization and interest distribution of village collective land, collective enterprises and infrastructure, so as to optimize the allocation of rural resources in a larger administrative scope. At the same time, on the basis of the division of economic, social and ecological functions and regional disparity between urban and rural areas, it should continue to increase government financial investment in the development of rural areas, bring rural residents into the social security system such as urban minimum living security, unemployment, medical care and pension, and gradually promote the improvement of the integrated social security system between urban and rural areas, so that urban and rural residents can have equal access to survival security services and promote the development of rural areas, and farmers' quality of life can be effectively increased.

5.3 Overall Situation of Urban–Rural Transformation

5.3.1 Economic Growth and Optimization of Urban–Rural Structure

Since the reform and opening-up, the domestic system reform and the policy of opening to the outside world have promoted the rapid development of labor-intensive processing and manufacturing, agricultural product processing and modern high-tech industries. To meet the needs of urbanization development and urban–rural infrastructure construction in China, the steel industry, building materials industry and machinery manufacturing industry also grow rapidly. From 1978 to 2016, the average annual growth rate of China's GDP was nearly 10.0%, and the average annual growth rate of per capita GDP was 8.7%. In terms of industrial structure, the added value of the secondary industry accounted for more than 40% of GDP. Urbanization development and the changes in social demand provided a lot of opportunities for the development of service industry. Therefore, the proportion of added value of the tertiary industry showed a trend of continuous rise, and surpassed that of secondary industry for the first time in 2012, which accounted for 45.5% of GDP. On the contrary, the proportion of the added value of primary industry in the national economy decreased year by year, from 27.7% in 1978 to 8.1% in 2016. Industrial differentiation in rural areas is an important embodiment of agricultural industrial structure adjustment and social demand changes. Specifically, the proportion of agricultural planting decreased significantly, while the proportion of animal husbandry continued to increase, and the agricultural production structure remained relatively stable after 2000. In general, China's industrial structure has realized the transformation from "secondary-tertiary-primary" to "tertiary-secondary-primary" since the reform and opening-up, the trend of industrial structure upgrading is obvious, and the unit labor productivity continues to increase (Fig. 5.2).

5.3.2 Improvement of Social and Living Conditions

Economic growth and the upgrading of industrial structure promote the transformation of urban and rural employment structure and the improvement of urban and rural living standards. In terms of the employment structure, the proportion of employed population in the secondary and tertiary industries showed an upward trend, while that of the primary industry showed a downward trend, from 70.5% in 1978 to 27.7% in 2016. However, the number of employed people in agriculture has been higher than that in the secondary and tertiary industries before 2010 (Fig. 5.3). In 2016, the proportion of employment in the secondary and tertiary industries reached 28.8% and 43.5% respectively, and the service industry became the main channel the main channel of agricultural surplus labor transfer. Compared with the structure of output value, the dual characteristics of China's urban and rural industrial structure are obvious, and the absorption of agricultural employment population in service industry is still the main trend of economic development. As to the urban and rural residents' income, the income level of urban and rural residents has been continuously improved. From 1978 to 2016, the per capita disposable income of urban households increased from 343 yuan to 33,616 yuan, and that of rural households increased from 134 to 12,363 yuan.

Promoted by the economic development, China's urban–rural infrastructure construction continues to improve, and the efficiency of public resource allocation continues to improve. With the popularization of nine-year compulsory education and the development of secondary and higher education, the number of illiterate populations in China has decreased year by year, and the average education level has been significantly improved. Taking rural areas as an example, the proportion of rural population with primary and secondary education level and below of population aged 6 and above is declining, while that of

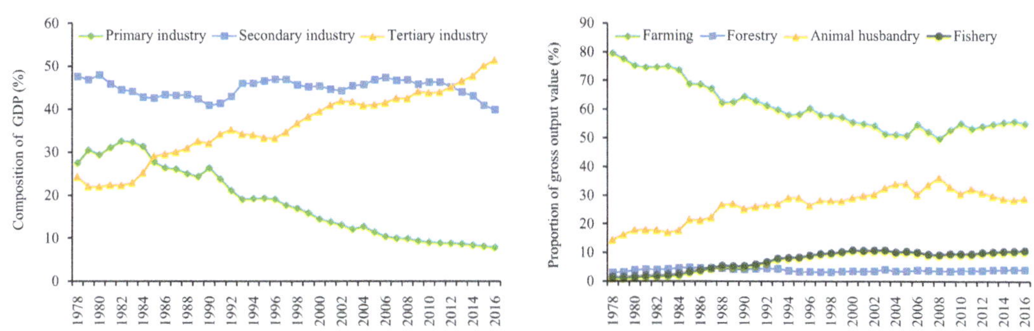

Fig. 5.2 Changes of urban and rural industrial structure from 1978 to 2016

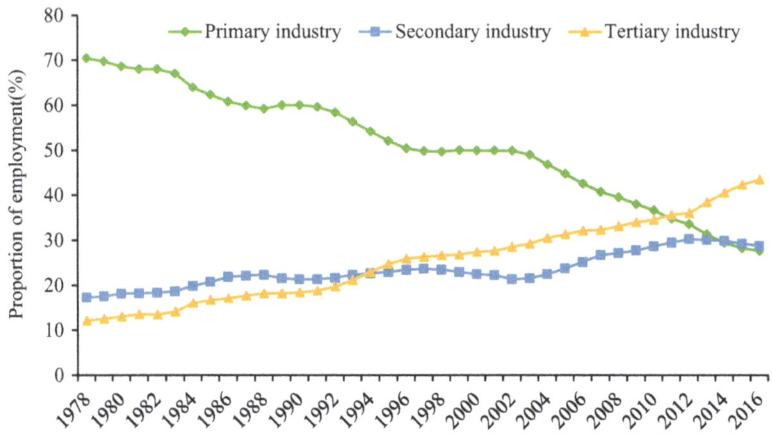

Fig. 5.3 Changes of employment structure in three industries from 1978 to 2016

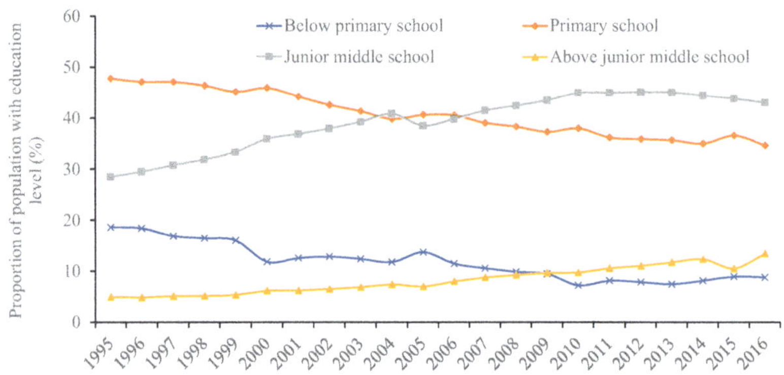

Fig. 5.4 Changes of the educational level of rural population aged 6 and above from 1995 to 2016

population with junior high school education or above is increasing. In 2016, the proportion of rural population aged 6 and above with education in junior high school and above increased from 33.47% in 1995 to 56.49% (Fig. 5.4).

5.3.3 Expansion of Urban Scale and Space

Under the promotion of industrial development and employment, rural population is moving rapidly to cities and towns and the urban space is expanding rapidly, which reflects intuitively in the change of land use types is the changes of urban and rural landscape pattern. From 1978 to 2016, the urbanization rate increased by 1.04 percentages per year, especially after 1995, the urbanization rate accelerated significantly, with an average annual growth of 1.35 percentages. The transfer of rural population to cities and towns has led to significant changes in the traditional urban and rural settlement patterns and spatial structure (Fig. 5.5).

Land use is the mapping of urban and rural socioeconomic development in geographical space. In general, the most obvious change in urban and rural land use structure, type and space is promoted by industrialization and urbanization, and is mainly manifested in the expansion of urban space and the construction of urban–rural road. In particular, the emergence of new urban space forms with various development zones as carriers has greatly promoted the expansion of urban space. From 1981 to 2016, the area of urban construction land in China has increased from 6720 to 52,761.3 km^2, with an increase of 7.85 times and an average annual growth of 1315.47 km^2 (Fig. 5.6). Compared with the urban residential land, although the rural

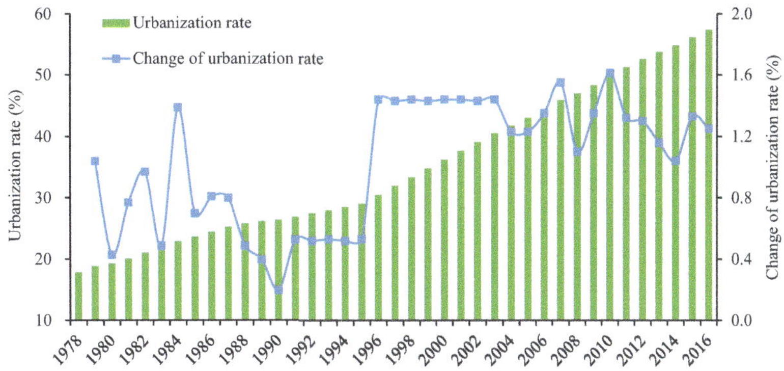

Fig. 5.5 Changes in urbanization rate and change rates of urbanization rate from 1978 to 2016

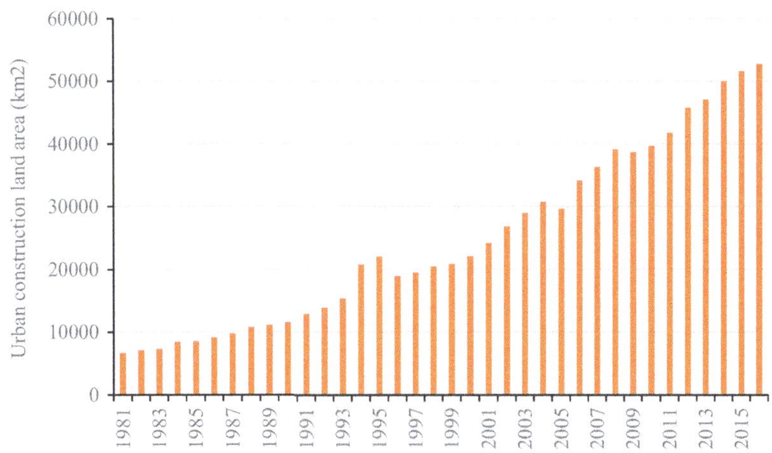

Fig. 5.6 Changes of urban construction land area from1981 to 2016

population is constantly flowing out, the differentiation of rural households and the outward expansion of newly built houses lead to the increase of rural residential land instead of decrease in the period of 1996–2008. Meanwhile, due to the rapid expansion of cities and towns, the proportion of urban construction land in urban–rural construction land is increasing (Fig. 5.7).

In order to alleviate the contradiction of land use, meet the demand of economic growth and urban expansion for construction land, and improve the level of intensive land use, different forms of rural residential renovation have been carried out in various regions, promoting the construction of rural communities and making the spatial form of rural settlements change significantly, especially in the surrounding areas of cities and developed areas. Under the promotion of urban expansion, facilities supporting, township merging and new countryside construction, Suzhou, one of the most developed areas in China, Suzhou has implemented the "three concentration" development strategy of agriculture to scale operation, industry to industrial park, and farmer's residence to town, leading to the appearance of different types of communities such as integrating into urban areas, removing and settling down, scale operation, investment and development, leisure and sightseeing and stock cooperation (Ju 2006). In terms of the spatial pattern, these communities show a trend of centralization. With a large number of rural populations flowing into cities and towns, many rural areas are declining or even disappearing. Take Shandong Province as an example, the number of subdistrict office in the whole province increased by 21.0 per year on average from 2000 to 2016, while the number of town and township decreased by 15.3 and 34.6 per

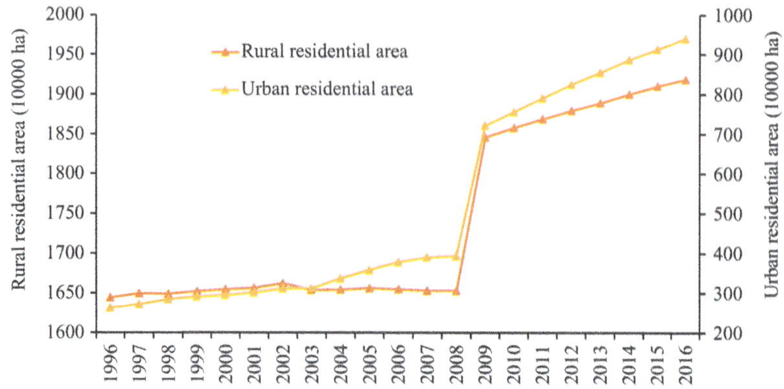

Fig. 5.7 Land change of urban and rural residential areas from 1996 to 2016

year respectively, and the number of administrative villages decreased by 1278.8 per year on average.

5.3.4 Outstanding Urban–Rural Development Problems

Urban–rural relationship is the relationship between urban and rural areas in economic and social aspects. In a broad sense, it includes economic, social, spatial, resource utilization, ecological environment and other aspects; in a narrow sense, it usually refers to the economic relationship between urban and rural areas. The contradiction between urban and rural development is directly related to the stage of China's economic growth, but it is also subject to the unique urban–rural dual system, as well as the obvious urban bias strategies and policies. In general, it is manifested in the changes of contradiction between urban and rural development with the change of economic growth and the imbalance between them. At the beginning of the founding of PRC, how to solve the problems of food and capital accumulation for industrialization were the first priority. To meet the accumulation of primitive capital for industrialization development, the government chose to divide the cities and villages through the household registration system, employment system and social security system. After the reform and opening-up, the household responsibility system, price mechanism reform and the development of township enterprises promotes the rapid development of grain production of rural employment transfer, which effectively solve the problems of food and industrialization funds. Furthermore, to solve the problems of rigid system and insufficient employment of state-owned enterprises, and accelerate urban development, urban–rural development in China began to enter the development mode with city as the core in the middle and late 1990s, and the core is to promote economic growth and solve the employment problem. Entering the twenty-first century, because of the continuous expansion of regional gap and urban–rural gap, coordinated development has become the key issue in China's urban–rural relationship.

The income difference between urban and rural residents is the direct representation of urban–rural development difference. After the reform and opening-up, the urban–rural income gap showed a fluctuating trend. Specifically, the urban–rural income gap showed an obvious narrowing trend in 1978–1984, while it turned to a trend of continuous expansion in 1985–1994; after a short period of narrowing, it continued to widen from 1997 to 2008, reaching the maximum of 3.33 times in 2009; after that, it showed a downward trend, but the gap was huge. Engel's coefficient is the proportion of residents' food expenditure in the total consumption expenditure, which is used to measure the economic and social development level of a country or region. From 1978 to 2016, both the Engel's coefficients of urban and rural residents showed a downward trend, dropping from 57.5% and 67.7% to 28.8% and 32.2%, respectively. Comparing the differences between the two changes, the trends of Engel's coefficient and income gap between urban and rural residents are roughly the same. In the years when the income gap between urban and rural residents increase, the difference of Engel's coefficient of urban and rural residents is also large, especially in the 1990s and the beginning of the twenty-first century (Fig. 5.8).

There are great differences in the level of economic and social development in different regions, mainly in the industrial structure, employment structure and spatial structure. This difference forms the difference of urban and rural regional evolution in the East, the West and the West, and shows the difference between urban agglomeration and non-urban agglomeration. According to the per capita disposable income of urban and rural households in 2016, the situation of regional urban and rural development is analyzed. The five provinces with the highest per capita disposable income of urban households are Shanghai, Beijing, Zhejiang, Jiangsu and Guangdong, and the lowest are Gansu, Heilongjiang, Jilin, Guizhou and Qinghai; The five provinces with the highest per capita disposable income of rural households are Shanghai, Zhejiang, Beijing, Tianjin and Jiangsu, and the five lowest provinces are Gansu, Guizhou, Qinghai, Yunnan and Tibet. It is obvious that the urban–rural development gap between the East, the Central and the West is significant, and the areas with high level of urban and rural development are concentrated in the eastern coastal area, while the low-level areas are concentrated in the western China (Fig. 5.9).

Promoted by export-oriented economy, township enterprises and urbanization, the industrial structure and employment structure in the eastern coastal rural areas transform rapidly, and the urban–rural space forms a network structure with central cities, small towns and rural settlements as nodes and roads and other infrastructure as channels. Thus, the trend of urban–rural integrated development is increasingly obvious. As an important ecological function area and grain production area in China, the carrying capacity of resources and environment in the central and western regions is weaker than that in the eastern regions. Although the national biased policies have supported the economic growth in the central and western regions since the twenty-first century, it was difficult for the central and western China to obtain the international and domestic environment of the eastern region in the period of reform and opening up. The eastern region is still the main region of inter-provincial population inflow and agglomeration, and labor-intensive industries are still concentrated in the eastern coastal region. In addition, with the rapid development of provincial capitals and prefecture-level cities in Central and Western China, and the scale of population agglomeration and urban expansion in these areas is much faster than that of the county areas, and the urban–rural gap is extremely significant between the eastern, central and western regions as well as between large, medium and small cities.

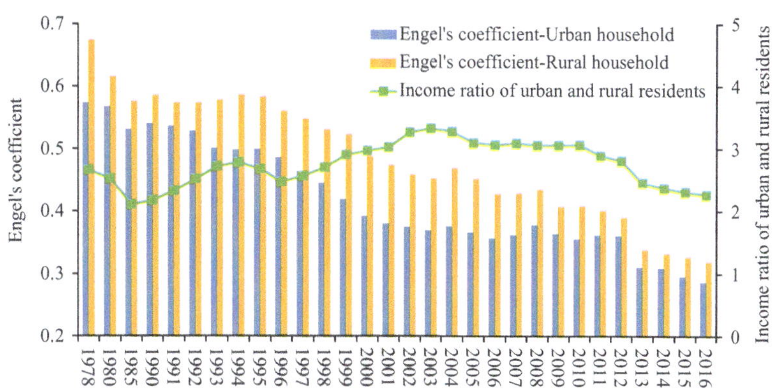

Fig. 5.8 Gaps and changes in living standards of urban and rural residents from 1978 to 2016

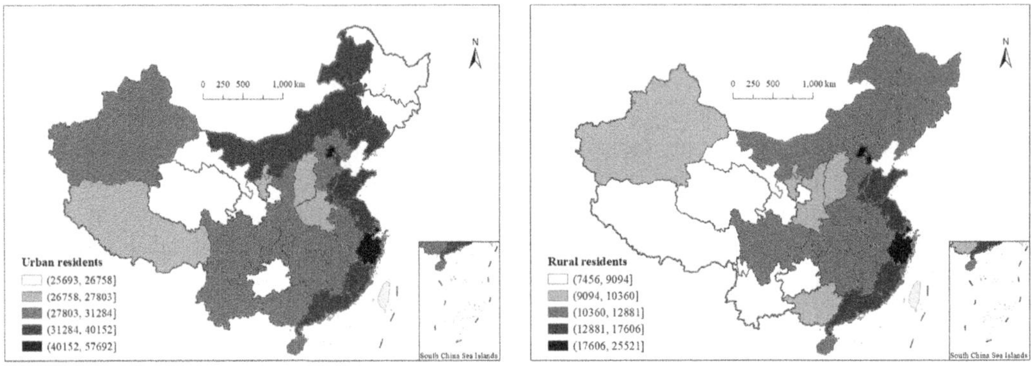

Fig. 5.9 Spatial pattern of the per capita disposable income of urban and rural households in 2016

5.4 Spatial Pattern of Urban–Rural Transformation in China

5.4.1 Design and Selection of Evaluation Index

1. **Analytical framework**

The state evaluation of urban–rural transformation is a comprehensive and systematic work, involving multiple factors and multi levels. The final evaluation result is to transform the theoretical research into practical decision-making to direct the overall development of urban and rural areas, so as to promote the transformation of urban and rural areas and the sustainable development of rural areas. Urban–rural economic growth, policy system and subject game jointly promote the coordinated evolution of urban and rural areas. Therefore, the evaluation of urban–rural transformation should integrate the stage of economic growth, institutional synergy and interest balance. In the fifth Plenary Session of the 18th Central Committee of the CPC, the development concept of "innovation, coordination, green, openness and sharing" was put forward. In the new era, urban–rural coordination is an important basis and guidance for the overall development of urban and rural areas, and is an important guide for changing the concept of emphasizing urban areas over rural areas and innovating the urban–rural integration system. The evaluation of urban–rural transformation can comprehensively reveal the basic stage, overall pattern and series of problems faced by urban and rural development. Driven by a series of factors such as economic growth, urbanization, capital accumulation and government management, urban–rural areas develop rapidly, which is manifested in the flow of urban and rural elements and the spatial redistribution of resources. Meanwhile, urban–rural areas are integrated into the industrial division system through the prenatal and postpartum linkages of enterprises. Based on the four dimensions of population transformation, industrial transformation, land transformation and social transformation, Liu et al. (2014b) evaluated the urban–rural development status and transformation trend. However, there is a research gap in the quantitative description of urban–rural relationship.

In October 2008, the Third Plenary Session of the 17th CPC Central Committee passed the "decision of the CPC Central Committee on several major issues of promoting rural reform and development", which clearly put forward "to establish a system to promote the urban–rural economic and social integrated development". Also, it stressed that "breakthroughs should be made in urban and rural planning, industrial layout, infrastructure construction, and integration of public services as soon as possible, so as to promote the balanced allocation of public resources and the free flow of production factors between urban and rural areas, thus boosting the integration of urban and rural economic and social development". Urban–rural equivalence is the advanced stage of urban–rural relationship and the ultimate goal of urban–rural relationship evolution. It is necessary to establish a scientific and reasonable evaluation index system of urban–rural equivalence development. In general, the determination of evaluation criteria can make a vertical comparison of the development trend of urban–rural equivalence in China or a certain region, and make a horizontal comparison of the development degree and evolution trend of urban–rural equivalence in various regions of the country. Through quantitative evaluation and analysis, we can find the main influencing factors of urban–rural equivalent development from the numerical value, and find the appropriate paths for the further integrated and coordinated development of urban–rural economy and society. Based on the existing researches, this paper proposes a comprehensive evaluation index system including three dimensions of transformation conditions, transformation subjects and transformation effectiveness (Fig. 5.10).

The three dimensions of urban–rural transformation are urban–rural socio-economic transformation, urban–rural structural transformation and urban–rural relationship transformation. The urban–rural socio-economic transformation is the basis and prerequisite of urban–rural development, which provides the initial conditions for the current urban–rural transformation and become the basis for the factor allocation and departments connection. Urban–rural structural transformation is usually consistent with economic and social transformation. In other words, the change of urban–rural structure is the response of urban–rural regional space to external environmental changes, including industrial structure, employment structure, population structure, land use structure, etc. For example, when the industrial structure, urbanization level and economic level are mutually appropriate, the scale effect of social and economic activities can be fully exerted to further attract the investment of production factors and facilities; while when the urban scale is excessively concentrated and the economic level is not high, the phenomenon of excessive urbanization, which is generally exists in Latin America, will occur. The urban–rural relationship is a symbol of the urban–rural transformation. When elements between urban and rural areas are reasonably allocated and urban–rural areas achieve coordinated development, the urban–rural relationship shows benign evolution; otherwise, it will continue to deteriorate.

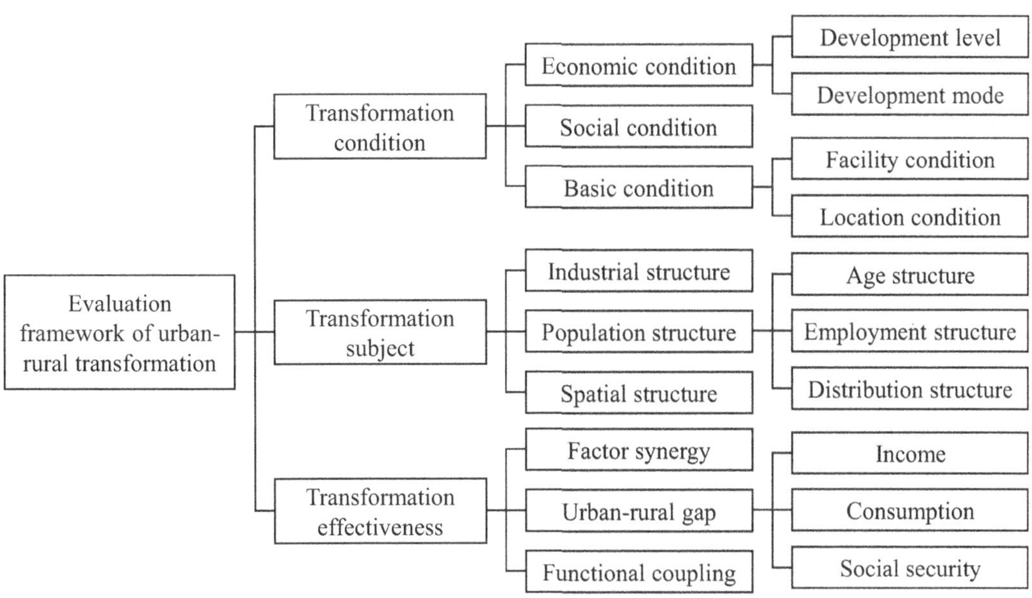

Fig. 5.10 Evaluation framework of urban–rural transformation

Transformation is the representation of condition, subject and effectiveness. Therefore, the overall situation of urban–rural transformation depends on the coordinated promotion of economic society, urban–rural structure and urban–rural relationship. In a specific period, due to the differential in the basis of regional development, social and economic changes vary in different dimensions. According to the general law of economic growth and urban–rural transformation, when urban–rural economy and society are in the low-level stage, urban and rural areas lack of growth momentum, and there is little factor flow and connection between urban and rural areas, which is manifested as the regional structure dominated by rurality, and urban and rural development is in the low-level stage of coordination. When urban–rural economy and society is in the stage of rapid development, rural elements accelerate to flow and gather to urban areas due to the promotion of industrialization and capital agglomeration. In this case, urban–rural areas are evolving from rural dominance to urban–rural conflict and urban dominance, and the urban–rural relationship is in the stage of conflict. If there is no policy to regulate and control, factors such as the loss of young and strong labor force and lack of capital investment result in the slow rural development, and some rural areas become hollow seriously. When urban–rural economy and society enter a higher-level stage, the radiation and driving effect of the city on the surrounding rural areas is strengthened, and the urban capital and population transfer to the small rural towns. At the same time, with the support of policies, the urban–rural structure tends to be stable and the urban–rural relationship will be continuously improved.

2. **Establishment of evaluation index system**

Tables 5.2, 5.3 and 5.4 show the detailed evaluation index system of the three dimensions to measure urban–rural transformation. Because the data of some indicators are difficult to obtain in some years, and the principle of representativeness is fully considered, the research focuses on selecting the key indicators of urban–rural transformation. The data are derived from China Statistical Yearbook, Tabulation on the Population Census of the People's Republic of China, China Regional Economic Statistical Yearbook, China Employment and Social Security Statistical Yearbook, etc. Specifically, the data of population, including total population, urban population and employment population are from the census yearbook. Land use data comes from the Earth System Science Data Sharing Network (https://geodata.nju.edu.cn/). According to the time limits of census data and urban–rural land remote sensing interpretation data, this study selected two years of 1990 and 2015 to analyze the change trend of urban–rural transformation in various regions from 1990 to 2015.

There are four levels in China's administrative systems, i.e., province, city, county and township. The county-level administrative unit is often used to spatial pattern of regional development. From the perspective of urban- rural regional evolution, the county-level unit has formed a village and town hierarchy system with small and medium-sized cities as the core, lacking the regional driving force of big cities. If the municipal district is regarded as a separate unit, the relationship between urban and rural units will be separated. Therefore, this study selects provincial units and city units as the evaluation units of urban–rural transformation in China. Beijing, Shanghai, Chongqing and Tianjin are taken as separate research units in the analysis of city units, and Hong Kong, Macao and Taiwan are excluded. In addition, due to the lack of time series data and some prefecture-level city data in Tibet and Qinghai, the provincial data is used to replace the missing data in these areas. Finally, a total of 337 prefecture-level units are obtained.

5.4.2 Regional Situation of Urban–Rural Transformation

The economic environment, institutional environment and natural environment of urban–rural development in different regions are quite different, forming the transition path and state of urban–rural development with regional characteristics.

Table 5.2 Comprehensive evaluation index system for urban–rural socio-economic transformation

Criteria layer (Weight)	Index layer (Weight/Index polarity)	Definition
Economic transformation (0.253)	Per capita GDP (1.000) (+)	GDP/Total resident population
Social transformation (0.483)	Income of rural residents (0.861) (+)	Per capita net income of rural households
	Income of urban residents (0.139) (+)	Per capita disposable income of urban households
Infrastructure and services transformation (0.264)	Education attainment (0.495) (+)	Average years of education
	Medical level (0.505) (+)	Number of beds per thousand people

Table 5.3 Comprehensive evaluation index system for urban–rural structural transformation

Criteria layer (Weight)	Index layer (Weight/Index polarity)	Definition
Population structure (0.264)	Urbanization rate (+)	Urban resident population/total resident population
Social structure (0.295)	Employment structure (+)	Non-agricultural employment/total employed population
Economic structure (0.224)	Industrial structure (+)	Added value of non-agricultural industry/GDP
Landscape structure (0.217)	Urban–rural land use structure (+)	Area of urban construction land/area of urban and rural construction land

Table 5.4 Comprehensive evaluation index system for urban–rural relationship transformation

Index layer (Weight/Index polarity)	Definition
Income gap (0.373) (−)	Per capita disposable income of urban households/per capita net income of rural households
Consumption gap (0.235) (−)	Per capita consumption expenditure of urban households/per capita consumption expenditure of rural households
Industry duality (0.392) (−)	Labor productivity of non-agricultural industry/labor productivity of agricultural industry

Relying on the advantages in economic base, location conditions, history and culture, the eastern coastal area in China have been established as the priority areas of national development after the reform and opening-up. Since 1984, Chinese government has successively established four special economic zones and 14 coastal open cities to speed up the pilots of reform and opening-up. At the same time, the Yangtze River Delta, the Pearl River Delta and the southern Fujian Delta have been determined as economic development zones, and then expanded to Shandong Peninsula, Liaodong Peninsula and Hainan Province. In 1990, the central government granted Shanghai Pudong New Area preferential policies for a period of 10 years to speed up social and economic construction, forming an all-round pattern of opening-up in coastal areas. County economic growth and rural construction supported by township enterprises also promote the rapid development of rural areas. The advanced market system and the first opportunity of opening to the outside world help the eastern region ahead of other regions in economic growth and urbanization, and promote the transformation of urban-rural structure and function.

Based on the comprehensive evaluation index system for urban–rural socio-economic transformation, urban–rural structural transformation and urban–rural relationship transformation, the situation of urban–rural transformation in each province is calculated. And according to the National Bureau of Statistics of China, the Chinese mainland can be divided into four parts, i.e., eastern, central, western, and northeastern regions. The results show that in 1990, the highest values of the three dimensions were all in northeast China, while the values of each dimension in western region were the lowest. The urban–rural transformation generally showed a downward trend from the northeast to the east, the central, and then to the west. In 2015, the eastern region surpassed the northeast region in urban–rural socio-economic transformation and urban–rural structural transformation, but in the dimension of urban–rural relationship transformation, the highest-value area was still the northeast China, and the values of all dimensions in the western region is still the lowest. In terms of numerical value, the gaps between the maximum and minimum value of urban–rural socio-economic transformation expands from 0.055 to 0.310, and those of urban–rural structural transformation and urban–rural relationship transformation expand from 0.176 to 0.310, 0.171 to 0.351, respectively. Obviously, the regional differences of urban and rural transformation showed a trend of expansion (Table 5.5).

In terms of the provincial scale, the spatial pattern of urban–rural socio-economic transformation and structural transformation in 2015 were very similar, showing a decreasing trend from the eastern coastal areas to the inland (Fig. 5.11). Except for Hebei province and Hainan province, other provinces in the eastern region had higher values of urban–rural socio-economic transformation and structural transformation. The dimension of urban–rural relationship transformation in the three northeast provinces had a high level. There was a relatively high economic and social level in Xinjiang, but the slow transformation of urban and rural industries, employment structure and spatial structure resulted in the low level of urban–rural structural transformation. The overall level of urban–rural socio-economic transformation and structural transformation in southwest and northwest China were relatively low, especially in Tibet, Guangxi, Yunnan, Guizhou, Ningxia and Gansu.

Unlike the spatial pattern of urban–rural socio-economic transformation and structural transformation, the high-value areas of urban–rural relationship transformation were distributed in Hebei, Jiangsu, Zhejiang, Fujian, Hainan and Tianjin in the eastern coastal areas, Heilongjiang and Jilin in the northeast, and Hubei, Jiangxi and Anhui in the central region. The secondary high-value areas were concentrated in Liaoning, Shandong, Henan and Sichuan, and the low-value areas were still concentrated in the western region. Unlike the dimensions of urban–rural socio-economic transformation and structural transformation, Hebei and Hainan, with lower level, had relatively high values of urban–rural relationship transformation, which indicated that urban and rural development in these areas was in a relatively low-level equilibrium state. However, Inner Mongolia, Shanxi and other provinces showed different situations. The overall economic and social level of these regions was relatively high, but the gap between urban and rural areas was higher than that in other regions since the strong dependence of economic development on energy and mineral resources and energy mining enterprises, the high proportion of state-owned enterprises in economic activities, and the lack of effective interaction between urban and rural areas.

Plain agricultural area is the main grain production area in China, and also the main area of economic growth and population agglomeration. In the process of rapid industrialization and urbanization, there are obvious social and economic differentiation in plain agricultural areas. Although the agricultural areas in Henan, Anhui, Jiangxi, Hebei and Hunan have obvious advantages in soil, climate, water resources and agricultural technology, it is difficult to realize large-scale production in these areas due to the small per capita cultivated land resources, and they do

Table 5.5 Urban–rural transformation of the four regions in 1990 and 2015

	1990			2015		
	Socio-economic transformation	Structural transformation	Relationship transformation	Socio-economic transformation	Structural transformation	Relationship transformation
Eastern region	0.116	0.295	0.388	0.548	0.670	0.682
Central region	0.109	0.236	0.323	0.284	0.421	0.643
Western region	0.088	0.239	0.293	0.238	0.360	0.380
Northeast region	0.143	0.412	0.464	0.375	0.429	0.731

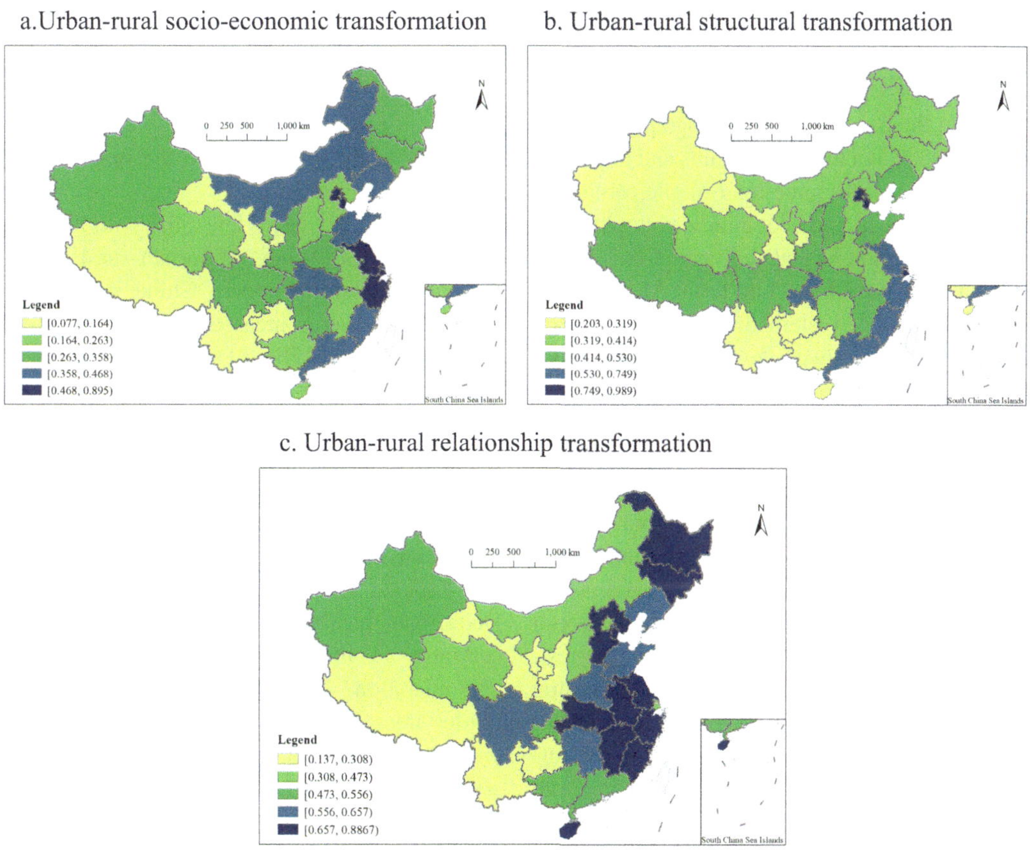

Fig. 5.11 Spatial pattern of China's urban–rural transformation at provincial scale in 2015

not have the economic foundation and location conditions of the eastern region. In addition, the development of township enterprises and small cities has insufficient driving capacity for rural labor transfer and employment. Therefore, the coordination between urban and rural areas in these areas is lower than that in eastern China. Heilongjiang and Jilin provinces in the Northeast are the most important industrial bases in China, with high level of industrialization and urbanization. At the same time, due to the large-scale use of agricultural machinery and advantages of rich land resources, agricultural production income is significantly improved. Thus, there is little difference between urban and rural development. However, due to the large number of resource-based cities and heavy chemical enterprises, the high proportion of state-owned enterprises, and the rigid management system, the transformation and development in northeast China is under great pressure. In recent years, the problems such as economic downturn and human resources outflow have emerged in northeast China, which makes the urban–rural transformation face severe challenges of regional social and economic development.

Due to the bad natural and geographical environment, such as a lot of mountains and hills, lack of water and soil resources and fragile ecological system, the social and economic development in the western region has lagged behind the eastern and central regions for a long time. Although the promotion of the western development strategy has effectively improved the regional transportation infrastructure and communication condition, the high transportation cost has restricted the local economic development due to the constraints of topography and soil and water conditions. Some regions rich in energy and mineral resources have high economic level. However, due to the lack of labor demand of energy and mineral enterprises, the profits of state-owned enterprises are difficult to benefit the local government and local residents, and the dynamic mechanism of urban–rural development and transformation is still insufficient, which ultimately slows down the transformation of urban–rural industrial structure and employment structure. In addition, restricted by topography and water and soil resources, most rural areas remain a low-level of agricultural modernization, and the phenomenon of rural population outflow is serious, which result in the insufficient endogenous growth power in rural areas and the continuous widening gap between urban and rural areas.

5.4.3 Spatial Pattern of Urban–Rural Transformation

Prefecture-level city is an important administrative unit and economic unit in China, which is composed of municipal districts and surrounding counties and cities. It has a relatively complete system from big cities to small towns, and has the location advantage of population and other factors gathering in a certain region. Because of the scale differences between prefecture-level cities and provinces, the comprehensive evaluation at the prefecture-level city scale can better reflect the spatial differentiation of urban–rural transformation within the province. The results show that the evaluation results of prefecture level city units are consistent with those of provincial units in the overall pattern, but there are still large differences among prefecture level cities within the province, and the spatial differences are more obvious.

In 1990, there were a few cities with high and relatively high value of urban–rural socio-economic transformation in China, mainly distributed in the core areas of the Yangtze River Delta and the pearl River Delta and the northeast region; while, there are many cities with low and relatively value, which show a spatial distribution characteristic of ubiquity in the central and western regions, except for some median-value cities in Northwest China. In 2015, high-value cities were mainly distributed in the core areas of the Yangtze River Delta and the Pearl River Delta, and the relatively high-value areas concentrated in the Yangtze River Delta urban agglomeration, Northwest China and northern border areas. Except for a few provincial capital cities with relatively high level of urban–rural socio-economic transformation, the spatial distribution of low value and relatively low value spatial units is wide in central and western China. Overall, the agglomeration of the high-value areas of urban–rural socio-economic transformation from 1990 to 2015 is more obvious in the eastern coastal areas, while the vast western region is the main distribution area of low-value areas (Fig. 5.12).

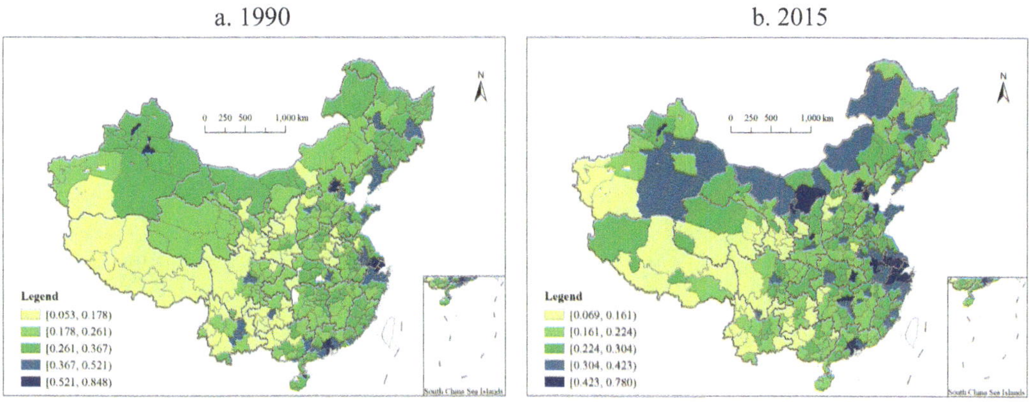

Fig. 5.12 Spatial pattern of urban–rural socio-economic transformation in 1990 and 2015

The level of urban–rural structural transformation in the eastern and northeastern regions was higher than that in the central and western regions. The high-value areas showed a spatial characteristic of scattered distribution, the level of urban–rural structural transformation in provincial capital cities was significantly higher than that in other regions, and the level of urban–rural structural transformation in northeast China decreased significantly in the process of urban–rural transformation. In 1990, the high-value areas were mainly distributed in northeast China and a few provincial capitals. While the low-value areas were mainly distributed in Huang-Huai-Hai Plain, southern hilly areas and southern Xinjiang. In 2015, the high-value and relatively-high-value areas showed the trend of spatial dispersion, but the scope of high-value units in the eastern region was expanded, and were distributed in the Yangtze River Delta and the Pearl River Delta; some areas changed from low-value areas to middle-value areas; and the western region was still dominated by low-value and relatively-low-value units (Fig. 5.13).

The spatial pattern of urban–rural relationship transformation was different from that of urban–rural socio-economic development transformation and urban–rural structural transformation. In 1990, the high-value and relatively-high-value areas of urban–rural relationship transformation were mainly distributed in the northeast three provinces, Inner Mongolia, Xinjiang and the southeast coastal areas, while the low-value and relatively-low-value areas were mainly distributed in the Huang-Huai-Hai Plain and the western region. By 2015, the high-value and relatively-high-value areas were mainly distributed in Xinjiang, the north of Heilongjiang, the eastern coastal areas and the central regions; while the low-value and relatively-low-value areas were still concentrated in the southwest and the Loess Plateau. In general, the spatial pattern of urban–rural relationship transformation in 2015 was similar to that in 1990, but the urban–rural synergy in Inner Mongolia, Jilin and other places decreased significantly (Fig. 5.14).

5.4.4 Trend of Urban–Rural Transformation

1. **Changes of urban–rural transformation**

Since the 1990s, the reform of economic system and the opening to the outside world have gradually deepened. Globalization, decentralization and marketization have become significant motivations for regional economic growth and regional differences (Wei 2001). From 1990 to 2015, the level of urban–rural socio-economic transformation and urban–rural structural transformation showed an increasing trend, which showed that the social and economic level of urban and rural areas was continuously improved, the centrality of urban and rural areas

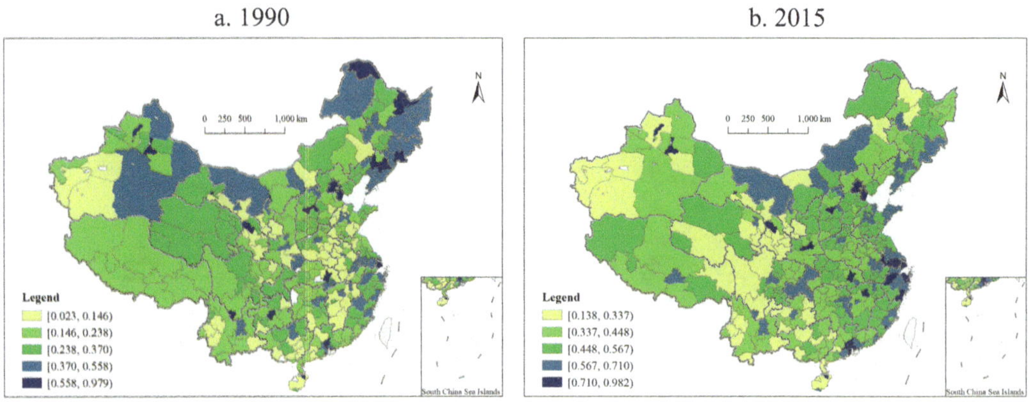

Fig. 5.13 Spatial pattern of urban–rural structural transformation in 1990 and 2015

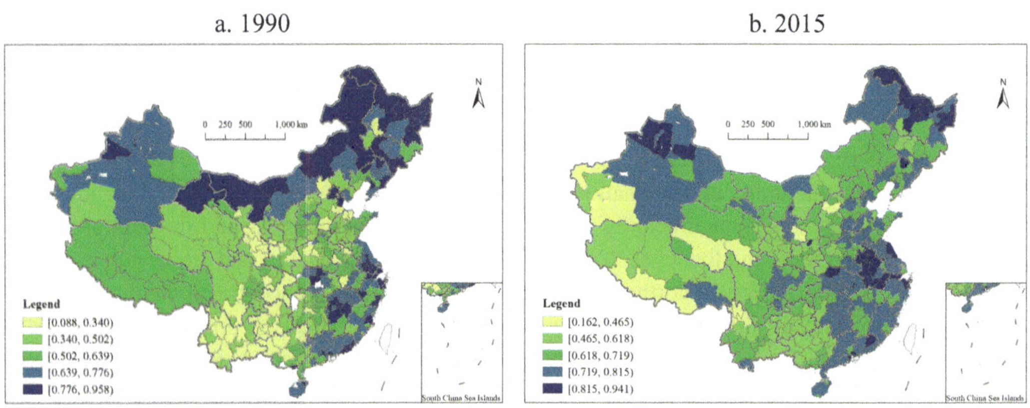

Fig. 5.14 Spatial pattern of urban–rural relationship transformation in 1990 and 2015

was constantly strengthened, and the output-value of non-agricultural industries, non-agricultural employment and the proportion of urban population increased. Due to the fact that the urban–rural dual management system has not been fundamentally broken, the elements flow between urban and rural areas is difficult to achieve equal exchange, the transferred rural migrant workers is hard to achieve citizenship, and the agricultural production security system and industrialization system are still not perfect, resulting in the widening gap between urban and rural areas. During this period, the urban construction land increased by 6.12 times, and the urban population increased by 1.55 times.

The spatial–temporal pattern of urban–rural socio-economic transformation indices shows an obvious trend of spatial agglomeration (Fig. 5.15). The high-value and relative-high-value areas of increase concentrated in the Yangtze River Delta, Pearl River Delta, Beijing-Tianjin-Tangshan, Shandong Peninsula in the eastern coastal areas, as well as central and western Inner Mongolia, eastern Xinjiang, northern Shaanxi and Ningxia. The medium-value of increase areas was distributed in the eastern coastal areas, the plain areas of central and western regions and the distribution areas of abundant energy and mineral resources in the northeast. The increase in southwest China, Qinghai-Tibet Plateau, southern Xinjiang, Liupan mountain area and Western Hunan is relatively small. Under the guidance of market and foreign capital, the township enterprises in the eastern coastal areas have obtained the opportunity to develop again through a series of measures, such as the reform of property rights system and the elimination of backward enterprises. Meanwhile, the rapid growth of urban economy helps the urban–rural industries establish close ties through the division of labor, which promotes the construction of development zones, and the scale effect of industrial agglomeration in cities and parks is also emerging. Except for some big cities, the main grain producing areas and ecologically fragile areas in the central and western regions are developing slowly, which makes it difficult to support the large-scale gathering demand of production elements. The monitoring report of population flow demonstrates that the rural population in these areas is still in the main trend of transferring to outside county.

Compared with urban–rural socio-economic transformation, the spatial–temporal pattern of urban–rural structural transformation show a different trend. The areas with rapid transformation of urban and rural structure are mainly concentrated in Yangtze River Delta, Pearl River Delta and some provincial capital cities, where the industrialization and urbanization have entered the second half of accelerated development. Thus, industrial transformation and upgrading, employment structure transformation and rural population transfer to cities and towns have become the obvious trends of urban–rural structure transformation. In some large urban areas, such as Beijing and Tianjin, where they are in the late stage of industrialization, the industrial structure changes little. However, with the increase of urban population and the expansion of urban scale, the spatial structure of urban and rural areas has changed significantly. Meanwhile, under the synthesis of multiple indicators, the change of urban–rural structure is less significant than that of other rapid urbanization areas. The low-value areas of the changes of urban–rural structural transformation were mainly distributed in Daxing'anling region, four prefectures in southern Xinjiang, Liupanshan Mountain area, western Sichuan Plateau, Yunnan-Guizhou Plateau and Hainan. Restricted by natural geography, history and culture and other factors, the economic growth in these areas is slow, and the power of the transformation of urban–rural structure is insufficient, resulting in the urban–rural structural transformation.

As to the urban–rural relationship transformation, there are 30.45% of the cities whose increase is less than 0, mainly distributed in northeast China, Qinghai-Tibet Plateau, southern Xinjiang, southern Jiangsu, Inner Mongolia and Jiangxi; and 69.55% of the cities have an increment of urban–rural relationship transformation greater than 0. Although there is a relatively high-degree of urban–rural relationship transformation in some developed areas, rural areas lag behind urban construction in the process of rapid industrialization and urbanization because of the lack of endogenous growth power cultivation. In some backward areas in the west, on the one hand, economic growth and structural transformation is slow, on the other hand, the driving effect of cities and towns on rural areas through technology diffusion and market demand is insufficient, resulting in the widening gap between urban and rural areas.

2. **Cold and hot spot analysis of the changes in urban–rural transformation**

To reflect the spatial heterogeneity of the changes in urban–rural transformation, this study used Getis-Ord G^* to identify different the hot and cold spots. Employing the ArcGIS software platform, the local G^* statistical value is calculated and divided into hot spot areas of 99% confidence interval, 95% confidence interval and 90% confidence interval, not significance, as well as cold spot areas of 90% confidence interval, 95% confidence interval and 99% confidence interval, respectively (Fig. 5.16). The hot spots of the increase of urban–rural socio-economic transformation are mainly distributed in the Yangtze River Delta and its peripheral areas, showing an obvious circle structure; while the cold spots are mainly distributed in the southwest. The hot spots of the changes of urban–rural structural transformation are mainly concentrated in Haixi area, t Bohai rim area, as well as theYangtze River Delta and its peripheral areas, while the cold spots are mainly distributed in Southwest China and southern Xinjiang. The hot

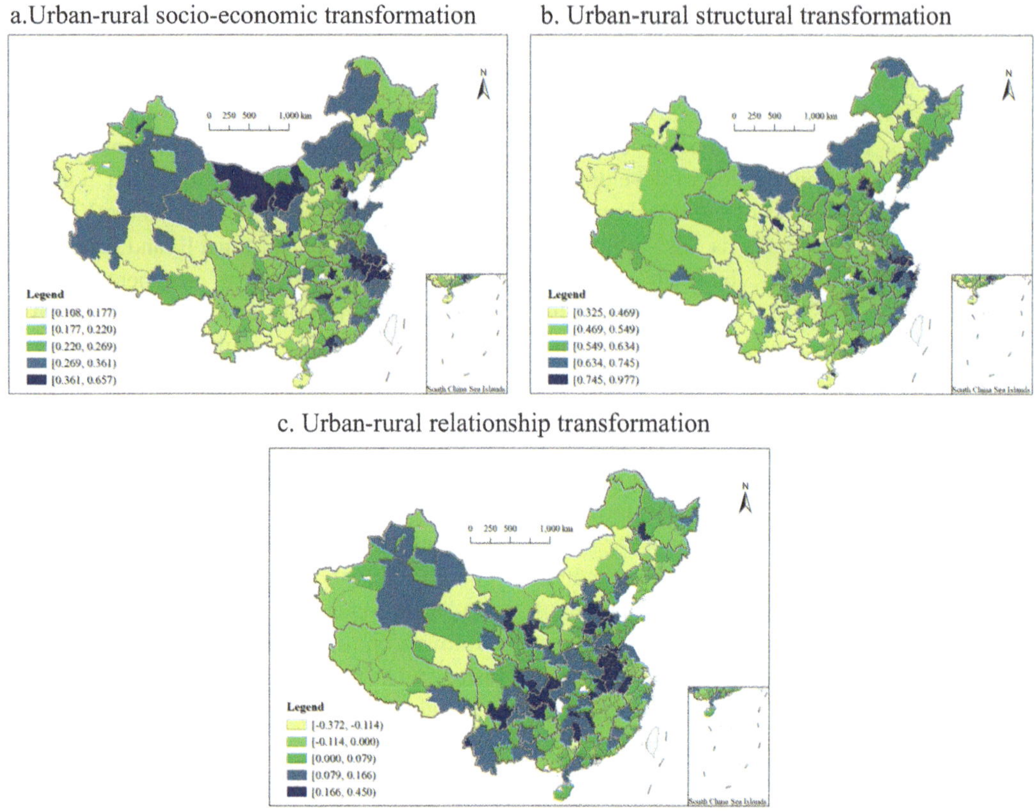

Fig. 5.15 Spatial pattern of the increment of urban–rural transformation in different dimension from 1990 to 2015

spots of the changes of urban–rural relationship transformation are mainly distributed in Sichuan Basin, the middle and lower reaches of Yangtze River Plain and Huang-Huai-Hai Plain, while the cold spots areas are mainly distributed in southern Xinjiang, Qinghai-Tibet Plateau, central of Northeast China and central Inner Mongolia.

5.5 Regionalization and Orientation of Urban–Rural Transformation

5.5.1 Comprehensive Regionalization of Urban–Rural Transformation

Regionalization is an important tool and means for geography to explore natural laws and guide national economic and social development. Based on China's natural geographical division, economic development, ecological environment and other basic conditions, this paper studies the macro and overall control of urban and rural transformation, and constructs the corresponding index system to cluster and overlay the transformation status of different dimensions. The regionalization index system of urban–rural transformation is an important basis for dividing the regional types of urban and rural development. Therefore, this study selects the natural and human factors that can fully reflect the urban and rural transformation in each region. Based on the comprehensive evaluation of urban and rural transformation in this Chapter, a comprehensive zoning index system is formed, which is composed of four dimensions, i.e., physical geographical environment, regional development conditions, urban–rural structure status and urban–rural relationship form (Table 5.6).

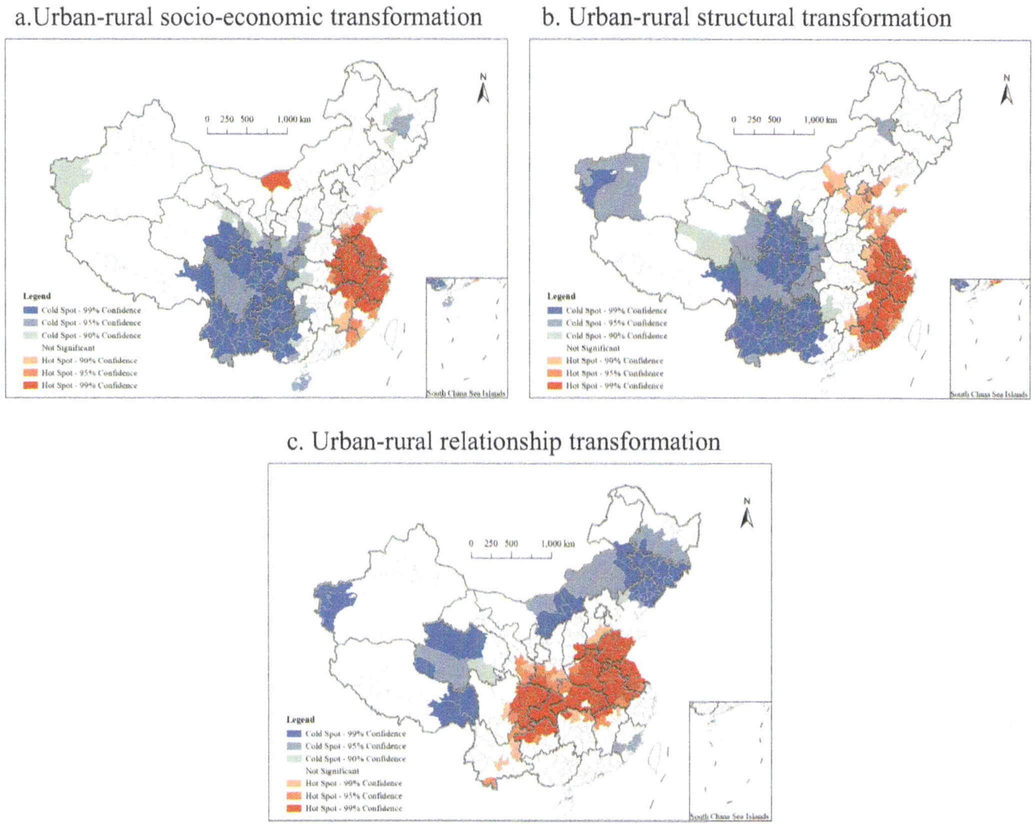

Fig. 5.16 Getis-Ord G* index of the changes in dimensions of urban–rural transformation in 1990 and 2015

Based on SPSS 18.0 and ArcGIS 10.2 software platform, this study takes the evaluation index system of urban and rural transformation as the clustering factor, and uses Ward cluster analysis method to divide the regional type of urban–rural transformation. After several times of quantitative adjustment, it is found that it is more appropriate to divide China into eight regional types of urban–rural transformation, which are basically consistent with the spatial pattern of economic growth, urbanization and urban–rural coordination degree (Liu and Yang 2012; Wang et al. 2016; Fang et al. 2015). Meanwhile, ArcGIS software is used to visualize the clustering results (Fig. 5.17). Further, based on the natural geographical environment, urban and rural development situation and the integrity and similarity principle of administrative division (Liu et al. 2018), the regionalization of urban–rural transformation is formed, which divides China into seven region and 19 sub-regions (Fig. 5.18). To represent the characteristics of the region, the level of economic and social development and the situation of urban and rural transformation are employed to name the region. Specifically, the level of economic and social development is divided into five categories, i.e., developed, relatively developed, medium, relatively backward and backward; the situation of urban and rural transformation can be divided into four categories, i.e., integration, composition, running-in and antagonism. Using the query function of ArcGIS, the statistical characteristics of the main indicators in each type area were obtained by sorting out the index data. Besides, combined with the existing agricultural comprehensive regionalization, economic regionalization, main function zoning, new urbanization planning and other achievements, the study analyzes the characteristics, main control factors

Table 5.6 Comprehensive regionalization index system of urban–rural transformation

Index	Maximum	Minimum	Average	Standard deviation
Per capita GDP (ten thousand yuan)	20.72	1.02	4.87	2.91
Per capita net income of rural residents (ten thousand yuan)	4.46	0.48	1.28	0.54
Per capita disposable income of urban residents (ten thousand yuan)	5.77	1.65	2.75	0.64
Education attainment (year)	11.71	3.47	8.75	1.10
Medical level (beds / thousand people)	19.54	1.45	5.00	1.64
Urbanization rate (%)	100.00	21.36	53.54	15.04
Employment structure	99.99	12.10	62.19	16.59
Industrial structure	99.96	51.37	86.48	8.65
Urban–rural land use structure	97.36	4.78	25.35	12.55
Urban–rural income gap	4.71	1.00	2.34	0.61
Urban–rural consumption gap	5.09	0.83	2.09	0.54
Industrial duality	20.61	0.22	5.18	2.75
Elevation (m)	5043	1	804	1029
Topography	5.96	0.00	1.22	1.29
Temperature (°C)	24	−6	12	6
Precipitation (mm)	2188	52	934	501

and obstacles of different urban–rural transformation regions.

Region I: Urban–rural running-in area with medium economic level in northeastern China

This region includes the three provinces in northeast China and northern Inner Mongolia. In general, the northeast is an important grain production base and forestry production base in China due to its vast territory with a sparse population and rich cultivated land resources. Here, large-scale production and specialized management are adopted to promote the cultivation of grain crops, unit labor productivity is much higher than other regions. For example, the average farming area of agricultural employees in Heilongjiang province is as high as 12 mu or 0.8 hectares. On the solid basis of agricultural production, the rapid development of agricultural processing, animal husbandry and logistics industry has promoted the transformation and upgrading of agriculture, and the development of urban and rural areas is highly coordinated. In recent years, Jilin Province and Heilongjiang Province have accelerated the development of agricultural product processing industry, which has become a regional leading industry and driven rural economic development to a certain extent.

As an old industrial base, northeast China enjoys relatively good conditions in industrial base and road infrastructure. Industrial development provides a large number of jobs for urban population and promotes economic growth and urbanization development. Among them, the industrial level and scale of Liaoning Province ranks first in the country. However, in the period of market economy construction, the northeast region is faced with problems such as low efficiency of state-owned enterprises, slow technological upgrading, difficulty in system transformation and brain drain. Although China has issued the strategy of revitalizing the northeast old industrial base to promote the transformation of traditional industries and the

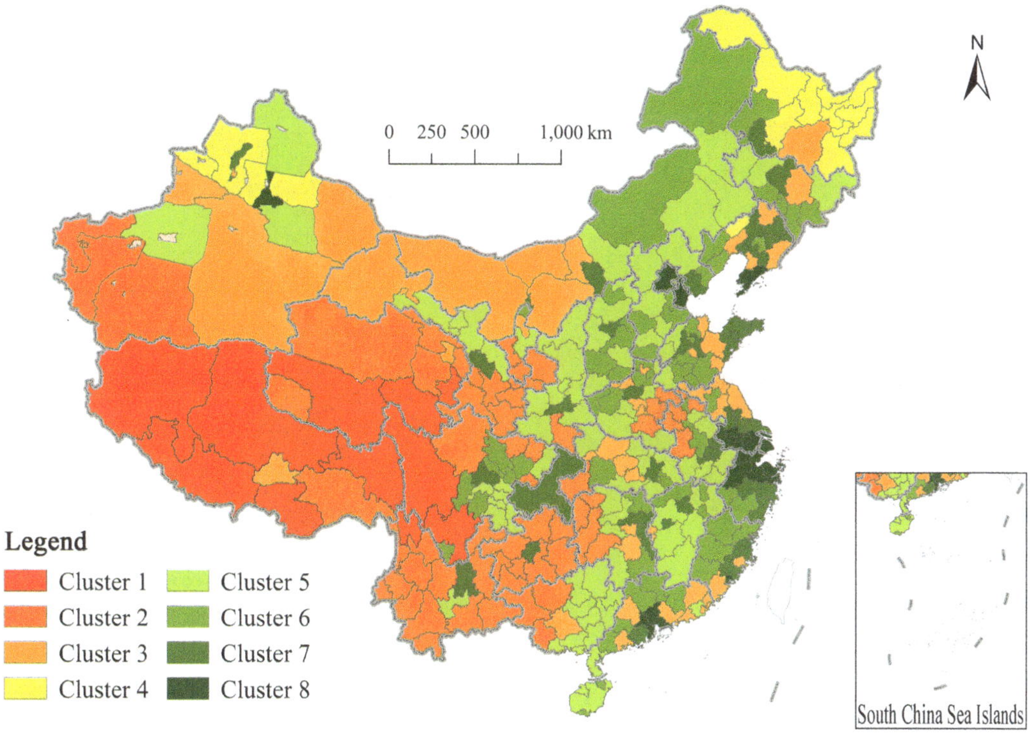

Fig. 5.17 Spatial distribution of the clustering of urban–rural transformation in China

improvement of regional infrastructure, the development of northeast China still faces problems such as ineffective management system and low efficiency of enterprises when compared with the eastern coastal developed regions and central region. Therefore, the urban–rural development is faced with severe regional functional decline, structural transformation difficulties and other challenges. From 1990 to 2015, the urban–rural economic growth and urban–rural structural change in northeast China were slower than those in other regions.

In terms of urban and rural economy and urban construction, this region can be divided into three subregions, namely, urban–rural integration area with a relatively developed economy in central and south of Liaoning Province, urban–rural integration area with a relatively developed economy in Harbin-Changchun urban agglomeration, and running-in area of urban–rural development with medium economic level in northeast agricultural area. In the long-term development process, northeast China has formed a T-shaped development spatial pattern of Harbin-Dalian and Harbin-Daqing, which further forms the main urbanization areas of central and southern Liaoning, Changchun-Jilin and Harbin-Daqing. Based on the primacy analysis of cities, the first cities in Jilin and Heilongjiang account for a high proportion in the provincial GDP and total population, but lack of small and medium-sized cities. And the economy in some counties and cities relies on the town economy, which has limited the driving capacity of cities on villages within the county.

Region II: Urban–rural integrated development area with developed economy in eastern coastal China

This region includes all or part of Beijing, Tianjin, Hebei, Shandong, Jiangsu, Shanghai, Zhejiang, Fujian, Guangdong, Hainan and other provinces. These areas are economically developed, with per capita GDP, per capita income of urban and rural residents ranking first in China.

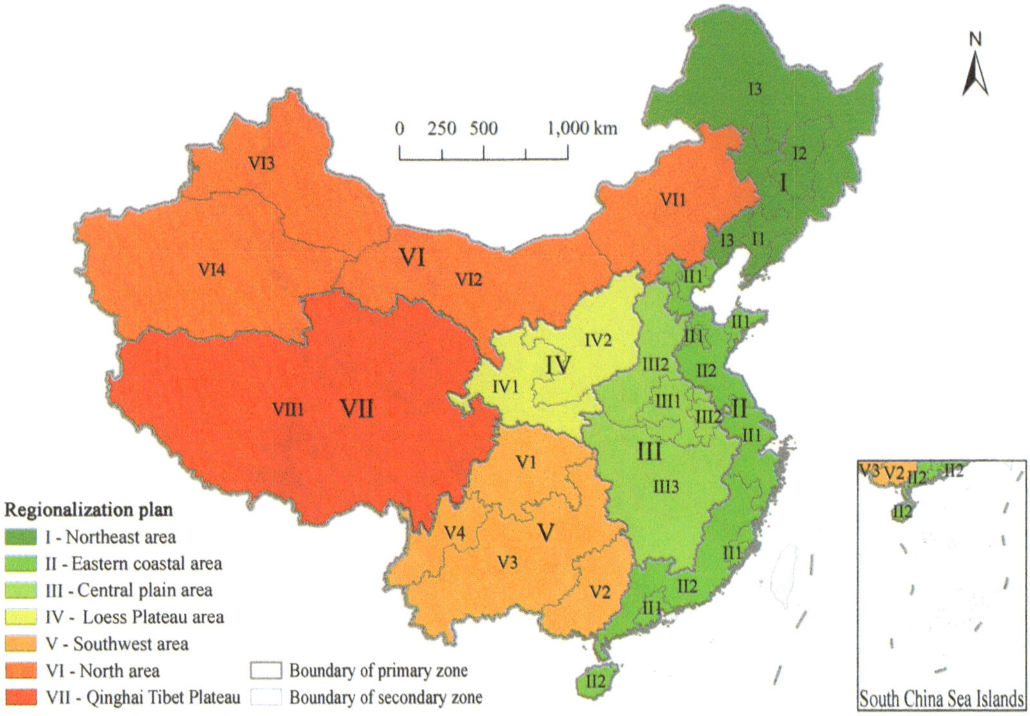

Fig. 5.18 Comprehensive regionalization of urban–rural transformation in China

Overall, they have entered the stage of post industrialization, and are the important economic growth engine and the main areas of population agglomeration. With 7.76% of the territory, these areas have gathered 33.18% of the total population and 47.92% of the GDP, which means the population density is the highest in China, reaching 618.62 people per square kilometer. The township enterprises, urban expansion and development zone construction in the eastern coastal areas have greatly promoted economic growth, and the structures of urban–rural industry and employment have entered a relatively advanced stage, which has attracted a large number of migrant workers and become the main area of population inflow across provinces in China. At the same time, the spatial pattern of urban and rural areas has changed greatly, which makes this region become a mature area of urban agglomeration and metropolitan area in China. Under the guidance of township enterprises and cities, traditional farming civilization and production and life style in rural areas are constantly transforming and reconstructing, forming a diversified development pattern of urban agriculture, agricultural industrialization, leisure industry and rural tourism. Meanwhile, the proportion of rural non-agricultural employment and agricultural labor productivity are very high, correspondingly, the income gap between urban and rural residents is small. However, under the traditional urbanization and industrialization mode, the rapid development of urban and rural areas causes serious problems such as the loss of cultivated land resources, the shortage of urban and rural reserve space, the shortage and pollution of water resources, and the human-earth relationship is tense.

According to the spatial structure and the degree of urban development, the eastern coastal area is divided into two subregions: the economically developed urban–rural integration area in the core city and the economically developed urban–rural integration area around the core city. The core urban area involves Beijing-Tianjin, Shandong Peninsula, Yangtze River Delta,

Fujian coastal areas and Pearl River Delta, forming urban agglomeration with Beijing, Tianjin, Qingdao, Shanghai, Shenzhen, Guangzhou and other mega-cities as the core. These areas have accelerated the construction of small towns in suburban counties to varying degrees, strengthened the investment in transport infrastructure, information network and environmental facilities, and promoted the spatial reconstruction of rural settlements. The peripheral areas of core cities are in the stage of rapid economic growth and urbanization development, and the urban–rural regional spatial patterns are changing significantly. These regions are close to developed areas, and have the advantage of space proximity in regional cooperation and industrial transfer, especially in the regional cooperation and urban–rural coordination within the province. For instance, the industrial transfer from the Pearl River Delta to the north and west of Guangdong has driven the economic growth and infrastructure construction in the eastern, western and northern regions of Guangdong.

Region III: Urban–rural running-in area with intermediate economy in central agricultural area

This region includes central and southern Hebei, and the central China such as Henan, Hunan, Anhui, Hubei, Jiangxi, and is mainly composed of Huang-Huai-Hai Plain and the middle reaches of the Yangtze River, and also includes some mountainous and hilly areas. The Huang-Huai-Hai Plain and the middle reaches of Yangtze River are important grain-producing areas with excellent agricultural climate, soil and topographic conditions and perfect agricultural irrigation facilities. Meanwhile, this region has good location conditions and natural geographical environment, and is an important area for bearing economic growth and population agglomeration. With 9.02% of the national land area, it carries 26.83% of the total population and 20.84% of the GDP, and the population density is second only to that of the eastern coastal area, reaching 430.74 people per square kilometer. Generally, the Central Plains Economic Zone with Zhengzhou as the center, Wuhan City Circle with Wuhan as the center, Changsha-Zhuzhou-Xiangtan Economic Zone and other population and economic agglomeration areas have been formed in this area, which have good foundations for economic development and good natural geographical environments for urban construction. Thus, they have become the advantageous areas of provincial factor agglomeration and industrial layout. However, as a traditional plain agricultural area, the economic development of counties and cities far away from big cities is relatively backward.

According to the location and economic development level, this region is divided into three sub-regions, i.e., the urban–rural antagonism area with backward economy in the transitional agricultural area of Henan and Anhui, the urban–rural running-in area with intermediate economy in agricultural area of Hebei-Henan-Anhui, and the urban–rural running-in area with intermediate economy in middle reaches of the Yangtze River. The first sub-region includes the southwest of Henan province and the north of Anhui province. It is far away from big cities, and its social and economic development lag behind, showing the characteristics of "large grain producing county, weak industrial county and poor financial county". Due to the lack of support from county economy and township enterprises, there are many national poverty-stricken counties in this sub-region, the most typical of which is Huaxian county in Henan Province. Generally, the grain production function of the central plain agricultural area is strong, and the grain production and economic development are competing fiercely for urban and rural regional space. However, due to the strict constraints of farmland protection and regional function orientation, economy in many counties develops slowly, and large-scale farmers go out for work, which leads to the lack of the endogenous mechanism for rural growth. With the continuous outflow of rural labor force and land system restriction, the phenomenon of rural hollowing is becoming more and more serious.

Region IV: Urban–rural running-in area with backward economy in Loess Plateau

This region mainly involves Shanxi and Shaanxi, the eastern part of Gansu, and the southern part of Ningxia. This region is located in the Loess Plateau, with complex terrain and fragile ecological environment. Relying on the unique natural conditions, it has developed characteristic crops such as economic forest and fruit and grains, which is an important farming pastoral area in China. Meanwhile, the region is also an important distribution area of coal, oil and other energies and mineral resources in China. Regional economic growth is heavily dependent on energy and mineral resources, and there are many energy and mineral enterprises and many state-owned enterprises, lack of small and medium-sized enterprises and energy substitution enterprises, and the market system is not perfect. Due to the outflow of resource income, resource-based enterprises have limited driving force on local economy and fiscal revenue, resulting in industrial duality and large income gap between urban and rural residents. In some areas, fossil energy is exhausted, and the economy is in a downward trend. Due to the ravines and weak infrastructure, it is difficult to expand the towns and industrial parks in this region. In rural areas, due to the barren land and poor agricultural production conditions, agricultural returns is low, the driving force of urban and economic development is insufficient, which results in the low income of rural residents and the low proportion of non-agricultural employment.

According to the difference of economic development level and urban–rural transition state, this region is divided into two sub-regions: the urban–rural antagonism area with backward economy in Qinba-Liupanshan mountain area and urban–rural running-in area with intermediate economy in Shaanxi-Shanxi Loess Plateau. The former is located in intersection area of Loess Plateau, Qinling Mountains and Liupanshan mountains, with big mountain, deep gullies, and rugged terrain. The human-earth relationship is tense and the regional economy is seriously backward, which makes this area become a concentrated and contiguous areas with special difficulties in China. The per capita net income of rural residents is only 4607 yuan, which is only 36.82% of the national average level. The latter is mainly distributed in Shanxi and Shaanxi, where the economic and social development is greatly constrained by topography and ecological environment. Since the implementation of the policy of returning farmland to forest protection, the ecological environment of the Loess Plateau has been improved continuously, and the area of soil erosion area has decreased year by year. In recent years, with the expiration of the policy, Shaanxi has carried out the gully control and land reclamation project with land consolidation as the main content in the Loess Plateau to coordinated grain production and ecological protection, and promotes rural spatial reconstruction through the unified planning of rural communities.

Region V: Urban–rural antagonistic area with backward economy in southwestern China

This region mainly involves Yunnan, Sichuan, Chongqing, Guizhou, Guangxi and other southwest regions. With a high average altitude, undulating terrain and fragile ecological environment, this area is an important ecological conservation area in China, and is also concentrated and contiguous areas with special difficulties in China. Restricted by lagging behind in social and economic development, the rural areas are still dominated by traditional agricultural production, and the conversion of urban and rural employment is insufficient. The proportion of non-agricultural industries and non-agricultural employment are at a relatively low level in China, and the urbanization rate is only 47.82%.

According to the difference of terrain and economic development, this region is divided into four sub-regions: urban–rural running-in area with intermediate economy in Sichuan Basin, urban–rural running-in and antagonistic area with relatively backward economy in eastern Guangxi, urban–rural antagonism area with backward economy in Yunnan-Guizhou Plateau, urban–rural antagonism area with backward economy in Yunnan-Sichuan Plateau. Generally, the Yunnan-Guizhou Plateau is characterized by complex terrain, large fluctuations, and backward economy, and the economic and social indicators of urban and rural areas are lower than those of other regions in China, and the urban development gap is large. Chongqing, eastern Sichuan and eastern Guangxi are dominated by plains and hills, and the natural geographical environment there is better than other areas in southwest China, forming developed areas such as Chengdu-Chongqing economic zone, Guangxi Liangjiang economic zone, which are concentration areas of production factors and rapid development of cities and towns.

Region VI: Urban–rural running-in area with intermediate economy in northern China

This region mainly distributes in Inner Mongolia, western Gansu, northern Ningxia, and Xinjiang, including Inner Mongolia Plateau and Taklimakan Desert. It is located in arid and semi-arid and arid areas, which is an important pastoral area and ecotone of agriculture and animal husbandry. Water resources and transportation facilities are the restricting factors for the development of this region. With 30.39% of the national territory, there are 5.13% of the national population concentrating here, and the population density is only 23.43 people per square kilometer. Due to its rich coal, oil and other mineral resources, the region has a high level of economy development and urbanization. At the same time, driven by characteristic agriculture such as cotton, fruit and animal husbandry, the average added-value of agricultural labor reaches 27,374 yuan, far higher than that in other western regions, and the gap between urban and rural development is small.

According to the economic level and topographical conditions, the region can be divided into four sub-regions: urban–rural running-in area with intermediate economy in Inner Mongolia and the pastoral areas along the Great Wall, urban–rural running-in are with relatively developed economy in the agricultural and pastoral areas of Inner Mongolia-Gansu-Ningxia, urban–rural running-in area with intermediate economy in northern Xinjiang pastoral area, and urban–rural antagonism area with backward economy in southern Xinjiang pastoral areas. Among them, northern Xinjiang is rich in per capita cultivated land resources, and the oasis agriculture is developed, becoming an important cotton and fruit producing area in China. Although the conversion of rural non-agricultural employment is slow, the average productivity of agricultural labor is high. Driven by the development of petroleum industry, the overall economic level of this region is relatively high. In terms of the southern Xinjiang, there is a vast area with few people, widespread deserts, serious shortage of water resources and poor ecological environment. Combined with the destruction of extreme religious forces, the economic development of this region lags behind, and the urbanization level is as low as only 26.80%.

Region VII: Urban–rural antagonistic area with backward economy in Qinghai-Tibet Plateau

The region mainly includes Tibet, Qinghai, northern Yunnan and western Sichuan. The Qinghai-Tibet Plateau is located in the first stage of the terrain, with high altitude, complex-varied terrain, dry-thin air, and fragile ecological environment, and the population density is only 5.37 people per square kilometer. The region is dominated by natural features, with low human influence. Agricultural production accounts for 10.48% of the local economy, and the level of agricultural labor productivity is low. More than 70% of the rural population being engaged in traditional plateau farming and animal husbandry. Affected by the natural environment, the level of economic development, the proportion of non-agricultural industries and the proportion of non-agricultural employment in this area are far lower than the national average level, and the urbanization rate is only 40.59%. The rurality occupies a dominant position in urban–rural regional space, and the level of population agglomeration is low. The income ratio of urban and rural residents reaches 3.00:1, which is far higher than that of other areas in China.

5.5.2 Mechanism of the Difference of Urban–Rural Transformation

The evolution of urban–rural regional system is influenced by many factors, mainly including industrialization, urbanization, regional investment, human capital, foreign investment and regional economy. However, due to the regional differences in natural conditions, development foundation and institutional environment between regions, as well as the stage of economic and social development, the dynamic mechanism of urban–rural transformation in different regions is quite different. Based on the comprehensive regionalization of urban–rural transformation, this section analyzes the main driving mechanisms of each region.

The physical geographical environment is the basis of the direction, speed and capacity of urban–rural transformation, especially in the primary stage of urban–rural transformation. With economic growth, industrial development and institutional changes, the changes of urban–rural regional system are more affected by human activities. This section mainly analyzes the influence of economic, social and institutional changes on the regional difference between urban and rural transformation and its internal mechanism. Referring to the researches of Ou et al. (2008) and Chen et al. (2009), the economic and social driving forces of urban–rural transformation are divided into administrative force, market force, endogenous force, outward force and agriculture-drive force.

Administrative force refers to the role of government power in promoting the transformation of urban and rural development, mainly involving urbanization and economic growth led by local government. The government promotes economic growth, employment transformation and urban–rural coordination through fiscal input, industrial layout, engineering projects and public service facilities. Both the national and provincial governments have strong administrative resources, especially the direct investment through industrial support policies and infrastructure construction and the direct transformation of rural areas to urban areas through administrative division. This study employs the per capita fiscal expenditure on agriculture, forestry, animal husbandry and fishery to represent administrative power, which reflects the local government's ability to promote rural development.

Market force refers to the urbanization and economic growth under the guidance of market mechanism, and is the basic driving force of regional urbanization and urban–rural regional transformation. It requires that the various production factors and their regional combination should be allocated effectively according to the law of market economy, which is manifested in the in the transfer of rural production factors from agricultural and rural areas to non-agricultural industries and cities under the comparative interests, which promotes the spatial agglomeration of production factors. In general, market force is represented by total retail sales of consumer goods.

Industrialization is the initial driving force of urban and rural economic growth and structural transformation. The development of township enterprises has promoted the rapid development of eastern coastal areas in China and has been the main reason for the differences in rural areas. With the construction of development zones, the division of traditional urban industry and township industry is meaningless, while the industrial agglomeration area with development zones as the carrier has become an important driving force for regional economic growth. Therefore, the output-value of the secondary industry is employed to represent the level of urban and rural industrialization in this study.

The outward factor mainly refers to the impact of globalization on urban and rural development. Since the implementation of China's reform and opening-up policy, foreign investment and foreign trade have become important driving factors of regional economic development. In China, the foreign investment is mainly distributed in the eastern coastal areas, which has obvious regional characteristics and affects the national import–export pattern. At present, foreign trade still plays an important role in local economic growth, population agglomeration and industrial upgrading. Here, the ratio of import and export trade in GDP is used to reflect the magnitude of outward force.

As the basic industry of urban–rural social and economic development, agriculture plays an irreplaceable role in employment promotion, social stability and rural development, and the productivity improvement and organizational development of agriculture provide a large number of labor forces for industrial development and urban construction. In underdeveloped areas, agriculture is still an important driving force of national economic development. Thus, the per capita gross output value of agriculture, forestry, animal husbandry and fishery is used to represent the agriculture-drive force.

Urban–rural transformation is the unity of regional conditions, transformation structure and coordinated effect. Here, the weighted sum of the three is used to represent the comprehensive state of regional urban–rural transformation. The calculation formula is as follows:

$$URT = 1/3 \times URD + 1/3 \times URS + 1/3 \times URE \quad (5.1)$$

where *URT* denotes the comprehensive state of urban–rural transformation, *URD*, *URS*, and *URE* represents the level of urban and rural economic and social development, the state of urban and rural structure, and the urban–rural coordinated effect, respectively.

Based on EVIEWS and partial correlation analysis, the driving forces are analyzed in six urban–rural transformation regions except for the urban–rural antagonistic area with backward economy in Qinghai-Tibet Plateau. Then, the partial correlation coefficient is obtained, and T-test is selected for bilateral test. The value of T-test less than 0.01 is defined as significant correlation, and those less than 0.05 and 0.1 are relatively significant correlation and weak correlation, respectively.

Results show that except for the administrative force, the other five driving forces hold a significant impact on urban–rural transformation at the national level (Table 5.7). Among them, the influences of market force and agriculture-drive force are the most significant, which indicates that the opening of domestic market and the establishment of market mechanism are conducive to urban–rural transformation. Agricultural modernization can alleviate the urban–rural contradiction, and then boost the transformation of urban and rural areas. Outward force has a relatively obvious influence, indicating that participation in global trade can narrow regional differences in urban–rural transformation. The endogenous force of industrialization also promotes urban–rural economic and social changes, but the partial correlation coefficient is significantly smaller than those of other driving factors.

By comparing the effects of different driving forces on each region, it can be found that the outward force is most obvious in the eastern coastal China. The endogenous force of industrialization is more obvious in the central and western China, especially in the western region. The driving force of market to all regions is very significant. The agriculture-drive force is the most significant factor in eastern China and Loess Plateau. Fiscal expenditure on agriculture varies greatly among different regions, which reflects the regional difference and complexity of the effect of the government's macro management in urban–rural transformation. Restricted by terrain, ecology and other factors, the cost of infrastructure construction and social services is high in central and western China, and the efficiency of enterprise capital utilization is also lower than that of developed areas due to the poor economic foundation. In this context, the fiscal expenditure in these regions is largely obtained through the central transfer payment, and the per capita fiscal expenditure is relatively high. For example, the per capita fiscal expenditure related to agriculture, rural areas and farmers in the Loess Plateau and southwest China reached 622 yuan and 600 yuan respectively in 2015, while the national average was only 568 yuan. Despite the support from the central government, the level of urban–rural transformation in these regions is still low, and it is facing prominent problems and contradictions in urban and rural development.

Table 5.7 Partial correlation analysis of driving forces in different urban–rural transformation regions

Region	Correlation and significance	Driving forces				
		Administrative force	Market force	Endogenous force	Outward force	Agriculture-drive force
China	Correlation (Partial correlation coefficient)	−0.086	0.742***	0.284***	0.567***	0.625***
	Significance (Bilateral test)	0.117	0.000	0.000	0.000	0.000
Region I	Correlation (Partial correlation coefficient)	−0.018	−0.067	−0.155	0.292*	0.190
	Significance (Bilateral test)	0.914	0.969	0.361	0.08	0.261
Region II	Correlation (Partial correlation coefficient)	0.268**	0.784***	0.201*	0.662***	0.634***
	Significance (Bilateral test)	0.018	0.000	0.077	0.000	0.000
Region III	Correlation (Partial correlation coefficient)	0.161	0.720***	0.336***	0.323***	0.426***
	Significance (Bilateral test)	0.174	0.000	0.004	0.005	0.000
Region IV	Correlation (Partial correlation coefficient)	−0.463***	0.838***	0.459***	0.309*	0.769***
	Significance (Bilateral test)	0.008	0.000	0.008	0.085	0.000
Region V	Correlation (Partial correlation coefficient)	−0.487***	0.547***	0.438***	0.243**	0.392***
	Significance (Bilateral test)	0.000	0.000	0.000	0.029	0.000
Region VI	Correlation (Partial correlation coefficient)	−0.204	0.634***	0.441***	0.300*	0.475***
	Significance (Bilateral test)	0.241	0.000	0.008	0.080	0.004

Note ***means significant at 0.01 level, **means significant at 0.05 level, and *means significant at 0.1 level

In different regions, there are significant differences in the driving mechanism of the leading forces of urban–rural transformation. Region I is weakly driven by the outward force, but not related to other driving forces. It may be related to the regional internal development environment differences. Although its urban–rural transformation is the result of multiple driving forces, they are not statistically significant. Region II are significantly affected by the five driving forces, especially the market force and outward force, which are consistent with the actual situation of economic and social development in the eastern coastal areas. Except for the administrative force, the impact of the other four forces is significant in region III, especially the market force and agriculture-drive force. This is mainly because agriculture still plays a very important role in the economic growth and social employment in the agricultural area of the central plain. The five forces play a significant role in region IV, among which the market force and agriculture-drive force are stronger than others. However, the administrative power holds a negative effect on urban–rural transformation, which means that the level of urban–rural transformation is higher in areas with low financial revenue. This may be related to the exploitation of regional energy and mineral resources since its weak driving effect on local social and economic development. Region V is significantly promoted by the five forces, the most obvious is market force and endogenous force, and the administrative power is the same as region IV, showing a significant negative impact. In addition to the administrative power, the role of the other four forces is significant in region VI, especially the market force and agriculture-drive force.

To accelerate the transformation of urban and rural areas, there are significant differences in the measures to be taken in different regions. In region V and VI, more attention should be paid to market reform, improving the role of market mechanism in urban–rural transformation, and agricultural modernization, forming characteristic agricultural production system and raising consumption level of urban and rural residents. The level of industrialization and agricultural modernization in region III should be further improved to accelerate the adjustment of industrial structure and increase the proportion of the secondary and tertiary industries. The regulation and control of local government is an important means to promote urban–rural transformation. In addition to region II, there is a lack of effective interaction between local government revenue and urban–rural development in other regions, such as low degree of urban–rural transformation and high government revenue or high degree of urban–rural transformation and low government revenue. Since Yunnan-Guizhou Plateau, Qinghai-Tibet Plateau and Qinba mountain area in central and western regions are important concentrated and contiguous areas with special difficult in China, it is necessary to increase financial transfer payment. Meanwhile, in the energy mining area, heavy chemical industry area and state-owned enterprise cluster area, it is urgent to straighten out the relationship between enterprise development and local government and residents, striving to solve the dilemma of "enriching areas but not the people".

5.5.3 Development Orientation of Different Urban–Rural Transformation Regions

Based on the comprehensive influence of physical geography, social economy and institutional environment, there are corresponding paths and models for urban–rural transformation in different regions. Based on the analysis idea of "obstacle factors, transformation basis, optimization countermeasures", this study puts forward optimization measures to promote urban–rural coordinated development in different regions according to local conditions (Fig. 5.19). The obstacle factors are divided into natural geographical environment constraints, economic and social constraints and Institutional constraints; the transformation basis is composed of regional characteristics, economic conditions and urban–rural relationship; and the optimization countermeasures focus on public resources allocation, industrial coordination, land use, and village-town pattern.

Region I: Urban–rural running-in area with medium economic level in northeastern China

The urban–rural transformation in northeastern China faces multiple unfavorable factors, which are mainly manifested in slow regional economic growth and rigid institutional mechanism. In terms of physical geographical environment, the climate in the east of northeastern China is cold and the period of agricultural production is short; the spatial distribution of water and soil resources does not match, the water resources in Jilin and the central and south of Heilongjiang are less, and the construction of farmland and water conservancy facilities is insufficient. Due to the excessive use of cultivated land resources, black land degradation and soil quality decline are serious, which are not conducive to the sustainable development of agriculture. In terms of economic development, the proportion of heavy industry and state-owned enterprises is too high, the local economy relies too much on heavy chemical industry and machinery manufacturing industry, resulting in low economic benefits, and many old industrial cities have

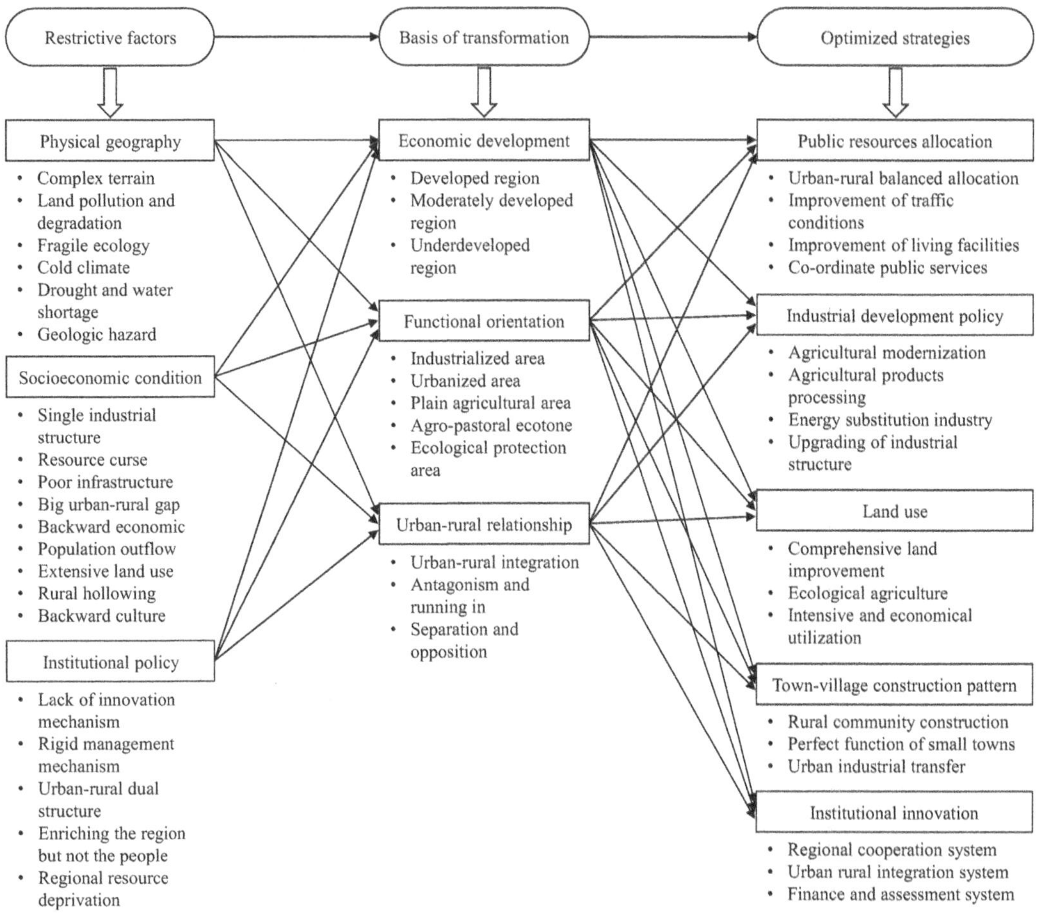

Fig. 5.19 Optimized path of urban–rural transformation in different regions

become resource-exhausted cities; due to the inflexible government management and enterprise management, lack of market competition consciousness and enterprise innovation culture, small and medium-sized enterprises are few, and new urban employment is different. In recent years, the economic growth of northeast China is slow, the economic growth rate is in the last few places of China, and the population outflow is serious. In terms of regional coordination, the central and southern Liaoning urban agglomeration and Harbin-Changchun urban agglomeration are the main agglomeration areas of machinery manufacturing and heavy chemical enterprises in northeastern China, while the cities and towns in the peripheral areas are small in scale and lack of industrial support, and are seriously insufficient in population agglomeration and public service supply; enterprises in some districts and counties are gradually moving to major cities, which weakens the development of county economy, thus there is no division of labor between central and peripheral regions. In terms of institutional obstacles, the development of rural cooperative organizations in northeast China develops slowly, the market system of agricultural products is not perfect, and the self-regulating function of grain sales and processing is lack, which is greatly affected by the national grain protection price and new policies; in addition, the structure of agricultural industry is single, the development of economic crops and animal husbandry is slow.

Stabilizing agricultural development and building national grain production base are strategic choice for rural development in north-eastern China, which requires effective protection of cultivated land and basic farmland. Driven by the strategy of revitalizing the old industrial base in northeast China, measures have been taken to integrate the heavy industrial structure and implement innovative development strategy. Therefore, the trend of non-heavy industry has been continuously strengthened, which provides a foundation for promoting the industrial development in rural areas. Correspondingly, the urban–rural coordinated development in the future should focus on the following aspects. First, it is necessary to accelerate the development of agricultural products processing industry, forming a pattern of mutual promotion between the production of high-quality and base-type commodity grain and the agricultural products processing. Second, it is urgent to improve the rural financial and agricultural socialized service system, and speed up the scale and organization of agricultural production. Third, it is needed to improve the public service functions to promote the development of small and medium-sized towns in the county, form small towns with the ability of factor aggregation, and develop a number of industrial, resource-based and ecological small towns according to local conditions, which will help to build a platform for two-way free flow of urban and rural elements. Forth, more attention should be paid to reforming the administrative management system, stimulate market vitality and form a competitive atmosphere. Fifth, it is urgent to implement the talent strategic innovation plan to enhance the regional innovation ability.

Region II: Urban–rural integrated development area with developed economy in eastern coastal China

The eastern coastal areas are the main areas of population agglomeration and economic activity agglomeration in China, where the intensity of land development is high, and the level of economic and urban–rural relationship is high. However, the sustainable development of urban and rural areas faces the following constraints. First, water resources needed by urbanization and agricultural production in the northern part of the region are seriously insufficient; the per capita cultivated land resources are lower than the national average level, and the cultivated land resources are seriously lost in the process of industrialization and urbanization; the development of industrial enterprises has caused serious environmental problems, especially the serious pollution of rivers and lakes. Second, with the economic growth and the spatial agglomeration of population, the demand for urban and rural construction land is strong, but the rapid development and disorderly spread of land space led to the lack of urban expansion space. Third, with the transfer of rural population to urban areas, the problem of rural hollowing is prominent, and the land use efficiency of rural residential areas is low. Forth, there is a large gap between the core cities and the peripheral areas, the industrial connection and industrial transfer in the peripheral areas are insufficient, which means there is a lack of coordination mechanism among regions.

The eastern region has entered the stage of urban–rural integration, but it still faces the problem of declining coordination between urban and rural areas. Therefore, this region should make full use of the advantages of good location and strong economic strength, promote rural development with cities and boost agricultural development with industry, and further enhance the level of urban–rural coordinated development. First, it is necessary to accelerate the transformation and upgrading of the industrial structure of core cities, improve their innovation abilities, and enhance their competitiveness of participating in the international division of labor. Also, the surrounding areas should actively undertake the industrial transfer of core cities, thus improving the level of regional division and cooperation. Second, the urban system should be improved, and the middle level urban nodes should be developed, forming a city-town-village system which is guided by big cities, relied on by regional central cities and secondary central cities, and supported by small towns, central villages and communities in the county, so as to build an urban–rural integration platform. Third, more attention should be paid to agricultural industrialization, promote the moderate scale operation of agriculture and the intensive processing of agricultural products, improve the social service ability of agriculture. At the same time, based on the needs of urban development, business format such as leisure agriculture and rural tourism should be actively developed to promote the multi-function of agriculture. Forth, it is necessary to establish the concept of intensive and economical land use under the background of rural population outflow, and promote comprehensive governance of hollow village and rural community construction through land consolidation and overall planning of urban and rural construction land.

Region III: Urban–rural running-in area with intermediate economy in central agricultural area

The main problems of urban–rural transformation in this region include the following three points. First, it is difficult to realize large-scale operation of agricultural production because the per capita cultivated land area is only one mu, and as the main grain producing area, the land demand of industrial enterprises and urban construction is restricted by the policy of cultivated land protection. Second, agriculture has a high proportion in both regional output value and employment, while the industrialization level and urbanization rate are low, which makes it difficult to effectively absorb the population transferred from agriculture. Therefore, it is urgent to solve the dilemma of large grain producing counties and poor fiscal counties. Third, with the population going out to work and settlements expanding outward, the phenomenon of rural hollowing is serious, resulting in the inefficient use of land resources. In the plain area of the middle reaches of the Yangtze River, the population migration has changed rice planting from two seasons to one season, and some cultivated land has been abandoned seriously. Under the comprehensive effect of these factors, the traditional agricultural areas have formed a cycle of "slow economic growth, rural population migration, weakening of rural subject, backward rural development", which hinder the sustainable development of agriculture and rural areas.

This region is an important grain production base in China. It is necessary to fully coordinate the relationship between agricultural industrialization, urbanization and industrialization based on clearing positioning, and promote rural development through the comprehensive role of internal and external forces. First, focusing on the positioning of main grain producing area, this region needs to actively seek the support of central finance, increase investment in infrastructure and public services, thus improving the support for regional economic and social development. Second, measures should be taken to strengthen the contact with regional central cities and the eastern coastal areas, actively undertake the industrial transfer in these areas, centralize the distribution of industries in counties and towns with a good economic foundation, develop agricultural products processing, and strengthen cooperation before, during and after production. Third, this region should make good use of national grain production supporting policies, improve the agricultural socialized service system, and promote moderate scale operation of agriculture. Forth, the renovation of hollow villages should be combined with community construction. Based on the objective reality of hollowing development of rural areas in Central China, it is urgent to promote village relocation and consolidation. In particular, rural settlements should be renovated based on community construction. Thus, it will help to optimize the spatial layout of settlements and improve the hierarchical system of county town, key towns, central towns and central villages (Liu et al. 2014a).

Region IV: Urban–rural running-in area with backward economy in Loess Plateau

The main problems of urban–rural development in this region mainly include the following four points. First, the complex topography and ravines crisscross in this area are not conducive to the development of agricultural production and urban construction; and this region is a typical ecotone of agriculture and animal husbandry in China, where the ecological environment is fragile and soil erosion is serious. These factors seriously restrict agricultural industrialization, industrial enterprise development and urban construction. Second, agricultural production in this region is extensive, production income is low, and the degree of agricultural industrialization is not high. Third, due to the poor infrastructure, the regional accessibility is low. Forth, economic growth is heavily dependent on energy and mineral resources such as coal and oil, and there is a lack of alternative industries. Under the background of the decline of international energy and mineral resources prices, the economic growth rate has declined rapidly.

Because of the fragile ecological environment, the Loess Plateau should develop ecological, environmental protection industries and ecological characteristic agriculture based on the functional localization, and reconstruct urban–rural space through measures such as land consolidation. First, it is necessary to accelerate industrial transformation and upgrading, actively develop coal chemical industry and petroleum intensive processing industry, and actively develop energy substitution industry. Forestry, fruit and coarse grain industry with plateau characteristics should further be developed to enhance the value of ecological and local agricultural products, and it is needed to accelerate the development of agricultural products processing industry and tourism industry. Second, focusing on the positioning of national production functional areas, this region should speed up infrastructure construction and improve the education and training system through financial transfer payment and special construction investment, thus enhancing regional accessibility and endogenous growth momentum. Third, Qinba-Liupanshan poverty-stricken areas should actively promote the implementation of targeted poverty alleviation strategy, make differential policies to boost rural development, and ecological migration should be adopted to improve farmers' production and living conditions in areas with poor ecological environment. Forth, it is necessary to strengthen the combination of ecological environment protection and comprehensive treatment of degraded land and inefficient land use, continue to encourage the project of returning farmland to forests and enriching the people with ecology, and improve farmland quality through land consolidation in gullies, thus promoting agricultural mechanization production. Meanwhile, measures should be taken to comprehensively renovate the abandoned pits, kilns and ditches under the background of rural population outflow, and reconstruct rural space by coordinating farmland construction, industrial development and rural community construction.

Region V: Urban–rural antagonistic area with backward economy in southwestern China

There are four practical problems in the development of urban and rural areas in Southwest China. First, the topography is complex and diverse, and its ecological environment is fragile, which is not suitable for large-scale agricultural development and urban construction. Second, the proportion of agricultural economy and employment is high, and the development of industrialization is slow due to the constraints of natural conditions and ecological environment. Third, the level of urbanization is low, which results in the slow spatial agglomeration of factors and development of service industry. Forth, regional accessibility is low since the lag of infrastructure construction.

Southwest China is an important ecological functional area, concentrated contiguous poverty-stricken areas with special difficulties, and minority areas. To solve the dual structure of urban and rural areas and industry and agriculture in this region, efforts should be made to promote regional economic development and urban–rural coordinated development by means of measures such as main functions, improvement, industrial transformation and improvement of public services. First, it is needed to speed up the development of high-efficiency, high-quality and ecological agricultural planting industry with plateau and mountain characteristics, accelerate the development of animal husbandry, forming a comprehensive three-dimensional ecological agriculture and improving the ternary structure of cash crops, feed crops and grain crops. Second, it is necessary to develop agricultural products processing industry according to local conditions, and encourage the development of clean and environment-friendly industrial enterprises. Third, measures should be taken to develop rural tourism and industries with national characteristics based on the natural scenery and national culture. Forth, this region should strive for the support of national finance and special finance based on the orientation of ecologically functional zones, thus improving the infrastructure and enhancing regional accessibility. Fifth, relying on large regional cities, it is urgent to accelerate the development of small and medium-sized cities and towns, improve the service functions of small towns, and attract farmers and herdsmen to gather in central villages and towns.

Region VI: Urban–rural running-in area with intermediate economy in northern China

The main problems of urban and rural development in the northern region are as follows. First, the climate in this region is mainly characterized by drought and semi-drought, and the shortage of water resources is the main constraint factor of regional land development and utilization, industrial development and urban construction. Meanwhile, the ecological environment is fragile, and excessive reclamation causes land desertification, overgrazing causes grassland degradation, and excessive irrigation causes river drying up. Second, the transportation infrastructure is poor, which causes the low regional accessibility and convenience. Third, economic growth is highly dependent on energy and mineral resources industry, and the proportion of state-owned enterprises is high. Forth, the per capita cultivated land area of rural residents is large, but the distribution of rural settlements is scattered, and the public service function of villages and towns is not strong or even seriously missing.

This region is an important agricultural and pastoral area and energy and mineral resources area in China, which is constrained by water shortage and ecological degradation. Based on the regional characteristics of soil and water resources and mineral resources endowment, the paths for urban–rural transformation in this region mainly include the following. First, efforts should be made to speed up the development mechanization of agriculture to ease the labor demand of agricultural production, vigorously develop water-saving and ecological agriculture, and develop modern characteristic animal husbandry industry to comprehensively improve the quality of agricultural products and build modern animal husbandry production base. Second, it is necessary to expand the local industrial chain and develop strategic industries such as desert exploration and sightseeing, new energy and characteristic cultural tourism. Third, this region should accelerate infrastructure construction and enhance the level of intensive and efficient utilization of regional factors. Forth, with the rural residents gradually gathering to cities and towns, it is necessary to carry out land reclamation timely in rural areas.

Region VII: Urban–rural antagonistic area with backward economy in Qinghai-Tibet Plateau

The main problems of urban and rural development in the Qinghai-Tibet Plateau include the following. First, due to the high altitude, cold climate, low air pressure, complex terrain and sparsely populated population, this region is not suitable for industrial production and large-scale urban construction. Second, the long-term unreasonable land development and resource utilization have caused significant damage to the regional natural environment and ecosystem. Third, the poor infrastructure and the low education level of local residents are not conducive to the establishment of local factor agglomeration and regional endogenous growth mechanism.

Forth, the distribution of rural settlements is scattered, and the construction of rural public services and infrastructure is high-cost and difficult. In particular, due to the lack of industrial development support, the construction of small and medium-sized towns and communities is seriously inadequate, and the promotion of new-type urbanization is still lack of long-term mechanism and endogenous driving forces.

The Qinghai-Tibet Plateau has clean air, water, soil, grassland and other natural resources and ecological environment, as well as contains profound regional culture and unique cultural resources. Therefore, it is necessary to strengthen regional characteristics and explore green development, efficient development and characteristic development paths of urban–rural transformation. First, it is necessary to optimize the structure of agriculture and animal husbandry, transform the development orientation of agriculture and animal husbandry from increasing production to increasing efficiency, strengthen the intra-industry alliance and the integration of industrial chain in agriculture, thus promoting the coordinated development of agriculture. Second, it is urgent to adhere to the concept of green, ecological and humanistic development and develop characteristic plateau and ethnic tourism based on natural and geographical scenery and human resources such as Tibetan culture, customs and religion. Third, based on the regional main function orientation, the region should actively strive for financial transfer funds, constantly improve infrastructure and social public service facilities, thus enhancing the human capital accumulation of local residents and the intensive ability of local economic self-development.

References

Chen M, Lu D, Zhang H. Comprehensive evaluation and the driving factors of China's urbanization. Acta Geogr Sin. 2009;64(4):387–98.

Fang C, Ma H, Wang J. A regional categorization for "New-Type Urbanization" in China. PloS One. 2015;10(8):e0134253.

Holmes J. Impulses towards a multifunctional transition in rural Australia: gaps in the research agenda. J Rural Stud. 2006;22(2):142–60.

Ju D. Rural community is the effective carrier of new rural construction. Shanghai Rural Econ. 2006;9:35–7.

Liu Y, Chen C, Li Y. The town-villages construction pattern under new-type urbanization in China. Areal Res Dev. 2014a;33(6):5–10.

Liu Y, Hu Z, Li Y. Process and cause of urban-rural transformation in the Bohai Rim Region, China. J Geog Sci. 2014b;24(6):1147–60.

Liu Y, Wang J, Guo L. The spatial-temporal changes of grain production and arable land in China. Sci Agric Sin. 2009;42(12):4269–74.

Liu Y, Yang R. The spatial characteristics and formation mechanism of the county urbanization in China. Acta Geogr Sin. 2012;67(8):1011–20.

Liu Y, Zhang Z, Wang J. Regional differentiation and comprehensive regionalization scheme of modern agriculture in China. Acta Geogr Sin. 2018;73(2):203–19.

Liu Z, Liu Y, Li Y. Extended warm temperate zone and opportunities for cropping system change in the Loess Plateau of China. Int J Climatol. 2019;39(2):658–69.

Long H. Land consolidation and rural spatial restructuring. Acta Geogr Sin. 2013;68(8):1019–28.

Ou X, Zhen F, Qin Y, et al. Study on compression level and ideal impetus of regional urbanization: a case of Jiangsu Province. Geogr Res. 2008;27(5):23–32.

Wang Y, Liu Y, Yan B, et al. Spatial patterns and influencing factors of urban–rural coordinated development in China. Sci Geogr Sin. 2016;36(1):20–8.

Wei YD. Decentralization, marketization, and globalization: the triple processes underlying regional development in China. Asian Geogr. 2001;20(1–2):7–23.

6 Spatial–Temporal Patterns of Urban–Rural Transformation in Bohai Rim Region

Abstract

The Bohai Rim region is the "third pole" of economic growth in China's except the Pearl River Delta and the Yangtze River Delta, and is currently in the stage of rapid industrialization and urbanization. Therefore, an in-depth study of the urban–rural transformation in Bohai Rim region is of great significance to accelerate the development of Beijing-Tianjin-Hebei Region and ensuring the coordinated development of China's eastern, central, western, and northeastern regions. This chapter analyzes the situation of urban–rural transformation and the pattern of coupling and coordination in Bohai Rim region from three aspects of population transformation, industrial transformation and land use transformation, and discusses the types and divisions of urban–rural transformation in Beijing Tianjin Hebei region. Then, the spatiotemporal pattern of urban and rural development in the Bohai Rim Region from 2000 to 2015 is revealed, and the coordination between the two is discussed. With the grid data and geostatistical approaches, the process and pattern of urban–rural transformation are explored, and the spatial centrality, growth and different laws of urban–rural transformation are analyzed. On this basis, the spatial–temporal patterns of urban–rural transformation in Bohai Rim region are depicted from four dimensions of population transformation, land transformation, industrial transformation and social transformation.

Since the late 1980s, the global factors of production have been rapidly moving eastward and agglomerating. Guided by the strategic vision of "two overall situations" put forward by Comrade Deng Xiaoping, the Pearl River Delta and the Yangtze River Delta took the lead in achieving rapid economic restructuring and upgrading. In the new reform period, China has been shifting its regional focus of reform, opening-up, and modernization northward and westward as a result of the rapid economic development and overall rise of the southeastern coastal areas. Because of the synergy of the eastward movement and the northward shift, the Bohai Rim Region, located in the coastal economic center of North China, has been growing into the "third pole" of China's economic growth in addition to the Pearl River Delta and the Yangtze River Delta. From the perspective of regional balance, urban–rural integration, and human-earth coordination, the effort to accelerate the economic integration process in the Bohai Rim Region is in line with regional economic development law and fulfills the strategic requirement for faster implementation of urban agglomeration. Moreover, it is conducive to coordinating the regional economic growth of China and promoting the balanced and

Y. Liu, *Urban-Rural Transformation Geography*, Sustainable Development Goals Series,
https://doi.org/10.1007/978-981-16-4835-9_6

sustainable development of the eastern, central, western, and northeastern regions.

An in-depth study of the urban–rural transition in the Bohai Rim Region is an important subject for scientifically balancing regional populations, resources, the environment, and economic development, exploring ecological progress models for urban and rural areas, and establishing a sound mechanism of regional coordinated development. It is also an effective way to accelerate the construction of the capital economic circle, improve the urban system and layout in the region, and promote the coordinated and innovative development of the Bohai Rim Economic Zone. Considering the actual social and economic development in the Bohai Rim Region, the key to urban–rural transition in the region lies in the coordinated promotion of infrastructure connectivity, industrial development complementarity and mutual help, exchange and flow of resource factors, the joint construction and sharing of public services, and collaborative protection of the ecological environment. Advancing scientific urban–rural transition in the Bohai Rim Region is of great practical significance in accelerating the coordinated development of the Beijing-Tianjin-Hebei Region and ensuring the balanced development of China's eastern, central, western, and northeastern regions. The Bohai Rim Region here includes Beijing, Tianjin, Hebei, Liaoning, and Shandong (hereinafter referred to the "three provinces and two municipalities").

6.1 Process of Urban–Rural Transformation in Bohai Rim Region

6.1.1 Urban–Rural Population Transformation

The Bohai Rim Region includes the Beijing–Tianjin–Hebei Urban agglomeration, with Beijing, Tianjin, and Tangshan as core cities; the Shandong Peninsula Urban Agglomeration, with Jinan and Qingdao as core cities, and consisting of Yantai, Weifang, Zibo, Dongying, Weihai, Rizhao, and other cities; and the Middle and Southern Liaoning Urban Agglomeration, with Shenyang and Dalian as core cities and consisting of Anshan, Fushun, Benxi, Dandong, Liaoyang, Yingkou, Panjin, Tieling, and other cities. Driven by these three urban agglomerations, the Bohai Rim Region is undergoing rapid urbanization. The urbanization rate is used to measure the level of urbanization and characterize the urban–rural population transition in a region. The urban–rural population transition in the Bohai Rim Region exhibits a trend of continuous rise (Fig. 6.1). In 1986, the region had a total population of 190.14 million, with an urban population of 43.4 million, and an urbanization rate of 22.83%. After 30 years of development, the region's total population increased to 255.3 million in 2016, with an urban population of 159.78 million—an increase of 116.38 million, with an urbanization rate increased to 62.6%. Especially since 2000, the pace of urbanization has grown even faster, with an average annual growth rate of 1.9 percentage points, making the Bohai Rim Region one of the fastest growing regions in China in terms of urbanization. Urban–rural population transition of over 50% in the region indicates that most parts of the region are experiencing a rapid transition from a rural society to an urban society, and are gradually entering the stage of urban–rural integrated development.

There are significant regional differences in the urban–rural population transition in the "three provinces and two municipalities" of the Bohai Rim Region. Since the reform and opening-up, the level of urban–rural population transition in Beijing, Tianjin, and Liaoning has been higher than the average level of the Bohai Rim Region in the same period, while the level of urban–rural population transition in Shandong and Hebei has been lower than the average level. The urbanization rate of Beijing, Tianjin, Liaoning, and Shandong in 1978 was 55.0%, 49.5%, 31.7%, and 8.8%, respectively, and that of Hebei was 13.93% in 1986. By 2016, the urbanization rate of Beijing, Tianjin, Liaoning, Shandong, and

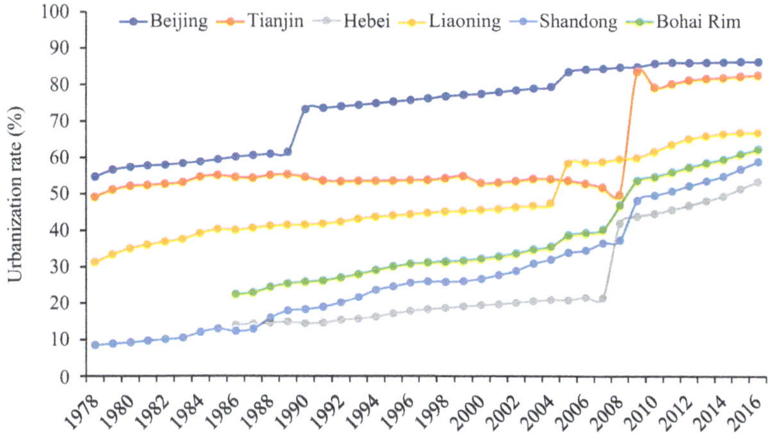

Fig. 6.1 Dynamic change in urban–rural population transition in the Bohai Rim Region as well as in provinces and municipalities

Hebei grew rapidly to 86.5%, 82.9%, 67.4%, 62.6%, and 53.5%, respectively. Beijing and Tianjin have entered a stage of urban–rural integrated development, with an urbanization rate of over 70%. Shandong and Liaoning are stepping into an urban–rural integration stage. Hebei still faces the arduous task of urban–rural population transition, where most traditional agricultural areas need a higher level of urbanization. Urbanization in the Bohai Rim Region is neither led by the government nor caused by an agricultural collapse. In general, it is driven by industrial development and upgrading.

Since 2000, the "three provinces and two municipalities" in the Bohai Rim Region have seen significant changes in the spatial pattern of population urbanization (Fig. 6.2). Generally, the level of urbanization continues to rise in counties. The trend is emerging that counties with a high urbanization rate are largely distributed across the Beijing-Tianjin-Hebei Urban Agglomeration, the Shandong Peninsula Urban Agglomeration, and the Middle and Southern Liaoning Urban Agglomeration. In 2000, 105 or 32.11% of the research units[1] had an urbanization rate of over 30.00%, which were mainly distributed in such core cities as Beijing, Tianjin, Shenyang, Dalian, Yantai, Qingdao, and Jinan as well as their surrounding areas. The urbanization level was low in most of the rest research units. Counties with a low urbanization rate in the vast traditional agricultural areas of the Huang–Huai–Hai Region were characterized by mean distribution. Compared with 2000, the level of county urbanization in the Bohai Rim Region has improved markedly in 2015, especially in the three urban agglomerations and their surrounding counties. The number of research units with an urbanization rate below 30% fell from 221 in 2000 to 18, accounting for 5.81% of the total. The number of research units with urbanization rates above 50% increased from 45 in 2000 to 111, accounting for 33.94% of the total, further expanding its coverage in space.

6.1.2 Urban–Rural Industrial Transformation

Regional economic development is mainly reflected in two aspects: economic aggregate increase, and structural transformation and upgrading. From 1952 to 2016, the evolution of industrial development in the Bohai Rim Region can be roughly divided into five stages (Fig. 6.3). The first stage was from 1952 to 1960. Thanks to the national strategy for the development of heavy industry, heavy industry developed quickly, and the proportion of output accounted for by secondary industry increased rapidly in all provinces and cities during the implementation

[1]For the sake of research, this chapter incorporated downtown areas of prefecture-level cities based on the Bohai Rim Region's county-level statistical units listed in the *China Statistical Yearbook (County Level)* and obtained 326 research units.

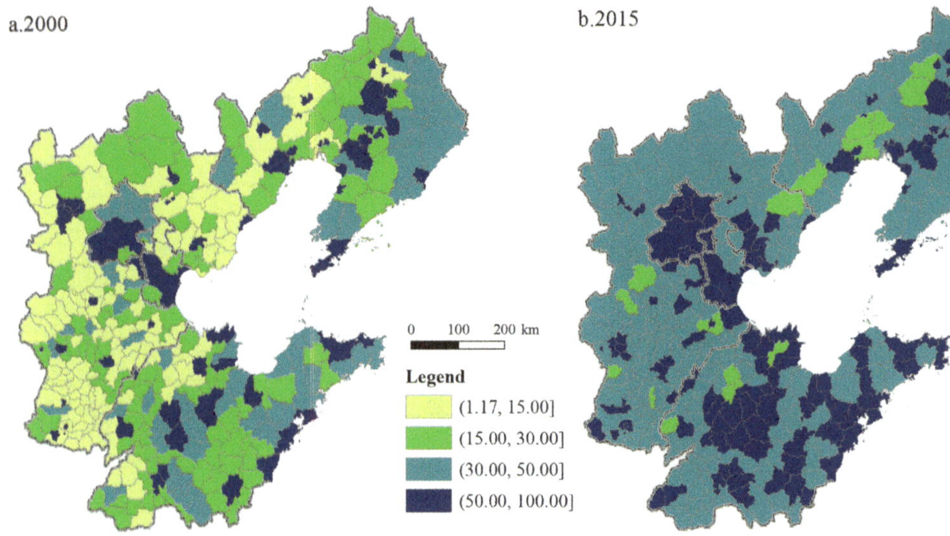

Fig. 6.2 Spatiotemporal process of urban–rural population transition in the Bohai Rim Region

of the first Five-Year Plan. In 1952, the shares of primary, secondary, and tertiary industries in the region were 48.09%: 30.23%: 21.69%. The larger share of the primary industry suggested the primary stage of industrialization. In 1960, the industrial shares were 16.23%: 61.52%: 22.24%. The share of secondary industry increased by 31.27% in ten years, while tertiary industry developed very slowly. The economic form was prominently biased toward heavy industry, and the industrial structure was imbalanced.

From 1960 to 1978, due to the Great Chinese Famine and the Cultural Revolution, economic development slowed down dramatically. Moreover, the economic development strategy biased on heavy industry remained unchanged. The proportion of output accounted for by secondary industry was still high (at 61.64% in 1978), and the development of tertiary industry lagged seriously (at 18% in 1978). From 1978 to 1984, the reform and opening-up policy spurred a change in the economic development strategy biased toward heavy industry, as well as constant adjustments and transformations of the industrial structure. In the few years before the reform and opening-up, rural labor productivity was continuously liberated under the effect of the institutional reform. The output of primary industry increased rapidly; tertiary industry began to develop steadily; secondary industry experienced a process of decline and adjustment. In 1984, the shares of the first, second, and third industries in the Bohai Rim Region were 25.64%: 51.31%: 25.64%. After 1984, all industries of the Bohai Rim Region developed in a mutually complementary way. The added value of secondary industry kept fluctuating around 50%, and the proportion of the tertiary industry kept increasing. After 2008, under the impact of the global financial crisis, the industrial structure was constantly adjusted in the Bohai Rim Region, featuring tertiary, secondary, and primary industries in order of decreasing proportion. To be specific, the share of primary industry continued to decline to only 6.6% in 2016; the share of secondary industry also decreased annually to 40.8% in 2016; the share of tertiary industry increased annually, surpassing secondary industry, gaining a dominant position in 2014, and reaching 52.6% in 2016. Although the industrial structure in the Bohai Rim Region has undergone a leapfrog transition from "secondary, tertiary, and primary industries" to "tertiary, secondary, and primary industries" in the middle and late stage of industrialization, the share of heavy chemical enterprises is still high and the

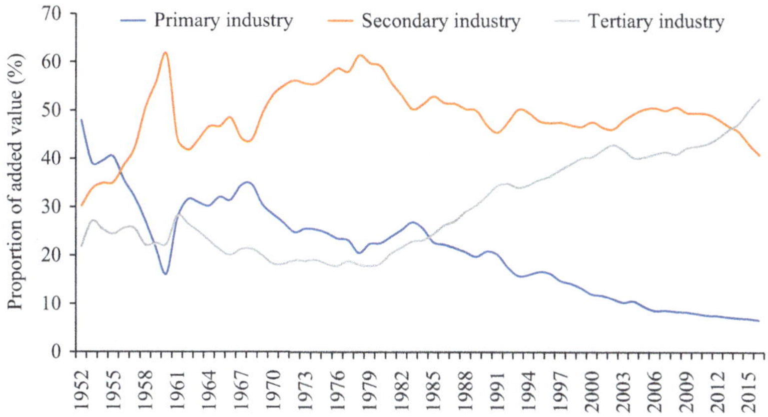

Fig. 6.3 Evolution of the industrial structure in the Bohai Rim Region

transition of industrial cities is slow, especially in Hebei.

As urbanization and industrialization continue to grow, industrial upgrading is constantly boosting urban–rural transition. According to the Petty–Clark theory of industrial evolution, industrial development in a region follows the trajectory of the gradual transfer and evolution of primary, secondary, and tertiary industries, and is an important basis for the division of urban–rural transition stages. In this study, the proportion of secondary and tertiary industries' added value to the gross national product (GNP) was used to represent the level of urban–rural industrial transition. At the current stage of economic development, regional heterogeneity is significant in industrial transition in the Bohai Rim Region (Fig. 6.4). In 2000, 116 research units (counties and districts) with an industrial transition of at least 80% were mainly concentrated in the downtown areas and surrounding areas of such prefecture-level cities as Beijing, Tianjin, Dalian, Jinan, Shenyang, and Qingdao, accounting for 35.58% of the total research units in the region. These areas made a small contribution to the agricultural output, while highly developed industry, commerce, and services dominate in the industrial development of these areas. Traditional agricultural areas, which are characterized by low industrialization and urbanization, have a single industrial structure, and the output of primary industry accounts for more than 20% in most of these areas that have difficulties in industrial transition and upgrading. In 2015, the industrial structure of counties in the Bohai Rim Region further moved toward secondary and tertiary industries. Specifically, 234 or 71.78% of the research units had an industrial transition level of at least 80%, an increase of 118 or 36.2 percentage points over 2000. Industrial transition is an important engine and component of regional urban–rural transition, and directly affects resource allocation as well as the interaction and circulation of urban and rural factors. Moreover, the agglomeration of urban and rural factors directly affects the transition of urban–rural patterns.

6.1.3 Urban–Rural Land Use Transformation

As a basic factor of production for economic development, land is the spatial support for industrial development and population agglomeration. It also changes with the evolution of economic and social systems. Economic and industrial transitions spur land-use changes. Urban–rural transition at the spatial level is represented by the land de-agriculturalization rate, and is examined from the perspective of land use. Since the 1980s, the land de-agriculturalization rate in the Bohai Rim Region has been increasing annually, from 1.96% in 1985 to 10.41% in 2015, with 7.63% of that growth occurring since 1995. The land de-agriculturalization rate in Beijing and Tianjin was much higher than in the other three provinces, reaching 17.57% and 20%,

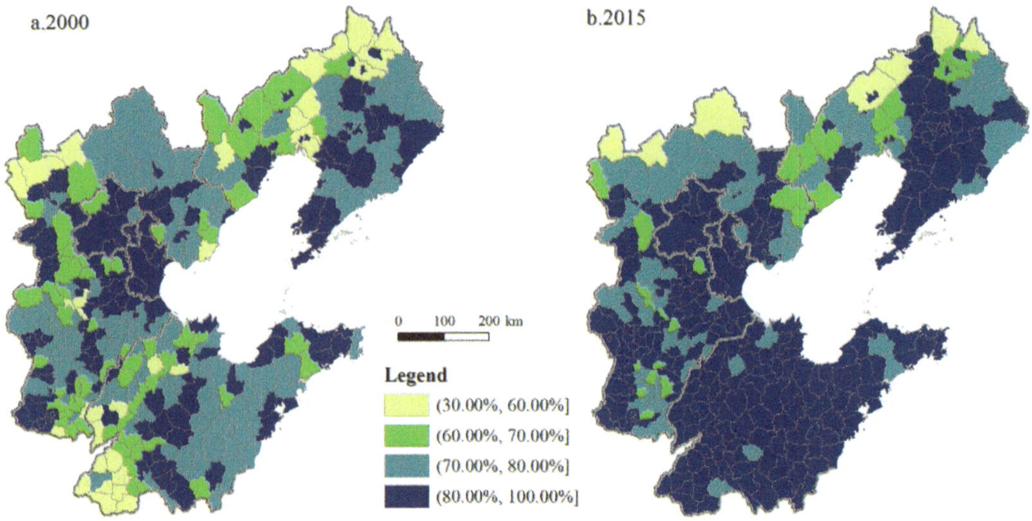

Fig. 6.4 Spatial characteristics of industrial transition in the Bohai Rim Region

respectively, in 2015 (Fig. 6.5). Due to the fast speed and large scale at which farmland was occupied for construction purposes at the national level, China introduced the strictest farmland protection policy in 1997. In that year, the examination and approval of land construction for non-agricultural purposes were strictly controlled and frozen. Since 2000, driven by the rapid urbanization rate of one percentage point per year, land de-agriculturalization has been intensified in terms of both speed and scale, resulting in a hidden occupation by land protectors. The rapid land de-agriculturalization directly threatens the country's food security and ecological security, and it is urgent to address the use and allocation of land, both connotative and intensive.

Urban and rural construction land characterizes the urban–rural spatial transition. In terms of land use, the urban–rural transition is mainly reflected in land urbanization. In this study, the proportion between the total area of urban construction land and the total area of county (district) land was used as the representation index of urban–rural spatial transition. Overall, the spatial heterogeneity of land urbanization in the Bohai Rim Region is significant, consistent with urban–rural population transition and industrial transition (Fig. 6.6). In 2000, the areas with a high land urbanization rate were mainly concentrated in Beijing, Tianjin, and Tangshan, where the level of spatial transition of urban construction land was over 20%. The land urbanization rate was generally lower than 5% in the Yanshan–Taihang mountainous areas in the north and west of Hebei, the mountainous areas in the west of Liaoning, and the Changbai mountainous areas in the east of Liaoning. Research units with a land urbanization rate higher than 20% had a larger spatial coverage, increasing from 27 in 2000 to 44 in 2015. They were mainly the downtown areas of prefecture-level cities, and were concentrated in Shandong. The research units with a land urbanization rate lower than 5% and those with a land urbanization rate between 5 and 10% had little change, falling by only seven and 10, respectively, showing no significant change in spatial distribution.

In terms of the urban–rural spatial pattern and the urban system, large cities in the Bohai Rim Region feature a strong agglomeration effect. Since the 1980s, large cities such as Beijing, Tianjin, Shenyang, Dalian, and Qingdao have significantly expanded their space (Fig. 6.6). In the traditional Huang–Huai–Hai agricultural areas, the land space for rural residential areas has been also expanded significantly. Rural construction land is used extensively, and there

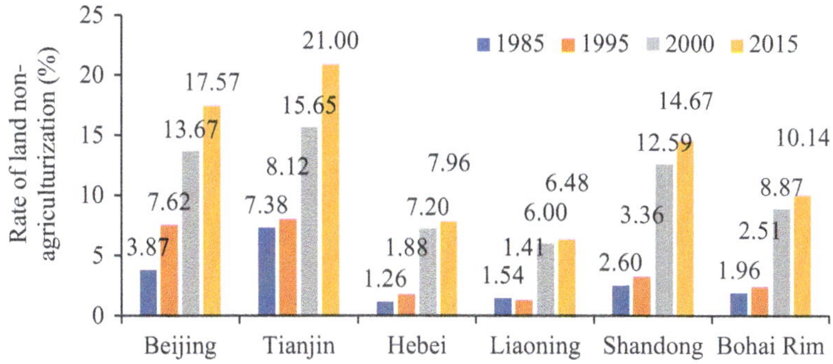

Fig. 6.5 Land urbanization transition process in the Bohai Rim Region

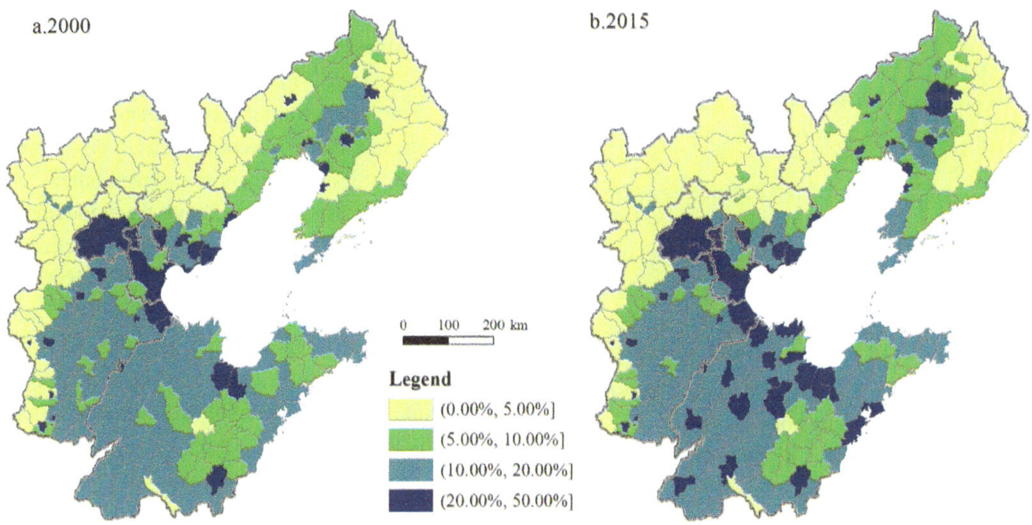

Fig. 6.6 Spatial–temporal patterns of urban–rural construction land transformation in the Bohai Rim Region

are even many "hollow" villages, stemming from efforts to build a reasonable city–town–village system and to take the path of scientific and sound urbanization, and spatially rely on cities, central towns, central villages, and new farmer communities. In practice, a platform is built for the comprehensive treatment of rural fields, waters, roads, forests, and villages, and major rural land projects are comprehensively regulated based on urban construction and new countryside construction. These actions are intended for improving the efficiency of land use, tackling difficulties in land use for urban and rural development, and ensuring the scientific, rational, and sustainable use of urban and rural land.

6.2 Measurement of Urban–Rural Transformation Based on the Perspective of "Population-Land-Industry"

6.2.1 Data Sources and Research Methods

1. **Indicator system of the evaluation of urban–rural transformation**

According to the connotation of space analytic geometry, if the states of the three subsystems of population (S_1), land (S_2) and industry (S_3) represent a three-dimensional space (Fig. 6.7), then a point (P) in this three-dimensional space

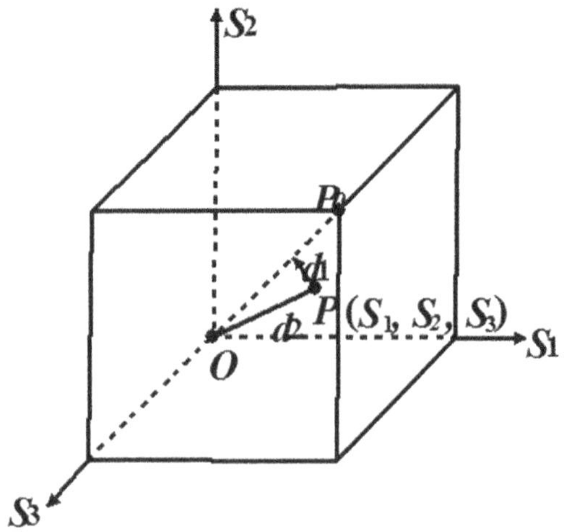

Fig. 6.7 The three-dimensional space built by three states of regional subsystems

represents one state of the regional system. If the states of the three subsystems have the same starting point (that is, the origin at O), the points on the line OP_0 ($S_1 = S_2 = S_3$) will represent the fully synchronous development of the three subsystems and a fully coordinated state of the regional system. Yet, if P, which represents the regional system state, does not often fall on line OP_0, then the vertical distance from point P to line OP_0 (d_1) represents the degree to which a regional system deviates from a fully coordinated state. The distance between the point P and origin O (d_2) represents the regional comprehensive level. The formulae are as follows.

$$d_1 = \sqrt{\frac{(S_1 - S_2)^2 + (S_1 - S_3)^2 + (S_2 - S_3)^2}{3}} \tag{6.1}$$

$$d_2 = \sqrt{S_1^2 + S_2^2 + S_3^2} \tag{6.2}$$

where $d_1 = 0$ represents a fully coordinated state of the regional system and $d_1 = 1$ represents a completely uncoordinated state of the regional system. The larger the value of d_1, the more uncoordinated is the regional system. d_2 represents the regional comprehensive level. The larger the value of d_2, the higher is the regional comprehensive level.

Considering data accessibility, quantification and method scalability, we adopted the three indexes of the population urbanization rate, land urbanization rate, and industrial non-agriculturalization rate to analyze the non-agricultural transition of population-land-industry, which is described in detail in Table 6.1. Population urbanization refers to the process of a rural population's transformation into the cities, the core of urbanization and a general increase in the proportion of the urban population accounting for the total population, which could be an important measure of the level of national socioeconomic development. Land urbanization refers to the increasing proportion of land with urban morphological features per the total construction land area in a region (Lu et al. 2007). Industrial non-agriculturalization refers to the flow of production factors from rural areas to urban areas that accompanies socioeconomic development and results in the transformation and development of industrial structure. As an increasing proportion of the population moves to the cities, more urban land and non-agricultural industries are needed. So, as P_i increases, L_i and I_i will increase accordingly. When land urbanization and industrial non-agriculturalization keep level with urban population growth, the three indexes of P_i, L_i, and I_i exhibit synchronous growth and the urban–rural system is in a fully coordinated state.

The three indexes of the evaluation system are positive for the subsystems. To make the various indexes comparable, we transformed all the indexes into comparable common units by normalizing all measures (P_i, L_i, I_i) and ranging the values of the cube diagonal from 0 to 1. When the three indexes are applied to the three-dimensional space in Fig. 6.7, the starting point O represents a completely agricultural society with human agricultural production activities taking place in the rural areas. At this point, all the values of the three indexes referring to P_i, L_i, and I_i are 0. Instead, the point P_0 represents a completely urbanized society with human non-agricultural activities taking place in the urban areas. This hypothesis implies a definition: an urban society is a society resulting from a process of complete urbanization. Although this

Table 6.1 Indexes for evaluating three major aspects of urban–rural transformation

Criteria layer	Index	Definition	Calculation formulae
Population	Population urbanization rate (P_i)	Proportion of urban resident population per total resident population	$P_i = UP_i/TP_i$ (6.3)
Land	Land urbanization rate (L_i)	Proportion of urban, industrial, and mining land per total construction land	$L_i = \frac{LU_i + LI_i}{LU_i + LI_i + LR_i}$ (6.4)
Industry	Industrial non-agriculturalization rate (I_i)	Proportion of the second and third industries' output per total GDP	$I_i = \frac{SGDP_i + TGDP_i}{GDP_i}$ (6.5)

Note i is the evaluated unit (county); P_i is the population urbanization rate of unit i; UP_i is the urban resident population of unit i; TP_i is the total resident population of unit i; L_i is the land urbanization rate of unit i; LU_i is the urban land area of unit i; LR_i is the rural residential land area of unit i; LI_i is the industrial and mining land area of unit i; I_i is the industrial non-agriculturalization rate of unit i; $SGDP_i$ is the second industry added value; $TGDP_i$ is the tertiary industry added value; GDP_i is the gross domestic product (GDP) of unit i

urbanization is virtual today, it will become real in the future (Haas and Westlund 2017; Lefebvre 2003). At this point, the values of P_i, L_i, and I_i are 1, with all rural residents being citizened and rural land being completely transferred to urban land. P_i, L_i, and I_i can be regarded as the measurement indexes of the three-dimensional spatial subsystems, whereas the length of the space vector about the three indexes (OP) could be considered as the coupling degree of urban–rural transformation. As expressed in Table 6.2, the accumulated values of P_i, L_i, and I_i refer to the length of the OP line in Fig. 6.7. The line represents the transformation coupling degree (TCD), which reflects the state of the peasants' citizenization, industrial non-agriculturalization, and land urbanization. Under the fully coordinated condition, the vector overlaps with the cube diagonal, which represents the fully coordinated line (OP_0), indicating that the three subsystems of population, land, and industry develop simultaneously and the urban–rural system is in a fully coordinated state. The vertical distance of the vector OP to the fully coordinated line (shown as d_1 in Fig. 6.7) is also named the deviation degree, which represents the coordinated deviation degree (CDD) and measures the matching status of each dimension. CDD reflects the difference between the real urban–rural transformation and its coordinated condition. The coordinated transformation degree (CTD) could be calculated by deducting the deviation degree from the accumulated value (Table 6.2) and is also the projection of the line OP onto the line OP_0 in Fig. 6.7, describing a comprehensive degree when we project the real urban–rural transformation process onto the fully coordinate state of the urban–rural system.

2. **Trajectory analysis method**

Trajectory-based detections of time series could be described by codes in the form of figures or letters for each unit in the vector layer to track the state changes. As numeric codes are convenient for operations and calculations in ArcGIS software for trajectory computing models, we applied digital coding to obtain the changing trajectory of urban–rural transformation. In this study, urban–rural transformation was divided into four types according to the values of CTD: Low-level type (Type I); Middle-level type (Type II); High-level type (Type III); Higher-level type (Type IV). Then, the numbers 1–4 were used to sequentially represent Types I–IV for each layer of each time node in the trajectory analysis. Trajectory codes for each unit were computed as below:

$$Y_i = (G_1)_i \times 10^{n-1} + (G_2)_i \times 10^{n-2} \cdots + (G_n)_i \times 10^{n-n} \quad (6.9)$$

where Y_i represents the calculated code of unit i in the trajectory-based detection layer; n represents the number of time nodes; $(G_1)_i$, $(G_2)_i$, and

Table 6.2 Proxy index system for evaluating urban–rural transformation

Proxy index	Index representation	Calculation formula
Transformation coupling degree (TCD)	Accumulated value	$TCD = \sqrt{\frac{P_i^2 + L_i^2 + I_i^2}{3}}$ (6.6)
Coordinated deviation degree (CDD)	Deviation degree	$CDD = \frac{1}{3}\sqrt{(P_i - I_i)^2 + (L_i - I_i)^2 + (P_i - L_i)^2}$ (6.7)
Coordinated transformation degree (CTD)	Comprehensive degree	$CTD = \sqrt{TCD^2 - CDD^2}$ (6.8)

$(G_n)_i$ represent the codes of the types of urban–rural transformation at each time node at the given unit.

The trajectory-based codes (e.g., 2222, 3344) of every unit were calculated automatically. For example, "2222" means that the urban–rural transformation had always stayed at the Middle-level during 2000–2015 and "3344" means that the urban–rural transformation was at the High-level in 2000 and 2005, then reached a Higher-level from 2010. According to the phylogenetic rules of a system's evolution from basic to advanced and from disorder to order, an urban–rural system tends to develop and upgrade to higher levels. Sometimes, the system might fluctuate at a lower level due to uncontrollable driving factors. However, it should be specially noted that urban–rural transformation often changes during successive and subsequent stages. For example, the urban–rural transformation may change from Type III (High-level) to Type II but rarely into Type I. Not only do the trajectory-based codes signify that the type can change for each unit and each time node but the trajectory for each unit through the time series can also be specified by the type classifications (Fig. 6.8). If there is no arrow between two consecutive time nodes, the trajectory is impossible.

3. **Data sources**

Considering the relative integrity of urban–rural economic and social activities in space, as well as the availability of the data, this study takes the county-level administrative unit as the spatial research scale. Monitoring the non-agricultural transformation of "population-land-industry" in the Beijing-Tianjin-Hebei (BTH) region was done for the years, 2000, 2005, 2010, and 2015, in the new millennium. The data on population and industry were obtained from the China Statistical Yearbooks (County-Level) of various years, China Statistical Yearbook for Regional Economy, Tabulation on the 2010 Population Census, China's Regional Statistical Yearbooks, and the Provincial Statistics. The data on land use and land cover (LULC) were collected from the Data Center for Resource and Environmental Science of the Chinese Academy of Sciences. The LULC data includes 6 first-level land categories and 25 two-level land categories where construction land includes urban land, rural settlement land, and other types of construction land (i.e., industrial and mining land). Because we aimed to describe and compare the data from 2000 to 2015 for the whole BTH region, we ensured that the data were unified and that the economic data had been based on comparable prices.

4. **Study area**

The BTH region (36°05′N–42°37′N, 113°11′E–119°45′E) lies in the North China Plain and at the core of China's Circum-Bohai sea region. BTH covers 134,735 km^2, occupies 1.9% of China's territory, and had a population of 111 million, which accounted for 8.08% of China's total population in 2014. As China's county administrative borders frequently change, we used the current (since 2015) borders of the BTH region

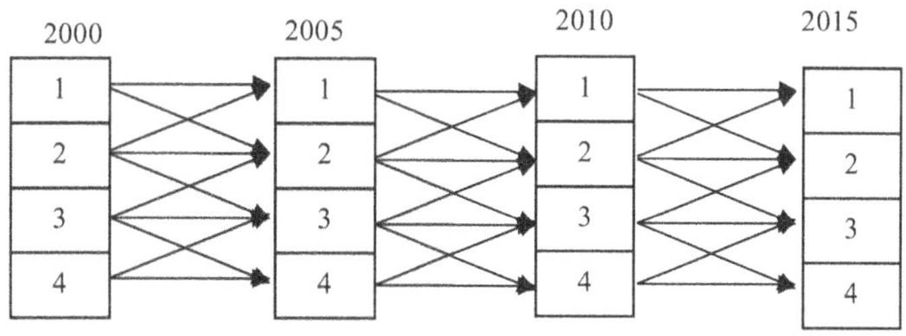

Fig. 6.8 Possible trajectories of type changes

as the study range, thereby ensuring comparability across different periods of time. The BTH region includes three administrative areas, Beijing, Tianjin, and Hebei, and contains 11 municipal cities, which are Shijiazhuang, Tangshan, Qinhuangdao, Handan, Xingtai, Baoding, Zhangjiakou, Chengde, Cangzhou, Langfang, and Hengshui. By the end of 2015, the BTH region included 202 distinct counties or districts with 16 in Beijing, 16 in Tianjin, and 170 in Hebei (Fig. 6.9).

6.2.2 Measurement of Urban–Rural Transformation in Beijing-Tianjin-Hebei Region

1. **Population-land-industry transformation**

This study adopted the equal interval method for the classification of the three-dimensional indexes to reflect population-land-industry transformation (i.e., P_i, L_i, I_i), which has five categories: lower-level transformation [0, 0.2), low-level transformation [0.2, 0.4), middle-level transformation [0.4, 0.6), high-level transformation [0.6, 0.8), and higher-level transformation [0.8, 1]. Figure 6.10 reveals the changes in P_i-L_i-I_i from 2000 to 2015 and reflects the marked increases in the mean values of P_i (0.34 → 0.30 → 0.49 → 0.55), L_i (0.33 → 0.36 → 0.38 0.40), and I_i (0.81 → 0.82 → 0.84 → 0.86). More than half of the counties had lower-level transformations of the population before 2005 and stepped into the low-level stage in 2010, then 42.57% of the counties entered middle-level transformation in 2015. It should be noted that 21.29% of the counties had higher-level transformations of the population. The above analysis reveals that since the turn of the new millennium, there has been rapid rural–urban migration in the BTH region, especially in central Beijing and Tianjin (Fig. 6.10), which presents a "Beijing-Tianjin" dual-core feature.

With the fast development of industrialization and urbanization, the demands on the national infrastructure, construction of township enterprises, and industry and commerce have caused aggravated changes in land usage toward non-agricultural uses, especially for urban, industrial, and mining purposes. During the study period, the proportions of low-level and middle-level land transformations increased with a corresponding decrease in lower-level land transformations while there was a slight increase in higher-level transformations and a small reduction in high-level transformations. Figure 6.11 shows the high-level and higher-level land transformations mainly distributed around the central Beijing and the "C" coastal zone of Tianjin. This change trend is similar to that of population transformation, signifying the profound effect of population mobility on changes in land use and land cover.

Presenting a very different change status compared with the population and land transformations, most of the counties' industry transformations are at the high-level and higher-level transformation stages with a decrease in the proportion of high-level and a corresponding increase in higher-level from 2000 to 2015. Among the total of 202 counties, there were 144 counties that reached the higher-level in 2015, occupying a very large proportion (71.29%). The counties with higher-level industry transformation are mostly distributed in the central BTH region while the ones at middle-level and high-level are located in the northern region because

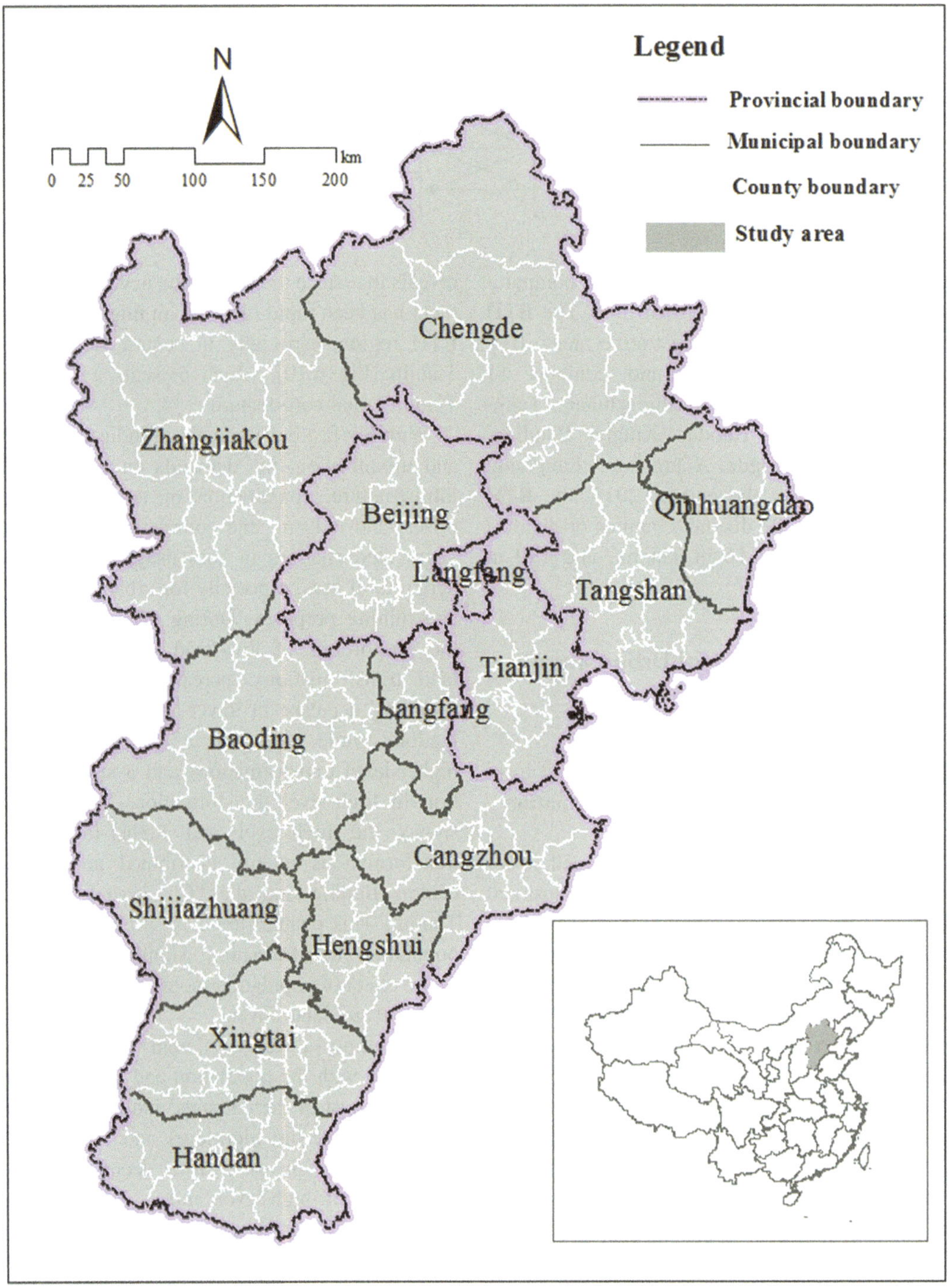

Fig. 6.9 Location of the BTH region, China

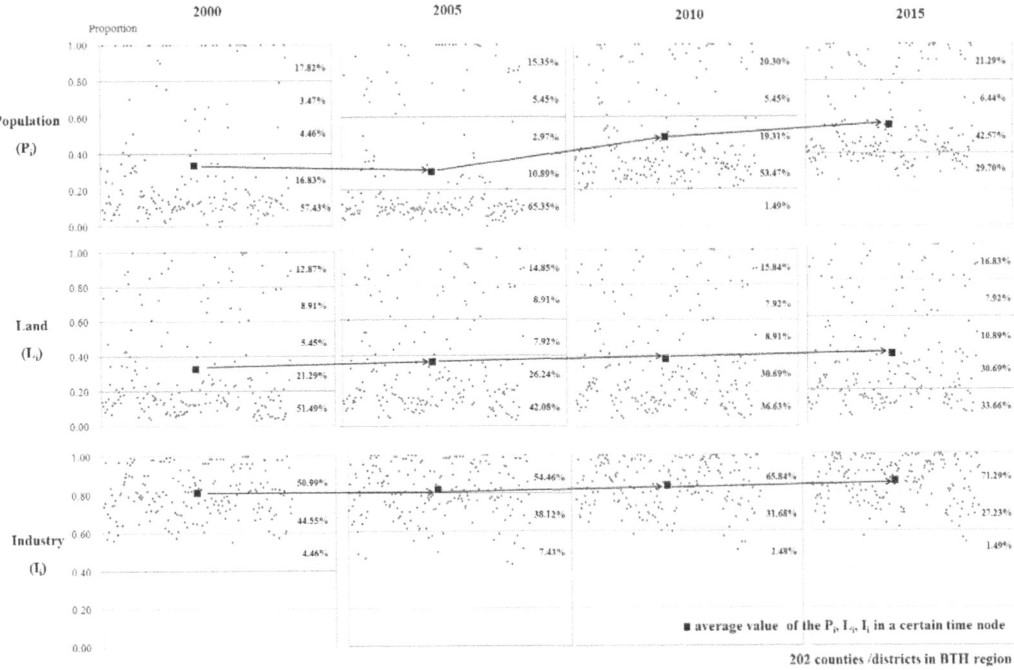

Fig. 6.10 Changes in P_i-L_i-I_i from 2000 to 2015

socioeconomic development in the northern region has lagged far behind that of other regions.

2. **Spatio-temporal patterns of urban–rural transformation**

The Transformation Coupling Degree (TCD) and Coordinated Transformation Degree (CTD) were calculated and five categories, which are lower-level [0, 0.2), low-level [0.2, 0.4), middle-level [0.4, 0.6), high-level [0.6, 0.8), and higher-level [0.8, 1], were subsequently classified by applying the equal interval method in ArcGIS 10.0 to reveal the degree division and spatial differentiation in BTH from 2000 to 2015. Meanwhile, this study adopted natural breaks (Jenks) for five classifications of the Coordinated Deviation Degree (CDD), i.e., lower-level deviation [0, 0.05), low-level deviation [0.05, 0.10), middle-level deviation [0.10, 0.15), high-level deviation [0.15, 0.20), and higher-level deviation (≥ 0.20).

From 2000 to 2015, there were no TCD and CTD values ranging from 0 to 0.2. The average TCD had a marked increase ($0.57 \rightarrow 0.57$ $0.62 \rightarrow 0.65$) while the average CDD had an overall decrease ($0.14 \rightarrow 0.14 \rightarrow 0.12 \rightarrow 0.12$), resulting in a corresponding increase in the CTD values ($0.55 \rightarrow 0.55 \rightarrow 0.61 \rightarrow 0.63$). The increasing trends of TCD and CTD reveal that the urban–rural system had evolved from basic to advanced driven by urbanization and industrialization while the decreasing trend of CDD indicated that the urban–rural system had changed from disorder to order. The results show that the three indexes of P_i, L_i, and I_i had experienced synchronous growths and the urban–rural system had gradually developed into a coordinated state. Accounting for the total number of counties, the proportions of high-level and higher-level TCD increased while the proportions of the low-level and middle-level TCD presented downward trends (Fig. 6.12). It is noteworthy that more than half of the counties were at the middle-level coupling. The overall CDD presented a decreasing trend. Accounting for the total number of counties, the proportion of high-level deviation showed a change trend of 51.49% $\rightarrow$ 49.50% $\rightarrow$ 32.67% $\rightarrow$ 27.23% while the

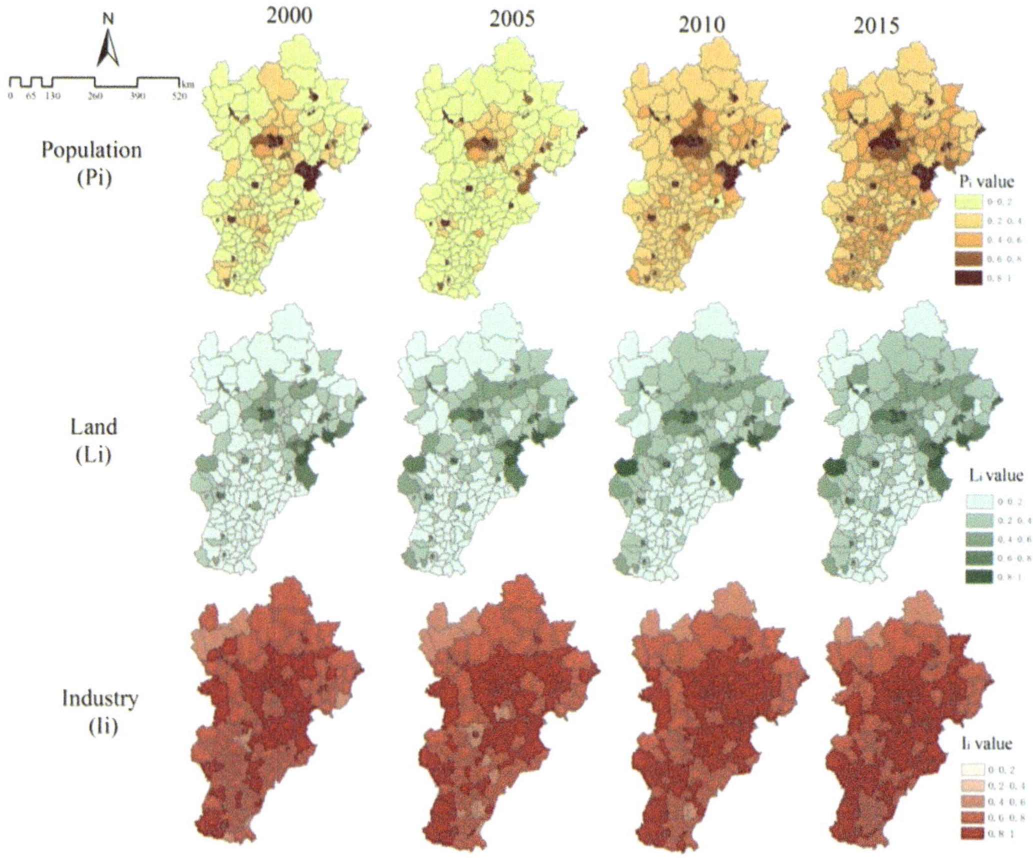

Fig. 6.11 Spatial–temporal changes of P_i-L_i-I_i in the BTH region

proportions of the lower-level, low-level, and middle-level deviations increased, reflecting a more coordinated system of population-land-industry over time. According to the change characteristics of the TCD and CDD variables, CTD showed a change trend similar to that of TCD as half of the counties remained at the middle-level transformation while the proportions of high-level and higher-level transformations increased over time. The results signify that the CTD level had gradually increased from 2000 to 2015 and there was a huge gap in these 202 counties due to their different population-land-industry developments.

Since 2000, the BTH region has witnessed fast development and significant changes in socioeconomic structure, which have exerted a profound influence on BTH's CTD. Figure 6.13 describes the spatio-temporal patterns of urban–rural transformation with a significant variation in the study area. There was a similar spatial distribution and development trend between TCD and CTD. The development cores of BTH, the urban districts in Beijing and Tianjin, as well as the surrounding high-level counties, have always had higher-level TCD and CTD. The two development cores expanded throughout 2000–2015, then merged into a belt around Beijing and Tianjin in high-level and higher-level coordinated transformations, revealing the order of socioeconomic development in the BTH region. A higher-level of CTD can be expected only when a region's economy, population, and land have experienced a major shift in industrial, demographic, and land-use structures, respectively. The regions with high CTD are mainly concentrated in central Beijing and Tianjin, due to their special advantages, such as location,

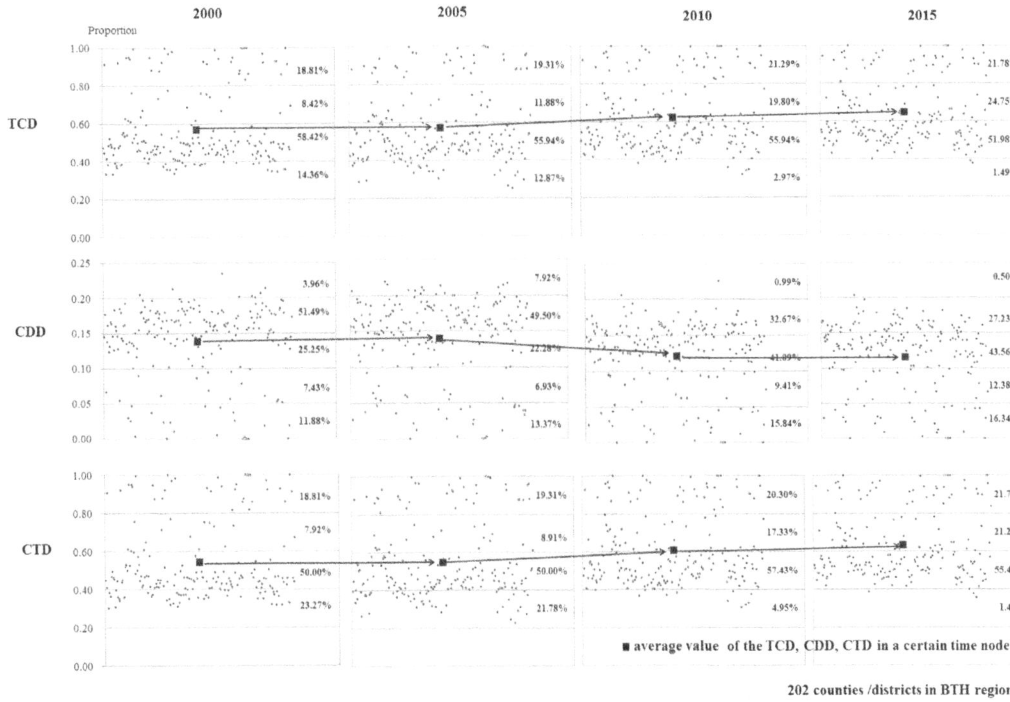

Fig. 6.12 Scattered plots of TCD, CDD, and CTD in the BTH region

socioeconomic factors, and optimal physical conditions, for urban–rural transformation development. These counties are more likely to move into a high-level stage with their fast socioeconomic development. In addition, urban districts in Shijiazhuang, Baoding, Xingtai, Zhangjiakou, Chengde, Tangshan, and Qinhuangdao have also had higher-level or high-level CTD. As for CDD, the northern area of BTH gradually had middle-level deviation while the south was scattered with high-level deviation, signifying that the urban–rural development in the northern counties had occurred in a more coordinated manner among the population-land-industry subsystems. The northern counties are located in the upper hand of wind and water of the BTH region with the prominent functions of ecological conservation and restricted industrial development, resulting in low-level CTD.

6.2.3 Types and Zones of Urban–Rural Transformation

1. **Types of urban–rural transformation**

This study combined all the indexes referring to P_i, L_i, I_i, TCD, CDD, and CTD in the 202 units during 2000–2015 into one large data set. After setting the county's number in ascending order of the CTD value in the 808 records, we obtained the fitting curves of the P_i-L_i-I_i, TCD, CDD, and CTD by adopting the moving average method with 20 iterations to eliminate the influence of numerical fluctuations. As shown in Fig. 6.14, the change trend of urban–rural transformation presents the "*S*" curve. As the CTD value is fundamentally determined by the three indicators of P_i, L_i, and I_i, the urban–rural transformation in the BTH region was divided into the following four types according to the ranges of CTD values

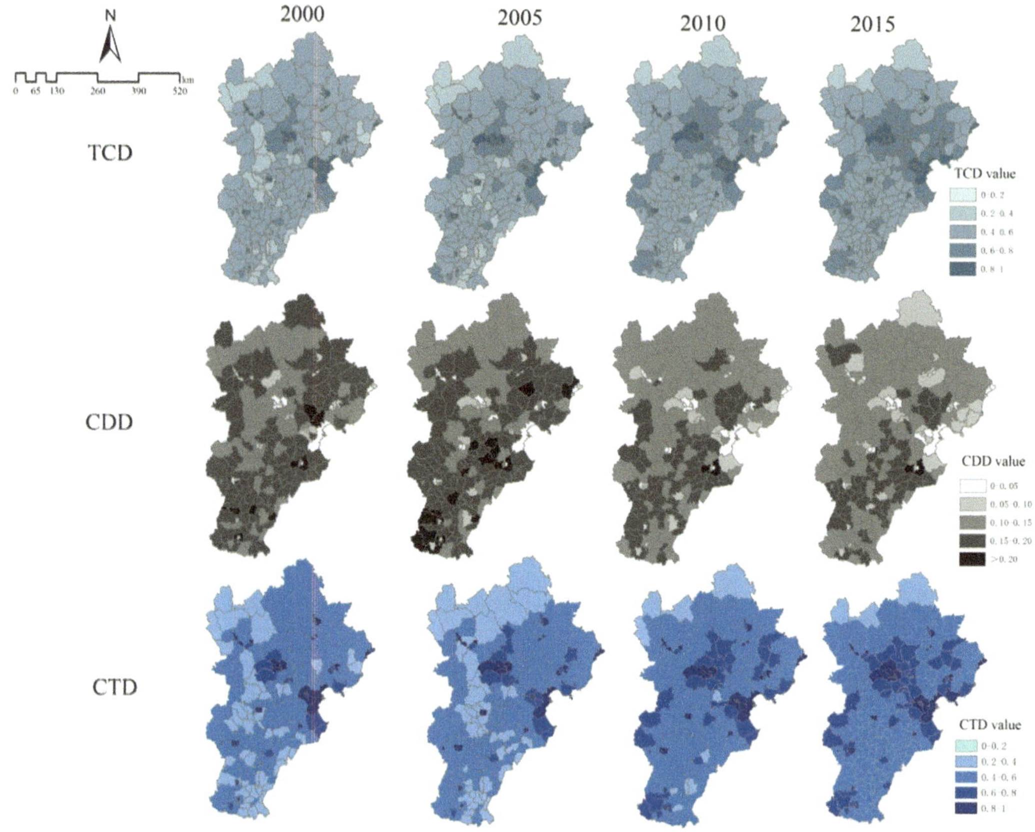

Fig. 6.13 Spatio-temporal patterns of urban–rural transformation in the BTH region

and the change trends of the three factors: (Type I) Industry driving type (Low-level type), CTD ∈ [0 to 0.4); (Type II) Population-land-industry driving type (Middle-level type), CTD [0.4 to 0.6); (Type III) Population-land driving type (High-level type), CTD ∈ [0.6 to 0.8); (Type IV) Uncoordinated population-land driving type (Higher-level type), CTD ∈ [0.8 to 1] (Fig. 6.14).

As there is no distribution of CTD ranging from 0 to 0.2, Type I could also be named as the low-level coordinated transformation stage. This type describes a continuously increasing industry transformation with rapid growth and a slight increase in both population and land transformations. The industrial non-agriculturalization rate (I_i) is far higher than the population and land urbanization rates (P_i, L_i) as most of the counties' I_i belongs to high-level transformation while all the counties' P_i and L_i are lower-level transformations for this type. This stage also witnessed a tiny increase in the CDD, signifying that the urban–rural system was increasingly uncoordinated with the transformation of population-land-industry. Meanwhile, the distance between the CTD and TCD curves gradually widened. At this stage, industry transformation has a great influence on the CTD and is the most important driving factor for urban–rural transformation. So, this stage could be considered as industry-driven urban–rural transformation.

Type II describes most of the population and land transformation being at low-level while industry transformation is at high-level or higher-level. The P_i, L_i, and I_i curves present similar change trends while population and land transformations develop rapidly along with rapid industrial non-agriculturalization at this stage.

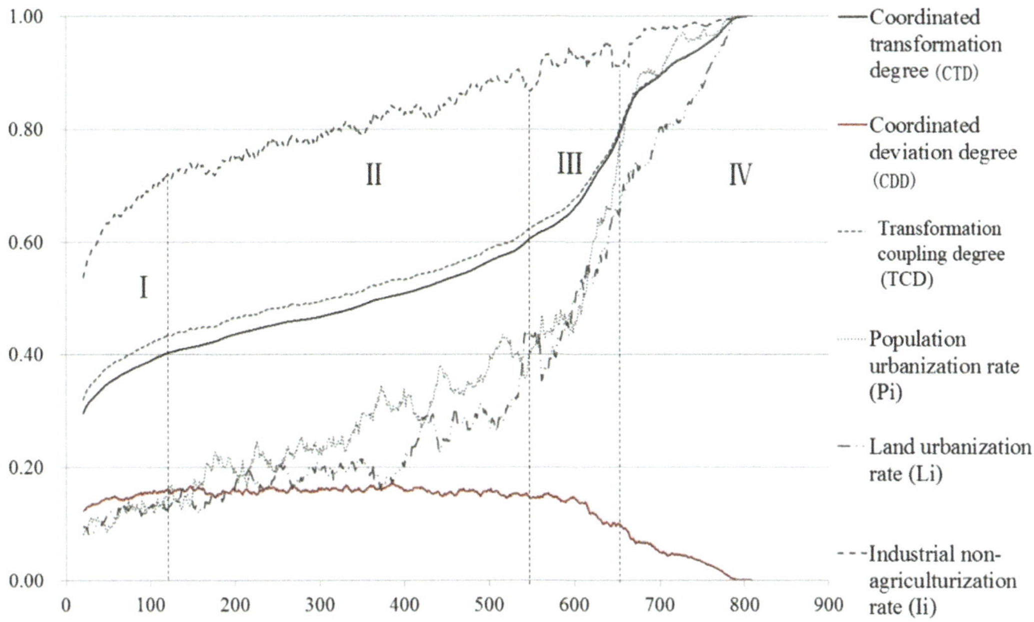

Fig. 6.14 Types of urban–rural transformations: presenting the "*S*" curve

The CTD curve keeps increasing at the middle-level transformation while CDD witnesses small fluctuations with a relatively stable state that reflects a relatively stable coordination system. In addition, it is noteworthy that the distance between the CTD and TCD curves is almost constant. In this context, urban–rural transformation is driven synchronously by the simultaneous effects of the three subsystems of population-land-industry at this stage, namely, population-land-industry-driven for Type II.

Type III describes a population-land driven stage for urban–rural transformation. For the third type, both population and land experiences rise sharply with most of P_i and L_i at the middle-level while industrial non-agriculturalization had a minor change and fluctuated at the higher-level. It is worth noting that the CDD curve begins to drop significantly during this stage, revealing that the urban–rural system becomes more coordinated with less coordinated derivation.

Type IV describes a sharp and continuing decrease in CDD until there is no coordinated deviation and the system is under a fully coordinated condition. The CTD curve overlaps with the TCD line, reflecting a negligible CDD at this stage. P_i and L_i keep increasing, but the gaps between their changes are obviously widened compared with Type III. P_i enters the higher-level during this stage while L_i is still in the high-level transformation and arrives at the higher-level. Although population and land are still important factors for the urban–rural transformations in these counties, there is an uncoordinated population-land driven stage.

Table 6.3 reflects the Pearson correlation coefficients between CTD and P_i-L_i-I_i for different types and reveals that CTD is positively correlated to I_i with the Pearson correlation coefficient of 0.852 for Type I. As for Type II, the Pearson correlation coefficients between CTD and P_i and L_i together increase to 0.572 and 0.511, respectively, while CTD still has a maximal relationship with I_i. The Pearson correlation coefficients between CTD and P_i-L_i-I_i for Type III and Type IV are similar. CTD is the most correlated with L_i, which is not too different from P_i and I_i. These results are consistent with the above analysis, indicating the different driving factors of urban–rural transformations for the four types.

Figure 6.14 presents the four types of urban–rural transformation in the "*S*" curve. The four

Table 6.3 Pearson correlation coefficients between CTD and P_i-L_i-I_i

Type	P_i	L_i	I_i
I	0.202*	0.259**	0.852**
II	0.572**	0.511**	0.722**
III	0.654**	0.824**	0.526**
IV	0.645**	0.824**	0.526**

Note *represents significant correlation at the 0.05 level; **represents significant correlation at the 0.01 level

types also represent the four stages of urban–rural transformation: early stage, middle-early stage, middle-later stage, end stage. We calculated the average values of P_i, L_i, and I_i by selecting the points that have the breakpoint values (CTD = 0.4, 0.6, 0.8) and obtained the following information: when CTD is 0.4, P_i, L_i, and I_i are 0.14, 0.14, and 0.72, respectively; when CTD is 0.6, P_i, L_i, and I_i are 0.38, 0.42, and 0.88, respectively; when CTD is 0.8, P_i, L_i, and I_i are 0.77, 0.67, and 0.92, respectively (Table 6.4).

2. **Function partition of urban–rural transformation**

Adopting the trajectory computing method, there are 19 categories of the types of changes in urban–rural transformations. Considering the change characteristics of different stages during the study period, the study area was delimited to the following four functional zones: dominant transition area includes 3334, 3444, 4344, 4434, and 4444 trajectory change; key transition area includes 2223, 2232, 2233, 2333, 3333, and 3433; potential transition area contains 1112, 1122, 1222, 2112, 2122, and 2222; restricted transition area contains 1111 and 2111.

The dominant transition area is mostly located in the central part of the BTH region, including 45 districts or counties with higher-level or high-level CTD (Fig. 6.15). This region is the earliest urban–rural transformation region and gradually forms the core area, which has the highest urban population density and is the most developed socioeconomic center in the BTH region.

The key transition area includes 44 districts or counties with no remarkable difference in the number of dominant transition areas. It is obvious that this key transition area is distributed around the dominant transition area because the rural residents in regions closer to the cities tend to have a higher proportion engaged in non-agricultural activities than do those in regions farther away from those cities. The counties or districts in this area have fluctuating CTDs, ranging from middle-level to high-level. Restricted by the limited land resources of the dominant transition area, the key transition area forms and develops while admitting a floating population for which land development must bear the increasing urban population and non-agricultural industrial development.

The potential transition area includes 111 counties or districts, accounting for almost 55% of the total study units. This region's CTD is generally developing from low-level to middle-level. Yet, the region has key driving forces, including population and industry. Many heavy industrial factories in Beijing or Tianjin have moved to Hebei according to the government's regional planning. These actions have brought about an advance in the regional economy and facilitated increasing numbers of rural dwellers to work in factories. Meanwhile, there are still many rural dwellers who have migrated to urban jobs for higher earnings than in rural areas and settled around Beijing, Tianjin, and Shijiazhuang. The citizenization of the new generation of migrant workers has become prevalent with the advance of urbanization. Thus, the main regional driving factors are population and industry transformations. However, serious problems, such as the weakening of agricultural businesses, village hollowing, and environmental pollution, that have emerged in recent years are limiting regional sustainable development as the

Table 6.4 P_i-L_i-I_i values at the breakpoint values

Index	CTD	P_i	L_i	I_i
Breakpoint value	0.4	0.14	0.14	0.72
	0.6	0.38	0.42	0.88
	0.8	0.77	0.67	0.92

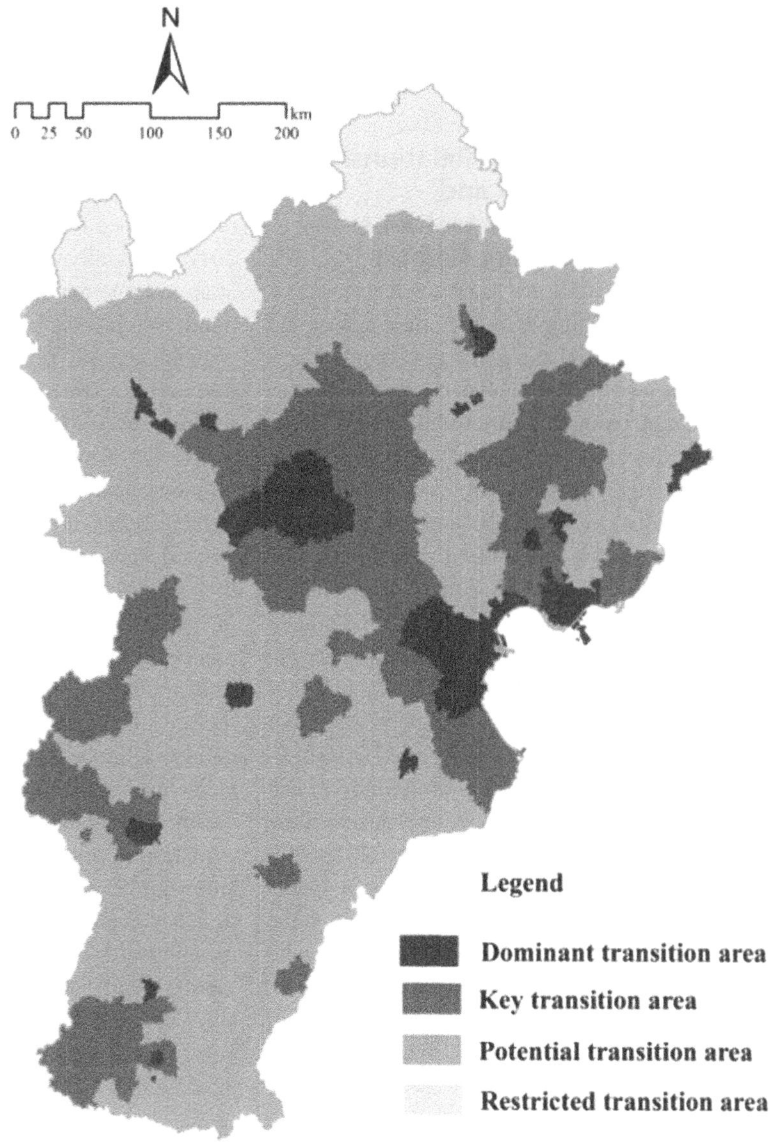

Fig. 6.15 Function partition of urban–rural transformation in the BTH region

region lacks a coordinated population-land-industry system. In this context, the building and development of sustainable rural areas will play an important role in coordinating urban–rural transformation.

The restricted transition area includes 3 counties whose CTD's are at a low level. Without the development of local industry, the population was easily attracted to Beijing and Tianjin, creating a poverty belt around Beijing.

The restricted transition area is a part of this poverty belt. Obviously, the rural development in the restricted transition area falls behind the development of other regions. With rich eco-tourism resources and greater cultural diversity, industry transformation will consider protecting the environment from being polluted and will promote rural development for regional tourism and leisure services.

6.3 Spatial Coupling Relationship of Population, Land, and Industrial Transitions in the Bohai Rim Region

Urban–rural transition is an important aspect and process of economic and social development in a region. As populations and industries move to cities, urbanization is manifested in the larger sizes of cities and the emergence of new cities in terms of regional space, and the increasingly higher proportion of urban population, higher proportion of non-agricultural industries, and higher proportion of urban construction land in terms of the relationship between regional urban–rural structures. The academic field of geography has carried out a series of studies on the relationship between urbanization and population de-agriculturalization, urbanization and industrialization, population and land urbanization, and urbanization and economic development, yielding fruitful results (Li 2008; Chen et al. 2010; Liu et al. 2008; Han et al. 2013). Urban–rural spatial and industrial transitions are closely related to the de-agriculturalization of population, land, and industry. The de-agriculturalization of population, employment, and industry drives constant changes in regional spaces. The rapid de-agriculturalization of land and cheap land input into the process of urbanization are depriving an increasingly large number of farmers of land. Some studies have shown that the speed of land urbanization in China is generally faster than population urbanization (Liu et al. 2013). The core of new urbanization lies in the urbanization of people. The continuous outflow of rural people, as well as urban living costs and institutional constraints, lead to a continuous increase in migrant workers living in both urban and rural areas. Theoretically, the coordinated linkage of population, land, industry, and employment de-agriculturalization is an objective requirement to form a reasonable and orderly urban–rural regional spatial pattern and ensure a sound urban–rural transition.

As noted above, the Bohai Rim Region is the third pole of China's economic growth after the Pearl River Delta and the Yangtze River Delta. Economic development has spurred fast urbanization in the Bohai Rim Region. The urbanization rate increased from 22.83% in 1986 to 56.02% in 2012. Especially since 2000, it has been increasing at an average annual rate of over one percentage point, making the region one of China's fastest growing regions in terms of urbanization during this period. Moreover, the region's land urbanization rate continues to increase, from 1.96% in the middle of the 1980s to 4.52% in 2010, and has grown by nearly two percentage points since 2000. In the process of urban–rural transition, the coordinated transition of economy, industry, population, and land is an important guarantee for the integration of urban and rural development. To explore the practical coordination of the urban–rural transition and the performance of different regions, this section, from the perspective of the de-agriculturalization of urban and rural system elements, sheds light on the characteristics of and existing problems in spatial coupling and the co-evolution of population, land, and industrial transition in the process of rapid urban–rural transition. It provides the basis for establishing a linkage mechanism for the de-agriculturalization of population, land, and industry in rapidly urbanized areas, as well as for regulating key aspects, and consolidates the theoretical basis for the optimal allocation of urban–rural land use.

6.3.1 Coupling Analysis Method for Multiple Factors

Population urbanization, industrial and employment de-agriculturalization, and land urbanization represent urban–rural transition from the perspective of the transition of urban–rural population, industry, and land use, respectively. The flow of population to cities, industrial upgrading, transition, and agglomeration, as well as land occupation for urbanization have constantly changed the pattern and form of urban–rural land use in a region. A scientific urban–rural transition should be supported by a coordinated transition in the structures of population, industry, and land. Excessive or lagging performance in any aspect may cause problems in urban–rural transition. For example, in Latin America, excessive population agglomeration and the absence of supporting industries may lead to slums. Land de-agriculturalization that is faster than population agglomeration and industrial transition may cause urban space expansion and inefficient land use. The coupling analysis of population, industrial, and land transition is helpful to identify the trends in regional urban–rural transition, and to capture the geographic spatial difference in transition using the GIS spatial analysis method. The three aspects of population, land, and industry in urban–rural transition are represented by three indicators: population urbanization rate, land urbanization rate, and industry de-agriculturalization rate. The calculation formula is shown in Table 6.1.

Population urbanization, industrial de-agriculturalization, and land de-agriculturalization are interrelated and mutually influenced, either positively or negatively, in a multilevel relationship. Therefore, the coupling coordination of the three aspects is the primary premise of smooth urban–rural transition. Population urbanization and industrial de-agriculturalization drive land de-agriculturalization and urbanization. Sound urbanization means that population, industry, and land use are in harmony, in which people move completely from rural to urban areas, settle down, and live and work in peace and contentment. Rational industrial transition and upgrading can boost employment and support efficient and intensive use of urban land. To explore how population urbanization, industrial de-agriculturalization, and land urbanization are coordinated in the urban–rural transition of the Bohai Rim Region, a model for measuring coupling coordination degree was introduced in this study. Physics developed the concept of coupling, where it means that the order and structure of two (or more) systems or forms of motion that reach a critical area are determined through various interactions and how they are coordinated. In other words, coupling determines the tendency of systems from disorder to order (Liu et al. 2005). With reference to the capacity coupling model in physics, the coupling degree model for the interaction of multiple factors is generalized, i.e.,

$$C_n = \left\{(u_1, u_2, \ldots, u_m)/\left[\Pi\left(u_i + u_j\right)\right]\right\}^{\frac{1}{n}},\quad u_i(i = 1, 2, 3, \ldots, m) \tag{6.10}$$

In this study, $n = 3$, so the coupling degree evaluation model for population urbanization, industrial de-agriculturalization, and land urbanization is:

$$C = \left\{\frac{U_i \times I_i \times L_i}{[(U_i + I_i + L_i)/3]^3}\right\}^{1/3} \tag{6.11}$$

To further represent the coupling coordination relationship between any two of population urbanization, industrial de-agriculturalization, and land urbanization, the model can be converted into:

$$C = \left\{\frac{U_i \times I_i}{[(U_i + I_i)/2]^2}\right\}^{1/2} \text{ or } \quad C = \left\{\frac{I_i \times L_i}{[(I_i + L_i)/2]^2}\right\}^{1/2} \text{ or } \quad C = \left\{\frac{U_i \times L_i}{[(U_i + L_i)/2]^2}\right\}^{1/2} \tag{6.12}$$

The coupling degree represents the intensity of interaction among the factors. To further explore the coordinated development level among the three factors, it is still necessary to conduct a comprehensive evaluation of the coupling coordination degree. For the coordination degree of interactive coupling among population urbanization, industrial de-agriculturalization, and land urbanization, the calculation model is shown in Formula 6.13.

$$D = \sqrt{C \times T}, T = \alpha U + \beta I + \lambda L \quad (6.13)$$

where C is the coupling degree; D is the coupling coordination degree; T is the comprehensive evaluation index of population urbanization, industrial de-agriculturalization, and land urbanization in a region; and α, β, and λ are the undetermined coefficients. In terms of the three aspects of regional urban–rural development, higher degrees of population urbanization and industrial de-agriculturalization, as well as a reasonable increase of the land urbanization rate are both the requirement and objective of the sound development of urban–rural regional systems. Based on this, the undetermined coefficients are: $\alpha = 0.40$, $\beta = 0.40$, $\lambda = 0.20$.

Coupling coordination relationship between any two of population urbanization, industrial de-agriculturalization, and land urbanization:

$$\begin{aligned} D = \sqrt{C \times T}, T = \alpha U + \beta I \quad \text{or} \\ T = \beta I + \lambda L \quad \text{or} \quad T = \alpha U + \lambda L \end{aligned} \quad (6.14)$$

To measure the degree of coupling coordination between population urbanization and industrial de-agriculturalization, the undetermined coefficient is 0.50. If land urbanization is involved in the measurement, the undetermined coefficient for land urbanization is determined as 0.4, and the undetermined coefficient for population urbanization and industrial de-agriculturalization is determined as 0.60.

6.3.2 Spatiotemporal Process of Population-Land-Industry Coupling Coordination

Population urbanization, land urbanization, and industrial transition and upgrading drive urban–rural transition. The coupling degree of non-agriculturalization of the three factors represents the intensity of interaction among the three population, land, and industrial subsystems. Rapid urbanization is accompanied by a steady movement of people from rural to urban areas. The continuous transition of rural areas to urban areas leads to a constant increase in the scale of urban construction land, the conversion of a great deal of agricultural land to urban land, and the non-agriculturalization and agglomeration of various production factors in urban areas, thus propelling regional economic and social development. Changes in economic aggregate and structure are further reflected in a higher degree of urbanization, and lead to a rigid increase in demand for urban construction land and an increase in the area of urban construction land. Land is not only a basic carrier to support industrial development, but also a spatial constraint. Its various functions support the continuous transition and evolution of industrial development. Changes in industrial structure directly spurs changes in land use and affect the spatial distribution of land use. The distribution characteristics and production efficiency of industries are restricted by land-use patterns and procedures, and they change land-use structures. Sound urban–rural transition requires the effective non-agriculturalization of population and employment. When land urbanization and population urbanization are coordinated, land urbanization is consistent with industrial transition. A healthy urbanization model features the linked and coordinated non-agriculturalization of population, land, and industry. The urbanization model with low-cost land input needs to be changed. Therefore, exploring a linkage mechanism and an allocation theory for the synchronous coordination of population, land, and industrial non-agriculturalization is of great importance for promoting coordinated and sustainable development in a region.

The Bohai Rim Region is now in a stage of rapid urban–rural transition, and population, industrial, and land structures are changing drastically in the region. Based on the comprehensive evaluation using the coupling model, the spatial differences in the intensity of interaction among population, land, and industrial non-agriculturalization in the region are significant.

(1) In 2000 and 2015, the areas with at least a 0.60 coupling degree of population, land, and industrial non-agriculturalization were mainly concentrated in Beijing and Tianjin, in the plains of Hebei, in the middle and southern Liaoning areas with Shenyang as the center, and in most areas of Shandong. Industrial agglomeration, transition, and upgrading in economically developed areas accelerate regional urbanization, and the land demand for industrial development, infrastructure, and living space continues to expand, resulting in the conversion of a great deal of farmland into construction land. The spatial expansion of land urbanization occurs extremely rapidly. As large cities have a diversified industrial composition, and mass population aggregations support the development of services at different levels, land-lost farmers in the suburbs of these large cities are more likely to shift to non-agricultural jobs after land acquisition, as compared with the land-lost farmers in less developed areas. Population urbanization, land urbanization, and industrial non-agriculturalization have consistent response sensitivity.

(2) The areas with a low coupling degree are mainly concentrated in the Yanshan–Taihang mountainous areas and the Changbai mountainous areas, where the coupling degree was generally lower than 0.600 in 2000 and 2015. These areas have poor natural and geographical conditions as well as a single industrial structure. The primary industry accounts for a large share, and industrialization is still in its infancy. Urbanization has a weak driving effect on industrial and employment transitions. Farmland is converted to construction land at a low scale and a slow speed. The correlation and interaction among population, land, and industrial non-agriculturalization are weak.

(3) From the perspective of the temporal dimension, the coupling degree of population, land, and industrial non-agriculturalization presented an overall rising trend in the Bohai Rim Region. The areas with a high coupling degree were further strengthened and had a larger coverage. The differences were still significant in the region (Fig. 6.16). Due to differential ground rent, land supply was concentrated in more advanced industries. From the perspective of land-use benefit, along with economic development and upgrading as well as rapid urbanization and industrialization, land resources in the region are bound to be supplied according to the demand for urbanized production, living, and ecological space; the scale of land urbanization is bound to increase; and industrial development is bound to drive population and employment non-agriculturalization in the process of a larger gross scale and a better industrial structure. After economic development enters the late (or developed) stage of industrialization, it follows that land use, industry, and employment will be adjusted gradually toward commerce and services. In the evolutionary process of industrial development in the region, the intensity of interaction, and the degree of correlation among the population, land, industry, and employment, industrial non-agriculturalization is bound to be enhanced, showing an overall rising trend in the coupling degree of population, land, and industrial non-agriculturalization. According to Lewis's theory of dual economic structure, when there is no difference in the income of urban and rural labor forces, the differential income effect of urban–rural factor flow may disappear, and the interaction and linkage among population, land, industry, and employment may come into balance.

A comprehensive study of whether the regional urban–rural transition is reasonable and orderly requires an analysis of how population, land, and industrial non-agriculturalization is coordinated, in addition to an integrated consideration of their interaction intensity and correlation degree. According to the results calculated from the coordination degree model, the spatial distribution pattern of the coupling coordination degree among population, land, and industry in the Bohai Rim Region is consistent with the coupling degree of population, land, and industrial non-agriculturalization. In hilly and mountainous areas with a low coupling degree, the coupling coordination degree tends to present a low degree of orderly coordination. The coordinated evolution of population, industrial, and

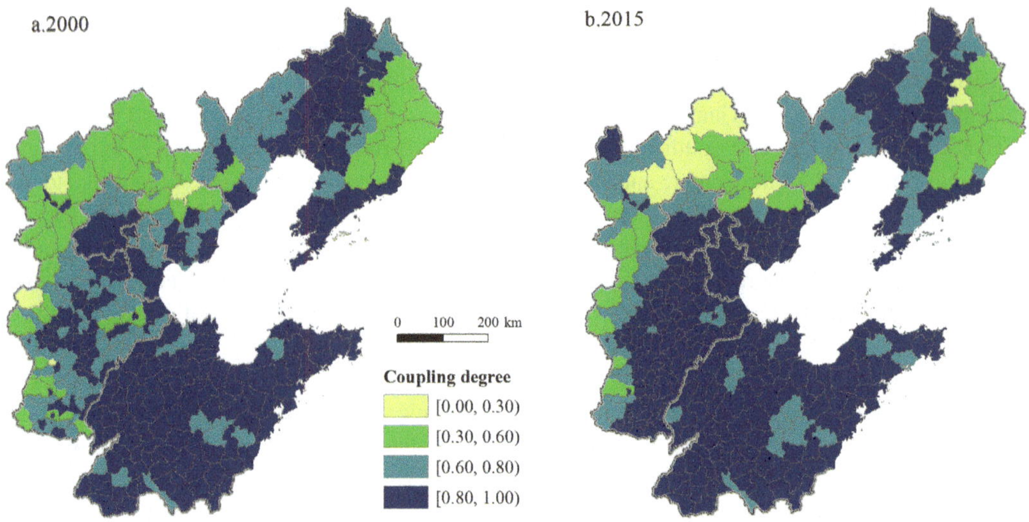

Fig. 6.16 Spatial distribution of coupling degrees of population, land, and industrial non-agriculturalization in the Bohai Rim Region in 2000 and 2015

land non-agriculturalization in the region is a kind of low-level synergistic evolution. Industrial development is mostly in the primary stage, and land urbanization and population non-agriculturalization have slow response sensitivity. In areas with a high coupling degree, the coupling coordination degree is also high. Population, land, and industrial non-agriculturalization manifests as high-level synchronous evolution (Fig. 6.17). From the perspective of time sequences and spatial patterns, as urban–rural transition advances, the areas with an improved coupling degree of population, land, and industrial non-agriculturalization in the region are mainly concentrated in the areas with high coupling degrees (0.80–1.00) (Fig. 6.18), while the areas with an improved coupling coordination degree are concentrated in the areas with medium and high coupling degrees (0.60–0.80) (Fig. 6.19). In 2000, 14.11% of the total research units in the Bohai Rim Region had a coupling degree lower than 0.60, and the coupling degree of most districts and counties was greater than 0.80, accounting for 58.59%. In 2015, the proportion of research units with a coupling degree lower than 0.40 decreased to 8.90%, and the proportion of districts and counties with a coupling degree greater than 0.80 increased to 73.01%. In 2000, 54.91% of the districts and counties had a coupling coordination degree between 0.4 and 0.6. In 2015, the proportion of districts and counties with a coupling coordination degree between 0.4 and 0.6 decreased to 29.45%, while the proportion of districts and counties with a coupling coordination degree between 0.60 and 0.80 increased to 52.15%. The coupling degree of population, land, and industry in the Bohai Rim Region was analyzed in groups using the cumulative frequency analysis statistical technique. With constant economic and social development as well as ongoing urban–rural transition in the region, the coupling degree of population, land, and industrial non-agriculturalization presents a wavelike and progressive evolution pattern from low to high levels. Considering the statistical curve of the evolution frequency of the coupling coordination degree, the crest of the evolution curve of the coordination degree keeps advancing and finally evolves into an approximately S-shaped curve. This is in line with the characteristics of a dissipative structure in which factors, structures, and functions within the open "population-land-industry" system interact with each other. System entropy decreases gradually in an open system, and the coordination degree gradually evolves to

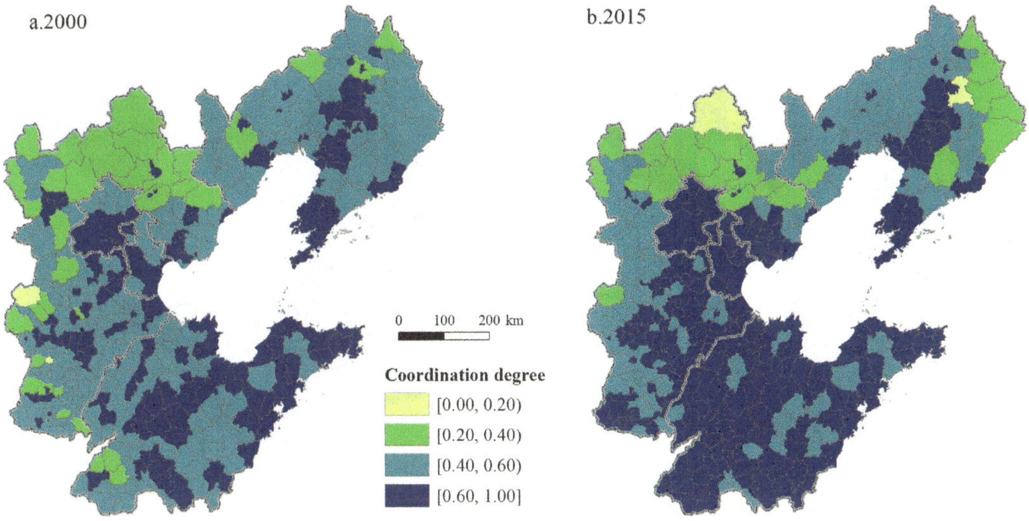

Fig. 6.17 Spatial distribution of the coupling coordination degree of population, land, and industrial non-agriculturalization in the Bohai Rim Region in 2000 and 2015

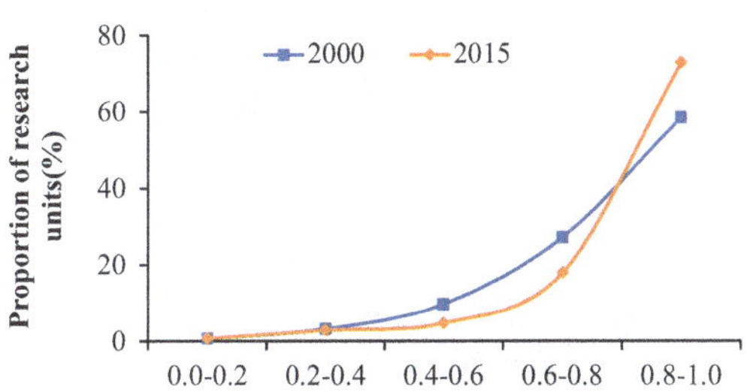

Fig. 6.18 Evolution curve of the coupling degree of population, land, and industrial non-agriculturalization in the Bohai Rim Region

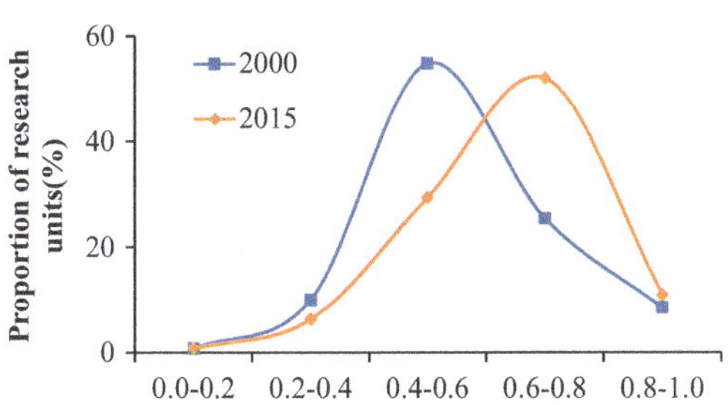

Fig. 6.19 Evolution curve of the coordination degree of population, land, and industrial non-agriculturalization in the Bohai Rim Region

a high coordination stage. After regional polarization and diffusion, the population, land, industry, and employment in the regional urban–rural system eventually reach a sound state of highly coordinated self-organization and self-regulation, and sub-spatial units as well as urban and rural areas within the system move toward a balanced and integrated development pattern.

6.3.3 Spatiotemporal Pattern of Coupling Coordination of Any Two of Population, Land, and Industry

The coupling degree of population, land, and industrial non-agriculturalization reflects the overall level of coordination, while the coupling coordination degree of any two of population, land, and industry further reveals the degree of coordination in the transition of the three systems. The extent and range of the coupling degree and coupling coordination degree of population–industrial non-agriculturalization and land–industrial non-agriculturalization are higher than those of population-land non-agriculturalization (Figs. 6.20 and 6.21). The long-term dual structure system of urban and rural land ownership, in terms of the land use allocation mechanism, is more presented as an administrative land allocation model dominated by the government,

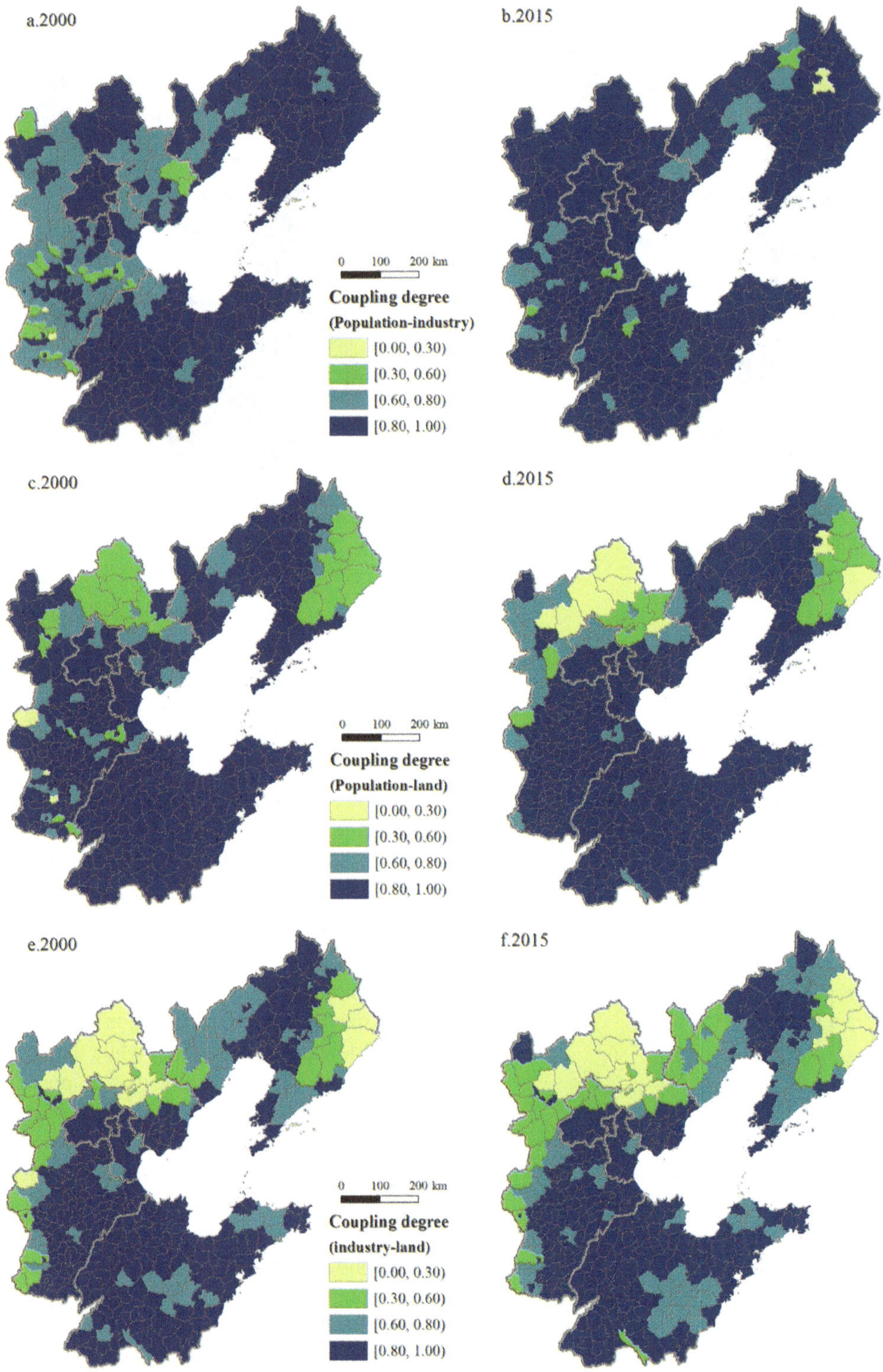

Fig. 6.20 Spatial distribution of the coupling degree of population–industrial non-agriculturalization, population–land non-agriculturalization, and land–industrial non-agriculturalization in the Bohai Rim region in 2000 and 2015

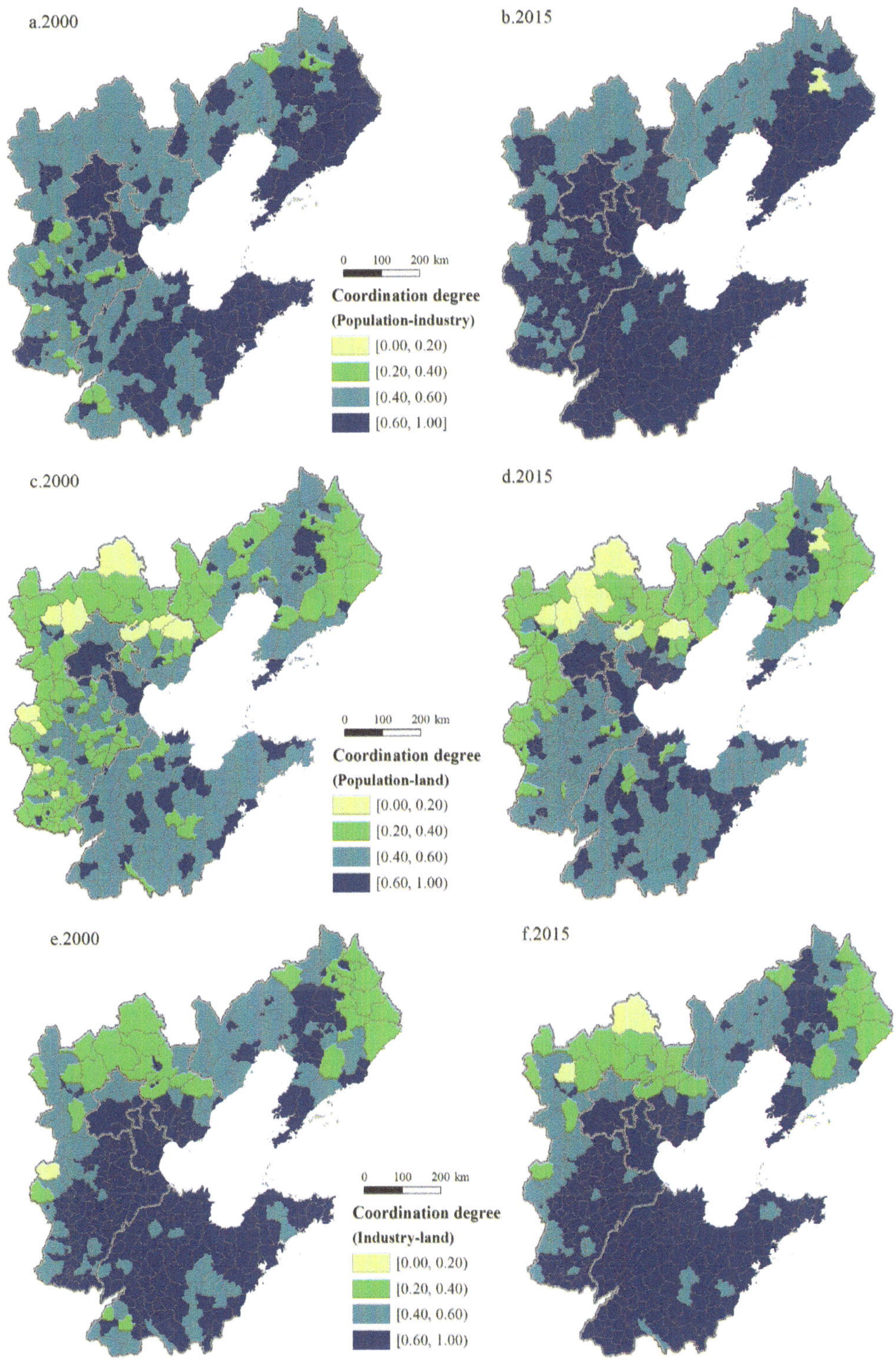

Fig. 6.21 Spatial distribution of the coupling coordination degree of population–industrial non-agriculturalization, population–land non-agriculturalization, and land–industrial non-agriculturalization in Bohai Rim in 2000 and 2015

resulting in a natural separation of urban and rural land allocation. In consideration of the spatial difference in social costs of land acquisition, governments—driven by the price scissors of economic benefits—are more willing to use administrative means to acquire rural land in urban suburbs. The low cost of land acquisition, as well as the monopolized secondary market of land, continuously promote the expansion of "land finance" and the sharp increase in land-lost farmers in the process of land non-agriculturalization. An urbanization model that excessively relies on land may lead to the rapid growth of land for urban construction, transportation, industry, and commerce, as well as the rapid expansion of urban space, leaving many "villages" in cities. Meanwhile, urbanization with low-cost land input may lead to an objective situation where land urbanization is far faster than population urbanization. Governments often use low land cost as the incentive to attract investment, forming an industrialization and urbanization model with low land cost. Although land urbanization drives industrial development and promotes the growth of economic aggregate in the region, many farmers lose their land and have not actually lived and worked in peace and contentment as urban residents.

The urbanization of land cost may promote urban–rural transition and the rapid development of the regional economy in the short term, but it is an unsustainable mode of land resource allocation, and is the root cause of the inefficient utilization of land resources as well as the low performance effect of land non-agriculturalization. From the evolution process of the coupling coordination degree of any two of population, land, and industrial non-agriculturalization, the development process of the coupling coordination degree of land–industrial non-agriculturalization is obviously faster than that of population–industrial non-agriculturalization as well as population–land non-agriculturalization (Figs. 6.22 and 6.23). In particular, population–land non-agriculturalization has the worst coordination degree and a slow evolution, proving that aggressive regional urbanization exists, i.e., an urbanization model that prefers land over population. In the process of rapid urbanization, urban space is usually utilized at low densities and in a decentralized manner. The scale of construction land expands; the hierarchical structure is unreasonable; the intensity of investment and development is low. As a result, the problems of idling and the extensive use of urban land are also prominent. The large-scale, long-distance, and migratory movement of population, to a certain extent, reflects defects in China's land and employment systems. In the urbanization model that prefers land over population, there is no mechanism linking land non-agriculturalization to population urbanization.

6.3.4 Policy Suggestions on Optimizing and Regulating Population, Land, and Industrial Non-agriculturalization

The optimal allocation of regional land use is an important means to address the imbalance in population, land, and industrial non-agriculturalization, as well as an essential condition and an important foundation for the sustainable development of regional economy and society. To a certain extent, it determines the patterns of regional economic development, land development, and the urban–rural relationship. In the new period of urban–rural transition, the contradiction between a more rigid demand for land resources and an insufficient supply of land intensifies, as does the exuberant demand for public service facilities land, industrial land, and residential land for urban municipal construction; the inefficient use and inertial increase of land also drive high-speed land non-agriculturalization. The inefficient use of urban land coexists with the abandonment and waste of rural land, leading to imbalances in population, land, and industrial non-agriculturalization, which is reflected in the fact that land urbanization is faster than population urbanization, and population urbanization lags behind industrial non-agriculturalization. First, in view of the problem that land urbanization happens faster than population urbanization, a unified trading platform for urban and rural land

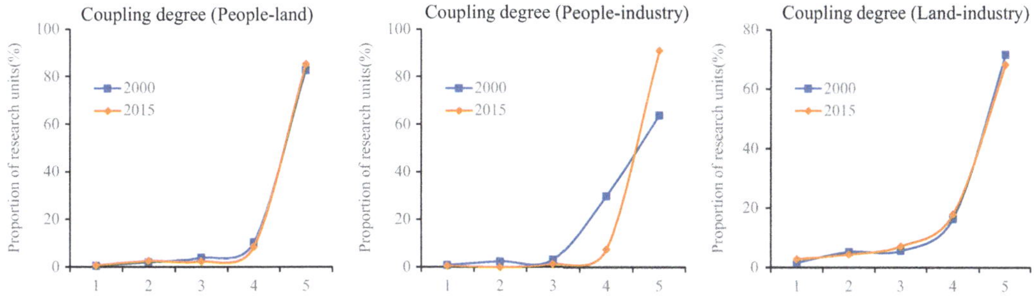

Fig. 6.22 Evolution curve of the coupling degree of population–industrial non-agriculturalization, population–land non-agriculturalization, and land–industrial non-agriculturalization in the Bohai Rim Region in 2000 and 2015

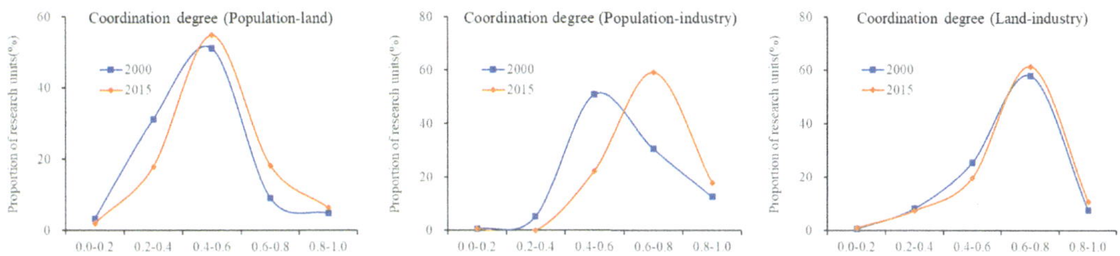

Fig. 6.23 Evolution curve of the coupling coordination degree of population–industrial non-agriculturalization, population–land non-agriculturalization, and land–industrial non-agriculturalization in the Bohai Rim Region in 2000 and 2015

should be built based on the inherent correlation and interaction between urban and rural development and in accordance with the principle of same land same price between urban and rural areas. Inhibiting the government's behavior of "low-cost acquisition and high-price sale" and replacing the government-led allocation mode with the market price allocation mechanism are conducive to improving the intensive and economical use of construction land in the process of urbanization. An innovative mechanism for overall urban and rural land planning should be created to make reasonable arrangements for urban and rural land in terms of time, space, and industry. It is important to integrate land acquisition in the process of urbanization with efforts to guide enterprises, develop industries, and expand employment. The mechanisms for basic land supply, planning and decision-making, market allocation, and income distribution should be improved, and a long-term mechanism that links urban and rural land, population and land, and land and industry should be established. Moreover, efforts should be made to promote the rational circulation of land used for urban construction and improve land-use efficiency. The contradiction between the rigid increase in demand for urban construction land and its insufficient supply should be solved, and "double constraints" on the intensity of urban land investment and the progress of land use should be strengthened. New ways should be actively explored to develop and reallocate idle and inefficient land, and a new coordination mechanism for land, population, and industry in urbanization should be established. Secondly, the industrial structure and development level in a region directly affect regional economic development, and determines the employment demand and structure in the region. Despite rapid urbanization, there is a misunderstanding of the theory of urbanization in which urbanization is the driving force for regional economic and social transition and stimulates regional demand, consumption, and economic development. Urbanization is essentially the result and process of economic and social development. According to local conditions, urbanization is supported by the efforts to reasonably plan market environment and resource endowment, steer industrial development structures, guide suitable enterprises, and stimulate the development of services. As the core driving force for urbanization, industrial development should be advanced to promote non-agricultural employment of population, facilitate a sound mechanism of "city–industry integration," and avoid unreasonable urbanization that has either "industry without city" or "city without industry." Systems of coordinated urban and rural housing, household registration, and social security should be gradually established, and rural labor forces that shift to non-farm jobs should be guided in a scientific and orderly way to live and work in peace and contentment, so as to ensure that rural labor forces no longer live in both urban and rural areas. It is also important to change the land-dependent urbanization model; reverse the rapid non-agriculturalization, idling, and inefficient use of land; reform the traditional urbanization model; and appropriately control the reasonable process and spatial pattern of urbanization. At the level of the land system, a multilevel land conservation and intensive use system should be established that integrates the basic land system, the core land system, and the security system.

6.4 Spatiotemporal Patterns of Urbanization and Rural Coordinated Development in the Bohai Rim Region

Urbanization is a dynamic process of rural–urban transition, including population migration, employment change, lifestyle transformation, and systematic evolution of spatial structure driven by the modernization of traditional agriculture-oriented economic structure. Urbanization is also an important pattern of manifestation for urban–rural transformation, demonstrating the decentralization to centralization transformation of population during the process of urban–rural transformation. Given the close link between demographic urbanization and the socio-economic development stage, urbanization, in a broad sense, contains both economic growth and structural transition, as well as social transformation. Rural transformation indicates the transformation of traditional industries, employment modes, and consumption structures in rural areas, which is helpful in transforming the traditional urban–rural isolated society to a harmonious integrated one. The essence is to promote the fundamental transformation of relationships between manufacturing and agriculture and between urban and rural areas. Obviously, urbanization is an important driving force for rural transformation. Constrained by industrial structures and urban–rural institutions, impacts of urbanization on rural transformation may vary in different regions. Thereupon, it has potential contribution to explore the coupling relationship between urbanization and rural transformation in summarizing their interaction and identifying regional differences of urban–rural transformation.

In recent years, the state successively issued some policies to support the construction of agriculture, rural areas, and peasants (*sannong*), issues like weakening of rural labors, limited increase of yield with increasing agricultural production, rapid non-agriculturalization of farmlands, rural hollowing, and difficult citizenization of rural migrants still exist with the long-term mode of urban-biased and economic-oriented development (Lu and Yao 2007; Liu 2010). Thus, it is difficult for rural areas to realize the effective development and restructuring with urbanization development, which consequently makes it difficult to synchronously realize rural prosperity and urban booming. The Third Plenary Session of the 18th CPC Central Committee facilitated the symbiotic development of middle-sized and small cities and towns, the integrative development of industries and towns, and the harmonious promotion of urbanization and new rural construction. In general, rural development is simultaneously driven by exogenous power of urbanization and endogenous power of rural construction. So that, it is necessary to explore rural development paths and urban–rural interactive modes and to improve relationships between industry and agriculture and between urban and rural areas under the urbanization background (Zhang and Liu 2008). With the constant development of urbanization, the traditional growth mode taking cities or towns as the center must be changed by focusing on the impact of urbanization on rural development (Antrop 2004). In this section, we construct a comprehensive evaluation index system of urbanization and rural development, quantitatively detect the spatial–temporal patterns of regional urbanization and rural development, explore the rural development trend under different urbanization levels and the harmonious status between the pair, and conclude with some policy suggestions on coordinating urbanization with new rural construction and balancing urban and rural development in the Bohai Rim region.

6.4.1 Data Sources and Research Methods

1. **Construction of the evaluation index system**

Urbanization is a necessary trend and an important indicator of regional development. It is also a crucial index to measure the social organizational degree and management level of a state or a region. Upholding principles of systematisms, scientific, dominance, and operability, indices from triple dimensions of agriculture, rural areas, and peasants were chosen to construct the rural development level index system based on the "element composition—structural relationship—benefit collaboration" (Table 6.5). The urbanization process is accompanied with population migration and land use transition. Under the people-centered urbanization philosophy, the harmonious development of "population-industry-land" should be highlighted. Thus, index selection ought to weaken the scale and strengthen the quality of urbanization. As Table 6.6 reported, a triple dimensional system of indices constructed with indicators regarding population, economy, and land. In the section, the population in 2000 and urban population data came from the fifth population census, while partially missing data in 2015 were polished by measuring the sixth population census. The sum between the urban settlement land and independent industrial and mining areas from the land-use alteration database was used to reveal the construction area. Outliers seldomly witnessed in some statistics were corrected by interpolation with data in years before and after.

2. **Measurement of urbanization and rural development level**

To eliminate the impacts of the subsystem index data dimension and make it comparable, standardization processing was conducted. Both subjective and objective approaches are employed to identify the weight of index. The subjective method evaluates the relative importance of indices according to the subjective cognition of evaluators; the objective method determines the weight in accordance with the amount of information provided by original values of indices. The empowerment information comes from the objective environment. As an objective weighting method, the entropy evaluation method calculates the weight according to the degree of uncertainty in the measurement system state. The more unbalanced system, the lower the information entropy, and the larger the index weight. We hereby apply the entropy evaluation method to endow weight for indices in the system. Moreover, a multi-target weighting sum model was conducted to measure the urbanization and rural development level index with the specific calculation method as follows:

Nondimensionalization of indices:

$$X'_{ij} = \left(X_{ij} - \min\left(X_j\right)\right) / \left(\max\left(X_j\right) - \min\left(X_j\right)\right) \tag{6.15}$$

Development level index:

$$F_i = \sum_{j=1}^{n} w_j X'_{ij} \tag{6.16}$$

Table 6.5 Evaluation index system of rural development level

Target layers	Structure indexes	Specific indexes	Description	Weights	Index nature
Rural development level	Agricultural production	Mechanical power input	Total agricultural machinery power/agricultural acreage (kw/hm^2)	0.1824	Positive
		Crop diversification	The proportion of non-grain acreage in the total seeded area (%)	0.0562	Positive
	Life of farmers	Rural residents' income	Per capita net incomes of rural residents (yuan/person)	0.1797	Positive
		Per capita electricity consumption	Rural electricity consumption/rural population (kW·h/person)	0.3258	Positive
	Rural economy	Labor productivity	Gross output of agriculture, forestry, animal husbandry, and fishery/agricultural employees (yuan/person)	0.2559	Positive

Table 6.6 Urbanization level evaluation index system

Target layers	Structure indexes	Specific indexes	Meanings of indexes	Weights	Index nature
Urbanization level	Population urbanization	Urban population proportion	Urban population/total population (%)	0.3633	Positive
	Economic urbanization	Proportion of secondary and tertiary industries	Output proportion of secondary and tertiary industries (%)	0.1578	Positive
		Per capita GDP	Per capita GDP (yuan/person)	0.1196	Positive
	Land urbanization	Economic density in construction areas	The output of secondary and tertiary industries/covered areas (ten thousand yuan/km^2)	0.1862	Positive
		Population density in construction areas	Urban population/covered area (person/km^2)	0.1732	Positive

Notes Construction land area denotes the sum of urban settlement land area and independent industrial and mining land area

where X_{ij} represents the original index; X'_{ij} denotes the dimensionless index; F_i is the urbanization or rural development level index of ith sample; and w_j refers to the weight value of jth index.

3. **Coordination index model**

Coordination is a concept that describes the benign interaction between objects, implying consistent harmony and proper coordination. Scholars generally use the coordination index to measure the degree of interaction and mutual impacts between multiple systems/elements. During the urbanization process, it is necessary to realize rural prosperity with urban development and synchronous promotion of urbanization and rural development. Hence, the coordination model between urbanization and rural development was constructed in this research.

$$C = \left[(UI \times RI)/((UI + RI)/2)^2\right]^{1/2} \quad (6.17)$$

where UI represents the urbanization level; RI refers to the rural development level; C is the coordination calculation model of urbanization and rural development. The value ranges from 0 to 1. The maximum was the best coordination state; on the contrary, the smaller value, the less coordinated.

6.4.2 Spatial Patterns of Urbanization and Rural Development

1. **Urbanization level pattern evolution in the Bohai Rim**

In 2000 and 2015, 214 and 193 counties in the Bohai Rim region were under the regional mean level of urbanization, accounted for 65.64% and 59.20% of total county-level units respectively. The standard deviations of the urbanization level were 0.1221 and 0.1151, suggesting the decrease of the number of counties with their levels of urbanization under the regional average. Moreover, the gap in urbanization level between counties is witnessing the narrowing trend. To clearly portray the spatial pattern of urbanization, the natural breaking point was employed to divide counties in the Bohai Rim region into different groups of low-level areas (LLA), relatively low-level areas (RLA), relatively high-level areas (RHA), and high-level areas (HLA) (Fig. 6.24). The findings indicates that: (1) The high and relatively high-level areas were concentrated in the Shandong Peninsula, mid-southern of Liaoning, and Beijing-Tianjin-Tangshan (BTT). Western Liaoning, western and northern Hebei, as well as western Shandong, had the relatively low urbanization level; (2) High level areas were scattered in the study area, including districts of prefectural cities; relatively high areas located surrounding their high counterparts; relatively low level areas located in mid-south of Shandong, northwestern Shandong, eastern Hebei, and southern Hebei; low level areas included southwestern Shandong and Yanshan-Taihang Mountains Area; (3) From 2000 to 2015, the dynamic degree showed that the urbanization level in the study area was present in growth at varying degrees. The faster growth areas were concentrated on surrounding counties centered on Beijing-Tianjin, southern Hebei, and the Shandong Peninsula. Urbanization growth in Liaoning Province was relatively slow.

2. **Spatial–temporal features of rural development level in the Bohai Rim**

The standard deviation of rural development level at the county-scale was 0.0427 and 0.0857 in 2000 and 2015 respectively, suggesting that the rural development gap in different counties was slightly increasing. Comparing with the standard deviation of the urbanization level, the rural development level gap was obviously narrower. Similarly, the natural breaking point was applied in dividing the counties based on the rural development level into four groups, namely low, relatively low, relatively high, and high (Fig. 6.25). The findings indicated that: (1) The rural development level in Beijing, Tianjin, and Shandong was undoubtedly higher than that in Liaoning and Hebei. The BTT region, Shijiazhuang, and Shandong Peninsula region were recognized as clusters of high and relatively high-level areas, while western and northern Hebei and western Liaoning were divided into hotspots of the low-level areas; (2) The dynamic degree of the rural development level in the Bohai Rim in 2000–2015 was positive. The BTT region and Liaoning had a higher dynamic degree, indicating the faster enhancement of rural development. Besides, western and northern Hebei had a relatively slower rate of rural development.

6.4.3 Coordination Analysis of Urbanization and Rural Development

1. **Statistical features of coordination**

As centers of regional development, towns have obvious advancement in element gathering and scale economy. City-oriented developmental strategies result in the great gap between urban and rural areas, as well as inconsistency between urbanization and rural development. As an important constituent part of the urban–rural territory system, the development of rural areas is

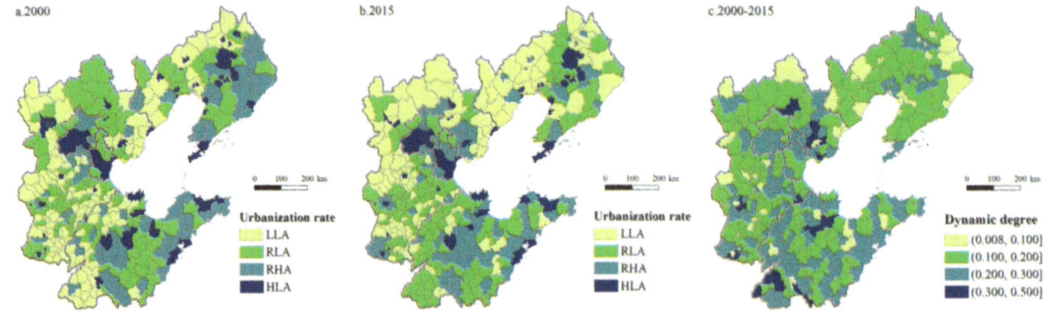

Fig. 6.24 Spatial pattern of urbanization level in the Bohai Rim from 2000 to 2015

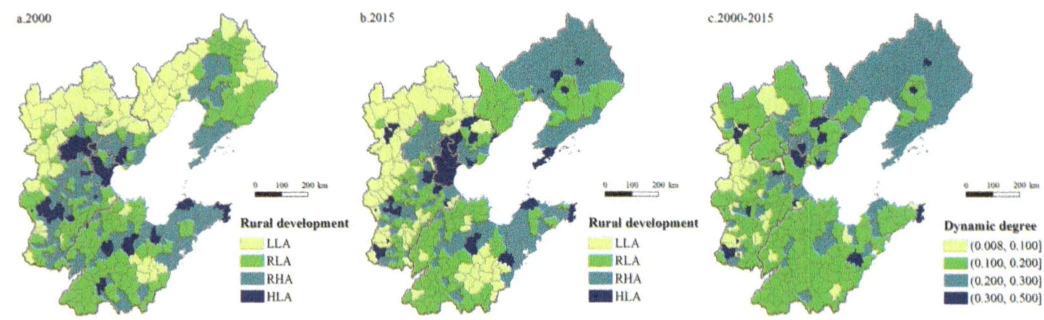

Fig. 6.25 Spatial pattern of rural development level in the Bohai Rim from 2000 to 2015

not only driven by the external urbanization progress, but also pushed by endogenous local factors. Hence, the analysis of coordination between urbanization and rural development ought to focus on the local response of rural development in urbanization progress, which can be reflected in agricultural labor migration, industrial and employment diversification, resident incomes increase, agricultural modernization, and overall enhancement of rural development. To explore coordinated relations between urbanization and rural development is good for a better understanding of rural response to urbanization and their interaction.

Through the correlation analysis at the county-scale, we found that urbanization and rural development were positively correlated (2000, $R = 0.356$, $P < 0.01$; 2015, $R = 0.458$, $P < 0.01$). The correlation coefficient before and after two years increased, implying that the association between rural development and urbanization was closer. Figure 6.26a represented the fitting result of nonlinear model between rural development and urbanization, which is significant at 1% level and provides quantitative support for promoting urbanization to facilitate rural development and narrow the urban–rural gap. However, there are no strict linear relations between urbanization and rural development. Rural development level in high urbanization level areas might be lower, so it is necessary to strengthen the driving ability of urbanization and endogenous power of rural development by coordinating urbanization with new rural construction.

To comprehensively unearth the relation between urbanization and rural development, the coordination degree model was further employed to detect the tightness degree of their mutual relations (Table 6.7). With the increase of the urbanization level, the coordination degree between urbanization and rural development decreased from 2000 to 2015 with the coefficient of variation constantly increasing, indicating that

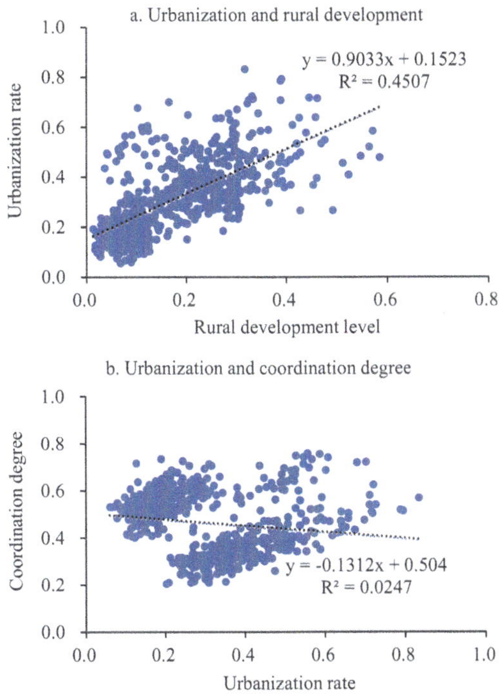

Fig. 6.26 Variation of rural development and coordination degree under different urbanization level in the Bohai Rim

with the enhancement of urbanization, the lower coordination degree between urbanization and rural development, the larger differentiation degree of rural development. In the initial stage of urbanization, the number of rural migrants in towns was limited, urban–rural elements flowed slowly, and coordination level between urbanization and rural development was relatively high. Moving to the speed-up stage, transfer of rural labors, diversification of agriculture, and urban–rural mobility of element accelerated, while the traditional city-centered model of urbanization excessively depended on rural lands, resources, and cheap labor forces. As the periphery with element outflowing, self-development power in rural areas was insufficient. The mutual separation between urbanization and rural development resulted in a reduction of their coordination.

Through correlation analysis of coordination degree and urbanization level at the county-scale in the Bohai Rim, we found a remarkably negative correlation between the pair ($R = -0.6880$, $P < 0.01$), implying that the higher the urbanization level, the lower coordination between urbanization and rural development. As represented in Fig. 6.26b, the goodness of fit is significant at 1%. In the low urbanization level stage, the coordination degree had a higher value and small variation. With the enhancement of the urbanization level, the coordination degree was remarkably decreased. Within the constraint of urban–rural binary mechanism and economic development orientation strategy, the rural development lagged behind the urbanization so that rural areas fell behind towns and the agriculture lagged behind the industry, resulting in "land urbanization" and rural hollowing. Turning into a town-centered development model and integrating the rural development into the urbanization progress are hereby argued as the key to coordinate the urban–rural relationship and facilitate rural development.

2. **Spatial pattern features of coordinating index**

To intuitively reveal spatial features of the coordinating index between urbanization and rural development in the Bohai Rim, we conducted spatial visualization on the coordinating index in 2000 and 2015 with the help of GIS software (Fig. 6.27). The quartile method was used to conduct grading for 325 counties in 2 time periods in the Bohai Rim. Three values at 25% (quartile), 50% (median), and 75% (third-quarters) were deemed as clinical values. In 2000, areas with high coordination degree and relatively high level were concentrated on central Hebei, Shandong Peninsula, Beijing-Tianjin Region, and Shenyang-Dalian Line. The low and relatively low areas mainly located in southwestern Shandong, western Liaoning, and Yanshan-Taihang Mountains Area. In 2015, areas with high and relatively high coordination degrees were concentrated on central and eastern Hebei, Shandong Peninsula, the BTT Region, and the Liaodong Peninsula. From 2000 to 2015, each county in the Bohai Rim realized enhancement on coordination degree of the rural development and urbanization, but the

Table 6.7 Coordination statistics between urbanization and rural development

Coordination	2000		2015	
	Mean	Coefficient of variation	Mean	Coefficient of variation
0.20–0.35	0.3052	0.1360	–	–
0.35–0.50	0.4004	0.1256	0.4587	0.1554
0.50–0.65	0.5411	0.2312	0.5636	0.0956
0.65–0.80	–	–	0.6958	0.1633

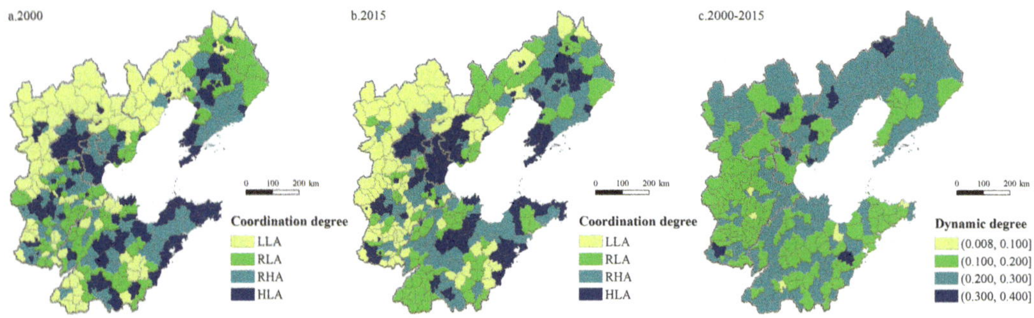

Fig. 6.27 Coordinating index pattern and variation between urbanization and rural development in Bohai Rim

enhancement speed in traditional agricultural areas was obviously slower than that in others.

Urbanization, rural development level, and the coordinating index in each county of the Bohai Rim showed the diverse spatial coupling features. To be specific, Beijing, Tianjin, Shandong Peninsula, and the mid-south of Liaoning had developed economy, optimal upgrading industries, higher urbanization level. Transfer of labor force employment, agricultural modernization, and rural tourist industry facilitated the rapid development in rural areas. These regions were therefore clusters of high and relatively high rural development levels areas. Compared with the rapid urbanization, the rural development degree was inconsistent, which made the differentiation of coordination degree obvious. As important major grain-producing areas in China, counties in western Shandong, southern Hebei, and mid-west of Liaoning were restricted by multiple factors including policies and locations and had some dilemmas, such as a high proportion of grain acreage, lagged agricultural diversification, slow adjustment to regional industrial structures, low rural income, slow urbanization, and insufficient rural development power. Counties in central and northern Hebei give priority to mountains and hills and are the most important ecological protective barriers in northern China, especially for the Beijing-Tianjin metropolitan. The central and local governments have constantly taken measures to limit resource development and industrial and agricultural production in these areas. Due to the impacts of the household registration system and administrative division, the high-speed development of economy in Beijing-Tianjin hasn't functioned well as a high radiation-driven core for surrounding peripheries. County-scale units in major grain-producing areas and ecological preservation areas have witnessed slow economic development and the coexistence of lagged urbanization and rural underdevelopment. The coordination degree of urbanization and rural development in these areas was relatively high at the low level. These areas should promote a combination between local resource environment endowment and national policy support and between market orientation and geo-economics advantages, and then take advantage of latecomers and explore the pathway of urbanization with local characteristics through internal and external interaction.

With the advancement of urbanization, two kinds of coordination declined. Coordination between urbanization and rural development had a significant differentiation across the geological space. Urban–rural element mobility and urban–rural association are the foundation of urban–rural development integration. New-type urbanization and urban–rural transformation should change the land-centered and city-biased development model, focus on promoting an equal exchange of urban–rural elements, reconstruct urban–rural relations, and rebuild the urban–rural system. What's more, it is essential to reinforce the combination between urbanization and new rural construction, to explore the household registration system and land system reform, to perfect the social welfare system, to coordinate the urban–rural industrial and employment mechanism, to gradually realize new rural construction community, moderate land operation scale, and agricultural modernization, as well as to promote basic urban–rural public service equalization and equivalence of urban–rural residents' living conditions. High urbanization level areas should balance the urban–rural development. Due to lagging economic development in major grain-producing areas and ecological preservation areas, areas with low urbanization and rural development levels should intensify their regional major functions, explore the valuation mechanism and long-effect protective mechanism of regional resources, intensify the regional collaborative mechanism, and explore the local pathway and model of urbanization.

6.5 Spatial Patterns of Urban–Rural Transformation in Bohai Rim Region

The urban–rural transformation is multi-level and multi-scale and is recognized not only as the transition of regional development patterns with the evolution of urban agglomeration and economic zone as the core but also as the socio-economic transformation in rural areas featured by changes of resource utilization, industrial structures, employment structures, and consumption structures under the impacts of industrialization and urbanization. As the dominant process of urban–rural transformation, urbanization is widely concerned by scholars and policy makers (Liu and Yang 2015). Urban spatial expansion undoubtedly results in land non-agriculturalization and continuous variation between urban–rural spatial structures and patterns. The current domestic urbanization studies lay particular emphasis on cognition of employment, residence, and sharing of urban social services on the urban–rural spatial structures and their differences by focusing on urbanization trends and spatial–temporal heterogeneities at different spatial scales, but seldom carry out the explorative studies on the steric configuration of urban–rural transformation in certain areas (Chen et al. 2009; Cao et al. 2011; Liu and Yang 2012). Particularly, geographic information science provides an effective approach to do urban–rural spatial type diagnosis, regional pattern cognition, and measurement study with the help of remote sensing and statistical models. The spatial configuration of urban–rural transformation is the core of urban–rural transformation studies. With the grid and geostatistical approaches, integrated GIS and RS were used to focus on the urban construction land and its spatial dynamics, to explore the hotspot clusters, expansion trajectories, and regional differences of urban–rural transformation, to explore and analyze the urban–rural transformation process and pattern, and to highlight the spatial centrality, growth and different laws of urban–rural transformation. These works may have potential in providing basis for scientific decision-making of optimizing the national land spatial development strategy and promoting urban–rural development integration in the Bohai Rim.

6.5.1 Research Methods of Spatial Exploration

1. **Grid analysis of urban–rural transformation**

The visualization of geographical things, phenomena, and processes is concerns of both domestic and overseas academic societies. To better portray the spatial dynamic and scenario prediction of urban–rural transformation, the grid spatial statistical analysis method was applied. The specific practice was to use grids at different scales to record the urban–rural construction land area and the dynamic variations, by which discrete spatial distribution attribute could be attributed to each grid to preliminarily explore and recognize the spatial clusters and dynamic expansion of urban–rural transformation and to make preparations for identifying the self-associated "cold spots" and "hot spots" in the further geostatistical analysis. Based on the grid statistical research advantages, it is convenient for data transition from vector to grid formats, spatial scene modeling simulation, and optimal simulation scale recognition (Fig. 6.28).

2. **Recognition of central clusters and spatial differences**

We generated 5 km × 5 km grid layer covering the whole study area with the help of ArcGIS platform, and then overlapped the urban construction land data with grid layer. Geostatistical analysis approaches were employed to detect urban spatial expansion clusters. The measurement model was shown in the following formulas.

Getis-Odr G_I^* was utilized to detect whether there were local highest or lowest areas of clustering in spatial statistics.

$$G_i^* = \frac{\sum_{j=1}^{n} w_{ij}x_j - \overline{X}\sum_{j=1}^{n} w_{ij}}{S\sqrt{\left[n\sum_{j=1}^{n} w_{ij}^2 - \left(\sum_{j=1}^{n} w_{ij}\right)\right]/(n-1)}} \quad (6.18)$$

$$\overline{X} = \sum_{j=1}^{n} x_j/n \quad S = \sqrt{\sum_{j=1}^{n} x_j^2/n - (\overline{X})^2} \quad (6.19)$$

where x_j was the attribute value of each grid; w_{ij} was the spatial weight of grids i and j. During the exploration process, the spatial searching radius method was applied to construct the spatial weight matrix by regarding 10 km as the correlation distance.

Urban–rural transformation affects the spatial configuration. The construction land expansion represents the growth features of urban areas. The urban–rural construction land expansion intensity index (L_i) was constructed in this thesis to represent the urban–rural construction land expansion speed and intensity of each unit grid within a certain period. Meanwhile, the mean construction land variation speed of a 5 km × 5 km grid was conducted the normalization processing, so that it had horizontal comparability. The measurement was shown in Formula 6.20.

$$L_i = \frac{\Delta U_i}{\Delta t \times TLA} \times 100\% \quad (6.20)$$

where ΔU_i was the increased area of construction land expansion during a period. Δt was the time span. *TLA* represented the total land area of research units. Based on measuring the dynamic intensity of urban–rural construction lands in each unit grid, Formulas 6.18 and 6.19 were adopted to deeply explore urban–rural transformation intensity differences.

3. **Spatial variation probability of urban–rural transformation**

The space of urban–rural transformation is mainly represented as the spatial differentiation pattern of urban–rural construction land. The spatial variation of urban–rural transformation echoes theories of central place in spatial configuration and pole-axis in regional development. Drawing a grid probability map is an effective way to predict the spatial variation trend of construction land. The precondition is using the stepwise logistic regression to model the spatial dynamics of urban–rural construction land and influencing factors. The measurement model was described in Formula 6.21.

$$\log\left(\frac{P_i}{1-P_i}\right) = \beta_0 + \beta_1 x_{1i} + \beta_2 x_{2i} + \cdots + \beta_n x_{ni} \quad (6.21)$$

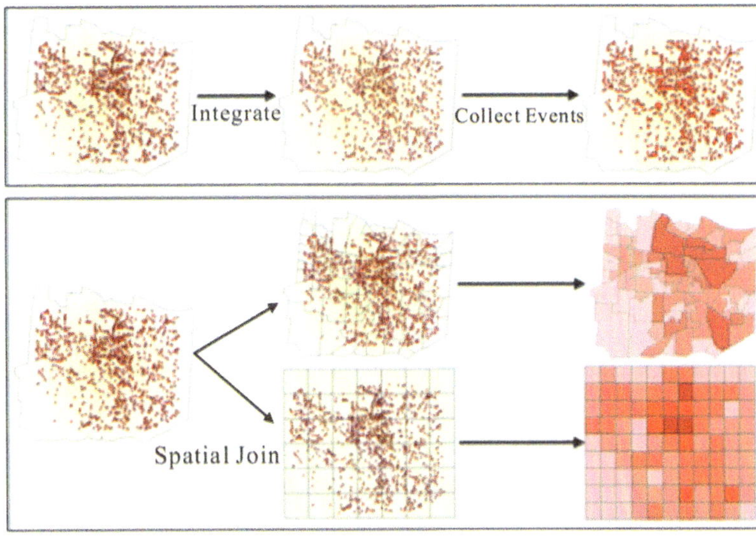

Fig. 6.28 The griding process of samples

where P_i represents the probability of occurring urban–rural construction land type i in each grid; x is an influence factor. ROC test method[2] is applied to verify logistic regression results. Also, the optimal simulation scale is explored and recognized to predict the spatial probability trend of urban–rural transformation. According to the regression coefficients and grid layer of influencing factors, the algebraic operation of weighting grids is conducted to ultimately get the spatial distribution probability map of land use structure.

6.5.2 Central Agglomeration of Urban Transformation

Cites are widely recognized as growth poles of regional development, making central agglomeration one of the important features. With urban–rural transformation, industrialization, and rapid urbanization, urban areas become clustering areas of production elements like assets, techniques, labor forces, and land. With the transformation and upgrading of industries, urban construction land witnessed the "core-gathering" expansion, which consequently affect the spatial structure form of regional urban–rural transformation. The Bohai Rim, as the third growth pole of China's economy, covers two megalopolises of Beijing and Tianjin and regional central cities like Jinan, Shenyang, Shijiazhuang, Qingdao, Tangshan, and Dalian. Based on the spatial pattern of urban construction land, the centrality of urban–rural transformation remains the spatial pattern expanded with the core of some central cities (Fig. 6.29).

Since the 1980s, the central clustering feature of metropolises in the Bohai Rim has been constantly enhancing, especially for increasingly prominent centrality of prefecture-level cities. Based on statistical units of 5 km × 5 km grids, the hot spots of spatial clustering in urban construction land indicated that the central regional differentiation characteristics of urban–rural transformation were constantly significant (Fig. 6.30). The Bohai Rim has been the most important foreign-oriented and multi-functional urban agglomeration that integrates with politics, economy, culture, and international exchange in northern China with Beijing and Tianjin as two centers, coastal open cities like Dalian, Qingdao,

[2]Receiver Operating Characteristic (ROC) curve is used for analysis evaluation of dichotomy judgment effect. The fundamental principle is to gain multi-sensitivity and 1-specificity by moving judgment points. By regarding sensitivity as the vertical axis and specificity as the horizontal axis, the curve is plotted by linking with each point to compare with straight lines at 45°. If the curve is closer, the judgment value of independent variables for dependent variables is worse, vice versa.

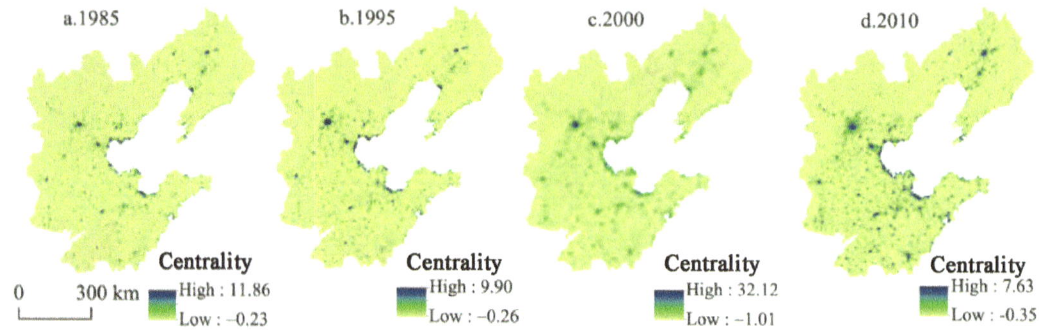

Fig. 6.29 The central agglomeration measure of cities and towns in the Bohai Rim in China

Yantai, Weihai, and Qinhuangdao as sectors, and provincial cities like Shenyang, Shijiazhuang, and Jinan as regional nodes.

Three major port clusters formed in the coastal areas. The first is the northeastern cluster with Dalian as the core and Huludao, Jinzhou, Yingkou, and Dandong as branch ports; the second is the Shandong cluster with Qingdao as the core and Longkou, Weihai and Yantai as branch ports; the third is the Tianjin-Hebei cluster with Tianjin Port as the core and Qinhuangdao Beijing-Tangshan, and Huanghua as branch ports. Convenient ocean shipping conditions will drive regional urban–rural industrial transformation and spatial transformation in hinterlands.

Urban construction land expansion in mountain areas and regional towns have unobvious central clustering features. Urbanization, industrialization and informatization can directly drive the spatial configuration evolution of urban–rural development. In advantageous areas of urbanization, industrialization and informatization, urban spatial clustering is constantly enhancing. Traditional agricultural and mountain areas have relatively lagging industrial development and are difficult to realize industrial transformation and upgrading. Non-agriculturalization of rural elements is characterized with "amphibious" mode of remote relocation working. The local non-agriculturalization and resettlement of rural labors are difficult, resulting in insufficient endogenous power of local urbanization, weak regional spatial clustering growth, and difficult formation of regional centers with certain radiating capacity.

Flat areas gradually form the basic pattern with provincial capitals as first-level clustering centers, prefecture-level cities as second-level centers, and counties and central towns as local centers. The BTT Region, mid-south of Liaoning, and Shandong Peninsula urban agglomerations are growing and constantly developing. The spatial configuration of urban–rural transformation varies under the framework of constantly enhancing regional centrality and town clustering, which shows the evolution characteristics of constantly expansion from the core to peripheries. According to the results of geostatistical exploration and recognition, the clustering centers of high hot spots concentrate in provincial capitals, as well as prefecture- and county-level cities. The central clustering features in counties and small and medium-sized towns is weak with limited hot spots of urban construction land.

From the perspective of optimizing the spatial configuration of urban–rural transformation, it is necessary to value the urban–rural land use optimization configuration, to focus on perfecting the "bottom-up" hierarchy of urban spatial system, to guide collaborative promotion of new rural construction and urbanization of small and medium-sized towns, to realize integrated development of regional population, land, and industries, and finally to promote the cultivation of characteristic regional industries and the integration between employment transformation and social services. During the synchronous

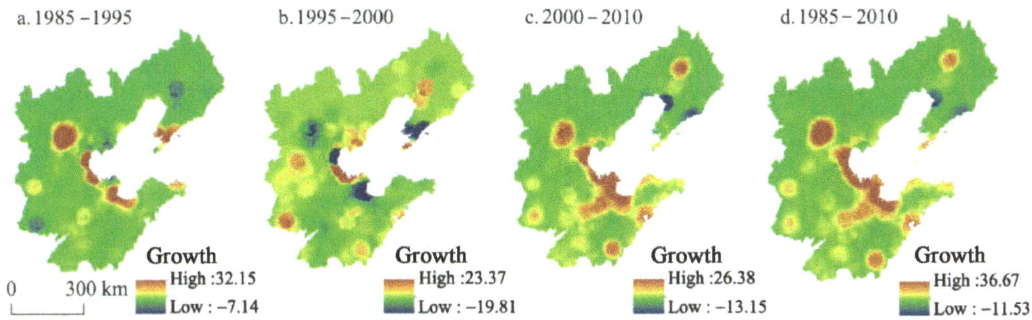

Fig. 6.30 The growth measure of cities and towns in the Bohai Rim in China

development of industrialization, urbanization, and agricultural modernization, the development pattern led by the manufacturing industry brings various drawbacks including development at the cost of resources overconsumption and environmental pollution, which had significant impact on land use. Central cities sprawled like a pancake, raising the cost of urban operation and management. And the radiating effect of industrial transformation that drives the development in peripheral areas was limited. The polar-nucleus development of megacities is not good for the integration of urban–rural transformation and hinders the formation of core-periphery urban–rural integrated interactive mode, but exacerbates the poverty in shadowed areas or even obstructs the regional urban–rural development integration and formation of the integrated pattern.

6.5.3 Spatial Heterogeneity of Urban–Rural Transformation

Urbanization is the dominant process of urban–rural transformation and directly results in the expansion of urban construction land, which also reveals the pathway and quality of urban–rural transformation and directly changes the structure and spatial pattern of regional land use. The majority of urban–rural spatial configuration therefore changes to construction land. The Bohai Rim has become the hot spots of China's urbanization since the reform and opening-up. With the constant speedup of industrialization and urbanization, the expansion of urban construction land is manifested as features of high density and high intensity.

Through the exploration of urban construction land hot spots in four stages (i.e., 1980s–1995, 1995–2000, 2000–2010, and 1980s–2010), we found that spatial expansion of urban–rural transformation in the Bohai Rim was present in the spatial features of "local gathering and global dispersion" (Fig. 6.31). The "gathering pattern" was continuously expanding outward with Beijing, Tianjin, Jinan, Shenyang, Dalian, Shijiazhuang, and Qingdao as cores. Driven by the coastal industrial development, the "C" shaped zone along the coastline with non-agriculturalization of its land has become the hot spots of urban–rural transformation in the Bohai Rim. From the perspective of territorial types, the regional differences in urban construction land are significant. In flat areas with densely distributed towns, the urban–rural spatial transformation configuration centered by prefecture-level cities is gradually forming. County-level cities and towns with limited radiating and gathering capacity grow slowly spatially and economy-wise. The spatial configuration of urban–rural transformation is generally led by the development of regional cores, indicating the "top-heavy" feature of urban system. Some practical problems like the disordered layout of small towns, limited carrying capability of industries and population employment, and underutilization of construction land and industrial hollowing in small towns are

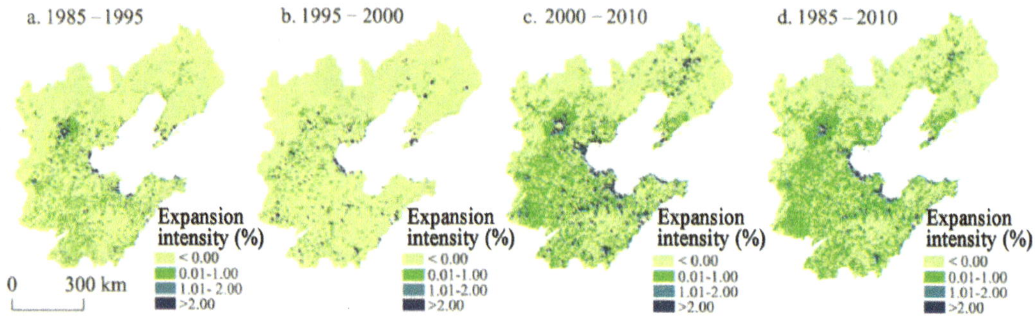

Fig. 6.31 The intensity difference of cities-towns construction land expansion in the Bohai Rim

urgent to be addressed in arranging the urban–rural development.

The spatial expansion and urban development of regional central cities bring a series of problems. Particularly, compression of regional life and ecological space result in a reduction of urban habitability and cost increase of urban system operation. Hence, optimizing the urban–rural spatial organization form becomes the strategic base for resource development, industrial development, and market expansion. Mountain areas suffer double constraints of natural conditions and economic locations with deficient urban hierarchy system and the unclear trend of spatial expansion. Meanwhile, lots of production elements gather in the BTT Region, which gives priority to a one-way mobility and interconnectivity of inter-region elements. The mutual feedback mechanism of elements consequently lacks. To date, the gathering mechanism of Beijing-Tianjin metropolitan is greater than the diffusive one, leading to an unreasonable pattern urban–rural development including the poverty of "shadowed areas" surrounding Beijing-Tianjin metropolitan.

6.5.4 Spatial Trend Sensitivity of Urban–Rural Transformation

Drawing upon the analysis of natural and cultural economic factors in urban–rural transformation, 16 dominant factors were selected in this research, including altitude (x_1), gradient (x_2), temperature (x_3), precipitation (x_4), distance from a provincial capital (x_5), distance from a prefecture-level city (x_6), distance from a county-level city (x_7), distance from a county (x_8), distance from a town (x_9), distance from a village (x_{10}), distance from a railway (x_{11}), distance from an expressway (x_{12}), distance from a national highway (x_{13}), distance from a provincial highway (x_{14}), distance from a county-level and village-level highway (x_{15}), and distance from a river (x_{16}). To reveal the correlation between urban–rural construction land and these dominant factors, the stratified sampling method and distance optimization algorithm were applied to do spatial rasterization processing of each factor. Considering the scale effect of regression precision, we conducted the study at nine spatial scales, namely 100 m, 300 m, 600 m, 1200 m, 1500 m, 1800 m, 2100 m, 2400 m, and 2700 m. Logistic regression was employed at different scales as well. The optimal regression scale of 1200 m was found with the ROC between regression prediction probabilities and actual observed values being 0.863. The Logistic regression model of urban construction land distribution probability was therefore described as follows:

$$\begin{aligned}\log(P/(1-P)) = &-0.005475x_1 - 0.075388x_2 + 0.000489x_3 \\ &- 0.000030x_4 + 0.000001x_5 - 0.000009x_6 \\ &+ 0.000005x_7 + 0.000001x_8 + 0.000036x_9 \\ &+ 0.000064x_{10} + 0.000004x_{11} \\ &- 0.000026x_{12} + 0.000005x_{13} \\ &- 0.000033x_{14} - 0.000085x_{15} \\ &+ 0.000048x_{16} - 2.131258\end{aligned} \tag{6.22}$$

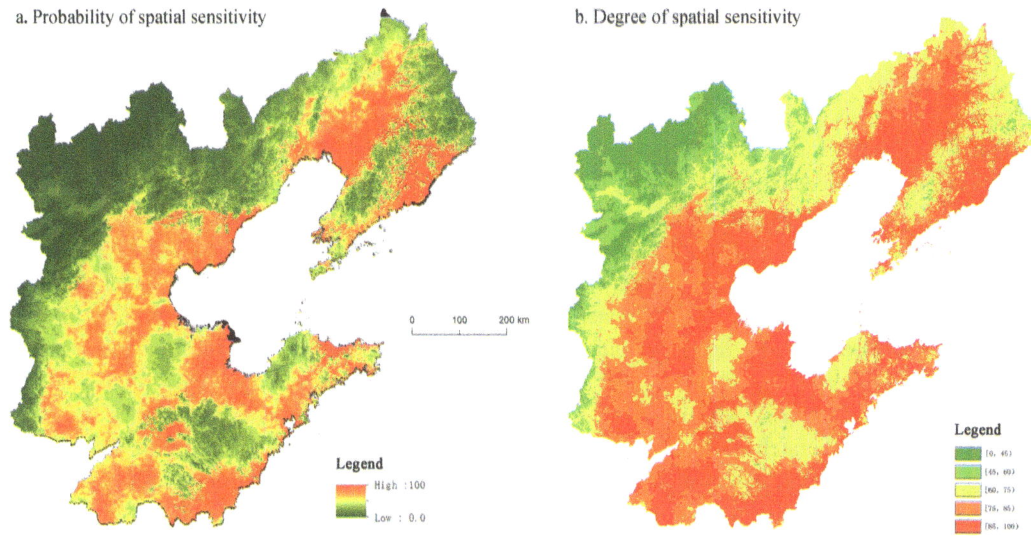

Fig. 6.32 The spatial sensitivity of rural–urban transformation and its grade distribution in the Bohai Rim

Employing the map algebraic approach, layers of influencing factors were substituted into the logistic regression model to get the spatial sensitivity probability map of urban–rural transformation (Fig. 6.32). Based on probability sensitivity degree differences of urban–rural transformation, five grades including extremely strong (85–100%), strong (75–85%), medium (60–75%), ordinary (50–60%), and weak (<50%) were divided (Fig. 6.32 right). The spatial ratios of 5 grades respectively reached 28.64%, 27.59%, 27.72%, 8.79%, and 7.27%.

Significant differences of urban–rural transformation have been witnessed between plains and mountain areas, coastal and inland, regional core and peripheries based on the configuration of point-axis-polygon. The difference is characterized by obvious sensitivity of urban–rural transformation in locationally advantage areas. The extremely sensitivity areas of future urban–rural transformation and integrated development is geographically distributed along the development axis of Beijing-Tianjin-Binhai New Area with Beijing-Tianjin-Hebei as the core and Liaodong and the Shandong Peninsula as two wings. The regional urban–rural transformation formed the overall regional framework with the core of Beijing-Tianjin-Hebei and sub-centers of Shandong Peninsula and mid-south of Liaoning. The region formed an integrated spatial pattern of regional urban–rural transformation driven by the core circle surrounding with secondary central cities as linking nodes, and bonds of consecutive economic corridors. Locally, the new-type urban system of small and middle-sized cities—central towns—central villages (communities)—general villages would be constructed to gradually form the village-town development space with the integration of rural life, production, and ecology. Facing practical problems of scattered, disordered, and hollow, it is necessary to optimize spatial structures of villages and the spatial layout of villages and towns, to form the central clustering with village population, industry, and settlement, and to promote life centralization, production scalarization, and ecological harmony (Liu and Yang 2015).

6.6 Comprehensive Patterns of Urban–Rural Transformation in Bohai Rim Region

6.6.1 Index System of Urban–Rural Transformation

Urban–rural transformation is the comprehensive transformation of the urban–rural territorial system, which is not only the transformation in population, industry, and society but also a systematic change in developing strategies and driving mechanisms. Urban–rural transformation in the Bohai Rim was studied from the perspectives of the multi-element layer and urban–rural coordinated layer in the above-mentioned sections. Hereby, based on the systematic and comprehensive perspective, urban–rural transformation in counties was comprehensively measured to master the geographical pattern of urban–rural transformation in the Bohai Rim. The county-scale data in 2000 and 2015 were collected. Analytic Hierarchy Process (AHP) was applied to construct the transformational index model that could comprehensively reveal urban–rural development status and transformation trends, and thereafter conduct the overall evaluation on urban–rural transformation level in the Bohai Rim. Based on indexes of population, industries, and lands, urban–rural transformation was divided into population growth and structure transformation, industrial growth and structure transformation, land use transition, as well as social development and service transformation. We then respectively construct four models of population growth index, land use change index, industrial evolution index, and social evolution index. Finally, a weighted urban–rural transformation index (*URTI*) was generated.

Urbanization rate, proportion of working age population, average years of education, and aging degree were selected to reveal the status of population transformation. Proportions of urban–rural construction land, cultivated land, and rural homestead were chosen to reveal the status of land use transition. Non-agricultural output structure and degree of agricultural mechanization were used to reveal the status of industrial transformation. Per capita GDP, per capita retail sales of consumer goods, per capita net incomes of rural residents, the number of beds in health services per thousand people, the number of beds in social welfare adoption per thousand people, and urban–rural resident income gaps were employed to reveal the status of economic transition. Specific indexes and weights were reported in Table 6.8.

Transformation is generated with deep development. When urban–rural regions develop in a certain stage and there are obstacles and bottlenecks, element transfer, mechanism shift, strategic change, and relational transformation of the urban–rural system would be triggered. For the sake of breaking through obstacles and realizing the new round development, the deep transformation occurs. To better understand and master the point of time transformation occurs, we define thresholds for index in the transformation evaluation index system. In other words, the transformation level evaluation index system can be set up to judge the reasons for realizing the urban–rural transformation degree and stage. Taking urbanization as an example, the process of urbanization enters the decelerating phase from acceleration when surpass 50%. The migration rate of population from rural areas to cities gradually abates. This can be deemed as an eigenvalue when the urban–rural regional development steps into the transformation stage. In terms of residents' education, when average years of education reach 9 years, it means that local population basically has completed the study of compulsory education and has owned the knowledge and skills of engaging in various basic applied tasks. This may largely lead to the upgrading and renewal of local employment structures and dominant industrial categories. This is therefore regarded as another symbolic feature of transformation. In the index of the population evolution model, a high or low aging degree is also an important representation that reveals whether urban–rural social development enters the transformation period. With the acceleration of the population aging process, local labor force proportion will decrease. Therefore, the "population bonus" undoubtedly disappear and then trigger the adjustment of regional industrial structures and production mode transformation, correspondingly bringing development and transformation of the urban–rural territorial system. Per capita GDP, as an index to measure social development transformation, also has the important representation significance for the occurrence and development of urban–rural transformation. Academic studies

Table 6.8 Evaluation index system of the urban–rural transformation level

Target layers	Index layers	Index computing methods	Weight	Polarity
Population transformation (*PTI*) (0.16)	Urbanization rate	Urban permanent resident population/total regional population	0.26	Positive
	Labor force proportion	15–64-year-old population/total regional population	0.25	Positive
	Mean schooling	Schooling years of specific age groups/total number of this age group	0.25	Positive
	Aging degree	Population above 65 years old/total regional population	0.24	Negative
Land transformation (*LTI*) (0.35)	Urban–rural construction land proportion	Construction land area/total regional area	0.42	Positive
	Cultivated land resource proportion	Cultivated area/total regional area	0.39	Negative
	House site proportion	Rural housing land area/construction land area	0.19	Negative
Industrial transformation (*DTI*) (0.15)	Non-agricultural output structure	Value added of secondary and tertiary industries/total regional production value	0.62	Positive
	Agricultural mechanical level	Total power of agricultural machinery/total cultivated area	0.38	Positive
Social transformation (*STI*) (0.34)	Per capita GDP	GDP/total regional population	0.24	Positive
	Living consumption level	Total retail sales of consumer goods/total regional population	0.18	Positive
	Income level of rural residents	Per capita net incomes of rural residents	0.16	Positive
	Medical health status of residents	The number of beds in health services for thousands of people	0.14	Positive
	Social security status	The number of beds in social welfare adoption for thousands of people	0.12	Positive
	Urban–rural resident income gaps	Per capita disposable incomes of urban residents/per capita pure incomes of farmers	0.16	Negative

argue that per capita GDP ranging from 560 to 1570 dollars indicates industrial starting stage, 1570 to 11,810 dollars denotes industrial realization stage, and 11,810 to 28,350 dollars means the developed stage (Qi et al. 2013). Different stages of per capita GDP correspond to the specific productivity level and urban–rural development state. Once the value of index rapidly increases, this undoubtedly is attributed to the entry of transformation stage.

6.6.2 Spatial–Temporal Pattern of Urban–Rural Transformation Index

Based on the aforementioned spatial–temporal pattern analysis of population transformation index (*PTI*), land transformation index (*LTI*), industrial transformation index (*DTI*), and social transformation index (*STI*), the weighted *UDTI* was calculated to analyze the overall pattern of urban–rural transformation in the Bohai Rim.

From 2000 to 2015, the overall *URTI* witnessed a growth trend. The mean *URTI* in all counties increased to 0.417 from 0.295. Index at the province (city) level also increased. Especially, *URTI* in Tianjin increased to 0.493 from 0.329, followed by Beijing increasing by 0.159. Mean *URTI* in Liaoning Province had the slowest growth, which increased to 0.430 from 0.330 (Table 6.9).

According to the spatial–temporal distribution of *URTI*, the low-value areas (LVA) and relatively low-value areas (RLA) in 2000 include 82 and 98 counties, accounting for 25.15% and 30.06% of total county units in the Bohai Rim respectively (Table 6.10). The number of counties with high *URTI* is 29, accounting for the least 8.90% of total units, locating in districts of Beijing, Tianjin, and other prefecture-level cities (i.e., Shijiazhuang, Tangshan, Jinan, Shenyang, Dalian, and Qingdao, etc.). 39 counties had relatively high-value of *URTI*, accounting for 11.95% and locating in Beijing, and cities in Liaoning (i.e., Dandong, Yingkou, Panjin, Fushun and its surrounding counties) and Shandong (i.e., Yantai, Weihai, Dongying, Weifang, Zibo, and Jining). The counties with moderate values mainly located in northern Hebei, Shandong Plain Areas, and Liaodong Peninsula, and scattered in southern and central Hebei. Counties with relatively low values mostly locate in northern and southern Hebei, western Liaoning, and hilly areas in central Shandong. Moreover, low value areas locate in western Liaoning, Bashang Plateaus in northwestern Hebei, and traditional agricultural areas in southern and central Hebei, and western Shandong (Fig. 6.33).

The spatial pattern of *URTI* in 2015 is similar with that in 2000 except some minor changes. The number of low-*URTI* counties changes limitedly and still accounts for the maximum proportion. The numbers of moderate, relatively high, and high counties however rise by 11, 12, and 4 respectively. The number of low counties reduced by 29. Spatially, the patterns of *URTI* in Beijing and Tianjin remained relatively stable. The high areas in eastern Liaoning obviously turned to relatively high areas. The north was present in the feature from low to relatively low areas. Though *URTI* increased in counties in Hebei at varying degrees, it remained relatively stable in the partition. However, the reduction of partition was relatively obvious in the north. The increase of *URTI* in counties in Shandong was obvious. The western Shandong was obviously changed from low areas to relatively low and moderate areas. The north had a significant change from the moderate to relatively high and high areas.

Based on aforementioned results, we have the whole idea on the status of urban–rural transformation in the Bohai Rim from 2000 to 2015 and the spatial distributions. The development transformation level in Beijing and Tianjin, as municipalities, reached the maximum and had increasingly significant impacts and radiation effects on surrounding areas. In the past decade, the rapid increase of the urban–rural transformation in the Beijing-Tianjin-Hebei Region should be attributed to the radiation driving role of municipalities. Liaoning had a higher urban–rural transformation level in 2000. The urban agglomeration of Shenyang-Fushun-Benxi, depending on their foundation of heavy industry,

Table 6.9 Mean variation of urban–rural transformation index in the Bohai Rim in 2000 and 2015

Years	Beijing	Tianjin	Hebei	Liaoning	Shandong	Bohai Rim
2000	0.399	0.329	0.284	0.330	0.284	0.295
2015	0.559	0.493	0.391	0.430	0.435	0.417
2000–2015	0.159	0.164	0.108	0.100	0.151	0.123

had dramatic enhancement on facilitating regional industrialization, absorbing rural residual labor forces, activating rural land transfer, accelerating urbanization construction steps, and promoting equality of urban–rural public service facilities. However, with the prominence of the institutional mechanism contradictions, the starting advantages of the urban–rural development were gradually losing and currently faced the prominent transformation issue, resulting in restricting enhancement of the urban–rural development in Liaoning. Determined by the development stage and level, the urban–rural transformation level in Shandong took the lead among three provinces. To be specific, Yantai, Weihai, and Qingdao in eastern coast of Shandong, together with Jinan and Zibo in the centeral had good developmental conditions. Fields and departments of urban–rural development had the fluent operational mechanism, ordered industrial structure and land optimization configuration, and therefore the higher urban–rural transformation level. Heze and Linyi in the south, combined with Binzhou, Dezhou, and Liaocheng in the west had relatively lagging overall development, resulting in a shortage of dominance and driving of the industrial sector. The overall urban–rural development level was relatively low. With the improvement of regional traffic conditions and driven by the industrialization and urbanization development of provincial central cities, the urban–rural transformation level in these lagging areas was rapidly enhanced. The urban–rural transformation in Hebei remained the relatively lagging position in the Bohai Rim, especially Zhangjiakou and Chengde in northwest mountainous areas, which had lagging economic development, lack of large-scale leading enterprises with dominant industries and influence. Consequently, this constrained the regional economic development or even led to invalid supply of urban–rural public service and social security, and relatively severe issues of land resource waste and idle.

6.6.3 Spatial–Temporal Pattern of Urban–Rural Transformation in Different Dimensions

1. **Spatial–temporal pattern of the evolution index in population dimension**

According to the mean *PTI*, Beijing had the biggest value of 0.468, followed by Liaoning. The smallest one was Hebei. In 2015, Beijing still ranked the first and reached 0.668. Besides, Tianjin had the most obvious increase, which was from 0.386 to 0.587. From 2000 to 2015, the mean *PTI* in counties of Beijing and Tianjin had the fastest growth, respectively increased by 0.200 and 0.202. The mean *PTI* in Liaoning had the minimum amplitude, which was only 0.053 (Table 6.11). Based on the comprehensive computing results of indexes, the natural breaking point in ArcGIS software was applied to divide indexes of population evolution into low areas, relatively low, mid-value, relatively high, and high groups (Fig. 6.34). In 2000, *PTI* was concentrated in counties with low and relatively low values, accounting for 31.90 and 34.05% of total counties respectively. The high areas of *PTI* had relatively fewer counties, only occupying 10.43% (Table 6.12). The spatial pattern also demonstrates that Beijing and Liaoning had a higher *PTI* index. Tianjin was slightly higher than the mean level of the Bohai Rim, while Shandong was slightly lower than the regional average. Hebei was the smallest. Specifically, high value areas basically had the fastest industrialization and urbanization development and the highest socio-economic development level, including districts of Beijing, Tianjin, and prefectures in Liaoning, Hebei, and Shandong (i.e., Shenyang, Tangshan, and Jinan, etc.). Relatively high areas basically referred to counties with the middle socio-economic development level or above, including suburbs of Beijing, main cities and surroundings of the eastern and central Liaoning, and scattered counties in Shandong. Mid-value areas mainly located in the mid-west and eastern Liaoning, and some areas in Shandong; relatively low values mainly located in the northern mountain areas of Hebei and traditional agricultural areas in the central and southern Hebei, Shandong Peninsula, and a few counties scattered in Liaoning; low values mainly got involved in western and southern Shandong, Bashang Plateaus in eastern and northwestern Hebei, and central and southern Hebei.

The *PTI* level in the Bohai Rim slightly increased to 0.492 in 2015 from 0.378 in 2000. As a whole, the numbers of low and relatively low counties decreased by 6 and 20. Mid-value and relatively high areas significantly increased to 105 and 54 counties from 53 and 24 counties, respectively. Based on the spatial pattern, high areas mainly located in districts of Beijing, Tianjin, and other prefectures in three provinces. The overall pattern was similar to that in 2000. The distribution of relatively high areas changed significantly. The shrinkage of this group in Liaoning was obvious. While Shandong and Hebei witnessed a significant expansion of this group, which mainly located in surrounding districts of major prefectures. The number of counties with mid-values increased significantly. Some counties in Liaoning changed from relatively high to this mid-value group, while those in Hebei and Shandong were evolved from the low value group. The coverage of relatively low areas obviously shrunk to western and northern Hebei, western Shandong, and central Liaoning. This indicates that Hebei and Shandong had a great improvement and enhancement regarding natural population structures, economic

Table 6.10 Grading statistics of URTI in the Bohai Rim in 2000–2015

Year	Low-value areas (LVA)		Relatively low-value areas (RLA)		Mid-value areas (MVA)		Relatively high-value areas (RHA)		High-value areas (HVA)	
	Counties	Proportion	Counties	Proportion	Counties	Proportion	Counties	Proportion	Counties	Proportion
2000	82	25.15	98	30.06	78	23.93	39	11.96	29	8.90
2015	53	16.26	100	30.67	89	27.30	51	15.64	33	10.12

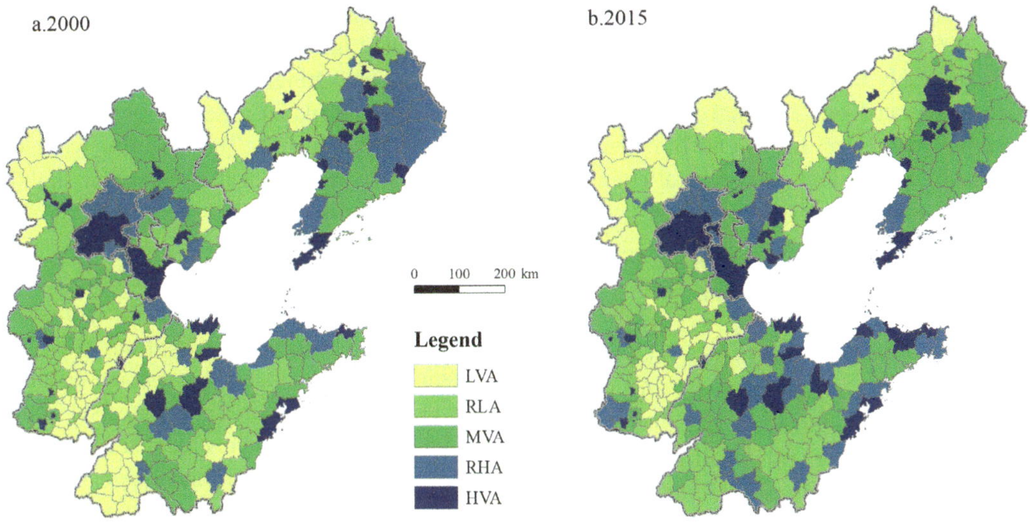

Fig. 6.33 Spatial pattern evolution of the urban–rural transformation index in the Bohai Rim

Table 6.11 Mean variation of population evolution index of the circum-Bohai Region in 2000 and 2015

Years	Beijing	Tianjin	Hebei	Liaoning	Shandong	Bohai Rim
2000	0.468	0.386	0.351	0.463	0.363	0.378
2015	0.668	0.587	0.497	0.516	0.458	0.492
2000–2015	0.200	0.202	0.146	0.053	0.095	0.114

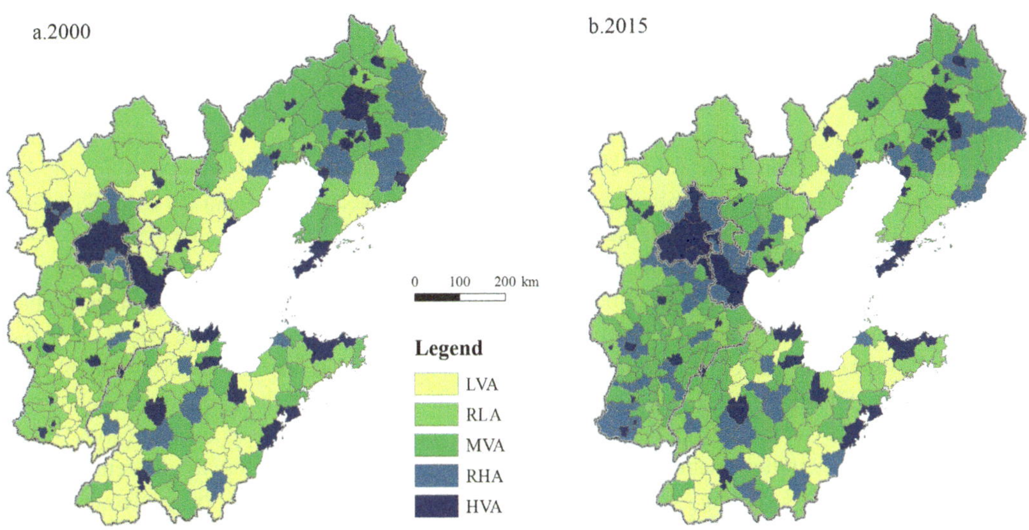

Fig. 6.34 Spatial pattern of population evolution index in the Bohai Rim

structures, and comprehensive quality structures from 2000 to 2015. The number of the low areas was remarkably reduced. The spatial distribution was concentrated on Zhangjiakou in the northwestern Hebei, Heze, Linyi, and Binzhou in the hinterland of Shandong, and several counties in western Liaoning.

2. **Spatial–temporal pattern of the evolution index in the land dimension**

In 2000, Beijing had the biggest mean of the county-scale land evaluation index (0.438), followed by Liaoning (0.338) and then Tianjin, Hebei, and Shandong with small gaps. In 2015, Beijing still ranked the top regarding the mean of land evolution index (0.497), followed by Liaoning (0.355). However, the gap between Tianjin, Shandong, and Liaoning was remarkably shrunk. As a whole, the county-scale land evolution index in the Bohai Rim remained the trend of increase but not significant. Beijing with the most significant change only had 0.059 of amplitude (Table 6.13). The natural breaking point in ArcGIS software was applied to divide indexes into different groups and got the low, relatively low, mid-value, relatively high, and high areas of land evolution (Fig. 6.35). In 2000, the study area was dominated by counties with relatively low and mid-values, accounting for 33.44% and 25.46% respectively, followed by relatively high and the low counties with the respective proportion of 18.71% and 16.26%. The high value counties occupied only 6.13%. To be specific, the high areas mainly located in districts of Beijing, prefectures in Liaoning (e.g., Dalian, Fushun, Anshan, and Benxi) and Hebei (e.g., Qinhuangdao, Shijiazhuang, and Baoding). Relatively high areas mainly located in eastern Liaoning, Liaodong Peninsula, Shandong Peninsula, northern and central Hebei, and the Beijing-Tianjin Region. Mid-value counties mainly situated in western and central Liaoning, Zhangjiakou, Xingtai and eastern Hebei, coastal Shandong, and Zibo and Tai'an in central Shandong. Relatively low areas concentrated on central Hebei (Shijiazhuang), central and northern Liaoning, and northwestern Shandong (e.g., Binzhou and Liaocheng), and southeastern Shandong (e.g., Qingdao and Linyi). The low areas mainly located in Hengshui and Xingtai (Table 6.14).

By 2015, the spatial pattern of land evolution index greatly changed. The majority was still relatively low areas with the proportion slightly increasing. Moreover, the proportion of low areas was increased by 3.68%, while the mid-value, relatively high, and high areas decreased at varying degrees. To be specific, the number of high counties reduced to 15 from 20. The degrading counties mainly located in Liaoning, while the high counties in Hebei remained stable. Relatively high areas mainly located in the BTT Region, while the rest of them remained the scattered distribution features in Dalian, Yantai, Qingdao, Jinan, and Cangzhou. Mid-value areas mainly located in eastern Liaoning, counties in Hebei and surrounding Beijing and Tianjin, as well as central and eastern Shandong. Relatively low counties mainly located in western Liaoning, and western, central, and southern Shandong, as well as Baoding and Shijiazhuang in Hebei. The low areas mainly located in northwestern Liaoning and central and southern Hebei.

3. **Spatial–temporal pattern of the evolution index in the industrial dimension**

In 2000–2015, the county-scale *DTI* in the Bohai Rim remained the continuous growth trend from 0.465 to 0.608. *DTI* at the provincial scale was slightly enhanced as well, especially for the largest amplitudes of 0.214 and 0.110 in Shandong and Hebei, respectively (Table 6.15). In 2000, the study area was dominated by counties with mid-value and relatively high values, which accounted for 30.67% and 27.61 respectively and followed by relatively low areas with the proportion of 19.02%. The proportion of counties with high *DTI* was 15.64% (Table 6.16). Specifically, the high areas concentrated on Beijing, Tianjin, districts of prefectures, and surrounding developed counties. Relatively high areas mainly located in Beijing, Tianjin, central and southern Hebei, and central, and eastern coastal Shandong. Mid-value areas frequently located in eastern Liaoning, central and southern

Table 6.12 Grading statistics of population evolution index in the Bohai Rim in 2000–2015

Years	LVA		RLA		MVA		RHA		HVA	
	Counties	Proportion	Counties	Proportion	Counties	Proportion	Counties	Proportion	Counties	Proportion
2000	104	31.90	111	34.05	53	16.26	24	7.36	34	10.43
2015	38	11.66	91	27.91	105	32.21	54	16.56	38	11.66

Shandong, and traditional agricultural areas in southern Hebei. Relatively low areas mainly located in western Liaoning, northern and western Hebei, as well as central and western Shandong. The low areas mainly located in Heze (Shandong), Zhangjiakou (Hebei), and northern Liaoning (Fig. 6.36).

In 2015, the number of counties with high and relatively high values obviously increased. Their proportion was respectively increased to 17.18% and 31.29% from 15.64% and 27.61%. The number of mid-value and relatively low counties largely reduced (Table 6.16). To be specific, the high areas were concentrated on prefectures and surrounding developed areas in Shandong, and Shijiazhuang, Handan, and Tangshan in Hebei. Only minority areas in Liaoning belonged to the high group. Limited relatively high areas located in Liaoning. The distribution of relatively high areas in Hebei was similar to that in 2000, while Shandong witnessed an obvious spatial expansion of relatively high areas, especially in western and eastern Shandong. Mid-value areas mainly located in central and southern Hebei, counties surrounding Beijing-Tianjin, central and northern Shandong, as well as some counties in central Liaoning. Relatively low areas mainly located in east and mid-west of Liaoning, and northern, central, and southern Hebei. The low areas were concentrated on western Liaoning, Tieling, and Zhangjiakou in Hebei (Fig. 6.36).

4. **Spatial–temporal pattern of the evolution index in the social dimension**

From 2000 to 2015, the social evolution index in the Bohai Rim continuously increased. The overall mean increased to 0.210 from 0.203. The mean county-scale *STI* at the provincial scale significantly increased and exceeded 0.200 except Hebei. The value in Tianjin increased to 0.529 in 2015 from 0.227 in 2000, showing the largest amplitude, followed by Beijing with an amplitude of 0.283. In three provinces, Shandong had the largest amplitude, which was increased to 0.461 from 0.203 (Table 6.17). To be specific, relatively low and mid-value counties had the biggest proportion in 2000, respectively reaching

Table 6.13 Mean variation of land evolution index in the Bohai Rim in 2000 and 2015

Years	Beijing	Tianjin	Hebei	Liaoning	Shandong	Bohai Rim
2000	0.438	0.293	0.256	0.338	0.254	0.274
2015	0.497	0.343	0.287	0.355	0.300	0.308
2000–2015	0.059	0.050	0.031	0.017	0.045	0.034

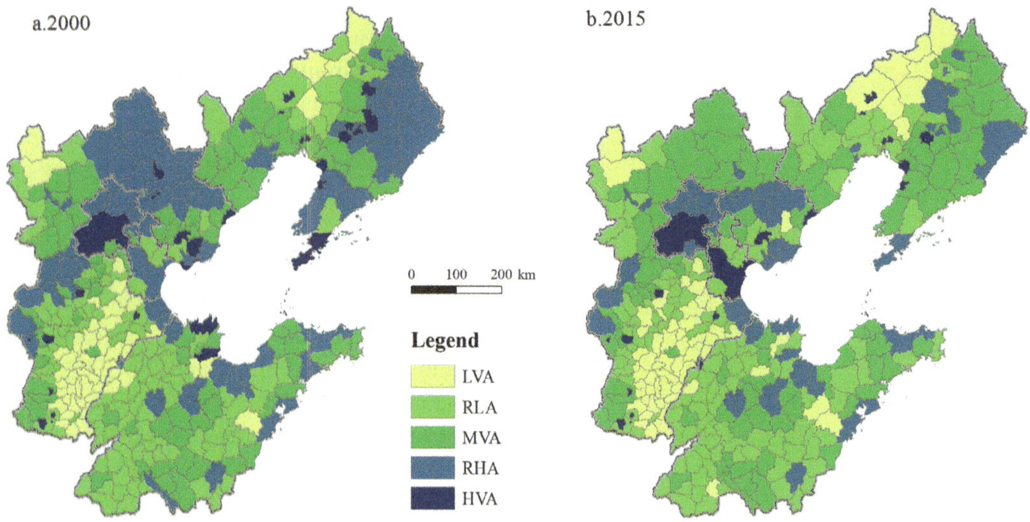

Fig. 6.35 Spatial pattern of the land evolution index in the Bohai Rim

35.58% and 31.90%. In 2015, they still had the biggest proportions, but the specific number respectively decreased to 30.06% and 26.38%. The proportion of the low counties increased to 20.86% from 9.20% (Table 6.18).

From the perspective of the spatial pattern, the high areas mainly located in districts of Beijing, Jinan and Dongying in Shandong, Baoding, Shijiazhuang and Tangshan in Hebei, Shenyang, Dalian, and Anshan in Liaoning in 2000. Relatively high areas were concentrated on Beijing-Tianjin, Anshan, Shenyang, Liaodong Peninsula, Yantai, Weihai, and Qingdao. Mid-value areas mainly located in the Shandong Peninsula, suburbs in the BTT Region, central Liaoning, Shijiazhuang, Handan, Xingtai, and Cangzhou. Relatively low areas involved in southern, western, and central Shandong, central and southern Hebei, and western Liaoning. The low areas mainly located in western and northern Hebei, and western Liaoning. The high areas were also widely distributed in prefectures of Shandong Peninsula, Shenyang, Dalian, Shijiazhuang, Baoding, and Tangshan, in addition to districts of Beijing and Tianjin. Relatively high areas mainly located in districts and counties of Beijing and Tianjin, northern and eastern Shandong, but scattered in Liaoning and Hebei. Mid-value areas concentrated in central and eastern Liaoning, eastern Hebei, central, and western Shandong. Relatively low areas mainly located in northern Liaoning, western and southern Shandong, Chengde, Baoding, Shijiazhuang, and Handan in Hebei. The low areas were concentrated in Yanshan-Taihang Mountain Areas, traditional agricultural areas in mid-southern, western and northwestern Hebei, as well as five counties in western Liaoning (Fig. 6.37).

Table 6.14 Grading statistics of land evolution index in the Bohai Rim in 2000–2015

Years	LVA		RLA		MVA		RHA		HVA	
	Counties	Proportion	Counties	Proportion	Counties	Proportion	Counties	Proportion	Counties	Proportion
2000	53	16.26	109	33.44	83	25.46	61	18.71	20	6.13
2015	65	19.94	117	35.89	89	27.30	40	12.27	15	4.60

Table 6.15 Mean variation of industrial evolution index in the Bohai Rim in 2000 and 2015

Years	Beijing	Tianjin	Hebei	Liaoning	Shandong	Bohai Rim
2000	0.554	0.587	0.487	0.409	0.453	0.465
2015	0.625	0.664	0.597	0.498	0.667	0.604
2000–2015	0.071	0.077	0.110	0.089	0.214	0.139

Table 6.16 Grading statistics of industrial evolution index in the Bohai Rim in 2000–2015

Years	LVA		RLA		MVA		RHA		HVA	
	Counties	Proportion	Counties	Proportion	Counties	Proportion	Counties	Proportion	Counties	Proportion
2000	23	7.06	62	19.02	100	30.67	90	27.61	51	15.64
2015	26	7.98	57	17.48	85	26.07	102	31.29	56	17.18

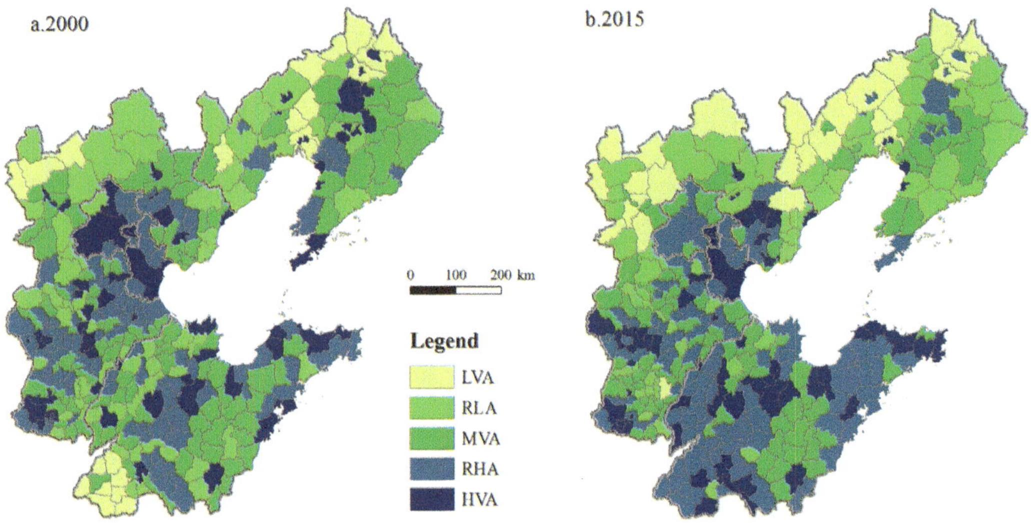

Fig. 6.36 Spatial pattern evolution of industrial evolution index in the Bohai Rim

Table 6.17 Mean variation of social evolution index in the Bohai Rim in 2000 and 2015

Years	Beijing	Tianjin	Hebei	Liaoning	Shandong	Bohai Rim
2000	0.259	0.227	0.191	0.225	0.203	0.203
2015	0.542	0.529	0.358	0.438	0.461	0.413
2000–2015	0.283	0.302	0.168	0.213	0.258	0.210

Table 6.18 Grading statistics of social evolution index in the Bohai Rim in 2000–2015

Years	Low areas		Relatively low areas		Mid-value areas		Relatively high areas		High areas	
	Counties	Proportion	Counties	Proportion	Counties	Proportion	Counties	Proportion	Counties	Proportion
2000	30	9.20	116	35.58	104	31.90	53	16.26	23	7.06
2015	68	20.86	98	30.06	86	26.38	47	14.42	27	8.28

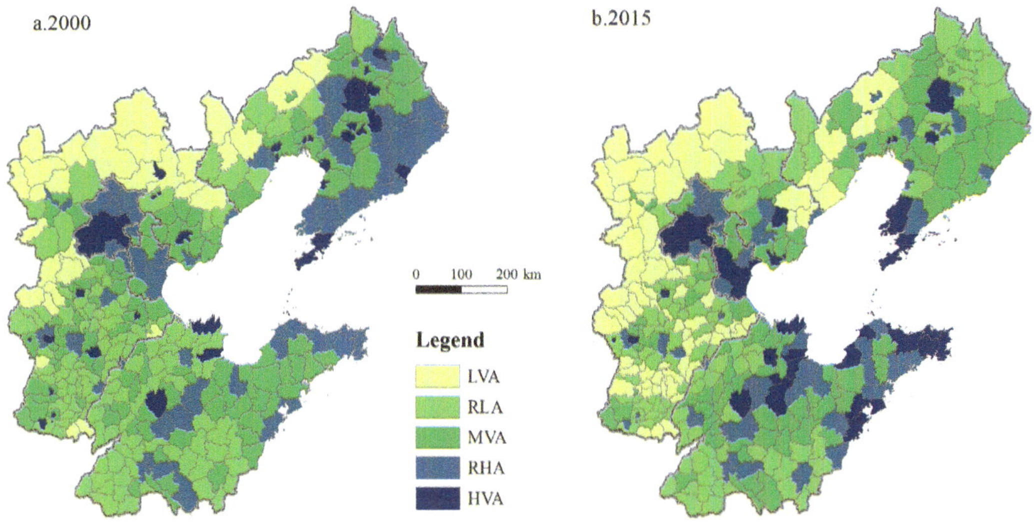

Fig. 6.37 Spatial pattern evolution of social evolution index in Bohai Rim

References

Antrop M. Land scape change and the urbanization process in Europe. Land Scape Urban Plann. 2004;67 (1–4):9–26.

Cao G, Bian X, Liu T. Comprehensive evaluation of the urbanization level in Yangtze river delta region: an index framework based on the population, economic structure and land use. Geogr Res. 2011;30(12):2139–49.

Chen M, Lu D, Zhang H. Comprehensive evaluation and the driving factors of China's urbanization. Acta Geographica Sinica. 2009;64(4):387–398.

Chen F, Zhang H, Wu Q, et al. A study on coordinate development between population urbanization and land urbanization in China. Hum Geogr. 2010;25(5): 53–8.

Han F, Wang Z, Yang H. The effect of industrial structure on intensive urban land use. Resour Sci. 2013;35(2): 388–95.

Haas T, Westlund H. In the post-urban world: Emergent transformation of cities and regions in the innovative global economy. In: Urban-Rural Relations in the Post-Urban World. Routledge, 2017:70–81.

Lefebvre H. The rban evolution. Minneapolis: University of Minnesota, 2003.

Li G. The analysis and evaluation of the compatible relationship between industrialization and urbanization in China. Areal Res Dev. 2008;27(5):6–11.

Liu Y. Progress and prospect on the study of rural hollowing in China. Geogr Res. 2010;29(1):35–42.

Liu Y, Yang R. The spatial characteristics and formation mechanism of the county urbanization in China. Acta Geogr Sin. 2012;67(8):1101–10.

Liu Y, Yang R. The spatial pattern measure of urban-rural transformation in the Bohai Rim region in China. Acta Geogr Sin. 2015;70(2):248–56.

Liu X, Zhang D, Chen B. Characteristics of China's town-level land use in rapid urbanization stage. Acta Geogr Sin. 2008;63(3):301–10.

Liu Y, Lu S, Chen Y. Spatio-temporal change of urban-rural equalized development patterns in China and its driving factors. J Rural Stud. 2013;32:320–30.

Lu D, Yao S. A scientific thought about urbanization progress in China. Hum Geogr. 2007;22(4):1–5.

Lu D, Yao S, Li G. Comprehensive analysis of the urbanization process based on China's conditions. Econ Geogr. 2007;27(6):883–7.

Qi Y, Yang Y, Jin F. China's economic development stage and its spatio-temporal evolution: A prefectural-level analysis. J Geogr Sci. 2013;23(2):297–314.

Zhang F, Liu Y. Dynamic mechanism and models of regional rural development in China. Acta Geogr Sin. 2008;63(02):115–22.

7 Resources and Environmental Effects of Urban–Rural Transformation in China

Abstract

With the rapid development of industrialization and urbanization, great progress has been made in China's urban and rural socioeconomic development, but it also brings a lot of resource problems and environmental pressure, resulting in issues such as the imbalance of factor input structure and the excessive consumption of water and land resources. From 2000 to 2009, there exists an inverted U-shaped relationship between non-agriculturalization of farmland and economic growth in China, and the rapid non-agriculturalization rate of agricultural land has slowed down. Then, this chapter analyzes the land use change in Beijing-Tianjin-Hebei region from 2000 to 2015, constructs a driving force index system of land use change, and discusses the land use pattern of Beijing Tianjin Hebei region in 2030 under the scenarios of business as usual, cropland protection and ecological security. Furthermore, the improved STIRPAT model is used to study the impact of China's urban–rural transformation on energy consumption, CO_2 emissions and industrial pollutant emissions, and some suggestions are put forward to reduce energy consumption, CO_2 emissions and industrial pollutant emissions. Finally, the theory of village transformation and its resources and environmental effects is discussed, and Beicun village in the suburb of Beijing is taken as an example to analyze the resources and environmental effects and the process, characteristics and internal mechanism of village transformation in the process of coordinated development of "planting, breeding, processing and tourism".

7.1 Analysis of the Resources and Environmental Effects of Urban–Rural Transformation

7.1.1 Resources and Environmental Carrying Capacity in a Region

Human social and economic activities are accompanied by the exploitation and utilization of natural resources, and the discharge of all kinds of production and living wastes, which bring about resource and environmental effects. With the increasingly serious resource and environmental issues in the process of economic growth, especially after the "limits to growth" was put forward in the early 1970s, sustainable development has become the focus of attention from all circles. Various resource and environmental factors have become constraints on human development. Research on carrying capacity started with grain restrictions on the population but has been extended to fields such as water, soil, and mineral resources, and has grown from a single factor to multiple factors.

Y. Liu, *Urban-Rural Transformation Geography*, Sustainable Development Goals Series,
https://doi.org/10.1007/978-981-16-4835-9_7

Resources and environmental carrying capacity are the degree of functional suitability of resource and environment conditions to human production activities in a specific region. In addition to the maximum population size that can be sustained and its maximum economic size, it also includes the ability to satisfy material needs, provide a quality living environment, maintain the stability of the ecosystem, and rationally exploit and utilize resources.

From the perspective of sustainable utilization, the resource and environmental carrying capacity explores the synergistic relationship between economic and social dimensions and the dimensions of resources, the environment, and ecology in a specific region. It has been extensively covered in the existing research, including studies on a single or comprehensive multi-index carrying capacity such as land resource carrying capacity, water resource carrying capacity, environmental carrying capacity, and ecological carrying capacity (Mao and Yu 2001; Liu and Fang 2008).

(1) **Land resource carrying capacity**. The research on land resource carrying capacity mainly focuses on the production capacity of farmland and the carrying capacity of population's grain consumption. In 1977, the United Nations Food and Agriculture Organization conducted a study on population support capacity in developing countries by measuring the size of the population that could be sustained by land in these countries.

(2) **Water resource carrying capacity**. As more water is used for production and living, the constraints of water resources on economic growth and urban construction are increasing, especially in the inland and arid areas where water resources are seriously deficient and polluted. Studies have been carried out on the carrying capacity of water resources in basins, lakes, oases, coastal areas, and other types of areas.

(3) **Environmental carrying capacity**. In the context of increasingly severe environmental pollution, such concepts as environmental self-purification, environmental capacity, and environmental carrying capacity have been proposed. The environmental carrying capacity is defined as "the threshold of an environmental system within a specific space and time range to support human social and economic activities in terms of resource supply, environmental accommodation, and ecological services on the premise that the environmental function and structure do not change" (Liu et al. 2009). Some studies have evaluated the supply of water, soil, and gas resources, pollution bearing, economic and social status, etc.

(4) **Ecological carrying capacity**. In a narrow sense, the ecological carrying capacity studies the maximum population that an ecosystem can sustain. In a broad sense, the ecological carrying capacity studies the adaptability of an ecosystem's self-sustaining and self-regulating capacity to socioeconomic intensity and population size. The ecological footprint theory has been widely applied in a broad sense to ecological carrying capacity, which was proposed by William Rees in 1992. The ecological footprint method calculates the biologically or ecologically productive area required for consumption and waste discharge in a specific region to represent the ecological load caused by development (the ecological footprint requirement), and uses the ecologically productive land area available in the region to represent the biological supply capacity or ecological supply capacity. The sustainability of economic system development in the region is then measured and analyzed by comparing the two.

China now has a large population base, a small per capita resource endowment, and a limited environmental carrying capacity. Because of the continuous growth of population as well as rapid economic industrialization and social urbanization, many needs of urban and rural residents in most regions are approaching or even exceeding the local resource support capacity and the environmental carrying capacity. The research on resource and environmental carrying capacity provides a reference for coordinating the relationship between population and land and in turn promotes sustainable development, as well as a basis for dividing regional functions and managing space. As the background and threshold for economic growth and urbanization in a region, the resource and environmental carrying capacity restricts regional development. With its economic growth and urbanization, China's economic growth, population agglomeration, and increasing demand are facing resource shortages, environmental damage, and ecological deterioration. Due to the differences in resources and the environment among regions, the dominant resource and environmental factors that restrict economic and social changes in these regions are different, and the human–land relationship also presents a functional relationship with the changes of geographical space.

7.1.2 Relationship Between Economic Growth and the Changes of Resources and Environment

Generally, the theory of development model and development stage is used to explain the relationship between economic growth and the changes of resources and environment. Development model, the type and path adopted by a country or region in economic development, reflects the development strategy and the mode of economic production. According to the theory of development model, the problems of resources and environment are related to the development mode adopted by a country or region, which can be generally divided into following two types: extensive and intensive development model. The extensive development model is based on a large number of labor and resources input, but the output efficiency is low, which shows the inefficient use of resources. If we do not pay attention to the problems of resources and environment, a lot of waste and pollution will be produced, which will eventually lead to environmental deterioration. According to the theory of development stage, environmental quality is closely related to the stage of economic development. In the low stage of economic development, the pollution and environmental damage caused by human activities are very weak, and the level of environmental quality is very high. In the middle stage of economic development, with the enhancement of human activities, the weakness of environmental awareness and the lack of corresponding administrative intervention in environmental protection, environmental pollution and damage gradually become serious. After the economic development reaches a higher level, moreover, the environment will be gradually improved through effective environmental management and higher social investment for environmental problems.

The law of economic growth's impact on the environment has always been the focus of academic attention. The relationship between the two can be divided into the following three points of view: First, economic growth will inevitably lead to the aggravation of environmental pollution, which is the contradictory relationship between the two; second, economic growth is beneficial to the environment, and the two are mutually promoting harmonious relationship; the third is the environmental Kuznetz curve hypothesis put forward by Geossman and Kureger in the 1990s and the core idea is that in the process of economic development, the environment will deteriorate first and then improve. After the hypothesis is put forward, it has been widely accepted, and has been widely used in the study of the relationship between resource environment and economic development, such as pollution emission, carbon emission, forest land degradation, biological protection and so on. At the same time, it also involves the cross-section, time series or panel data of typical countries or regions, the verification of the inverted U-shaped curve hypothesis of environmental quality evolution and the prediction of its turning point threshold (Grossman and Krueger 1995; Cropper and Griffiths 1994; Panayotou et al. 2000; Liu et al. 2008a; Liu and Chen 2009).

In EKC hypothesis, the inverted U-shaped relationship between economic development and environmental pollution is described, which explains the difference of environmental impact in different stages of economic development (Fig. 7.1). In the economic take-off stage, with the rapid economic growth and large-scale exploitation and utilization of natural resources, industrial development takes heavy chemical

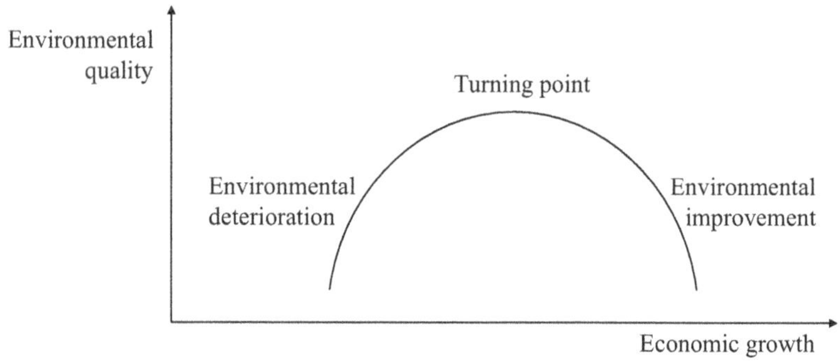

Fig. 7.1 Environmental Kuznets curve of economic growth

industry with high energy consumption and high pollution as the main content, which puts great pressure on the ecological environment. After the economic development to a certain stage, the industrial structure is gradually optimized, from resource intensive to technology intensive. In the manufacturing industry, the proportion of raw material industry with high energy consumption, high consumables and high pollution has gradually decreased, while the proportion of deep processing industry and technology intensive industry is increasing. Also, the emission of industrial pollution has been absolutely reduced, industrial pollution has been effectively controlled, and low pollution knowledge intensive industries have become the driving forces of economic growth.

Through the environmental curve, the long-term law between economic growth and environmental pollution is revealed. However, the economic type, institutional environment and other factors restrict the environmental situation of developing countries or regions to a certain stage, which does not necessarily lead to environmental improvement. If the ecological environment degenerates to exceed the ecological threshold, environmental degradation will become an irreversible process. If the industrial structure and industrial production efficiency of a country or a region cannot be smoothly transformed, the ecological environment will still face great pressure. Based on the experience of developed and developing countries, there are often conflicts between economic growth and environmental protection. The government is concerned about the demand for economic growth, and the environmental protection related laws are lagging behind, which will inevitably affect the relationship between the economy and the environment.

7.1.3 Comprehensive Resources and Environment Effect in Urban–Rural Transformation

The effect is the comprehensive characterization of effect and response. The process of urban–rural transformation has changed the allocation of resources and environmental conditions, resulting in a new comprehensive geographical effect. As for land use change, it is the spatial reflection of human economic and social activities on the surface of the earth. Land is the carrier of space. Through land use change, human activities lead to land degradation, inefficient use, ecological space compression, and decline of soil and water conservation and climate regulation functions. Taking population, industry and society as the three subsystems of urban and rural development transformation, in this paper, land change is regarded as a direct effect of urban and rural development transformation. And if the transformation of urban and rural development covers population, industry, society and land use, land use change will be regarded as the medium of resource and environment effect of economic and social change. The effect of resources and environment is the effect and response of resources and environment system under the influence of human activities, which can be divided into resource effect and environment effect.

Although the rapid progress of industrialization and urbanization has made great progress in China's urban and rural economic and social development, it is deeply affected by the tilt of system, institution, strategy and policy. China also faces the serious imbalance of urban and rural structure, industrial structure, regional structure and factor input structure, especially the imbalance of factor structure transformation. As for the imbalance of factor input structure, it is reflected in high resource consumption, increasing environmental pressure, and increasingly prominent constraints of resources and environment on economic growth, urbanization and agricultural transformation. In China, the proportion of consumption of major resource products in the global consumption is significantly greater than that of GDP in the global economy, the resource consumption per unit product is significantly higher than that of developed countries, the consumption of water and land resources is too fast, and the ecological environment cost of economic development is huge. In particular, in the process of industrialization, the local government takes economic growth as the guidance, resulting in the homogenization of industrial structure, and the relatively low-level industries of township enterprises and counties develop in scale. The competition between local governments for enterprise resources has brought many environmental problems and pressures, such as the substandard discharge of industrial "three wastes" in some areas, farmland pollution and food safety caused by sewage irrigation, lake eutrophication, river water quality decline and frequent environmental events. Besides, the phenomenon of polluting enterprises transferring from the lower reaches of the river basin to the upper reaches and from the city to the countryside has also appeared. The ecological and environmental problems have become more and more serious social problems.

Obviously, the impact of urban and rural development transformation is obviously comprehensive, and the trend of effect representation is closely related to the development mode. From the perspective of human land relationship of urban–rural regional system and the comprehensive effect of urban–rural transformation, this paper focuses on the stress and pressure of urban–rural system (including population, economy, society, land use and other subsystems) on the resource and environmental system (covering resources and environment subsystem), and reveals the mechanism, types and coupling law of urban–rural system and resource-environment system (Fig. 7.2). On the one hand, this paper analyzes the mechanism and coupling law between urban and rural system factors and resources and environmentalal factors using one-to-one two-way analysis; makes directional calculation of the influence coefficient of single factor X on single factor Y; quantitatively expresses the interaction equation and coupling curve between single factors. On the other hand, this paper analyzes the interaction mechanism and coupling law between urban and rural system and resource and environmental system factors, constructs the function equation of multi-factor influence, and explores the relationship between the two systems, so as to provide quantitative scientific basis for optimizing regional human land relationship and urban–rural relationship.

7.2 Coupling Analysis of Urban–Rural Transformation and Land-Use Change

7.2.1 Theoretical Analysis: Relationship Between Urban–Rural Transformation and Land Non-agriculturalization

1. **Relationship between farmland non-agriculturalization and socio-economic development**

Land-use changes are largely caused by human activities, and changes in land-use patterns can reflect on the transition across socio-economic systems (Long 2012). From 1996 to 2016, land area dedicated to construction in China increased

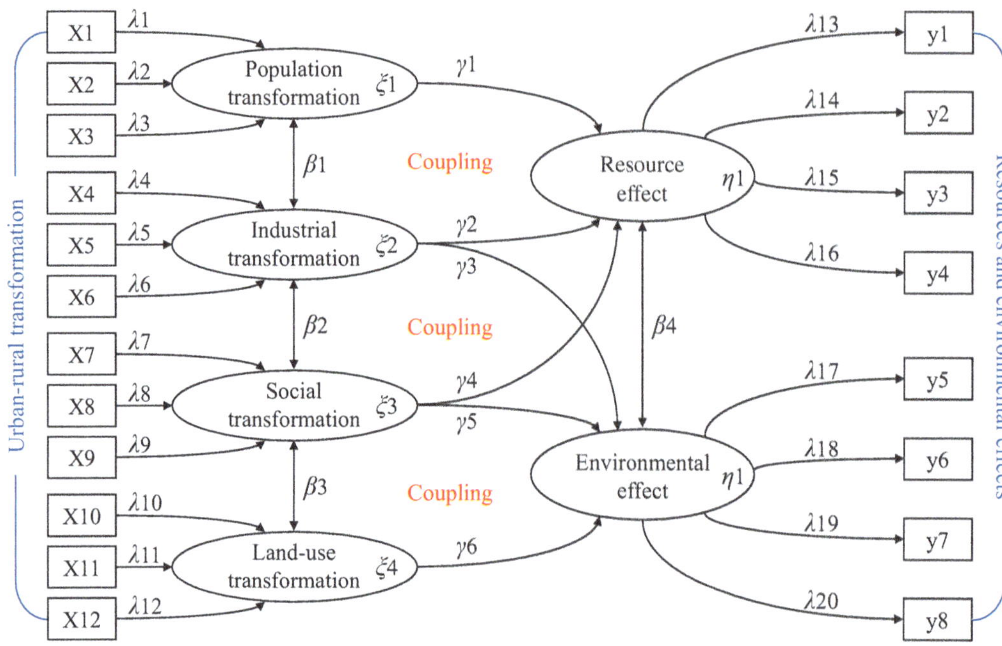

Fig. 7.2 Structural model for the diagnosis and evaluation of the resources and environmental effects of urban–rural transformation

from 29.6 million hm^2 to 39.1 million hm^2. Unlike the rapid urbanization in Latin America and Southeast Asian countries, urbanization in China has led to excessive non-agriculturalization of farmland. From 2000 to 2016, the urban built-up area increased by 2.42 times; the growth rate of built-up are is much higher than that of urban population during the same period. Rural population decreased by 219 million people, while the rural residential land area increased by 41,168,100 mu. Excessive non-agriculturalization of farmland leads to wastage of large amounts of cultivated land resources, intensifies the extensive land use, and does not ensure sustainable land use.

Since 1991, the environmental Kuznets curve has been proposed and widely used to study the relationship between economic growth, resource utilization, and waste emissions, considering forest changes, CO_2 emissions, oil extraction, and biodiversity (Li et al. 2014a). Land-use changes are closely related to economic growth and urbanization of population; studies have confirmed the inverted U-shaped relationship between economic growth and farmland changes since the economic reforms and opening-up of the economy in China (Li and Wu 2008). The agglomeration of urban population and development of non-agricultural industries have created social demands for non-agricultural land use, such as building infrastructure, increasing industrial and commercial use of land, and expanding residential land area. The stage of economic development, social needs, and industrial structure of an economy impact its land use. It is generally believed that socio-economic development affects environmental quality through scale effect, structural effect, and technological effect (Grossman and Krueger 1995). At the initial stage of socio-economic development, due to the difference in the marginal rate of return between the agricultural and non-agricultural sectors, the scale effect promotes a rapid increase in the rate of non-agriculturalization of agricultural land. With the gradual development of an economy to the intermediate stage, the economic structure transforms from a stage of being heavily reliant

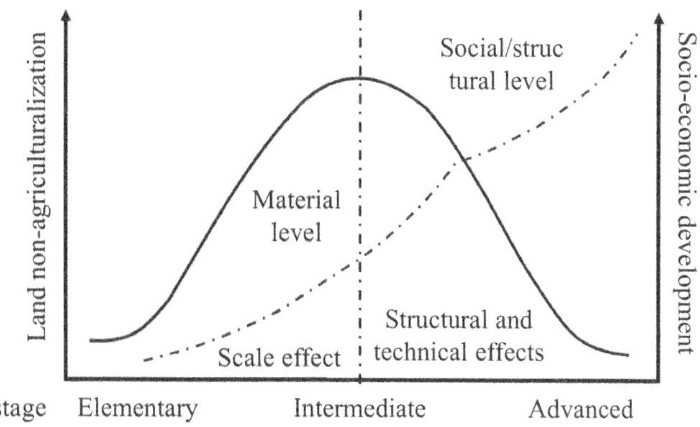

Fig. 7.3 Stages of economic development and the process of non-agriculturalization of cultivated land

on land input to a stage of being dominated by land-intensive industries. With this transition, the rate of non-agriculturalization of agricultural land begins to slow down. Finally, at the third stage of maturity, technological advancement promotes knowledge and innovation to replace large-scale land input and to improve land-use efficiency (Liu et al. 2008b). Antweiler et al. (2001) showed that structural and technological effects of change in output and those of pollution reduction (or resource reduction) were the main reasons for the downward slope of the environmental Kuznets curve after reaching the turning point. Due to the limitation of land resources, the development and utilization of cultivated land should be compatible with socio-economic development, and occupations dependent on cultivated land should follow the inverted U curve.

The development history of humans has provided a proof of existence of the relationship between non-agriculturalization of land and socio-economic development. In the past two centuries, the rapid industrialization and urbanization of developed countries have promoted the demand for land. When human society advances, urbanization is no longer manifested as the continuous expansion of metropolises; it is rather seen as the development of small and medium-sized cities and small towns under the influence of urban industry and population transfer (Leeuwen et al. 2006). Urbanization of developed countries has crossed the stage of population agglomeration in urban areas and entered the stage of urban civilization spreading to the countryside. Through this process of urbanization, violent urban expansion began to ease, and social inclusiveness became the theme of humans' socio-economic development. In the late 1980s, France implemented its first social inclusion policy to help new immigrants adapt to their new homes. In 2000, the European Union (EU) initiated a 10-year plan to improve the social inclusion of EU member states in the fields of employment, education and training, medical care, and housing (Ga and Huang 2008). Upon entering the advanced stage of urbanization, the social inclusion agenda plays an important role in curbing social exclusion and social isolation and provides equal opportunities, resources, and social welfare for residents in the new cities.

Based on the foregoing analysis, the relationship between non-agriculturalization of land and socio-economic development can be described by an inverted U curve (Fig. 7.3). In the initial and intermediate stages, the influx of population and the development of non-agricultural industries have converted a large amount of agricultural land into urban built-up land, and the scale of cities has expanded rapidly. At the onset of maturity, social inclusiveness and effective land use in urban and rural areas begin to dominate.

2. **Non-agriculturalization of agricultural land in China during the transition period**

Since the 1990s, industrialization and urbanization in China have caused the demand for industrial, infrastructure, housing, and commercial land to grow rapidly. From 1990 to 2016, the urban built-up area in China increased from 11,608 to 52,761 km^2, with an average annual expansion of 1,582.81 km^2 (Liu et al. 2013). Despite the substantial increase in area, urban land is characterized by low-density utilization. The per capita urban built-up area exceeds 130 m^2, which is much higher than current standard (82.4 m^2) for developed countries and that for developing countries (83.3 m^2). In addition, since 1992, large-scale construction of economic development zones at the provincial- and village-level in China has caused low land-use efficiency. As of 2018, there were 2,543 economic development zones in China, with a total area of nearly 36,000 km^2. However, about one third of them were currently unused, showing the characteristics of being occupied but not being used, being approved but having no building activity, or being more occupied but being used less.

With the advancement of urbanization, the large-scale transfer of agricultural labor to urban non-agricultural sector has exacerbated the hollowing of countryside, which is manifested in the vacancy, idleness, abandonment, and collapse of village houses. "One household, one house" policy system, homestead withdrawal mechanism, and lack of rural planning in China have become key in the formation of hollow villages. In the late 1980s, the Chinese rural baby boomers began to form nuclear families instead of living in the traditional main family (also known as "extended family" in which parents live with their married children). With this change, the new nuclear families (parents and unmarried children living together) built many houses on the outskirts of villages. Based on household statistics, the size of Chinese households has been declining since the 1980s, from 4.41 in 1982 to 3.96 in 1990 to 3.1 in 2015. In the twenty-first century, the number of migrant workers has increased, but it is difficult for migrant workers to settle in cities. Instead, they regularly remit money to their homes and build new houses in their original villages. Due to the fact that migrant workers go out to work all-year-round, the utilization efficiency of newly built houses in rural areas is very low, and some even remain idle throughout the year. With the rapid hollowing of villages, the rural population is decreasing but the homesteads are increasing (Fig. 7.4). The per capita homestead area is 229 m^2, which is much higher than the national maximum of 150 m^2.

With socio-economic development, agricultural land has been continuously used as construction land, accompanied by urban expansion, construction of development zones, and growth of rural homesteads (Fig. 7.5). In recent decades, the rapid economic growth in China has caused drastic changes in the urban–rural land-use patterns. However, the current urban expansion, construction of development zones, and rural residential land-use models are inefficient. With the transformation of economic structure from being industry-led to being tertiary industry-led in China, in order to improve land-use efficiency and intensity, measures such as land consolidation, linking the increase and decrease of urban–rural construction land, and structural adjustment of land use have been adopted. These measures are likely to help curb non-agriculturalization of land and wastage of land resources. It is estimated that

Fig. 7.4 Changes of rural population and housing land in China from 1996 to 2016

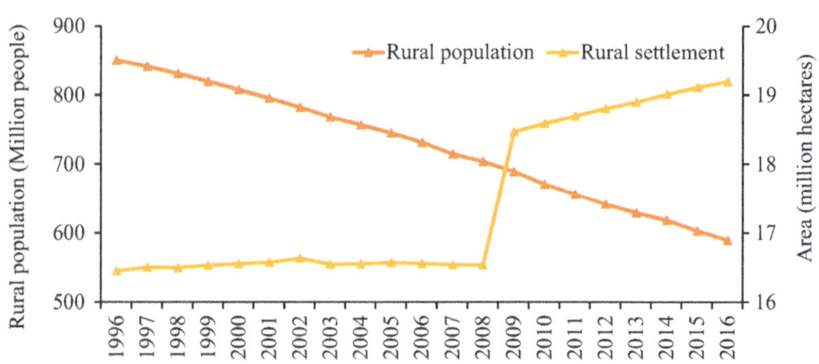

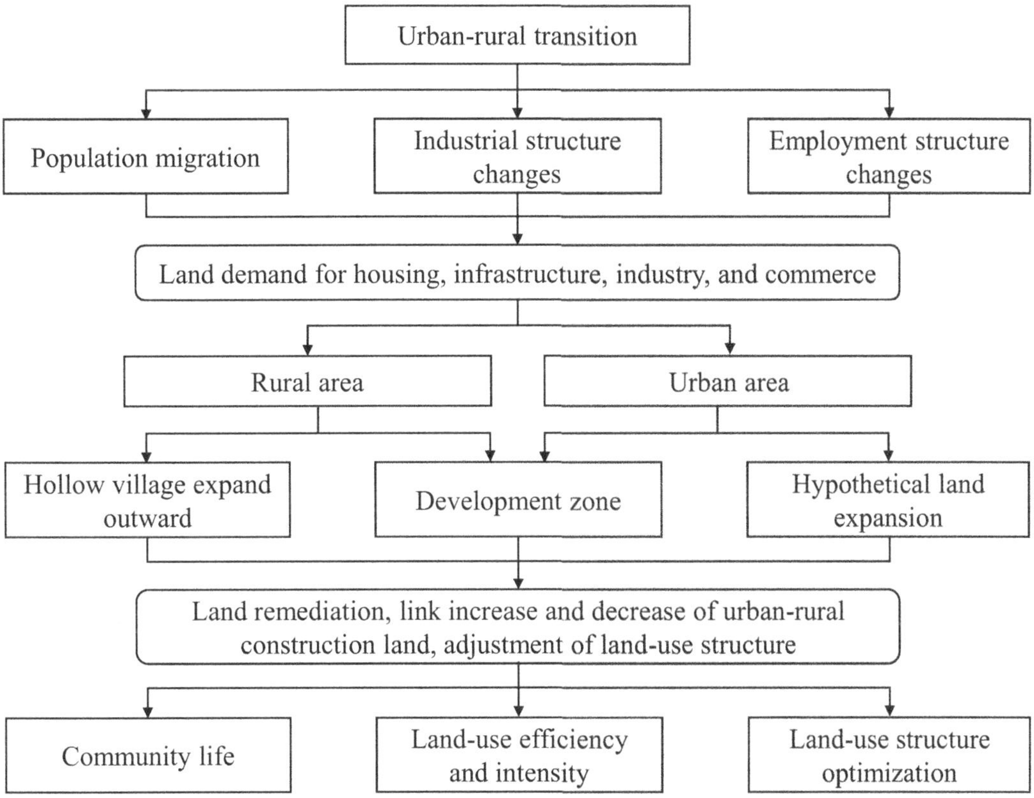

Fig. 7.5 Urban–rural structural transformation and non-agriculturalization of farmland in China

7.58 million hectares of construction land in rural areas of China has been converted into agricultural land through comprehensive land consolidation (Liu et al. 2013).

7.2.2 Empirical Methods and Usage Data

The current research focuses on analyzing the impact of economic growth and industrial transformation on agricultural land conversion as well as the relationship between agricultural land conversion and economic growth. We are drawing on the related research of Grossman and Krueger (1992) assuming that agricultural land conversion process has an environmental Kuznets curve relationship since the economic reforms and opening up of the economy in China. Based on existing research results, this article adopts panel data analysis method, controlling population density variables and using the following benchmark model for measurement:

$$FC_{it} = \theta_{it} + \alpha_1 GGDP_{it} + \alpha_2 PGDP_{it}^2 + \alpha_3 Pop_{it} + \mu_{it} + \varepsilon_{it} \tag{7.1}$$

If $\alpha_1 > 0$ and $\alpha_2 < 0$ both regress, there is an inverted U-shaped relationship between agricultural land conversion and economic growth.

In the second step, we substitute the urbanization and proportion of tertiary industry into the benchmark model to verify the influence of these factors on the relationship between economic growth and agricultural land conversion. We also incorporate land-use efficiency into the regression model and change the benchmark model as follows:

$$FC_{it} = \theta_{it} + \alpha_1 GGDP_{it} + \alpha_2 PGDP_{it}^2 + \alpha_3 Urba_{it} + \alpha_4 Urba_{it} \cdot PGDP_{it} + \alpha_5 Pop_{it} + \mu_{it} + \varepsilon_{it} \quad (7.2)$$

$$FC_{it} = \theta_{it} + \alpha_1 GGDP_{it} + \alpha_2 PGDP_{it}^2 + \alpha_3 Terti_{it} + \alpha_4 Terti_{it} \cdot PGDP_{it} + \alpha_5 Pop_{it} + \mu_{it} + \varepsilon_{it} \quad (7.3)$$

$$FC_{it} = \theta_{it} + \alpha_1 GGDP_{it} + \alpha_2 PGDP_{it}^2 + \alpha_3 Effici_{it} + \alpha_4 Effici_{it} \cdot PGDP_{it} + \alpha_5 Pop_{it} + \mu_{it} + \varepsilon_{it} \quad (7.4)$$

where t is the year, and α_i (i= 1…5) is the undetermined regression coefficient; FC_{it} and $PGDP_{it}$ respectively represent the non-agricultural land and per capita GDP of province i in year t; Pop_{it} represents the population density of province i in year t; ε_{it} is the error term; θ_{it} and μ_{it} are the cross-sectional effect and time effect.

FC_{it} is the amount of agricultural land occupied by urban–rural construction land in each province. $PGDP_{it}$ represents the per capita GDP of each province. $Effici_{it}$ characterizes the efficiency of construction land, which is measured by the proportion of non-agricultural industry construction land. $Urba_{it}$ represents urbanization, expressed as the proportion of urban population to total population. $Terti_{it}$ represents the level of industrial structure upgradation, measured by the proportion of tertiary industry. Pop_{it} is calculated by dividing population by area. Table 7.1 shows the statistical values of these variables. The available panel data covers 30 provinces, municipalities, and autonomous regions (except Tibet, Taiwan, Hong Kong, and Macau) from 2000 to 2009. The data on cultivated land occupation for urban–rural construction land is obtained from the "China Land and Resources Statistical Yearbook" and the other data is obtained from the "China Statistical Yearbook" and "China Regional Statistical Yearbook".

Figure 7.6 shows per capita GDP and the average level of non-agricultural land in different provinces from 2000 to 2009. Overall, the per capita GDP of the eastern coastal regions is higher than that of the central and western regions, and the scale of non-agricultural land in the eastern regions is also much higher compared to the other regions. In the central and western regions, the non-agriculturalization of farmland has declined with the decrease of per capita GDP. This means that there is a close connection between agricultural land conversion and economic growth in China.

7.2.3 Impact of Urban–Rural Transition on Farmland Non-agriculturalization

Panel data analysis results obtained using Hausman's test show that the fixed effects model is more appropriate. The regression results are shown in Table 7.2. According to the estimated coefficients of *PGDP* and *PGDP'* in all regression models, we found that there exists an inverted U-shaped environmental Kuznets curve relationship between the non-agricultural land

Table 7.1 Statistical characteristics of different variables

Variable name	Unit	Observed variable	Mean	Standard deviation	Minimum	Maximum
FC	km^2	300	699.64	6813.04	72.36	46,708.77
PGDP	Yuan per person	300	16,722.57	13,109.73	2759	77,205
Urba	%	300	0.45	0.16	0.23	0.89
Effici	GDP/km^2	300	0.79	0.9	0.1	6.25
Terti	%	300	0.4	0.07	0.29	0.75
Pop	$Person/km^2$	300	378.62	432.09	7.41	3486.25

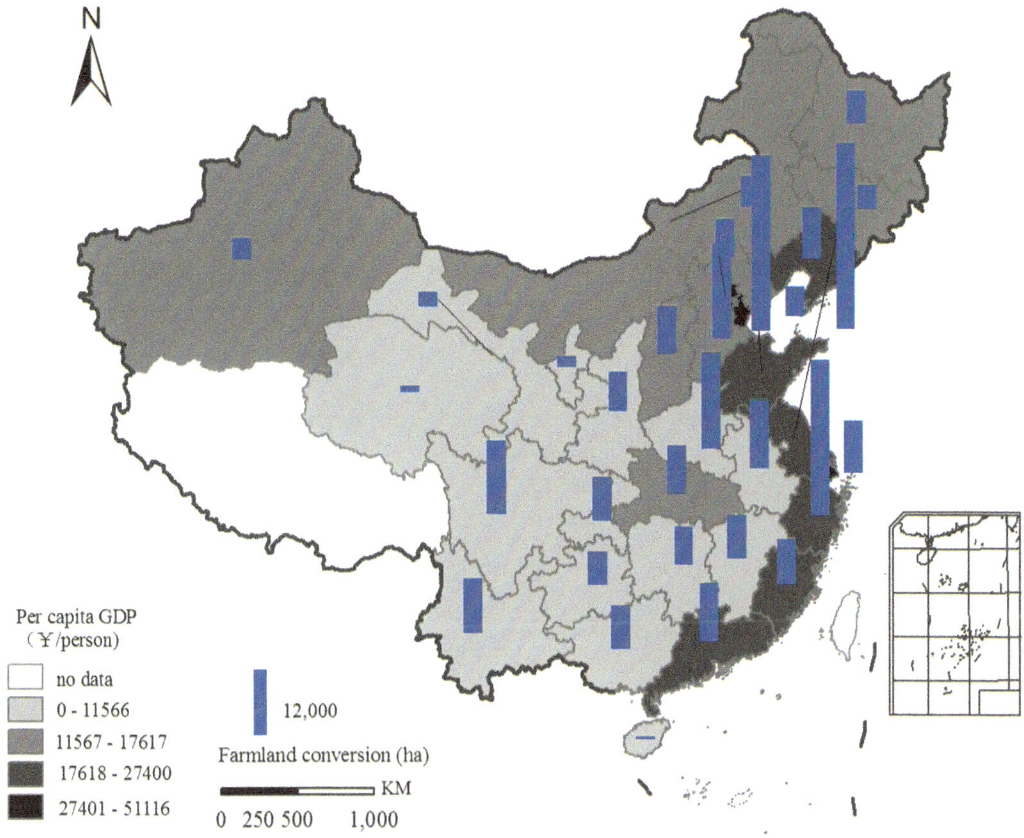

Fig. 7.6 Level of non-agriculturalization of farmland and per capita GDP in each province (2000–2009)

and economic growth (Table 7.2) during the study period. By deriving the *PGDP* in Model 7 at the turning point where the derivative is zero, we get the following equation:

$$\frac{\partial FC}{\partial PGDP} = \alpha_1 + 2\alpha_2 PGDP' = 0 \quad (7.5)$$

$PGDP^*$ at the turning point is 2,3660.7 yuan/person, which is almost equivalent to the value of per capita GDP (2,3707.7 yuan/person) in 2008 announced by the National Bureau of Statistics in China. From this, we can conclude that the turning point in China appeared in 2008, which was graphically at the intersection of economic growth and non-agricultural land conversion.

In Model 2, *Urba* and *Urba•PGDP* are not significantly different. Furthermore, in Model 5, when the land-use efficiency factor is added to the regression equation, the two variables are not significantly different. It shows that the migration of rural population to cities did not affect the relationship between economic growth and agricultural land conversion during the study period. This result can be attributed to the implementation of the policy of linking the increase and decrease for urban–rural construction land. Since 2000, the State Council has encouraged the relocation and integration of out-of-population villages. As a result of allocation of construction land quotas in urban and rural communities for intensive housing construction, the original rural construction land has been reclaimed as agricultural land. Through this process of urban–rural construction land replacement, the increase of urban construction land can be realized through the reduction of rural construction land, which

Table 7.2 Regression analysis results of all samples

Independent variable	Regression coefficients						
FC	1	2	3	4	5	6	7
PGDP	0.56*** (0.18)	0.59*** (0.22)	0.58*** (0.19)	0.43** (0.18)	0.48** (0.2)	0.6*** (0.22)	0.53** (0.23)
PGDP^2	−6.4e −06*** (1.79e−06)	−5.8e −06** (2.84e −06)	−6.6e −06*** (2.3e−06)	−6.7e −06** (3.45e −06)	−1.08e −05** (4.28e−06)	−8.76e −06** (3.74e−06)	−1.12e −05** (4.45e−06)
Urba		1091 (5119.39)			−2487.35 (5205.46)		−2607.6 (5318.7)
Urba•PGDP		−0.1 (0.39)			0.4 (0.39)		0.43 (0.42)
Terti			6455.36*** (1528.73)			4167.2*** (1344.3)	6100** (2346.2)
Terti•PGDP			−0.35*** (0.11)			−0.55** (0.31)	−0.68*** (0.23)
Effici				−6212.31* (3260.3)	−7918.84** (3351.72)	−5917.5** (3467.2)	−8279.5** (3497)
Effici•PGDP				0.09* (0.05)	0.097* (0.056)	0.097* (0.057)	0.11* (0.06)
Pop	7.96** (3.78)	7.81** (3.84)	8.11** (3.84)	7.54** (4.03)	4.73 (4.04)	7.6* (4.07)	4.72** (2.02)
Estimation method	FE	FE	FE	FE	FE	FE	FE
Pro-specific effect	YES	YES	YES	YES	YES	YES	YES
Year-specific effect	YES	YES	YES	YES	YES	YES	YES
Pro. num	30	30	30	30	30	30	30
Observations	300	300	300	300	300	300	300
R^2	0.525	0.378	0.49	0.57	0.32	0.48	0.69

Note The standard errors are in parentheses. *, **, and *** represent significant differences at the levels of 10%, 5%, and 1%, respectively

effectively reduces the large-scale occupation of agricultural land due to urban expansion.

In Model 3, *Terti* and *Terti•PGDP* have a significant difference at the level of 1%. By deriving the *PGDP* in Eq. 7.3 at the turning point where the derivative is zero, we get the following equation:

$$\frac{\partial FC}{\partial PGDP} = \alpha_1 + 2\alpha_2 PGDP' + \alpha_4 Terti = 0 \tag{7.6}$$

$$PGDP' = -\frac{\alpha_1 + \alpha_4 Terti}{2\alpha_2} = \left|\frac{\alpha_1}{2\alpha_2}\right| - \left|\frac{\alpha_4}{2\alpha_2}\right| \cdot Terti \tag{7.7}$$

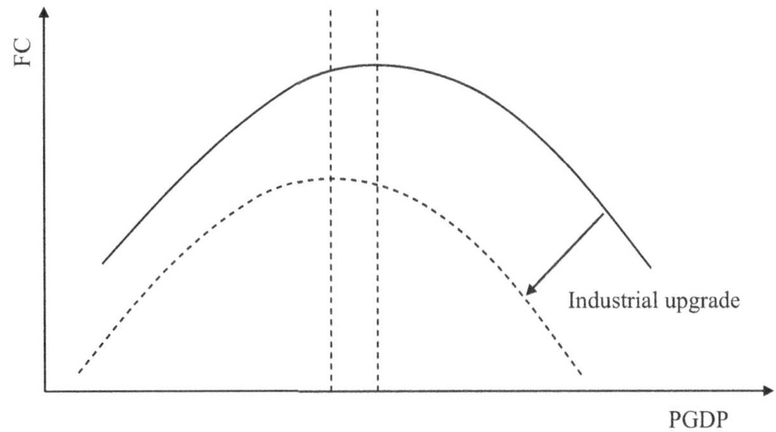

Fig. 7.7 Impact of industrial upgradation on the inverted U curve

Therefore, when the value of *Terti* increases, the value of *PGDP** at the turning point decreases, that is, the inverted U curve shifts downward (Fig. 7.7). It means that if the proportion of the tertiary industry rises, the turning point of agricultural land conversion will occur early. At the same time, the *Terti•PGDP* coefficient is negative, indicating that the impact of economic growth on agricultural land conversion can be attributed to the result of industrial upgradation. The rapid development and rising proportion of the tertiary industry can significantly reduce the demand for agricultural land conversion.

In Model 4, the estimated coefficients of *Effici* and *Effici•PGDP* are significantly different at the 10% level. By deriving the *PGDP* in Eq. 7.4 at the turning point where the derivative is zero, we get the following equation:

$$\frac{\partial FC}{\partial PGDP} = \alpha_1 + 2\alpha_2 PGDP' + \alpha_4 Effici = 0 \tag{7.8}$$

$$PGDP' = -\frac{\alpha_1 + \alpha_4 Effici}{2\alpha_2} = \left|\frac{\alpha_1}{2\alpha_2}\right| - \left|\frac{\alpha_4}{2\alpha_2}\right| \cdot Effici \tag{7.9}$$

Therefore, when the value of *Effici* becomes larger, the value of *PGDP** at the turning point also becomes larger, and the inverted U curve shifts upward (Fig. 7.8). This means that when land-use efficiency increases, the turning point will occur later. The negative performance of the *Effici•PGDP* coefficient also shows that the impact of economic growth on the non-agriculturalization of agricultural land increases

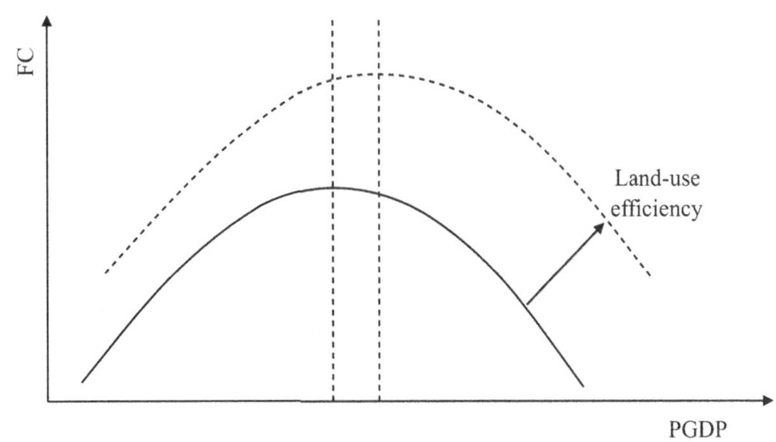

Fig. 7.8 Impact of land-use efficiency on the inverted U-shaped curve

with the improvement of land-use efficiency. Since *Effici* is measured by the proportion of urban–rural construction land in non-agricultural industries, a higher *Effici* means a higher non-agricultural output from per unit urban–rural construction land. In view of these findings, it can be expected that as the gap between non-agricultural and agricultural production increases, China will have more agricultural land for non-agricultural industries. At the same time, 70% of the 300 observations in this study showed that the proportion of the secondary industry was higher than that of the tertiary industry. This means that the secondary industry has more influence on the non-agricultural output rate of per capita construction land; it also generates more demand for non-agricultural land.

The regression results of Models 6 and 7 reveal that the relationship between non-agriculturalization of farmland and economic growth during the study period has a significant difference between *Terti* and *Effici*. According to these results, *Urba* and *Urab•PGDP* are not significantly different. By comparing the regression coefficients in Model 7 with the significance levels of *Terti* and *Terti•PGDP* as well as *Effici* and *Effici•PGDP*, we found that in terms of the inverted U-shaped relationship between non-agriculturalization of farmland and economic growth in China, industrial upgrading has more impact than land-use efficiency.

Based on the above research findings, there exists an inverted U-shaped relationship between non-agriculturalization of farmland and economic growth from 2000 to 2009 in China, and the rapid non-agriculturalization rate of agricultural land has slowed down. This process is accompanied by the transition from extensive land use to intensive land use. The coupling relationship between economic growth and non-agricultural land use tends to positive development, and the relationship between man and land has improved. The transition of the inverted U-shaped relationship is closely related to the upgradation of industrial structure, technological upgradation, and the implementation of the policy of linking the increase and decrease of urban–rural construction land. As the tertiary industry occupies less land than the secondary industry, the upgradation of the industrial structure and the development of the tertiary industry will make the non-agricultural land conversion curve fall instead of rise. Due to the implementation of the policy of linking the increase and decrease of urban–rural construction land, the reclamation of the inefficiently used construction land in rural areas along with the increase in urban construction land have made the transfer of rural population to cities and towns have limited impact on agricultural land conversion.

7.3 Scenario Simulation of Land System Change Under Urban–Rural Transformation

7.3.1 Materials and Methods

7.3.1.1 Study Area

The Beijing-Tianjin-Hebei (BTH) region (113° 04′–119°53′E, 36°01′–42°37′N) is located in the northeastern area of China, which includes Beijing Municipality, Tianjin Municipality and Hebei Province, with the total land area of 218,000 km^2. This region is characterized by a typical warm-temperate semi-humid and semi-arid continental monsoon climate. The terrain of the whole area slopes from northwest to southeast, which connects with Yanshan Mountain in the north, Taihang Mountain in the west, and Bohai Bay in the east, so there are mountainous, hilly, and plateau areas in the northwest, with basins and valleys distributed among them, while vast plains dominate the central and southeast. As one of the three major urban agglomerations of China, the BTH region contributes to 9.44% of China's GDP with 113 million people inhabiting here in 2018. The vast plain in the central and south of this region is a major grain and cash crop production base and there are many important ports and manufacturing bases distributed along the upper eastern coast. The northwest of this region has a high coverage rate of forests and grasslands, which plays an essential role in maintaining regional ecological security.

The BTH region is a highly developed zone characterized by significant internal imbalance. In 2018, BTH's urbanization rate reached 65.8%, with 86.5% in Beijing, 83.2% in Tianjin and only 56.4% across Hebei province. Beijing and Tianjin are developed metropolitan zones with relatively concentrated populations, capital and technology. However, cities located in the poverty belt around Beijing and Tianjin, such as Chengde and Zhangjiakou, have been confronted with challenges of boosting economic development and improving people's livelihood. Therefore, promoting regional sustainable development on the basis of ecological protection must be the primary focus for policy making in this region.

7.3.1.2 Data Source and Processing

The data used in this study included land use data, Digital Elevation Model (DEM) data, Net Primary Productivity (NPP) data, high/medium yield cropland data, national ecological function reserve data, national nature reserve data, basic geographic information data, meteorological data and relevant socio-economic data. Based on these data, simulations of land system changes in 2030 were carried out. Land use data pertaining to the years 2000, 2005 and 2015 were obtained from the Landsat TM/OLI remote sensing images through artificial visual interpretation, with accuracy of no less than 90% (Liu et al. 2014a; Ning et al. 2018). Spatial resolution of land use data is 100 m and these data were classified into six categories: arable land, forest, grassland, water, built land and unused land (Liu et al. 2003). DEM data were downloaded from the Geospatial Data Cloud website (http://www.giscloud.cn), with spatial resolutions of 90 m, and elevation and slope data were generated from DEM data using ArcGIS. NPP data for croplands, forests and grassland were calculated based on the Vegetation Photosynthesis Model (VPM) method (Niu et al. 2016). High and medium yield cropland distribution data were generated based upon the differentiation of cropland productivity represented by mean values of time-series NPP data (Ji et al. 2015). The national ecological function reserve and national nature reserve data, gridded gross domestic production (GDP) and population data were derived from the Resource and Environment Data Cloud Platform, Chinese Academy of Sciences. Both the raster datasets of gridded GDP and population were generated based on the China statistical yearbook at county level in 2005 and the resolution was 1000 m. Basic geographic information data were downloaded from the National Geomatics Center of China (http://www.ngcc.cn), including provincial boundaries, administrative centers at the city and county/district levels, highways and railways, etc. Annual average precipitation and temperature data were downloaded from the China National Meteorological Information Center (http://data.cma.cn).

7.3.1.3 Methods

1. **Dyna-CLUE model**

The Dyna-CLUE model is composed of two distinct modules: a non-spatial demand module which is designed to predict future land use demand, and a spatial allocation module designed to conduct land use pattern simulations. Required inputting elements of the Dyna-CLUE model include four parts: spatial policies and restrictions, land use conversion, land use demand and location suitability. ① Spatial policies and restrictions indicates areas where land use changes are prohibited due to regional policy and planning, such as the allocation of nature reserve status or permanent farmland, etc. ② The land use conversion section determines conversion elasticity and provides the conversion matrix of the defining rules of conversion between land use types. ③ The land use demand section calculates annual area changes for each land use type by either simple extrapolation or through complex multi-sectoral models. ④ Location suitability is estimated through spatial associations of current land use with associated driving factors using logistic regression model.

Stepwise logistic regression was performed in SPSS (IBM SPSS, version 22) to evaluate the relation between each land use type and the associated potential driving factors. The mathematical expression is as follows:

$$Log\left(\frac{P_i}{1-P_i}\right) = \beta_0 + \beta_1 x_{1,i} + \beta_2 x_{2,i} + \cdots + \beta_n x_{n,i} \quad (7.10)$$

where P_i is the spatial distribution probability, X_i represents the driving factor of i, β_i is the regression coefficient of each factor. Relative operating characteristic (ROC) was adopted as a quantitative measure to evaluate the fit of the logistic regression model (Pontius and Schneider 2001). The value of ROC ranges from 0.5 to 1.0. An ROC above 0.70 shows that the driving factors had a great explanatory power for certain land use type (Pontius et al. 2008).

The Dyna-CLUE model simulates land use distribution based on raster dataset, requiring the grid cell size, grid cell number, and spatial extent of all input data to be identical. The question of choosing an appropriate resolution for modelling has always been a challenge since the origin of the land system simulation research (Verburg et al. 2019). As the resolution improves, more land use information is involved, but the model's accuracy is likely to decrease (Pontius et al. 2008) and it may exceed Dyna-CLUE software's data processing capability. Thus, the spatial resolutions of most previous case studies were set between 100 and 1000 m (Lima et al. 2011; Lu et al. 2015; Verburg et al. 2002; Xu et al. 2013). Considering the spatial scale of the study area, the resolution of the available data and all the limitations above, the resolution of all input data were aggregated to 500 m.

2. **Markov model**

The Markov model was adopted in the process of predicting land use demand. The Markov chain is a non-aftereffect process, which is in accord with the characteristic of land use change. The later state (land use type) is only related to its beginning state and transition step size, but not to any other previous states (Fischer and Sun 2001; López et al. 2001). Prediction processes were as follows: firstly, the transition probability matrix was computed from the land use maps at the previous and current time node; secondly, the future land use demand was then produced by multiplication of transition probability matrix and the area of each land use type at the current time node (Guan et al. 2011). Transition probability matrix can be calculated through the Markov module in the IDRISI software package (Clark Labs, Worcester, Massachusetts) and its user's guide is highly useful (Myint and Wang 2006). Hence, detailed procedures and mathematical expression will not be discussed in this paper.

3. **Assessing simulation accuracy**

The method of three map comparison (Pontius et al. 2008, 2011) was adopted in this study to validate the simulation model by comparing the observed change with predicted change in the land use map. The maps used for the three-map comparison included the reference map of time 1, the reference map of time 2 and the simulated map of time 2. Observed change reflects the real dynamic changes of the land use, which were generated by overlaying the maps of reference time 1 and reference time 2. Predicted change reflects the land use changes in a simulated situation, which were generated by overlaying the map of reference time 1 and simulated time 2.

Model accuracy was determined by five types of indicators: null successes (accurate prediction due to observed persistence predicted as persistence), true hits (accurate prediction due to observed change predicted as right gaining category), partial hits (inaccurate prediction due to observed change predicted as wrong gaining category), misses (inaccurate prediction due to observed change predicted as persistence) and false alarms (inaccurate prediction due to observed persistence predicted as change). Thus, the figure of merit (FoM, Eq. 7.11) and total error (T, Eq. 7.12) were calculated using the indicators above (Chen and Pontius 2011).

$$FOM = \frac{H_1}{H_1 + H_2 + M + F} \times 100\% \quad (7.11)$$

$$T = H_2 + M + F \quad (7.12)$$

where H_1, H_2, M and F represent the true hits, partial hits, misses and false alarms; T refers to the total error.

Total error reflects the model inability both in the quantity prediction and in the spatial allocation. The FoM indicates the overall accuracy of the simulation model, ranging from 0% (no overlap between observed and predicted change) to 100% (perfect overlap between observed and predicted change), and if FoM value reaches 100%, the simulation is completely correct. This measure enables more realistic assessments of the cell-to-cell coincidence between simulated and real maps than more commonly used metrics, such as the kappa index, which are usually calculated using the entire surface area (Santé et al. 2010).

4. **Scenarios and parameters**

Based on comprehensive consideration of regional resource endowments, comparative advantages and different development orientations, scenario simulation has been largely applied in the formulation and assessment of land use planning and land use policy (Cairns et al. 2016; Jenerette and Wu 2001). Generally, several major aspects must be taken into account, including economic development, food production and ecological protection. Thus, combining the historical land use experience and future demand, three scenarios in 2030 were developed for the BTH region, including business as usual scenario (scenario 1), cropland protection scenario (scenario 2), and ecological security scenario (scenario 3). Then, different land use demand, spatial restriction policies and parameters were input into the simulation models for corresponding scenarios (Gong et al. 2018).

① **Conversion rules**

The conversion rules are composed of land use conversion sequences and conversion elasticity (ELAS). Land use conversion sequences were determined by the conversion matrix with a value of 0 or 1, with 0 indicating impossible conversion and 1 indicating possible conversion. By analyzing the historical land use conversions among land categories from 2000 to 2015 in the BTH region, we could conclude that all land use conversions are possible and therefore, we define each ELAS value of these six land categories as 1. Although conversion from built land to other types is generally difficult, it in deed happens in urbanization-dominant areas (Batisani and Yarnal 2009; Gong et al. 2018).

Land use conversion elasticity (ELAS) defines the relative conversion cost for a land use type converted to another, ranging between 0 and 1. The higher the elasticity value is, the more difficult the existing land use type can be transformed. Conversion elasticity should be defined based on the knowledge of real situation, but can also be adjusted during the calibration of the model (Verburg and Overmars 2009; Verburg et al. 2002). In this study, the ELAS for model validation in 2015 was preliminarily set based on the physical characteristics of land use type, and considering the ratio of gross loss to initial total for each land use type from 2000 to 2015 as the higher ratio indicates a lower conversion cost. For example, high ELAS were assigned to build land and forest, due to their low likelihood of being converted to other land use types. Subsequently, the ELAS for model validation was tuned during multiple tests and calibrations and finally defined to achieve satisfactory simulation accuracy.

In the BAU scenario, the ELAS was set to be the same as that for model validation in 2015. As for the CP and ES scenarios, the ELAS was adjusted according to corresponding protection measures on specific land use types. In the CP scenario, we increased the value of ELAS of arable land to prevent its conversion to other types. Accordingly, the values of ELAS of forest, grassland and unused land were reduced. In the ES scenario, the ELAS values of forest, grassland and water were raised to strengthen ecological protection while the ELAS value of arable land is moderately reduced (Table 7.3).

Table 7.3 Conversion elasticity

Scenarios	Arable land	Forest	Grassland	Water	Built land	Unused land
BAU	0.5	0.7	0.6	0.5	0.8	0.1
CP	0.9	0.5	0.4	0.5	0.8	0.0
ES	0.3	1	0.7	0.9	0.8	0.0

② Spatial distribution suitability

Spatial distribution suitability is determined by location preferences of land use types, and calculated through logistic regression models exploring the quantitative relationship between land use types and driving factors. Thus, only driving factors for which theoretical relationship with land use is known are taken into consideration, in order to avoid spurious correlations (Verburg et al. 2002). In this study, the choice for potential driving factors of land use change was based on prevalent knowledges (Lambin et al. 2003, 2001; Turner et al. 1995; Wu and Webster 1998; Xie et al. 2017). Therefore, based on the studies above and considering the availability and usability of data, the following biophysical and socio-economic driving factors were included in the modelling process: elevation, slope, population density, GDP density, distance to the nearest railway, distance to the nearest highway, distance to the nearest city/county center, annual average precipitation, and annual average temperature (Fig. 7.9).

③ Spatial policies

All land use changes were restricted in the designated area where spatial policies were implemented. In this study, the maps of high/medium-yield cropland (Fig. 7.10a), national nature reserve and ecological function reserve (Fig. 7.10b), and high *NPP* value area of grassland and forest (Fig. 7.10c) were selected as the basic data to create restriction areas in the scenarios of cropland protection and ecological security.

Spatial policies in the three scenarios were defined as follows: (1) In the BAU scenario, no spatial policies were implemented. (2) In the CP scenario, according to the requirements underlined in the "Overall land use planning for Beijing-Tianjin-Hebei coordinated development (2015–2020)", the high and medium-yield cropland in the study area were classified into the restricted area to prevent the non-agricultural transformation of high-quality cropland (Fig. 7.11a). (3) In the ES scenario, the BTH region is designed to be a demonstration region of eco-environmental conservation. Therefore, medium and high-yield cropland, national nature reserve and ecological function reserve, high *NPP* value area of grassland and forest were all included in the restricted area (Fig. 7.11b) to coordinate the relationship between cropland protection and ecological construction.

④ Land use demand

For model validation, land use demand in 2015 was simulated using Markov model based on the land use data in 2000 and 2005. The linear interpolation method was utilized to create annual land use data from 2006 to 2014.

For future land use simulation in 2030, land use demands in the three scenarios were defined in different ways. Land use demand in the BAU scenario was simulated by using Markov model based on the land use changes during 2005–2015 and the land use data from 2016 to 2029 was created by adopting the linear interpolation method. As for the CP and ES scenarios, land use demands were defined based on corresponding strategic requirements and data from regional planning. Since the regional planning for 2030 has not been available for the public by now, we have to use the latest data in 2020 planning as a reference. According to "Beijing overall land use planning (2006–2020)", "Tianjin overall land use planning (2006–2020)" and "Hebei overall land

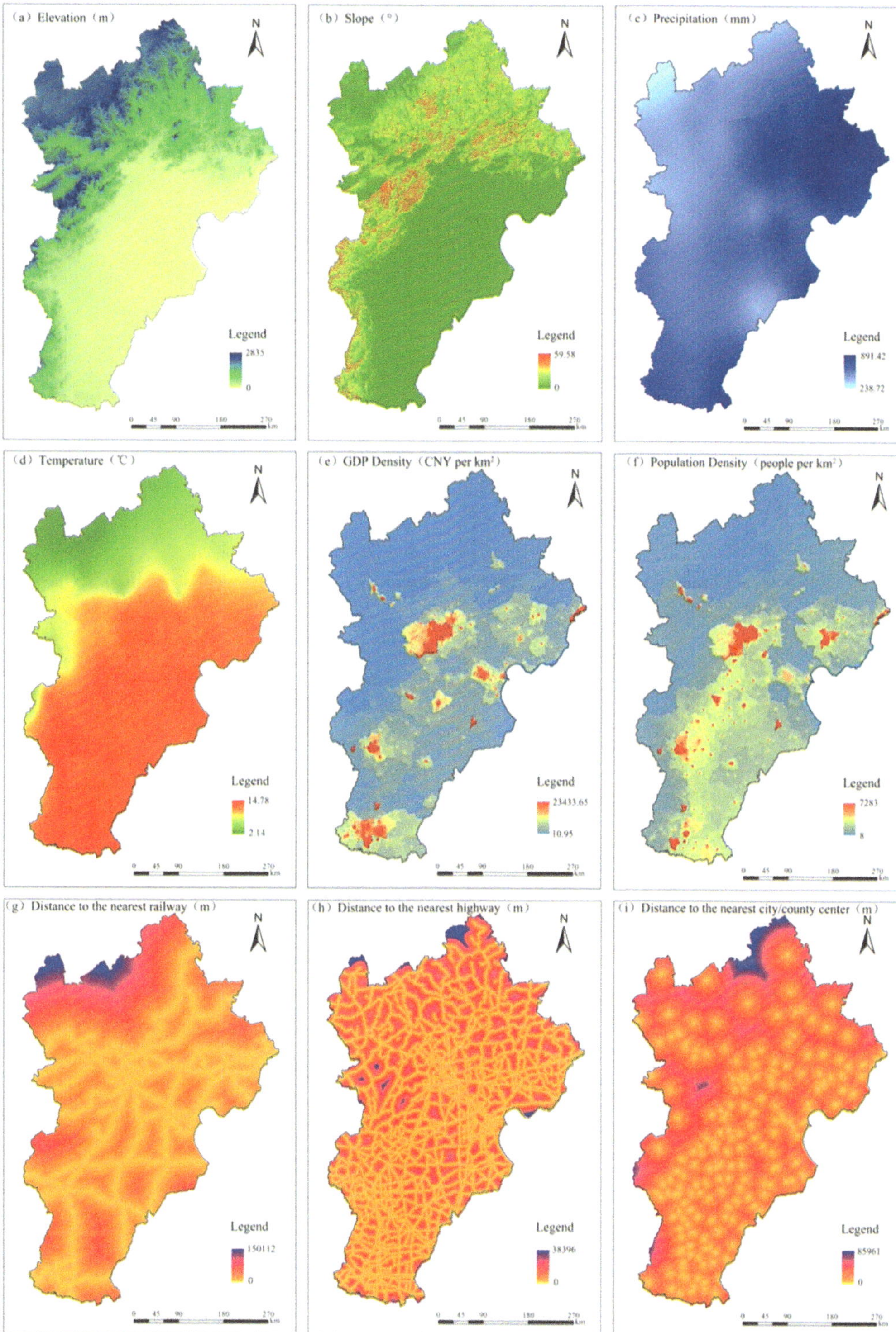

Fig. 7.9 Map of driving factors

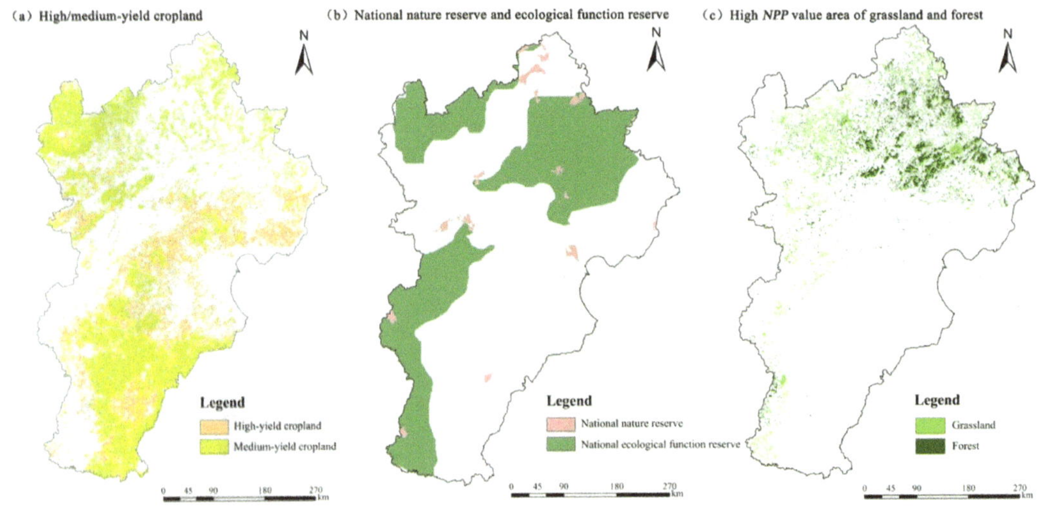

Fig. 7.10 Maps of high/medium-yield cropland (**a**), national nature reserve and ecological function reserve (**b**), high *NPP* value area of grassland and forest (**c**)

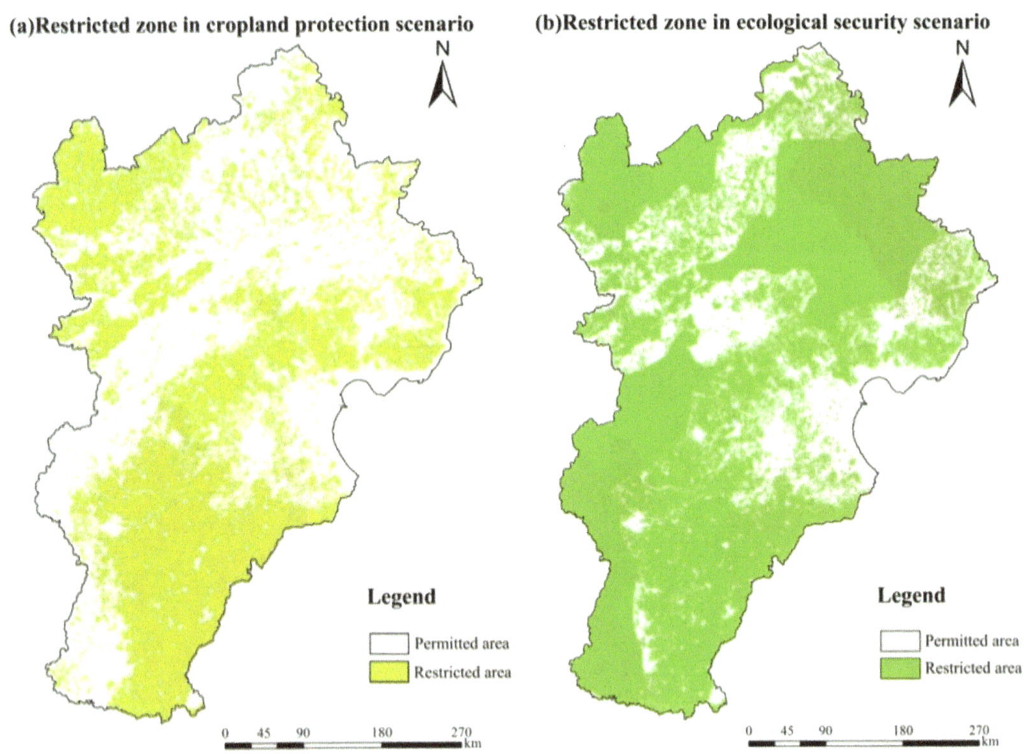

Fig. 7.11 Map of restricted zones

use planning (2006–2020)", the area of arable land in BTH region is planned to be at least 69,567 km^2 and the area of built land is planned to not exceed 26,965 km^2 in 2020. According to the "Framework agreement on promoting forestry ecology in the Beijing-Tianjin-Hebei region", the forest coverage is expected to reach 35% (75,482 km^2) in 2020. Thereby, according to the planning areas of the three land categories mentioned above, adjustments were made to the transition probability matrix of the BAU scenario to create land use demands for CP and ES scenarios.

In the CP scenario, we aim to simulate the possible influences and the environmental effects of cropland protection policies and cropland reclamation activities. The area of arable land should not be less than the maximum of the area in 2015 and area of high/medium yield cropland (83,059.31 km^2). Besides, the expansion of built land should be restricted and its area should not exceed the planned area in 2020. Therefore, one of the feasible methods to adjust the transition probability matrix is determined: the aggregate probability of arable land transferring to other types was reduced by 10%, and the probabilities of forest, grassland and unused land transferring to arable land were raised by 10%, respectively.

In the ES scenario, ecological protection policies and the effects of large-scale afforestation projects were intended to be simulated. The area of arable land should at least reach the maximum of the planned area in 2020 and area of high/medium yield cropland (83,059.31 km^2), and the area of forest should be at least the planned area in 2020. The area of water should not be less than the area in 2015, and the area of built land should not exceed its planned area in 2020. So, the transition probability matrix was adjusted as follows: the aggregate probability of forest transferring to other types reduced by 10% while the probabilities of arable land, built land transferring to forest increased by 10%, respectively, and the probability of grassland and unused land transferring to forest land increased by 50%, respectively.

7.3.2 Land Use Change and Simulation Under Urban–Rural Transformation

7.3.2.1 Land Use Changes from 2000 to 2015

During the period from 2000 to 2015, the land use conversions in the BTH region were relatively intensive with the most predominant characteristic of significant net gain of built land (Table 7.4). There was an evident increase in the built land by 2,744.96 km^2, which was almost two times larger than the central urban area of Beijing (consisting of Dongcheng, Xicheng, Chaoyang, Haidian, Fengtai and Shijingshan districts). However, other land use types demonstrated decreases in varying degrees during this time period, and the arable land experiences the largest decrease by 2,053.83 km^2.

The Sankey diagram was utilized to visualize the conversion pattern between land use types

Table 7.4 Land use conversion from 2000 to 2015 (km^2)

2000	2015							
	Arable land	Forest	Grassland	Water	Built land	Unused land	**Initial total**	**Gross loss**
Arable land	106,928	63.27	9.35	169.13	2224.42	2.57	109,396.74	2468.74
Forest	18.37	44,477.23	5.02	11.14	122.73	0.11	44,634.60	157.37
Grassland	81.33	33.34	34,929.82	43.93	156.11	0.24	35,244.77	314.95
Water	233.51	20.86	13.08	5975.53	262.48	13.67	6519.13	543.60
Built land	25.18	2.37	5.00	23.09	17,727.98	0.76	17,784.38	56.40
Unused land	56.52	2.59	2.64	30.83	35.62	1954.19	2082.39	128.20
Final total	107,342.91	44,599.66	34,964.91	6253.65	20,529.34	1971.54	215,662	3669.15
Net change	−2053.83	−34.94	−279.86	−265.48	2744.96	−110.85		

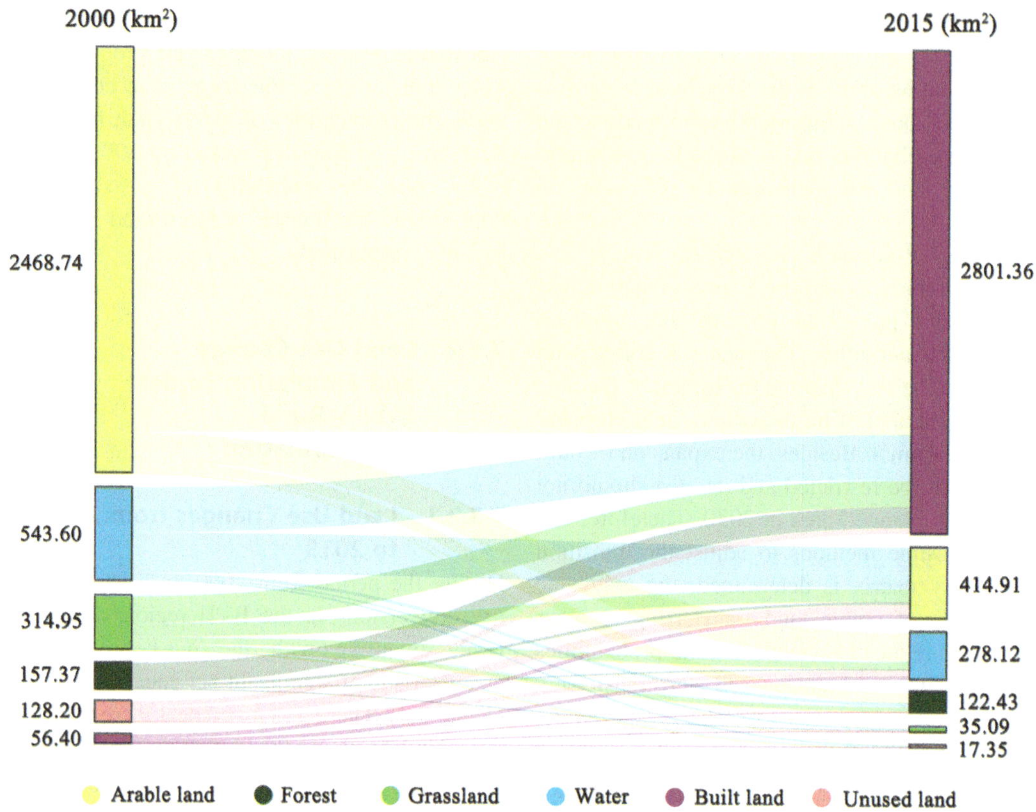

Fig. 7.12 Sankey diagram for land use conversion from 2000 to 2015

(Fig. 7.12). About 1.70% of the total study area (3,669.15 km^2) experienced intensive land use conversion, and the built land and arable land were the most active types. As for the flow-out process, arable land was the largest loss category during this period, which was mostly converted to the built land. Water and grassland were the second and third largest loss categories, respectively, which were mainly targeted to the built land and arable land. As for the flow-in process, built land was the largest gain category and the smallest loss category, which indicates that the built land grows at a high speed and tends to be more stable than other types after developed.

7.3.2.2 Regression Analysis of Land Use Change

All the land use types were taken as the dependent variables for logistic regression and nine driving factors were chosen as the independent variables according to regional conditions and data availability. Then, regression coefficients for each independent variable (Table 7.5) were calculated using a random sample of 10% of all the 862,648 cells, to reduce the potential influence of spatial autocorrelation (Verburg et al. 2002). Regression coefficient reflects the influence intensity that the independent variable exerts on the distribution probability of certain land use type. To be specific, if regression coefficient is positive, probability increases as the value of the independent variable increases; if regression coefficient is negative, probability decreases as the value of the independent variable increases. ROC test statistics for each land use type are all above 0.70 and the ROC values of four land use types exceed 0.8, which reveals that the selected driving factors performs well in explaining the land use pattern of the study area.

Table 7.5 Regression coefficients and ROC values for each land use type

	Y_1	Y_2	Y_3	Y_4	Y_5	Y_6
X_1	−9.11E−04	1.11E−03	8.57E−04	−3.02E−03	−1.70E−03	−2.26E−03
X_2	−1.69E−01	1.09E−01	2.71E−02	−1.13E−01	−1.75E−01	−2.42E−01
X_3	−3.30E−04	−3.97E−03	−1.23E−03	−1.33E−03	5.46E−04	−8.13E−04
X_4	−5.46E−05	–	−7.71E−04	1.40E−05	5.19E−05	−1.56E−04
X_5	3.84E−06	−1.44E−05	–	−7.45E−06	−6.44E−06	–
X_6	−3.84E−05	3.21E−05	−1.96E−05	−1.54E−05	–	3.71E−05
X_7	−2.24E−05	2.83E−05	7.24E−06	1.07E−05	−4.92E−06	–
X_8	−5.09E−02	7.43E−02	2.46E−02	–	–	6.58E−02
X_9	−1.92E−02	−4.95E−02	–	−3.52E−01	−1.01E−01	−6.75E−01
Constant	4.14	−5.61	−3.16	2.11	−0.48	0.99
ROC value	0.835	0.905	0.746	0.771	0.844	0.870

Note X_1, X_2, X_3, X_4, X_5, X_6, X_7, X_8 and X_9 represent to elevation, slope, population density, GDP density, distance to the nearest railway, distance to the nearest highway, distance to the nearest city/county center, annual average precipitation, and annual average temperature, respectively. Y_1, Y_2, Y_3, Y_4, Y_5 and Y_6 represent to arable land, forest, grassland, water, built land, and unused land, respectively. All variables significant at $p < 0.05$. The mark "–" denotes the factors removed

7.3.2.3 Model Validation

As shown in Fig. 7.13, the simulated 2015 land use map, reference land use maps of 2005 and 2015 were overlaid to create the map of prediction accuracy and error. Over the entire study area, the proportion of null successes and true hits was 78.37% and 18.58%, respectively, which indicates that 96.95% of the simulated map was consistent with actual land use map. Furthermore, the total error only accounted for 3.05%, including 0.48% partial hits, 1.15% misses, and 1.42% false alarms. This indicates that the quantitative prediction and spatial allocation in the modelling process is highly accurate. The FoM was 85.89%, which is higher than those of previous case studies (García et al. 2012; Pontius et al. 2008; Yang et al. 2017). This proves that the model has a satisfactory accuracy and can be used to conduct scenario simulations in the BTH region.

7.3.2.4 Multi-scenario Simulation Analysis

As shown in Table 7.6 and Fig. 7.14, the land use structures and spatial pattern of the study area differ significantly among scenarios, as a result of

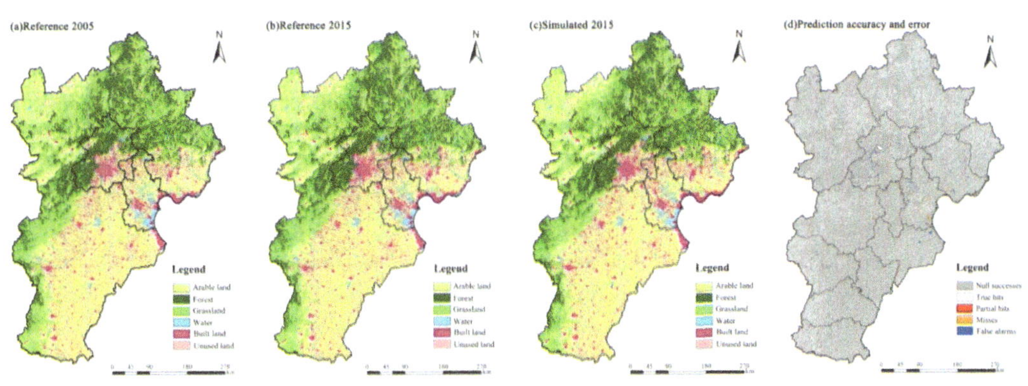

Fig. 7.13 Reference map for 2005 (**a**) and 2015 (**b**), simulated map for 2015 (**c**) and map of prediction accuracy and error (**d**)

Table 7.6 Area and proportion of land use types in 2015 and three scenarios (km^2, %)

		Arable land	Forest	Grassland	Water	Built land	Unused land
2015	Area	107,342.91	44,599.66	34,964.91	6253.65	20,529.34	1971.54
	Proportion	49.77	20.68	16.21	2.90	9.52	0.91
BAU	Area	96,910.19	41,196.15	32,172.71	7135.06	36,552.88	1694.99
	Proportion	44.94	19.10	14.92	3.31	16.95	0.79
CP	Area	115,857.19	36,976.65	28,917.83	6398.93	25,998.33	1513.07
	Proportion	53.72	17.15	13.41	2.97	12.06	0.70
ES	Area	87,712.59	75,550.47	24,281.95	6385.36	20,651.64	1079.99
	Proportion	40.67	35.03	11.26	2.96	9.58	0.50

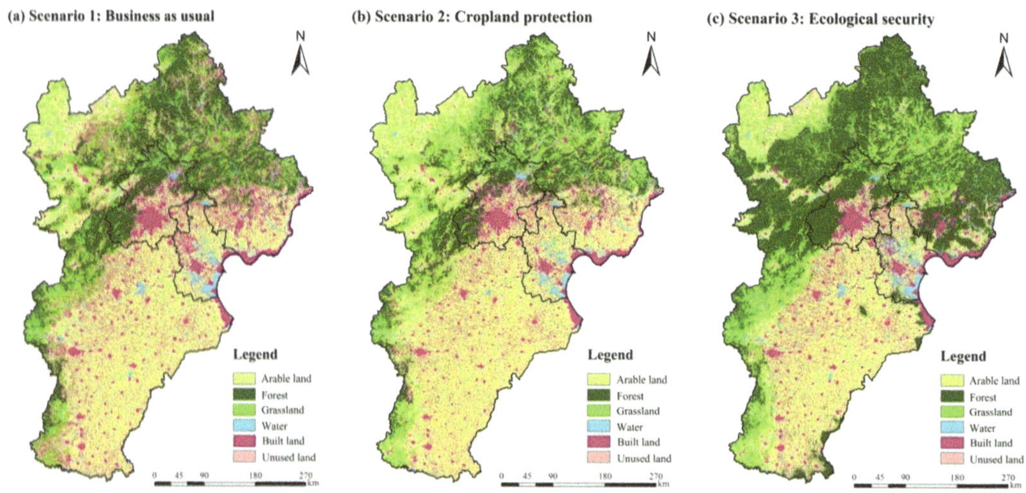

Fig. 7.14 Simulated land use map under three scenarios for 2030

different simulation objectives. Furthermore, spatial changes of dominant land use types from 2015 to 2030 are shown in Fig. 7.15, including built land, arable land and forest. In order to deeply explore the spatial distribution differences, the selected case study area on the four maps was enlarged to a uniform scale (Fig. 7.15).

In the BAU scenario, the study area experiences intensive land use changes during 2015–2030, characterized by the sustained increasing trend in built land. The area of built land surges by 16,031.18 km^2, with its percentage increasing from 9.52% in 2015 to 16.95% in 2030. Meanwhile, arable land experiences the largest decrease by 4.83%, while forest, grassland and unused land decrease moderately by 1.58%, 1.29% and 0.12%, respectively. As for the spatial pattern of land use changes (Fig. 7.14a), built land sprawl is characterized by spatial extension of the built-up area, and layout of major cities tends to be concentrated and well-organized in this scenario. New construction land will be mainly distributed in Beijing, Tianjin, Chengde, Zhangjiakou, Tangshan, Qinhuangdao and the eastern part of Baoding, Shijiazhuang, Xingtai and Handan (Fig. 7.15). Furthermore, there are two sites deserving special mention. Firstly, the Wuqing District of Tianjin, which is located in the intersection of Beijing, Tianjin and Hebei, demonstrates substantial built land expansion (areas surrounded by dotted-line circle A in Fig. 7.15a). This is mainly because Wuqing

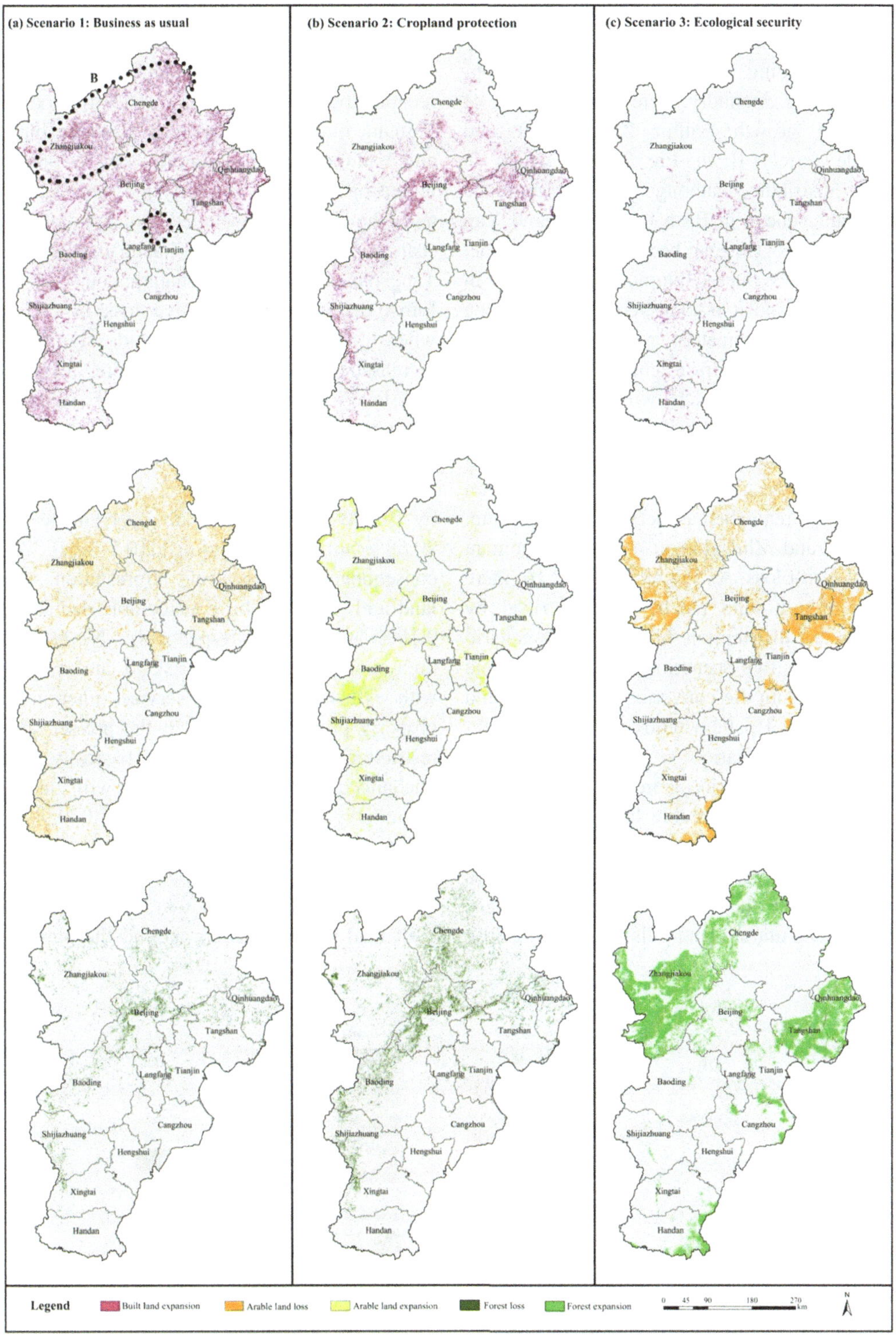

Fig. 7.15 Land use changes of built land, arable land, and forest from 2015 to 2030 under three scenarios

District is planned to be an important high-tech innovation base and place for regional industrial transfer according to the coordinated development planning. Secondly, there is a large scale of built land growth within Zhangjiakou and Chengde (circle B in Fig. 7.15a), which is mainly distributed among the intermontane basins and valleys in Bashang plateau and Yan mountain areas. This can be attributed to the fact that the 2022 Winter Olympics will be held in Zhangjiakou, and it is expected to promote the development of recreation, retirement and tourism industries and infrastructure construction in the nearby areas, which will create a huge demand for construction land on relatively flat terrain. The area of arable land loss is mainly distributed in the northern and southwestern part of the study area, whereas Beijing, Shijiazhuang, Chengde, and Zhangjiakou are the dominant areas of forest loss. As shown in Fig. 7.16, there are more built land patches characterized by concentrated distribution in the BAU scenario than the other two scenarios, while arable land and forest in situ are converted to the built land and their spatial pattern tends to be more fragmented in this scenario.

In the CP scenario, there is a remarkable increase by 3.95% in arable land from 2015 to 2030 with the area growth of 8,514.33 km^2, while the built land still shows obvious increase by 2.54%, and the area percentage of forest, grassland and unused land falls by 3.53%, 2.80% and 0.21%, respectively. As for the spatial pattern in this scenario (Fig. 7.14b), existing arable land mostly remains unchanged, while newly reclaimed arable land is mainly distributed over Taihang mountain in the east, Bashang plateau and Yan mountain in the north (Fig. 7.15b). The layout of built land is nearly the same as 2015, and the newly developed construction land scatters over Beijing, Shijiazhuang, Baoding, Tangshan, Chengde and Zhangjiakou. However, forest demonstrates intensive loss in the area where arable and built land significantly increased (Fig. 7.16b). Furthermore, forest and grassland show characteristics of significant fragmentation in spatial distribution due to the occupation by increasing arable land patches.

In the ES scenario, the area of forest increases by 30,949.81 km^2, with the percentage from 20.68% to 35.03% during 2015–2030. As a consequence of large-scale afforestation, arable land and grass exhibit considerable decrease by 9.10% and 4.95%, respectively. Nevertheless, high-quality cropland, grassland and wetlands with essential ecosystem service functions are properly protected, owing to the pre-set spatial restrictions. Built land and water basically remain unchanged, while unused land will decrease remarkably from 1,971.76 km^2 to 1,079.99 km^2 during 2015–2030. As for the spatial pattern in this scenario (Fig. 7.14c), built land and water is basically the same as that in 2015, while new built land mainly scatters around the peripheral

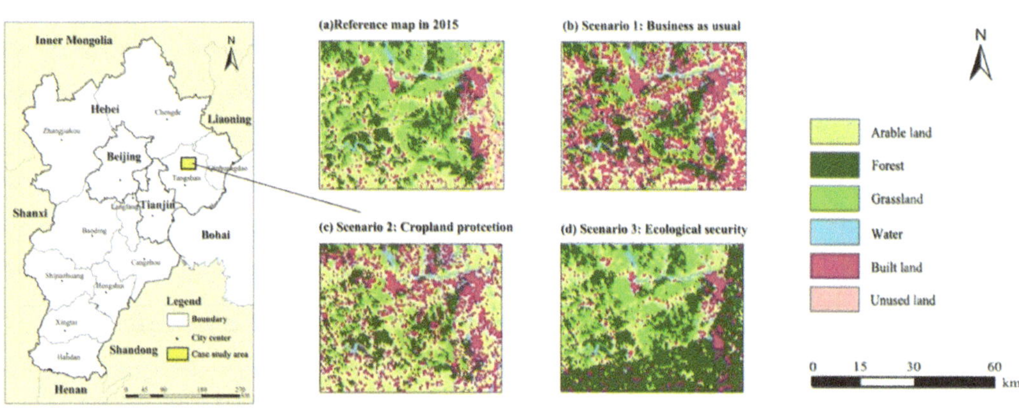

Fig. 7.16 Land use pattern of the case study area under three scenarios

areas of the major cities. The majority of forest, grassland, part of the waters and wetlands are mainly distributed in the ecological function reserves, including the Beijing-Tianjin water conservation ecological function reserve, the northern foot of Yinshan Mountain-Hunshandake Desert sand-fixing ecological function reserve, Taihang Mountain water and soil conservation ecological function reserve and west Liao River water conservation ecological function reserve, and 17 national nature reserves such as Beijing Songshan Mountain National Nature Reserve and Tianjin Ancient Coast and Wetland National Nature Reserve. The forest expansion is mainly distributed over plateau and mountain areas in Zhangjiakou and Chengde, and peripheral area of Qinhuangdao and Tangshan, where arable land loss takes place simultaneously (Fig. 7.15). Furthermore, forest demonstrates concentrated distribution trend in this scenario as revealed in the case study area (Fig. 7.16).

7.3.3 Optimization of Land Use Pattern in BTH Region Under Urban–Rural Transformation

The BTH region is one of the most important engines of China's economic development and scientific innovation, with an urbanization rate surging from 19.63% in 1980 to 65.8% in 2018. Driven by the rapid growth of urban land and population in this region, the contradiction between resources, environment and socio-economic development has significantly intensified since the implementation of the reform and opening-up policy in 1978 (Tan et al. 2005), with the most prominent characteristic of unbalanced development between regions (Huang and Lin 2017; Yao et al. 2008). In order to further resolve problems related to the intricate man-earth relationship in this region, the coordinated development of Beijing-Tianjin-Hebei has been positioned as one of the major national development strategies by the central government, aiming to take Beijing, Tianjin and Hebei as a whole to narrow the development gap between regions and solve the "megacity diseases" in Beijing. Based on this strategic background, the non-capital function transfer of Beijing should be the primary task at the current stage (Mao 2017). Thus, more effort will be taken in the construction of Hebei Xiong'an new district and Beijing city sub-center, to promote regional industrial restructuring and optimize urban–rural configuration (Lu 2015). Focusing on the traditional inefficient and extensive land use problems and aiming to find out the socio-economic and environmental effect of different future policies implemented to this region, land use patterns under three tailored scenarios were simulated in this study.

In the BAU scenario, the BTH region experiences remarkable built land growth. The built land expansion in Wuqing District, Zhangjiakou and Chengde is considered to be reasonable according to the analysis above. Therefore, this simulation result predicts the probable location and scale of construction activities taking place in the next 15 years. However, the total amount of built land under this scenario far exceeds the planned area in 2020 and it will lead to extensive destruction of cropland, forest and grassland, which violates the prerequisites of incremental construction land control and ecological land protection according to the overall land use planning for coordinated development. In the BTH region, area that can be used for future industrialization and urbanization is only about 2,000 km^2, accounting for about 1% of the total study area (Zhang and Zhao 2016). Therefore, the amount of land suitable for future development in this area is extremely limited, posing severe restrictions on expansion of urban–rural development space and industrial configuration in the BTH region.

In the CP scenario, although arable land exhibits significant growth, the newly reclaimed arable land is mainly distributed in areas where soil, irrigation, terrain, transportation and many other conditions are relatively poor, such as Taihang Mountain, Bashang Plateau and Yan Mountain. Consequently, more efforts need to be taken to improve tillage conditions through land

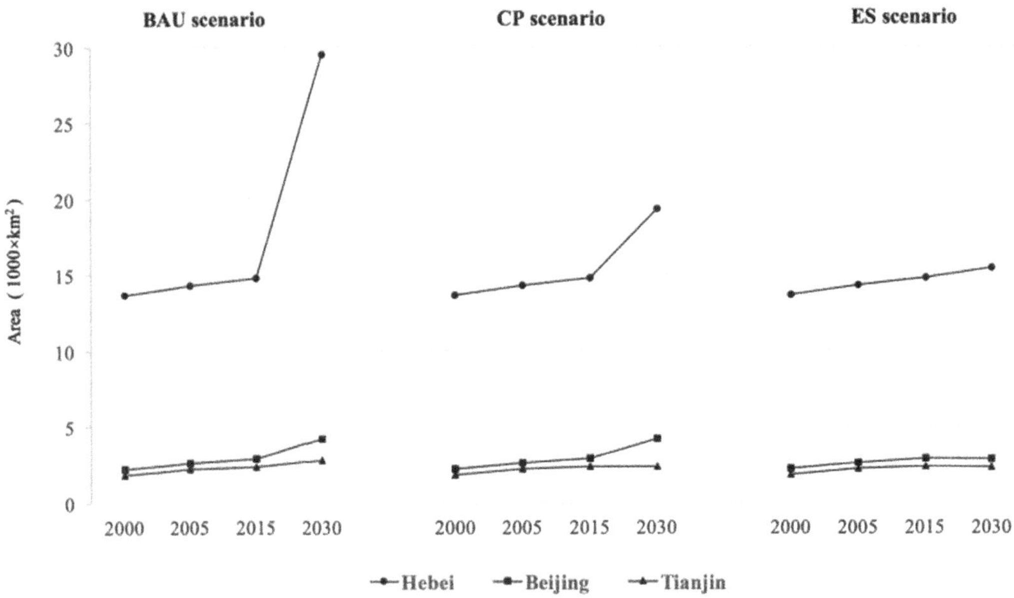

Fig. 7.17 Built land changes of Beijing, Tianjin and Hebei from 2000 to 2030 under three scenarios

consolidation and land engineering, so the input–output efficiency is worth seriously weighing and considering. In addition, under this scenario, there are still excessive built land growth and massive destruction of forest and grassland. Therefore, the land use pattern under this scenario, if it were to manifest, would be detrimental to both communities and the eco-system.

In the ES scenario, concentrated and contiguous high-quality cropland has been strictly protected, and the total amount of arable land has also been controlled within a reasonable range. In this way, regional food security can be certainly guaranteed. Meanwhile, the expansion of new construction land has been restricted, and the layout of existing built land tends to be better-organized in this scenario (Fig. 7.14c). Furthermore, forest, grassland, and water with important ecological functions are properly protected, and forest coverage reaches 35% in the study area, which indicates that the ecological environment has been effectively improved. In addition, it can be inferred that the problem of unbalanced regional development is expected to be effectively resolved under the ES scenario according to the characteristics that differ from other two scenarios in terms of built land changes (Fig. 7.17). The built land growth in Beijing and Tianjin is suspended in this scenario and it enters a new stage of stock land renovation and reclamation. Meanwhile, the built land growth in Hebei demonstrates a stable and sustained trend in the future, which is essential for Hebei to further boost economic growth and narrow the gap. Furthermore, incremental construction land for pollution intensive and high resource consumption industries should be strictly restricted in this scenario, while high-tech and environmental-friendly industries such as new materials, new energy, and advanced manufacturing should be vigorously supported through more favorable land use policies (Zhou et al. 2018). In conclusion, land allocation by overall planning and other administrative measures should be the key strategy in coordinating the relationship between high-quality economic development and ecological protection.

Through tradeoff, the land use mode in ES scenario is considered to be the optimal solution under the background of coordinated development at the current stage. Under this land use mode, the BTH region will make breakthroughs in ecological construction and regional rebalance through reallocation of land resources, which is expected to lay a solid foundation for developing into a world-class urban agglomeration (Han et al. 2015; Wang et al. 2014a). In terms of regional land use policy, the local government should give priority to ecological construction in the future coordinated development. Firstly, a long-term mechanism for comprehensive ecosystem management should be established, which is considered the bottom line, guaranteeing regional eco-friendly economic development. Secondly, we must utilize administrative measures such the targeted Grain-for-Green policy (Feng et al. 2005) to further expand ecological spaces and establish a more effective multi-level ecological network in this region. Furthermore, more efforts should be made by governments at all levels in optimizing the urban–rural production-living-ecology space structure through the two engines of new-type urbanization and rural revitalization, which has a vital role for social harmonious and sustainable development in the BTH region (Liu 2018; Zhou et al. 2017).

7.4 Urbanization, Economic Growth, and Carbon Dioxide Emissions in China

7.4.1 Data and Methods

7.4.1.1 Data Source

Urban population and built-up area are generally used as the proxy indicators of demographic urbanization (Bloom et al. 2008) and land (landscape) urbanization (Bai et al. 2012), respectively. We also chose built-up area (BA) and urban population (UP) as the proxy indices of land urbanization and demographic urbanization, respectively, and used real GDP per capita (pGDP) as economic indicator, and used CO_2 emissions as the environmental indicator (Table 7.7). All provincial data covering the period 1997–2010 are collected from the China Statistical Yearbooks (CSY) and China City Statistical Yearbook (CCSY). The CO_2 data over the period from 1997 to 2010 are obtained from the results calculated by Guan et al. (2012), in which the data on CO_2 emissions for all provinces are estimated from the publicly available energy statistics from Chinese authorities following the Intergovernmental Panel on Climate Change (IPCC) emission accounting approach. In the present study, all the variables are expressed in natural logarithms so that they may be considered elasticity of the relevant variables. The GDP is calculated at a constant price (1997 prices) and GDP per capita is calculated from GDP divided by the year-end population. The data for China do not include data for Hong Kong, Macao and Taiwan, China.

7.4.1.2 Econometric Analysis

This study utilized panel unit root, cointegration and causality analysis to investigate the relationship between urbanization level, economic growth and CO_2 emissions. The empirical modeling framework in the present study consists of four stages. First, the presence of unit roots in the all variables was tested. The LLC and IPS methods were used to test the presence of the panel unit root (Levin et al. 2002; Im et al. 2003). In addition to these methods, Maddala and Wu (1999) and Choi (2001) provided two

Table 7.7 Data description and sources

Indicators	Unit	Abbreviation	Meaning of indicators	Sources
Built-up area	km^2	BA	Land urbanization	CCSY
Urban population	persons	UP	Demographic urbanization	CSY
GDP per capita	Yuan RMB	pGDP	Economic growth	CSY
CO_2 emissions	million tons	CO_2	Pollutant emissions	Guan et al. (2012)

nonparametric unit root tests, the Fisher-ADF and the Fisher-PP statistics.

Second, because each of the variables contained a panel unit root, the heterogeneous panel cointegration test developed first by Pedroni (2004) was performed to examine whether there was a long-term equilibrium relationship between the variables. The Pedroni panel cointegration tests included two types, one was the panel cointegration test and the other was the group mean panel cointegration test. The former was based on the within dimension approach, which included the following statistics: Panel-v, Panel-rho, Panel-ADF and Panel-PP. The latter was based on the between dimension approach, which included three statistics: Group-rho, Group-PP and Group-ADF (Pedroni 2004; Mahadevan and Asafu-Adjaye 2007). It is generally accepted that the Panel-ADF and the Group-ADF statistics had better small sample properties than the other statistics, which made them more reliable.

Third, two techniques, i.e., the fully modified ordinary least squares (FMOLS) estimator and the dynamic ordinary least squares (DOLS) estimator were used to further estimate the long-run equilibrium relationships among the variables. The FMOLS estimator is believed to eliminate endogeneity in the regressor and serial correlation in the errors (Pedroni 2001). The DOLS estimators had a normal asymptotic distribution and their standard deviations provided a valid test for the statistical significance of the variables (McCoskey and Kao 1999). In general, the DOLS technique is more reliable than the panel OLS estimation for panel data (McCoskey and Kao 1999; Pedroni 2001). The FMOLS estimator can be calculated in the following:

$$w_{i,t} = \alpha_i + \beta_i x_{i,t} + \sum_{k=-K_i}^{K_i} \gamma_{i,k} \Delta x_{i,t-k} + \varepsilon_{i,t} \quad (7.13)$$

where $w_{i,t}$ and $x_{i,t}$ are cointegrated with slope β_i, and β_i may or may not be homogeneous across i. $-K_i$ and K_i are leads and lags.

$$\beta^*_{FMOLS} = \frac{1}{N}\sum_{i=1}^{N}\left[\left(\sum_{t=1}^{T}(x_{i,t}-\bar{x}_i)^{-1}\right)\left(\sum_{t=1}^{T}(x_{i,t}-\bar{x}_i)w^*_{i,t} - T\gamma_i\right)\right] \quad (7.14)$$

$$w^*_{i,t} = w_{i,t} - \overline{w}_i - \frac{\Omega_{2,1,i}}{\Omega_{2,2,i}}\Delta x_{i,t} \quad \text{and}$$
$$\gamma_i = \Gamma_{2,1,i} + \Omega^0_{2,1,i} - \frac{\Omega_{2,1,i}}{\Omega_{2,2,i}}(\Gamma_{2,2,i} + \Omega_{2,2,i}) \quad (7.15)$$

where $\xi_{i,t} = (\varepsilon_{i,t}, \Delta x_{i,t})$ and $\Omega_{i,t} = \lim_{T\to\infty} E\left[\frac{1}{T}\left(\sum_{t=1}^{T}\xi_{i,t}\right)\left(\sum_{t=1}^{T}\xi_{i,t}\right)'\right]$ is the long-run covariance for this vector process which can be decomposed into $\Omega = \Omega^0_i + \Gamma_i + \Gamma'_i$, where Ω^0_i is the contemporaneous covariance and Γ_i is a weighted sum of auto-covariance.

The panel DOLS estimator is defined as:

$$\beta^*_{DOLS} = \frac{1}{N}\sum_{i=1}^{N}\left[\left(\sum_{t=1}^{T} Z_{i,t}Z'_{i,t}\right)^{-1}\left(\sum_{t=1}^{T} Z_{i,t}w_{i,t}\right)\right] \quad (7.16)$$

where $z_{i,t} = \left[x_{i,t} - \bar{x}, \Delta x_{i,t-K_i}, \ldots, \Delta x_{i,t+K_i}\right]$ is vector of regressors, and $w_{i,t} = w_{i,t} - \overline{w}_i$.

Lastly, confirming the existence of cointegration between the variables, the next step was to examine both short- and long-run causality by performing Granger causality test based on vector correction model (VECM). The empirical models are specified as follows:

$$\Delta y_{i,t} = \theta_j + \sum_{k=1}^{K}\theta_{1k}\Delta y_{i,t-k} + \sum_{k=1}^{K}\theta_{2k}\Delta y_{i,t-k} + \mu\beta_{i,t-1} + \varepsilon_{i,t} \quad (7.17)$$

where Δ denotes the difference of the variable; K is the lag length; $\beta_{i,t-1}$ is the error correction term with lag 1; $\varepsilon_{i,t}$ is the residuals of the mode; μ is the coefficient of the error correction term $\beta_{i,t-1}$. The significance of causality results is determined by Wald F-test. In the short-run, the x does not Granger cause y where H_0: $\theta_{2k} = 0$ for all *i* and *k*, while the long-run causality can be established if $\mu = 0$.

Furthermore, we employed a heterogeneous panel Granger causality analysis, recently proposed by Dumitrescu and Hurlin (2012), to further verify the short-term Granger causal relationship between urbanization, economic growth and CO_2 emissions. The testing procedure is superior to former panel Granger causality tests since it can give efficient results even in panels with small sample sizes (Tugcu 2014). The Dumitrescu and Hurlin (2012) test of Granger non-causality for heterogeneous panel is based on the stationary fixed-effects panel model:

$$y_{i,t} = \theta_j + \sum_{k=1}^{K} \beta_i^{(k)} y_{i,t-k} + \sum_{k=1}^{K} \delta_i^{(k)} x_{i,t-k} + \varepsilon_{i,t},$$
$$i = 1, 2, \ldots, N;\ t = 1, 2, \ldots, T \tag{7.18}$$

where x and y are two stationary variables observed for N provinces in T periods. $\beta_i^{(k)}$ denote the autoregressive parameters, and $\delta_i^{(k)}$ are the regression coefficient's slopes; $\delta_i = \left(\delta_i^{(1)}, \ldots, \delta_i^{(k)}\right)$; individual effects θ_i are assumed to be fixed; K is the lag length. By definition, x causes y if and only if the past values of the variable x observed on the i-th province improves the forecasts of the variable y for this province i only. The test is based on the null hypothesis of homogeneous non-causality (HNC), there is no causal relationship from x to y for all the provinces of the panel $\left(\delta_i = \left(\delta_i^{(1)}, \ldots, \delta_i^{(k)}\right)' = 0, \forall i = 1, \ldots, N\right)$. Under the alternative hypothesis, there exists a causal relationship from x to y for at least one province of the sample. The test statistic is given by the cross-sectional average of individual Wald statistics defined for the Granger non-causality hypothesis for each province:

$$W_{N,T}^{HNC} = N^{-1} \sum_{i=1}^{N} W_{i,T} \tag{7.19}$$

where $W_{i,T}$ is the individual Wald statistic for the ith cross-section unit. Under the null hypothesis of non-causality, each individual Wald statistic converges to a chi-squared distribution with K degrees of freedom for $T \to \infty$.

The standardized test statistic $Z_{N,T}^{HNC}$ for fixed T samples is as follows (Dumitrescu and Hurlin 2002):

$$Z_{N,T}^{HNC} = \sqrt{\frac{N}{2K} \times \frac{T-2K-5}{T-K-3}} \times \left[\frac{T-2K-3}{T-2K-1} W_{N,T}^{HNC} - K\right] \to N(0,1) \tag{7.20}$$

7.4.1.3 The EKC Model

According to the EKC hypothesis, the long-term relationship between economic growth and CO_2 emissions can be expressed as a logarithmic cubic function of the income (Grossman and Krueger 1995). The simplified EKC model is given by:

$$\ln(CO_2)_{it} = \alpha_0 + \beta_1 \ln(\text{pGDP}_{it}) + \beta_2 \ln(\text{pGDP}_{it})^2 + \beta_3 \ln(\text{pGDP}_{it})^3 + \beta_4 \ln(\text{BA}) + \varepsilon_{it} \tag{7.21}$$

where *i*-province (i = 1, 2,…31), *t*-year (t = 1, 2, …T); α_0 represents cross-section effect; ε_{it} is random disturbance; β_1, β_2, β_3 and β_4 is the estimated coefficients; pGDP is real GDP per capita; BA is built-up area. Equation (7.6) allows us to test the various forms of environmental-economic linkages: $\beta_1 > 0$, $\beta_2 < 0$ and $\beta_3 > 0$ indicating an N-shaped relationship; $\beta_1 < 0$, $\beta_2 > 0$ and $\beta_3 < 0$ indicating an inverse N-shaped relationship; $\beta_1 < 0$, $\beta_2 > 0$ and $\beta_3 = 0$ indicating a U-shaped relationship $\beta_1 > 0$; $\beta_1 > 0$, $\beta_2 < 0$ and $\beta_3 = 0$ indicating an inverse U-shaped relationship, representing the EKC hypothesis, the turning point of the EKC is computed by $\omega = \exp(-0.5\beta_1/\beta_2)$; $\beta_1 > 0$, $\beta_2 = 0$ and $\beta_3 = 0$ indicating a monotonically increasing linear relationship; $\beta_1 < 0$, $\beta_2 = 0$ and $\beta_3 = 0$ indicating a monotonically decreasing linear relationship; $\beta_1 = \beta_2 = \beta_3 = 0$ indicating a level relationship.

7.4.2 China's Urbanization, Economic Growth and CO_2 Emissions

The variation in land urbanization, demographic urbanization, GDP and CO_2 emissions for China between 1997 and 2010 are shown in Fig. 7.18. The number of China's inland cities had increased from 193 in 1978 to 657 in 2010, and its urban population increased from 172.45 million to 681.13 million with an annual average growth rate of 4.53% (NBSC 2011; Fig. 7.18a). According the latest National New-type Urbanization Plan released by the central government, China's urbanization level was projected to reach 60% by 2020. Urban population growth in China was characterized by rural-to-urban migration (Gong et al. 2012). To accommodate this massive influx of population onto cities, China's urban area had expanded rapidly. Specifically, China's urban built-up area has increased from 13,685 km^2 in 1997 to 41,244 km^2 in 2010 (Fig. 7.18b). During the same period, China's urban area had expanded by 201%, whereas the urban population had only increased by 105%, indicating that the rate of China's land urbanization is almost twice as fast as population urbanization. China's real GDP had rapidly increased from 7,897 million Yuan in 1997 to 26,723 million Yuan in 2010 with an average annual increase of 10.7% (Fig. 7.18c). Large-scale industrialization and urbanization have also made China the largest CO_2 emitter in the world (Minx et al. 2011; IEA 2012a). The annual CO_2 emissions in China had been increasing from 2,887 million tons in 1997 to 8,175 million tons in 2010 (Fig. 7.18d).

Linear regressions were further used to examine the changes in each variable for China's 31 provinces during the period 1997–2010. The slopes of the lines of best fit reflected the changes in the variables, and a greater slope indicated greater changes in the variables. Figure 7.19 displays the changes in urban population, built-up area, per capita GDP and CO_2 emissions for 31 provinces between 1997 and 2010. The top 5 provinces with the highest growth rate of urbanization population were Guangdong, Jiangsu, Shandong, Henan and Hebei (Fig. 7.19a). In contrast, Tibet, Qinghai, Heilongjiang, Ningxia and Hainan were the provinces with the lowest growth rate of urban population. Similar to the urban population, the growth rate of built-up area in Guangdong province was the highest, followed by Shandong, Jiangsu, Zhejiang and Henan (Fig. 7.19b). The growth rate in GDP per

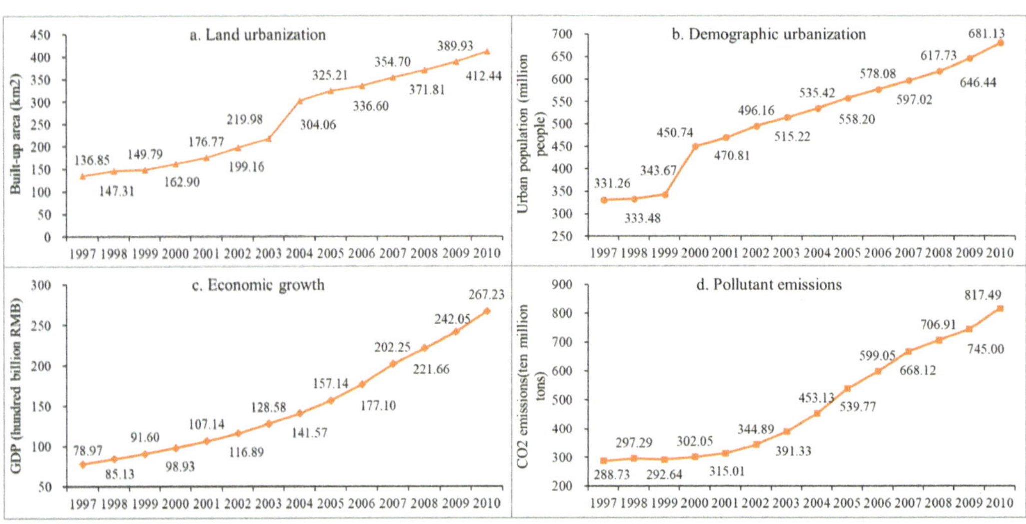

Fig. 7.18 The changes in land urbanization (**a**), demographic urbanization (**b**), GDP (**c**) and CO_2 emissions (**d**) for China between 1997 and 2010

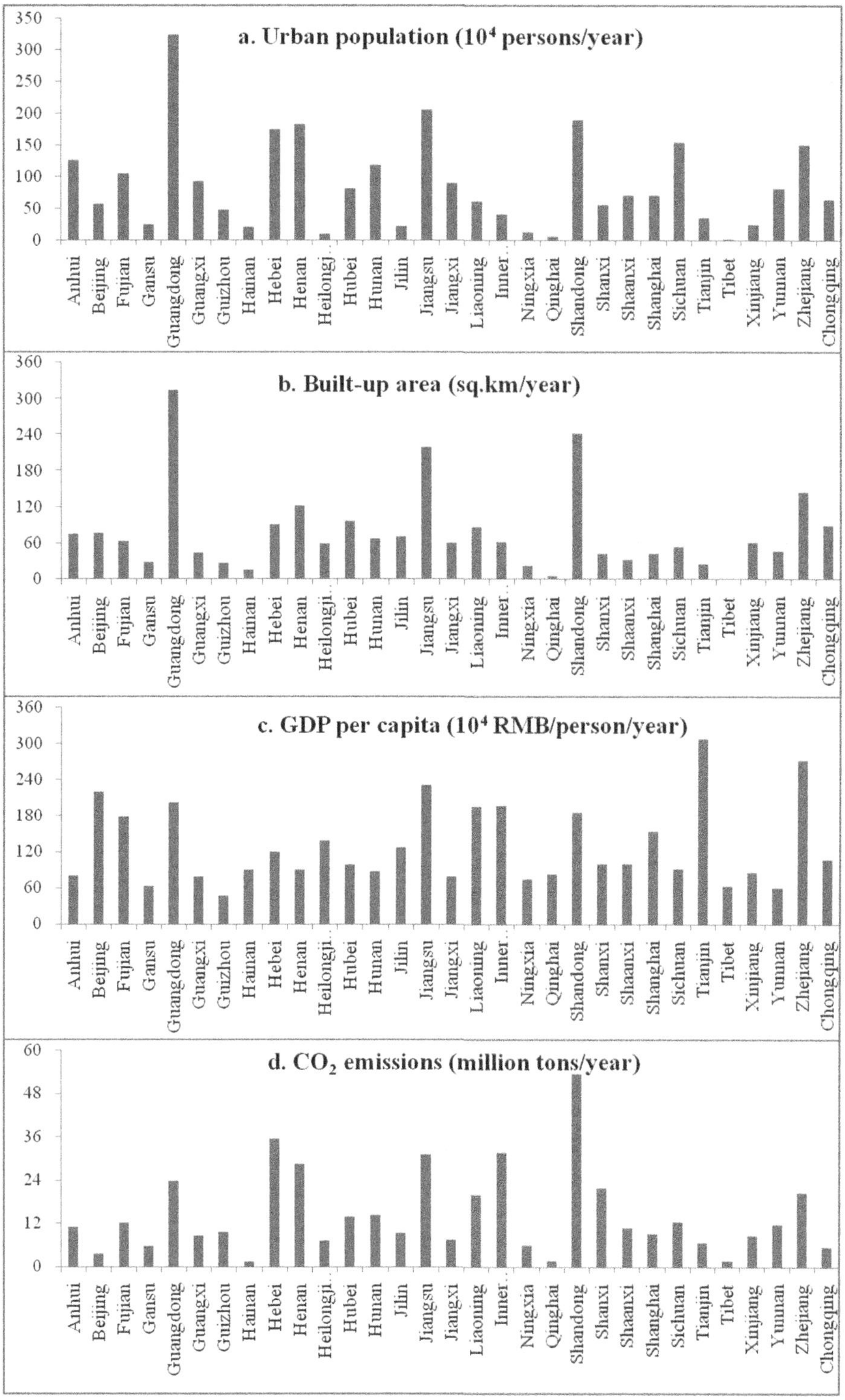

Fig. 7.19 The growth rate in urban population (**a**), built-up area (**b**), GDP per capita (**c**) and CO_2 emissions (**d**) for China's 31 provinces

capita in eastern China was relatively rapid over the past fourteen years (Fig. 7.19c). For CO_2 emissions, the growth rate in Shandong province (53.25 million tons per year) was the highest, followed by Hebei (35.35 million tons per year), Inner Mongolia (31.45 million tons per year), Jiangsu (31.15 million tons per year) and Henan (28.54 million tons per year) (Fig. 7.19d).

7.4.3 The Relationships Among Urbanization, Economic Growth and CO_2 Emissions in China

7.4.3.1 Panel Unit Root Tests

It is advisable to check the presence of unit roots in the all variables before proceeding to any econometric analysis because using the conventional ordinary least squares (OLS) estimator with non-stationary variables might result in spurious regressions. Table 7.8 provides the panel unit root test results with and without a trend term. Results show that many variables are non-stationary in their levels, but most of them become stationary at the 5% significance level after taking first differences. These results indicate that built-up area, urban population, real GDP per capita, built-up area and CO_2 emissions in all panels were integrated at order one, suggesting a possible long-run cointegration relationship among these variables.

7.4.3.2 Panel Cointegration Tests

If the series are integrated of the same order one can proceed with the cointegration test. Table 7.9 presents the panel cointegration test results for all panel datasets. For Panel A, all seven panel cointegration tests rejected the null hypothesis of no cointegration at the 10% significance level

Table 7.8 Panel unit root test results

Levels	Variable	LLC	IPS	Fisher-ADF	Fisher-PP
Intercept	BA	−2.75***	4.04	18.88	26.12
	UP	−4.68***	1.19	61.71	112.52***
	pGDP	5.863	12.1	9.23	4.24
	CO_2	1.65	7.18	8.4	6.45
Intercept and trend	BA	−1.50*	−1.47*	47.28	52.87
	UP	−9.11***	−0.27	67.02	27.63
	pGDP	−10.31***	−1.68**	91.47***	68.99
	CO_2	−6.39***	−2.32**	89.47**	74.63
First differences					
Intercept	BA	−6.39***	−4.65***	115.01***	224.74***
	UP	−11.32***	−6.40***	149.70***	220.52***
	pGDP	−4.51***	−0.29	55.29**	52.95***
	CO_2	−6.12***	−4.11***	106.59***	195.71***
Intercept and trend	BA	−5.54***	−1.65***	75.35***	202.62***
	UP	−18.90***	−9.45***	203.53***	309.78***
	pGDP	−2.16***	0.42	59.99**	105.29***
	CO_2	−2.59***	0.10**	53.40**	158.00***

Note Newey-West bandwidth selection using Bartlett kernel. Automation selection of lags was based on SIC. Levin, Lin and Chu test (LLC) Null: unit root (assumes a common unit root process); Im, Pesaran and Shin W-stat test (IPSW), ADF-Fisher Chi-square test (ADF) and PP-Fisher Chi-square test (PP) Null: Unit root (assumes an individual unit root process). BA, UP, pGDP and CO_2 are built-up area, urban population, and per capita GDP and CO_2 emissions, respectively. The null hypothesis of the LLC, IPS, Fisher-ADF and Fisher PP tests examines non-stationary. ***, ** and * indicates statistical significance at the 1%, 5% and 10% significance level, respectively

Table 7.9 Panel cointegration test results

Statistics	Panel A	Panel B	Panel C	Panel D	Panel E
Panel v	1.51*	1.23	3.22***	−0.03	3.70***
Panel rho	−1.75**	0.10	−3.25***	0.71	−1.93**
Panel PP	−2.40***	−1.17	−7.18***	0.24	−5.31***
Panel ADF	−2.98***	−1.29*	−8.12***	−1.19	−8.10***
Group rho	1.48	2.77	−0.35	2.67	0.13
Group PP	−0.36*	0.97	−5.74***	0.96	−5.63***
Group ADF	−2.04**	−1.06*	−7.61***	−1.26	−8.80***

Note Panel A (built-up area—pGDP); Panel B (urban population—pGDP); Panel C (built-up area—CO_2); Panel D (urban population—CO_2); Panel E (pGDP—CO_2). Statistics are asymptotically distributed as normal. All tests contain only the intercept and not the trend term. The variance ratio test is right-side, which the others are left-sided. The null hypothesis is that the variables are not cointegrated. Lag length selected based on AIC automatically with a max lag of 2. ***, ** and * reject the null of no cointegration at the 1%, 5% and 10% significance level, respectively

except the Group rho-statistic, meaning a long-run equilibrium relationship between land urbanization and economic growth, which is in agreement with a previous study (Bai et al. 2012). For Panel B, most of the panel cointegration test statistics were statistically insignificant at the 5% level expect the Panel ADF and Group ADF, which indicates that no long-run stable relationship exists between demographic urbanization and economic growth. The Pedroni test results revealed the existence of cointegration between urbanization and CO_2 emissions only in Panel C, suggesting a long-run equilibrium relationship between land urbanization and CO_2 emissions. All the panel test statistics in Panel D were statistically insignificant at the 10% level, implying that no a long-run equilibrium linkage exists between demographic urbanization and economic growth. With the exception of the Group rho-statistics, the other six test statistics in Panel E reject the null hypothesis of no cointegration at the 5% significance level, suggesting a long-term cointegrating relationship between CO_2 emissions and economic growth. It can be concluded that constant long-run equilibrium relationship exists between land urbanization, economic growth and CO_2 emissions in China for the period 1997–2010.

7.4.3.3 Panel Cointegration Estimation

Once the cointegration relationship is established, the next step is to estimate the long-run parameters. Table 7.10 provides the results of the whole China and its three regions (i.e., eastern, central and western regions[1]) based on the panel FMOLS and DOLS estimations. The cointegration coefficients between urban population, pGDP and CO_2 emissions are not estimated since the lack of cointegration among them. Two estimators produced almost identical results, suggesting that the estimates were not sensitive to whether the FMOLS or the DOLS method was used. Often the values of the DOLS estimators are determined under the assumption of one lead, one lag or tow leads, two lags in the change of the regressors (Li et al. 2011). The DOLS estimators were thus sensitive to the choice of number of lags and leads. But the most coefficients from DOLS estimation in our sample varied only slightly for different lags and leads. For the whole China, all estimated coefficients were positive and statistically significant at the 1% level when using pGDP, CO_2 and built-up area as dependent variables. The results suggest that urban expansion had a positive impact on economic growth and CO_2 emissions in China.

[1]The eastern region includes Liaoning, Beijing, Tianjin, Hebei, Shandong, Jiangsu, Shanghai, Zhejiang, Guangdong, Fujian and Hainan provinces; the central region includes Heilongjiang, Jilin, Shanxi, Henan, Anhui, Jiangxi, Hubei and Hunan provinces; and the western region includes Xinjiang, Gansu, Qinghai, Inner Mongolia, Ningxia, Shaanxi, Sichuan, Chongqing, Guizhou, Yunnan and Guangxi provinces (Zhong et al. 2011).

Table 7.10 Panel cointegration coefficients by FMOLS and DOLS for China and its eastern, central and western regions

Whole China							
Variable	Dependent variable: pGDP			Variable	Dependent variable: BA		
	DOLS (1,1)	DOLS (2,2)	FMOLS		DOLS (1,1)	DOLS (2,2)	FMOLS
BA	0.90*** (−27.29)	0.81*** (−21.66)	0.94*** (−27.58)	pGDP	0.66*** (−9.64)	0.22 (−0.47)	0.93*** (−43.33)
Variable	Dependent variable: CO_2			Variable	Dependent variable: CO_2		
	DOLS (1,1)	DOLS (2,2)	FMOLS		DOLS (1,1)	DOLS (2,2)	FMOLS
pGDP	0.83*** (−16.22)	0.60*** (−2.12)	0.94*** (−61.07)	BA	1.19*** (−36.78)	1.16*** (−29.56)	1.15*** (−34.88)
Obs	341	279	403	Obs	341	279	403
Eastern region							
Variable	Dependent variable: pGDP			Variable	Dependent variable: BA		
	DOLS (1,1)	DOLS (2,2)	FMOLS		DOLS (1,1)	DOLS (2,2)	FMOLS
BA	1.02*** (−50.73)	0.99*** (−6.63)	1.03*** (−38.21)	pGDP	0.66*** (−9.64)	0.43 (−0.51)	1.06*** (−26.07)
Variable	Dependent variable: CO_2			Variable	Dependent variable: CO_2		
	DOLS (1,1)	DOLS (2,2)	FMOLS		DOLS (1,1)	DOLS (2,2)	FMOLS
pGDP	0.90*** (−20.11)	0.57 (−0.77)	0.96*** (−42.37)	BA	0.98*** (−75.43)	0.99*** (−10.35)	0.95*** (−45.29)
Obs	121	99	143	Obs	121	99	143
Central region							
Variable	Dependent variable: pGDP			Variable	Dependent variable: BA		
	DOLS (1,1)	DOLS (2,2)	FMOLS		DOLS (1,1)	DOLS (2,2)	FMOLS
BA	1.17*** (−26.27	1.91*** (−52.62)	1.21*** (−26.97)	pGDP	0.90*** (−8.34)	0.30 (−1.4)	0.83*** (−25.16)
Variable	Dependent variable: CO_2			Variable	Dependent variable: CO_2		
	DOLS (1,1)	DOLS (2,2)	FMOLS		DOLS (1,1)	DOLS (2,2)	FMOLS
pGDP	0.64*** (−10.6)	0.40*** (−4.28)	0.84*** (−28.26)	BA	0.99*** (−38.35)	1.03*** (−28.1)	0.98*** (−24.42)
Obs	88	72	104	Obs	88	72	104
Western region							
Variable	Dependent variable: pGDP			Variable	Dependent variable: BA		
	DOLS (1,1)	DOLS (2,2)	FMOLS		DOLS (1,1)	DOLS (2,2)	FMOLS
BA	1.24*** (−32.66)	1.35*** (−49.14)	1.29*** (−28.75)	pGDP	0.53*** (−5.90)	0.25 (−1.92)	0.79*** (−24.24)
Variable	Dependent variable: CO_2			Variable	Dependent variable: CO_2		
	DOLS (1,1)	DOLS (2,2)	FMOLS		DOLS (1,1)	DOLS (2,2)	FMOLS
pGDP	0.64*** (−7.24)	0.63*** (−1.36)	0.97*** (−29.77)	BA	1.27*** (−43.92)	1.29*** (−41.09)	1.24*** (−27.96)
Obs	88	72	104	Obs	88	72	104

Notes BA and pGDP are built-up area and per capita GDP, respectively. The *t*-values are in parentheses. The panel method was grouped estimation. A panel data model with fixed effects was adopted. All tests were performed on the natural logarithm of the dependent and independent variables. Obs is observations. *, ** and *** indicate the estimates are statistically significant at the 10%, 5% and 1% level, respectively

In turn, economic growth contributed to the expansion of urban built-up area, and more people clustered in cities, which increases energy consumption and accordingly generates more emissions (Cole and Neumayer 2004). More specifically, a 1% expansion in urban built-up area increases per capita income by approximately 0.81%–0.94%. A 1% increase in per capita income contributes to the expansion in urban built-up area by 0.66–0.93% when using built-up area as the dependent variable. When using CO_2 emissions as the dependent variable, every 1% rise in land urbanization rate increases CO_2 emissions by approximately 1.15%–1.19%, and every 1% increase in per capita income increases CO_2 emissions by 0.60%–0.94%.

At the regional level, most estimated coefficients were also positive and statistically significant at the 1% level when using pGDP, CO_2 and built-up area as dependent variables. The urban expansion had a positive and significant impact on economic growth for the three regions, but the impact in the western region was slightly higher than that in the central and eastern regions. A 1% expansion in the build-up area contributes to the increase in per capita income by approximately 1.24%–1.35% in the western region, 1.17%–1.91% and 0.99%–1.03% in the central and eastern regions, respectively. In turn, economic growth also promoted to the expansion in urban built-up area. Specifically, every 1% rise in income per capita contributes to the expansion in built-up area by 0.66%–1.06% in the eastern region, 0.83%–0.90% and 0.53%–0.79% in the central and western regions, respectively. Meanwhile, both economic development and built-up area expansion also promoted to carbon emissions at the regional scale. A 1% increase in per capita income contributes to the increase in CO_2 emissions by approximately 0.90%–0.96% in eastern China, 0.40%–0.84% and 0.63%–0.97% in central and western China, respectively. Every 1% rise in urban expansion increases CO_2 emissions by about 0.95%–0.99% in the eastern region, 0.98%–1.03% and 1.24%–1.29% in the central and western regions, respectively. The impact of urban expansion to CO_2 emissions in the western region was larger than that in the eastern and central parts, whereas the effect of economic growth to emissions in the eastern region was larger than that in the western and central regions. This could be explained by the fact that in the eastern region, the increasing urban population, the scarcity of urban land, the competitive pressures of markets and advanced technology encourage the substitution of traditional energy sources by more flexible and reliable energy sources, which contribute to the reduction in proportion of coal use in energy consumption. On the other hand, the level of economic development in the eastern region was higher that than in the central and western regions. The economic development was positively associated with energy consumption (Bai et al. 2012). Therefore, the rapid economic growth and more energy consumption in the eastern region inevitably emit more CO_2 than that in the western and central regions.

The relationship between the growth rate in per capita income, built-up area and CO_2 emissions of China was displayed in Table 7.11. The correlation between per capita income and built-up area growth was statistically insignificant at the 5% level or higher. But the growth rate in economy and urban expansion had a significant and positive impact on CO_2 emissions in China, and economic growth had a larger effect on emissions than urban expansion. A 1% increase in economic growth contributes to the increase in CO_2 emissions by 1.76%–3.22%, and every 1% rise in urban expansion increases emissions by about 0.30%–0.80% in the eastern region.

7.4.3.4 Granger Causality Tests

1. **Granger causality tests based on panel VECM**

Table 7.12 lists the Granger causality test results based on panel VECM for the whole China and its three regions. Bidirectional long-run causalities between land urbanization (built-up area expansion), economic growth, and CO_2 emissions existed at the national level. Whereas there are only unidirectional short-run causal linkages running from: economic growth to land

Table 7.11 Panel cointegration coefficients by FMOLS and DOLS for China based on the growth rate of all variables

Variable	Dependent variable: pGDP growth			Variable	Dependent variable: BA growth		
	DOLS (1,1)	DOLS (2,2)	FMOLS		DOLS (1,1)	DOLS (2,2)	FMOLS
BA growth	0.07 (1.29)	0.04 (0.46)	0.06 (3.16)	pGDP growth	−0.06 (−0.09)	2.52 (1.13)	0.62 (2.30)
Variable	Dependent variable: CO_2 growth			Variable	Dependent variable: CO_2 growth		
	DOLS (1,1)	DOLS (2,2)	FMOLS		DOLS (1,1)	DOLS (2,2)	FMOLS
pGDP growth	2.58*** (5.47)	3.22** (2.00)	1.76*** (7.01)	BA growth	0.53*** (3.99)	0.80*** (2.25)	0.30*** (4.97)
Obs	310	248	372	Obs	310	248	372

Notes BA and pGDP are the growth rate of built-up area and per capita GDP, respectively. The *t*-values are in parentheses. The panel method was grouped estimation. A panel data model with fixed effects was adopted. All tests were performed on the natural logarithm of the dependent and independent variables. Obs is observations. ** and *** indicate the estimates are statistically significant at the 5% and 1% level, respectively

urbanization; land urbanization to CO_2 emissions; and economic growth to CO_2 emissions (Fig. 7.20). Specifically, economic growth contributed to urban expansion, but not vice versa (Panel A), which is not in agreement with a previous study showing long- and short-run unidirectional causality between economic growth and land urbanization in China (Bai et al. 2012). Further, it can be found that both land urbanization and economic growth were the Granger cause of CO_2 emissions, but not vice versa (Panel B and C).

At the regional scale, a bidirectional long-run causality between land urbanization and economic growth was found in the eastern and central regions (Panel D and G). But there exist only unidirectional long- and short-term causal linkages running from: land urbanization to CO_2 emissions (Panel E and F); and economic growth to CO_2 emissions in the eastern and central regions (Panel H and I). Both no long- and short-term causal relationship between land urbanization, economic growth and CO_2 emissions was detected in the western region (Table 7.12).

2. **Heterogeneous panel granger causality tests**

Table 7.13 provides the heterogeneous panel Granger causality test results for the whole China and its eastern and western regions.[2] These findings further verified the presence of unidirectional short-term causal linkages between urbanization, economic growth and CO_2 emissions at the national level. The short-run causalities from economic growth to land urbanization were detected in both Lag 1 and Lag 2 models. Land urbanization was the short-run Granger cause of CO_2 emissions in both Lag 1 and Lag 2 models, but not vice versa. In addition, the short-run Granger causality from economic growth to CO_2 emissions was also found both in Lag 1 and Lag 2 models, but the causal relation from CO_2 emissions to economic growth was detected only in Lag 2 model. These results indicated that both land urbanization and CO_2 emissions have little or no short-run impact on economic growth in China between 1997 and 2010. At the regional scale, we also further discovered the unidirectional short-term causal linkages running from land urbanization to CO_2 emissions and from economic growth to CO_2 emissions in the eastern region. Meanwhile, we also detected the presence of the unidirectional short-term causality between economic growth and land urbanization in the eastern region.

[2]Heterogeneous panel Granger causality test results for the central region was not given due to the number of cross-section in this region less than 9. Dumitrescu and Hurlin (2012) causality tests require the cross-sectional number greater than or equal to 9.

Table 7.12 Wald F-test statistics based on panel-based vector error corrected models for the whole China and its eastern, central and western regions

Whole China	Panel	Causal	Result	F-statistic value	
				Short-run causality	Long-run causality
	A	pGDP	BA	15.237 (0.00)	192.31 (0.00)
		BA	pGDP	7.204 (0.12)	6421.924 (0.00)
	B	BA	CO_2	18.446 (0.00)	313.338 (0.00)
		CO_2	BA	17.518 (0.10)	194.408 (0.00)
	C	pGDP	CO_2	9.019 (0.05)	6331.291 (0.00)
		CO_2	pGDP	4.444 (0.34)	305.076 (0.00)
Eastern region	D	pGDP	BA	0.635 (0.73)	14.571 (0.01)
		BA	pGDP	3.562 (0.17)	28.332 (0.00)
	E	BA	CO_2	5.004 (0.05)	8.156 (0.00)
		CO_2	BA	6.806 (0.16)	19.042 (0.15)
	F	pGDP	CO_2	3.373 (0.04)	10.885 (0.02)
		CO_2	pGDP	2.385 (0.30)	59.190 (0.10)
Central region	G	pGDP	BA	0.708 (0.70)	12.490 (0.01)
		BA	pGDP	3.116 (0.21)	110.951 (0.00)
	H	BA	CO_2	10.981 (0.00)	19.285 (0.00)
		CO_2	BA	3.562 (0.16)	10.308 (0.12)
	I	pGDP	CO_2	2.640 (0.05)	13.521 (0.00)
		CO_2	pGDP	12.367 (0.14)	133.69 (0.10)
Western region	J	pGDP	BA	0.231 (0.89)	2.928 (0.71)
		BA	pGDP	3.988 (0.13)	53.577 (0.10)
	K	BA	CO_2	2.112 (0.35)	6.898 (0.22)
		CO_2	BA	3.018 (0.22)	6.749 (0.15)
	L	pGDP	CO_2	1.749 (0.42)	4.501 (0.34)
		CO_2	pGDP	1.086 (0.58)	169.38 (0.34)

Notes The null hypothesis is non-causality. BA, pGDP and CO_2 are built-up area, per capita GDP and CO_2 emissions, respectively. Cases with probability levels (shown in parentheses) lower than 0.05 reject the null hypothesis

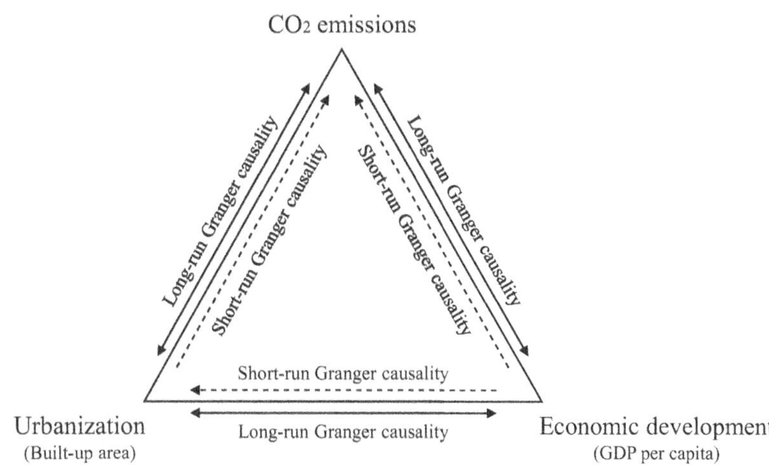

Fig. 7.20 Long- and short-run Granger causality between land urbanization, economic growth and CO_2 emissions in China

Table 7.13 Heterogeneous panel Granger causality test results for the China and its eastern and western regions

Whole China	Panel	Causal	Result	Wald-static	Zbar-static	Probability
	A	pGDP	BA	Lag 1: 2.052	Lag 1: 2.063	0.04
				Lag 2: 5.458	Lag 2: 3.460	0.00
		BA	pGDP	Lag 1: 3.743	Lag 1: 6.411	0.10
				Lag 2: 3.207	Lag 2: 0.530	0.60
	B	BA	CO_2	Lag 1: 10.190	Lag 1: 22.990	0.00
				Lag 2: 10.792	Lag 2: 10.403	0.00
		CO_2	BA	Lag 1: 1.537	Lag 1: 0.739	0.46
				Lag 2: 3.742	Lag 2: 1.226	0.22
	C	pGDP	CO_2	Lag 1: 7.138	Lag 1: 15.143	0.00
				Lag 2: 12.426	Lag 2: 12.531	0.00
		CO_2	pGDP	Lag 1: 1.789	Lag 1: 1.386	0.17
				Lag 2: 4.363	Lag 2: 2.035	0.04
Eastern region	Panel	Causal	Result	W-static	Zbar-stat	Porb
	D	pGDP	BA	Lag 1: 1.729	Lag 1: 0.735	0.46
				Lag 2:3.052	Lag 2: 0.196	0.84
		BA	pGDP	Lag 1: 3.643	Lag 1: 3.665	0.00
				Lag 2: 7.598	Lag 2: 3.721	0.00
	E	BA	CO_2	Lag 1: 9.610	Lag 1: 12.806	0.00
				Lag 2: 7.604	Lag 2: 3.725	0.00
		CO_2	BA	Lag 1: 1.514	Lag 1: 0.404	0.69
				Lag 2: 3.502	Lag 2: 0.554	0.59
	F	CO_2	pGDP	Lag 1: 0.953	Lag 1:−0.454	0.65
				Lag 2: 3.680	Lag 2: 0.683	0.49
		pGDP	CO_2	Lag 1: 6.039	Lag 1: 7.336	0.00
				Lag 2: 15.939	Lag 2: 10.189	0.00
Western region	Panel	Causal	Result	W-static	Zbar-stat	Porb
	J	pGDP	BA	Lag 1: 2.006	Lag 1: 0.987	0.32
				Lag 2:5.319	Lag 2: 1.666	0.10
		BA	pGDP	Lag 1: 4.519	Lag 1: 4.271	0.10
				Lag 2: 5.262	Lag 2: 1.628	0.10
	K	BA	CO_2	Lag 1: 7.959	Lag 1: 8.765	0.09
				Lag 2: 11.269	Lag 2: 5.601	0.25
		CO_2	BA	Lag 1: 1.566	Lag 1: 0.412	0.68
				Lag 2: 1.266	Lag 2:−0.014	0.31
	L	CO_2	pGDP	Lag 1: 1.643	Lag 1:0.513	0.61
				Lag 2: 4.694	Lag 2: 1.252	0.21
		pGDP	CO_2	Lag 1:8.808	Lag 1: 9.873	0.35
				Lag 2: 13.559	Lag 2: 7.114	0.68

Notes The null hypothesis is homogeneous non-causality. Cases with probability levels lower than 0.05 reject the null hypothesis. Lag 1 and Lag 2 represent the test models of the Dumitrescu and Hurlin (2012) causality tests of lag order 1 and 2, respectively

The short-run Granger causality relationship between land urbanization, economic growth and CO_2 emissions in the western region were not found.

7.4.3.5 Validation of the EKC Hypothesis in China

To further validate the EKC hypothesis in China, the Hausman test was performed to determine which one should be selected from two models: random effect and fixed effect models. Based on the assumption of the random effect model, the null hypothesis should be rejected for both quadratic and cubic models at the 1% significance level, which means that the fixed effect model may be more suitable than the random effect (Table 7.14).

Table 7.15 shows the estimates from the panel OLS estimator. For quadratic model, economic growth and urbanization had a positive and significant impact on CO_2 emissions. But the estimated coefficient on income squared was statistically insignificant at the 5% level. For cubic model, all the coefficients in the panel OLS equation were statistically significant at the 1% level or lower. Furthermore, the Wald test was also performed to choose the most appropriate one between the quadratic and cubic models. Results showed that the null hypothesis (the quadratic model) be rejected at the 1% significance level, indicating that the cubic function was more preferable to be accepted. From the sign of the parameters, there existed an inverse N-shape relationship between economic growth and CO_2 emissions in China. This demonstrates that as economic develops, CO_2 emissions fist decrease, and then rise after the left turning point and it will decline at last when arrive at the right turning point. By calculation, the left turning point was quite low and real GDP per capita was approximately 127 Yuan (1997 prices) and the right turning point was approximately 10,201 Yuan. These results further confirmed that land urbanization had a positive and significant impact on CO_2 emissions in China, implying that China's urbanization, especially land urbanization, does contribute to CO_2 emissions in the long-run.

Our empirical findings do not support the EKC hypothesis, which is in agreement with

Table 7.14 Hausman test results

Test summary	Chi-Sq. statistic	
Chi-Sq. statistic	Quadratic	Cubic
	40.304	41.210
Prob	0.000	0.000
Accept model	Fixed effects	Fixed effects

Table 7.15 Estimation results in pGDP and CO_2 emissions based on the panel OLS estimator

Dependent variable [ln(CO_2)]	Quadratic model			Cubic model		
	Coefficient	t-statistic	Prob	Coefficient	t-statistic	Prob
ln(pGDP)	0.674	3.054	0.000	−6.336	−2.402	0.017
$\ln(pGDP)^2$	−0.003	−0.184	0.854	0.997	2.657	0.008
$\ln(pGDP)^3$	–			−0.047	−2.666	0.008
ln(BA)	0.293	8.239	0.000	0.286	8.089	0.000
Constant	−1.717	−2.247	0.025	14.589	2.368	0.018
Turning point	–			(127.41, 10,201.29)		
F-statistic	597.542		0.000	589.035		0.000
Adjusted R^2	0.980			0.980		
Wald test	H_0: the quadratic model; H_1: the cubic curve					
Wald statistic	7.110***					

Note Fixed effect OLS estimator was used. The number of samples was 434. "***" indicates the estimator of a parameter is significant at the 1% level

Table 7.16 Comparison with the other studies

Source	Data type	Method	Result
Jalil and Mahmud (2009)	China; time series (1975–2005)	ARDL, quadratic model; VECM; EKC hypothesis	Inverted U-shaped, GDP → CO_2
Wang et al. (2011)	China's 28 provinces; panel data (1995–2007)	Pedroni cointegration; Panel VECM; EKC hypothesis	U-shaped curve; GDP → CO_2 (long-run)
Du et al. (2012)	China's 28 provinces; panel data (1995–2009)	Quadratic and cubic models; EKC hypothesis; GMM estimator	Inverted U-shaped is not strongly supported
Wang et al. (2012a)	Beijing; time series (1997–2010)	STIRPAT; OLS	Not support for EKC
This study	China's 31 provinces; panel data (1997–2010)	Pedroni cointegration; Panel VECM; EKC hypothesis; OLS; cubic model	Long-run: BA ↔ GDP; BA ↔ CO_2; GDP ↔ CO_2; short-run: GDP → BA; BA → CO_2; GDP → CO_2; inverted N-shaped curve

Note ARDL refers to the auto regressive distributed lag; GMM represents the generalized method of moment; OLS is the ordinary least square; and STIRPAT refers to the stochastic impacts by regression on population, affluence and technology. The symbol " ↔ ", " → " represent the bidirectional and unidirectional Granger causality, respectively

previous studies (Wang et al. 2011, 2012a; Du et al. 2012). In addition to the differences as summarized in Table 7.16, a major difference is that the models we employed are different from those used by Jalil and Mahmud (2009), which may have an impact on the validity of EKC hypothesis. The quadratic model in the regression equations was used in the study by Jalil and Mahmud (2009). More importantly, the studies by Wang et al. (2012a), and Jalil and Mahmud (2009) were based on the time series data while our study performed a panel data analysis using the China's provincial data. It is generally acknowledged that panel data models have several major advantages over conventional cross-sectional or time series data models (Wang et al. 2014a). Panel data models allow controlling for individual heterogeneity, as well as identifying effects that cannot be detected in simple time series or cross-section data (Du et al. 2012). Our results further suggest that the relationship between economic growth and CO_2 emissions in China do not support the EKC hypothesis.

7.5 Assessing the Impact of Population, Income and Technology on Energy Consumption and Industrial Pollutant Emissions in China

7.5.1 Theoretical Background and Methodology

7.5.1.1 Theoretical Base

Figure 7.21 showed the relationship between human activities and environmental impacts. With the accelerating urbanization and industrialization, rapid population growth and economic development may need to consume more energy and emit more pollutants, exerting pressure on the environment. However, the impact of human activities on the environment may be mitigated when the level of economic development reaches a certain threshold.

The effect of demographic, economic and technological factors on the environment is

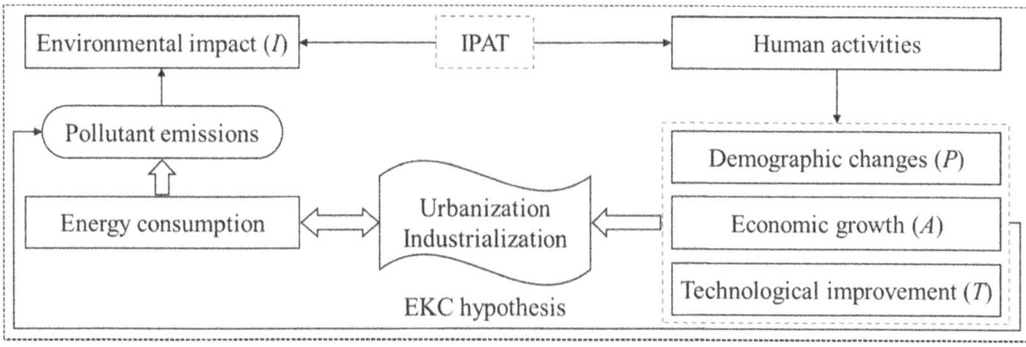

Fig. 7.21 Population, economy, technology and environment

mainly postulated in the IPAT model (I = PAT), which is firstly proposed by Ehrlich and Holdren (Ehrlich and Holdren 1971). The IPAT equation specifies the anthropogenic environmental impacts (*I*) as a function of population (*P*), affluence (*A*, usually proxied by per capita GDP) and technology (*T*, generally identified as the impact of per unit of economic activity).

The IPAT model is limited because it analyzes a problem by changing a factor while keeping others constant, resulting in proportionate impacts on the dependent variable (Dietzand Rosa 1994; Shi 2003). To overcome these weaknesses, Dietz and Rosa (Dietzand Rosa 1994) reformulated the IPAT model into a stochastic model (STIRPAT), which can statistically assess non-monotonic or non-proportional impacts of driving forces on the environment. York et al. (2003) refined the STIRPAT model and expanded it by incorporating additional factors, such as quadratic terms or different components of *P*, *A* or *T*. The STIRPAT model has been successfully utilized to analyze the effects of driving forces on various environmental impacts (York et al. 2003; Poumanyvong and Kaneko 2010; Wang et al. 2012, 2013). The STIRPAT model can be given as following equation:

$$\mathrm{I_{it}} = \alpha P_{it}^{b} A_{it}^{c} T_{it}^{d} \varepsilon_{it} \tag{7.22}$$

After taking logarithms, the model takes the following form:

$$Ln(I_{it}) = a + bLn(P_{it}) + cLn(A_{it}) + dLn(T_{it}) + \varepsilon_{it} \tag{7.23}$$

here, suffixes *i* and *t* refer to provinces and years, respectively; *P* denotes population size; *A* refers to real GDP per capita; *T* represents for technology; and the dependent variable *I* stands for pollutant emissions; ε_{it} denotes the error term, *a* is the constant term; *b*, *c* and *d* are, respectively, the coefficients of *P*, *A* and *T*.

7.5.1.2 Improved STIRPAT Model

The STIRPAT model allows other factors to be added to explore their effects on environmental parameters. In the STIRPAT model, factors *P* and *A* are decomposable, as is *T* (York et al. 2003). To further examine the impacts of population, income and technology on energy consumption and industrial pollutant emissions in China, population density, economic structures and industrial energy intensity were incorporated into an extended STIRPAT model. Liddle and his collaborators have demonstrated that population density was a better indicator than urbanisation in regional human-earth studies (the share of population living in urban areas) (Liddle 2004, 2014; Liddle and Lung 2010). Thus, this study used population density as a proxy indicator of demographic changes. Following the study of York et al. (2003), in this study, technology improvement was measured by three variables: the share of the added values of industry and tertiary industry in the real GDP and intensity energy intensity. Additionally, York's method was followed to establish the quadratic models for the affluence factor to verify whether the EKC hypothesis operated in China (York et al. 2003). The improved STIRPAT model was changed as follows:

$$Ln\left(\frac{EC_{it}}{P_{it}}\right) = a_0 + a_1 Ln(POPD_{it}) + a_2 Ln(pGDP_{it}) + a_3 Ln(INDU_{it}) + a_4 Ln(TERT_{it}) + \varepsilon_{1it} \quad (7.24)$$

$$Ln\left(\frac{EC_{it}}{P_{it}}\right) = b_0 + b_1 Ln(POPD_{it}) + b_2 Ln(pGDP_{it}) + b_3[Ln(pGDP_{it})]^2 + b_4 Ln(INDU_{it}) + b_5 Ln(TERT_{it}) + b_6 Ln(IEI_{it}) + \varepsilon_{2it} \quad (7.25)$$

where P is the total population and POPD is the population density; pGDP is the real per capita GDP; INDU and TERT are the shares of the added values of industry and tertiary industry in the GDP, respectively; IEI is the industry energy intensity (expressed as industry energy use per unit GDP); EC is per capita energy consumption; and I represents the environmental impact (with three types of industrial pollutant emissions per capita as proxies, i.e., exhaust gases, wastewater and solid waste). ε_{it} is the error term, a_0 and b_0 are the constant terms; and a_i and b_i are the estimated coefficients of the dependent variables. The industrial energy intensity is included in only the emission models because it contains part of the dependent variable.

The EKC theory states that environmental degradation initially increases with income during the early stages of economic growth and then decreases, with income arriving afterward at a turning threshold (Grossman and Krueger 1995). According to the EKC hypothesis, an inverted U-shaped relationship between per capita income and pollutant emission exists if the coefficient b_2 is statistically significant and positive, while the coefficient b_3 is statistically significant and negative (i.e., $b_2 > 0$ and $b_3 < 0$) and the turning point of the EKC is computed by $\omega = \exp(-0.5b_2/b_3)$. A U-shaped linkage exists between per capita emissions and income if the coefficient b_2 is statistically significant and negative and the coefficient b_3 is statistically significant and positive ($b_2 < 0$ and $b_3 > 0$) (Grossman and Krueger 1995; York et al. 2003; Liddle 2004).

7.5.1.3 Model Estimation

Two steps were performed to examine the impact of population, income and technology on energy consumption and pollutant emissions for the entirety of China and for the three regions (i.e., eastern, central and western China, Fig. 7.22). First, the presence of unit roots in all variables was verified before proceeding to any econometric analysis because using the conventional ordinary least squares (OLS) estimator with non-stationary variables may have resulted in spurious regressions. The Pesaran (2007) CIPS test, as one of the second-generation panel unit root tests, was performed to test the stationarity of the all variables. The second-generation panel unit root test is valid for both unbalanced panels and panels in which the cross-sectional and time dimensions are of the same order of magnitude (Pesaran 2007; Liddle 2015). The results of that test suggested that many variables are non-stationary in their levels, but most of them become stationary at the 5% significance level after taking first differences, which indicated that the variables in all panels are integrated of order one or $I(1)$ (Table 7.17).

Second, the impact of human activities on energy use and pollutant emissions for the whole sample and for the three regions was estimated using three methods: fixed effects (FE), the feasible generalized least squares (FGLS) and the linear regression with Driscoll-Kraay standard errors (DK), generating forty-four models. Specifically, the FE estimation was used to alleviate heterogeneity bias. The autocorrelation in the fixed effect models was verified by the Wooldridge test and the autocorrelation in the fixed effect models was found (Wooldridge 2002). Moreover, we confirmed the presence of groupwise heteroskedasticity in the all modes using the modified Wald statistic (Greene 2000). We also detected the existence of cross-sectional dependence by Pesaran test in the fixed effect

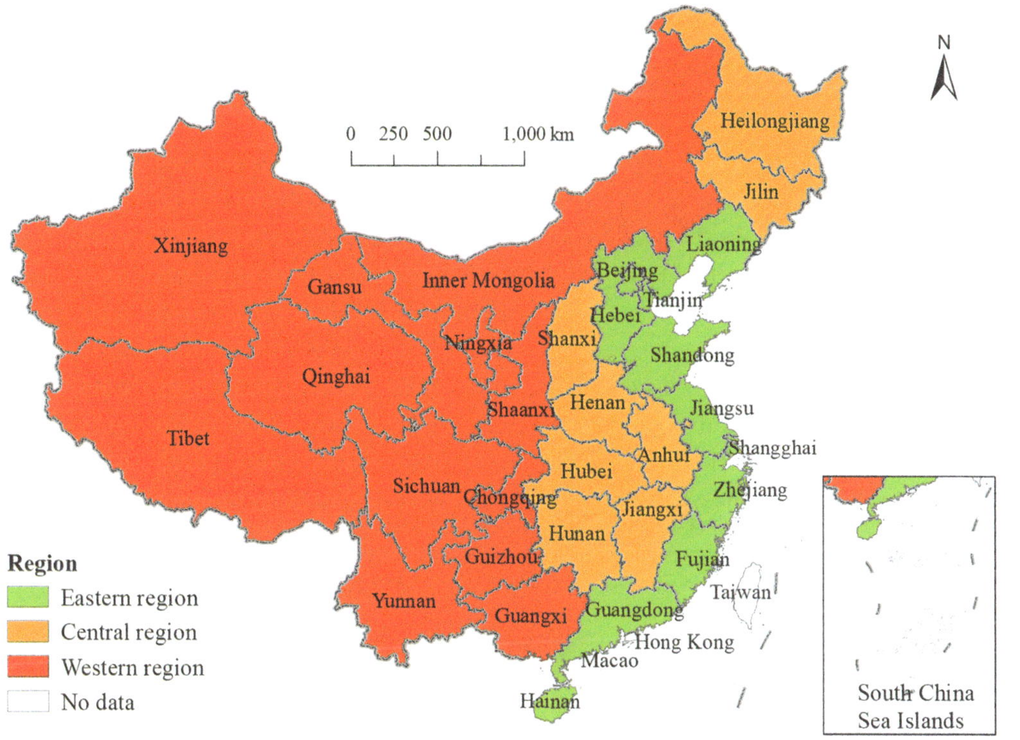

Note: The eastern region includes Liaoning, Beijing, Tianjin, Hebei, Shandong, Jiangsu, Shanghai, Zhejiang, Guangdong, Fujian and Hainan provinces; the central region includes Heilongjiang, Jilin, Shanxi, Henan, Anhui, Jiangxi, Hubei and Hunan provinces; and the western region includes Xinjiang, Gansu, Qinghai, Inner Mongolia, Ningxia, Shaanxi, Sichuan, Chongqing, Guizhou, Yunnan and Guangxi provinces. It is generally accepted that the eastern area of China belongs to the developed region, and the western and central regions are the underdeveloped or less developed region.

Fig. 7.22 Eastern, central and western regions of China

models (Pesaran 2004). The FGLS estimation was employed to address those issues. However, the FGLS method is infeasible if the panel's time dimension T is smaller than its cross-sectional dimension N, and it tends to produce unacceptably small standard error estimates (Beck and Katz 1995). To solve the issues, the Driscoll-Kraay estimation was applied. Besides being heteroscedasticity consistent, the Driscoll-Kraay standard error estimates are robust to general forms of cross-sectional and temporal dependence (Hoechle 2007; Zhang and Lin 2012). Because the DK estimation was robust to cross-sectional dependence, the emphasis was placed only on the estimation; the interpretations focused only on the results based on the DK models. The estimated results for FE and FGLS models were listed in Tables 7.18 and 7.19. Additionally, the multicollinearity among the explanatory variables in the DK models was tested by the variance inflation factors (VIF). The VIF values are all less than 10, indicating no multicollinearity (Table 7.20).

7.5.2 Data Source and Description

7.5.2.1 Data Sources

This study makes full use of a balanced panel dataset of 30 provinces in China over the period 1990–2012 (Hong Kong, Macao, Taiwan and Tibet are excluded for lack of data). The

Table 7.17 Comparison with the other studies

No. lags	Variables in levels						Variables in first differences					
	Constant			Constant and trend			Constant			Constant and trend		
	1	2	3	1	2	3	1	2	3	1	2	3
Ln(EC)	0.289	0.772	0.953	0.863	0.939	0.917	0.000	0.006	0.008	0.000	0.174	0.182
Ln(IEI)	0.000	0.004	0.041	0.917	0.586	0.722	0.000	0.000	0.031	0.000	0.000	0.674
Ln(FQ)	0.353	0.931	0.847	0.166	0.761	0.687	0.000	0.000	0.083	0.000	0.072	0.667
Ln(FS)	0.792	0.949	0.815	1.000	1.000	1.000	0.000	0.105	0.786	0.000	0.927	1.000
Ln(FW)	0.916	0.987	0.930	0.612	0.993	0.989	0.000	0.000	0.051	0.000	0.268	0.960
Ln(POPD)	0.864	0.920	0.998	0.999	0.997	1.000	0.012	0.029	0.843	0.009	0.071	1.000
Ln(pGDP)	0.953	0.994	1.000	0.984	0.997	1.000	0.009	0.028	0.823	0.005	0.019	0.396
Ln(INDU)	0.024	0.451	0.101	0.259	0.871	0.202	0.000	0.000	0.000	0.000	0.025	0.063
Ln(TERT)	0.660	0.976	0.713	0.449	0.990	0.370	0.000	0.008	0.014	0.000	0.308	0.934

Notes Ln() denotes natural logarithms; EC is energy consumption; IEI is industrial energy intensity; POPD is population density; pGDP is GDP per capita; FQ, FS and FW are industrial exhaust gas, wastewater and solid waste emissions, respectively; INDU and TERT are proportion of industry and tertiary industries, respectively. p-values shown for null hypothesis of $I(1)$

Table 7.18 Estimation results for energy use by FE and FGLS models

Variable	Whole of China		Eastern region		Central region		Western region	
	FE (1)	FGLS (2)	FE (3)	FGLS (4)	FE (5)	FGLS (6)	FE (7)	FGLS (8)
Ln(POPD)	0.24**	0.21***	−0.02	−0.07***	−0.79*	−0.28***	0.49	−0.21***
Ln(pGDP)	0.61***	0.56***	0.59***	0.46***	0.67***	0.32***	0.62***	0.43***
Ln(INDU)	0.59***	0.52***	0.99***	0.94***	0.05	1.43***	0.56***	1.38***
Ln(TERT)	−0.120	0.96***	0.48***	1.44***	−0.61***	0.86***	−0.02	0.73***
Constant	−7.75**	−8.47**	−10.0**	−11.89***	−1.100	−9.11***	−8.72***	−9.73***
R^2	0.904	–	0.915	–	0.912	–	0.924	–
AC test	F(l, 29) = 167.36***		F(l, 10) = 56.2***		F(l, 7) = 32.05***		F(l, 10) = 105.83***	
CD test	20.91***		7.03 ***		9.17***		6.31***	
HK test	x^2 (30) = 1402.75***		x^2 (11) = 449.44***		x^2 (8) = 65.11***		x^2 (11) = 520.11***	
Obs	690	690	253	253	184	184	253	253

Notes Ln() is natural logarithms; POPD is population density; pGDP is real GDP per capita; INDU and TERT are proportion of industry and tertiary industries, respectively. AC is autocorrelation; CD is cross-sectional dependence; HK is heteroskedasticity; Obs is observations. FE (fixed effects); FGLS (feasible generalized least squares)
*Indicate significance at the 10% level
**Indicate significance at the 5% level
***Indicate significance at the 1% level

provincial data on three types of industrial pollutant emissions, i.e., exhaust gases (denoted by FQ), wastewater (FS) and solid waste (FW) are collected from the China Environmental Yearbook. The data of the provincial population size, GDP, population density, industrial output, added values of the industry and tertiary industry are derived from the China Statistical Yearbook

Table 7.19 Estimation results for industrial pollutant emissions by FE and FGLS models

Variable	Whole of China		Eastern region		Central region		Western region	
	Industrial exhaust gases							
	FE (9)	FGLS (10)	FE (11)	FGLS (12)	FE (13)	FGLS (14)	FE (15)	FGLS (16)
Ln(POPD)	0.08	−0.21***	−0.29	−0.09***	2.47***	−0.09***	1.68***	−0.07***
Ln(pGDP)	0.33	0.06	−0.67	−0.20	−2.24***	−3.02***	−1.87**	−2.41***
Ln $(pGDP)^2$	0.04***	0.06***	0.09	0.08***	0.18***	0.22***	0.18***	0.22***
Ln(INDU)	0.72***	0.55***	0.98***	0.89***	0.29	1.85**	0.06	0.47***
Ln(TERT)	−0.46***	0.19***	−0.08	0.30***	−0.79***	1.24***	−0.18	0.40***
Ln(IEI)	0.10***	0.33***	0.18***	0.59**	0.01	0.12***	0.17***	0.51***
Constant	2.69**	3.12***	6.71***	1.246	3.62	8.58***	5.99**	11.07***
R^2	0.885	–	0.897	–	0.907	–	0.907	–
AC test	F(l, 29) = 76.52***		F(l, 10) = 45.92**		F(l, 7) = 24.59***		F(l, 10) = 23.99***	
CD test	20.36***		6.25***		4.96***		4.60***	
HK test	$x^2(30)$ = 323.93***		$x^2(11)$ = 20.32***		$x^2(8)$ = 156.8***		$x^2(11)$ = 141.26***	
Obs	690	690	253	253	184	184	253	253
Variable	Industrial wastewater							
	FE (17)	FGLS (18)	FE (19)	FGLS (20)	FE (21)	FGLS (22)	FE (23)	FGLS (24)
Ln(POPD)	−1.56***	0.13***	−2.75***	0.21***	0.55	−0.14***	1.33***	0.12***
Ln(pGDP)	−1.46***	0.16***	−2.23***	1.03***	−1.55***	−1.59***	−3.31***	−0.66**
Ln $(pGDP)^2$	0.09***	0.01	0.14***	−0.06***	0.09***	0.09***	0.19***	0.06***
Ln(INDU)	0.98***	0.19***	1.45***	1.07***	0.18	−0.36***	0.35	−0.46***
Ln(TERT)	−0.05	−0.57	0.78***	−0.11**	−0.91***	−0.38***	0.11	−0.61***
Ln(IEI)	0.12***	0.14***	0.19***	0.07***	0.06***	0.00	−0.04	0.25***
Constant	13.57***	1.64***	20.82***	−5.98***	8.78***	13.04***	9.47***	6.82***
R^2	0.361	–	0.685	–	0.581	–	0.167	–
AC test	F(l, 29) = 128.35***		F(l, 10) = 42.64***		F(l, 7) = 25.60***		F(l, 10) = 60.03***	
CD test	3.73 ***		6.25***		4.78***		4.60***	
HK test	$x^2(30)$ = 1009***		$x^2(11)$ = 80.41***		$x^2(8)$ = 128.8***		$x^2(11)$ = 1230.31***	
Obs	690	690	253	253	184	184	253	253
Variable	Industrial solid waste							
	FE (25)	FGLS (26)	FE (27)	FGLS (28)	FE (29)	FGLS (30)	FE (31)	FGLS (32)
Ln(POPD)	−0.31*	−0.34***	−1.14***	−0.42***	2.08***	−0.25***	2.06***	−0.06***
Ln(pGDP)	−0.33	−0.42***	−3.06***	−2.83***	−2.41***	−3.10***	−3.43***	−3.76***
Ln $(pGDP)^2$	0.06***	0.06***	0.21***	0.21***	0.18***	0.21***	0.25***	0.27
Ln(INDU)	0.86***	0.97***	0.89***	1.14***	0.10	1.30***	0.46	1.27
Ln(TERT)	−0.13	0.34***	0.02	0.77***	−0.70***	1.23***	0.39	0.69
Ln(IEI)	0.08***	0.15***	0.07	0.67***	0.01	0.11***	0.13**	0.21

(continued)

Table 7.19 (continued)

Variable	Whole of China		Eastern region		Central region		Western region	
	Industrial exhaust gases							
	FE (9)	FGLS (10)	FE (11)	FGLS (12)	FE (13)	FGLS (14)	FE (15)	FGLS (16)
Constant	−2.99*	—3.93***	13.52***	3.28***	−2.08	3.02***	−1.24	5.31
R^2	0.305	–	0.792	—	0.879	–	0.827	–
AC test	F(l, 29) = 102.46*** k*		F(l, 10) = 42.04***		F(1, 7) = 1.65***		F(l, 10) = 163.37***	
CD test	20.39***		2.95***		1.91***		3.43***	
HK test	x^2 (30) = 3284***		x^2 (11) = 1143.99***		x^2 (8) = 48.27***		x^2 (11) = 513.85***	
Obs	690	690	253	253	184	184	253	253

Notes Ln() is natural logarithms; POPD is population density; pGDP is real GDP per capita; IEI is industrial energy intensity; INDU and TERT are proportion of industry and tertiary industries, respectively. AC is autocorrelation; CD is cross-sectional dependence; HK is heteroskedasticity. Obs is observations. FE (fixed effects); FGLS (feasible generalized least squares)
*Indicate significance at the 10% level
**Indicate significance at the 5% level
***Indicate significance at the 1% level

Table 7.20 VIF tests for multicollinearity

	Model 1	Model 3	Model 5	Model 7	Model 9	Model 11	Model 13	Model 15	Model 33	Model 35	Model 37	Model 24
POPD	1.25	1.38	1.07	1.11	1.35	1.38	1.08	1.5	1.72	1.46	1.18	1.52
pGDP	2.14	2.34	2.82	2.98	3.52	4.61	4.09	5.15	5.2	8.25	4.43	7.37
INDU	1.80	1.78	2.06	2.21	1.83	1.81	2.17	2.32	2.18	4.14	3.59	2.42
TERT	2.34	2.97	2.29	1.71	2.44	3.08	2.32	1.87	2.45	3.34	2.71	1.87
IEI	–	–	–	–	2.21	2.67	1.81	2.38	2.57	4.08	2.10	3.25
Mean	1.88	2.12	2.06	2.00	2.27	2.71	2.29	2.64	2.76	4.48	2.96	3.21

Notes The VIF values are all below than 10, indicating that there is no multicollinearity (all variables in natural logarithms)

and the China Compendium of Statistics. The data on total energy consumption are obtained from the China Energy Statistical Yearbook and Statistical Yearbook of the various provinces, municipalities and autonomous regions. To eliminate the influence of the price index, the GDP is calculated at a constant price (1990 prices) and the proportion of industry and tertiary industry output in GDP are calculated accordingly. Table 7.21 provides a detailed description of all variables used in this study.

7.5.2.2 Data Description

Summary statistics and correlations, and cross-section independence (CD) tests among the variables are displayed in Table 7.22. Population density was negatively correlated with energy consumption and industry pollutant emissions except for wastewater, but the correlation was weak. The per capita GDP and the share of the added value of the industrial sector in GDP were positively correlated with energy consumption and pollutant emissions. The three types of industrial pollutant emissions were also positively associated with energy consumption but negatively correlated with industry energy intensity. In addition, the results of the CD test clearly indicated the presence of cross-sectional dependence for all variables.

Simple linear regressions were used to examine the changes in population density, energy use and pollutant emissions for 30

Table 7.21 Description of the variables used in the study over the period 1990–2012

Variable	Definition	Unit of measurement
Population density (POPD)	Ratio of a population to a given unit of area	Person per sq. km
GDP per capita (pGDP)	GDP divided by population at the end of the year	Yuan per capita
Proportion of industrial output (INDU)	Percentage of the added value of industrial sector in GDP	Percent
Proportion of tertiary industry output (TERT)	Percentage of the added value of tertiary industry in GDP	Percent
Energy use per capita (EC)	Sum of energy consumption divided by total population at the end of the year	Tce per capita
Industry energy intensity (IEI)	Total industry energy use divided by GDP	Tce per ten thousand yuan (RMB)
Exhaust gas emissions per capita (FQ)	Total volume of pollutant-containing gas emitted into the atmosphere divided by the total population	Hundred million cubic meters
Waste water discharge per capita (FS)	Total volume of waste water discharge by industrial enterprises divided by the total population	Ton per capita
Solid waste production per capita (FW)	Total volume of solid waste that is not regarded to be hazardous divided by the total population	Ton per capita

provinces in China during the period 1990–2012. The slopes of the lines of best fit reflected the changes in the variables, and a greater slope indicated greater changes in the variables. Over the period 1990–2012, the growth rate of the population in eastern China was relatively higher than those in the central and western regions (Fig. 7.23a). The five provinces with the highest population growth rates were Shanghai, Beijing, Tianjin, Hainan and Guangdong. For energy consumption, the growth rate in Shandong province was the highest, followed by Guangdong and Jiangsu (Fig. 7.23b). These results showed that the rapid population growth in most provinces of eastern China was accompanied by a significant increase in energy consumption over the past two decades.

For industrial exhaust gas emissions, Hebei province had a significant increase of 291 billion m^3 per year over the past 23 years, which was followed by the exhaust gas emissions from the provinces of Shandong, Jiangsu, Henan and Guangdong (Fig. 7.23c). The provinces with relatively rapid growth rates in generation of solid waste were Hebei, Henan, Shanxi, Inner Mongolia and Shandong (Fig. 7.23d). The rapid growth of solid waste production in these areas may be related to their abundant energy sources, especially coal and oil. For wastewater discharge, the trend increased in most provinces of eastern China, whereas both the central and western areas exhibited decreasing trends (Fig. 7.23e). The fast growth rate of industrial pollutant discharge in the eastern region may be attributed to its advanced industries and high population density. These results indicated that China's environmental management measures should consider the causes and the regional characteristics of environmental issues.

Table 7.23 lists population density, energy consumption and pollutant emissions per capita in eastern, central and western China in 1990 and 2012. Population density in the eastern region Increased from 636.98 persons per square kilometer (sq. km) in 1990 to 981.66 persons per sq. km in 2012. The density in the central region increased from 276.2 persons per sq. km in 1990 to 305.51 persons per sq. km in 2012, and the

Table 7.22 Summary statistics and correlations (all variables in natural logarithms)

Variable	Obs	Mean	Std. dev	Min	Max	CD test			
						CD statistics			Abs (corr)
EC	690	0.43	0.66	−1.68	2.07	94.69***			0.95
FQ	690	9.63	0.82	7.11	12.46	86.99***			0.89
FS	690	2.74	0.53	1.18	4.61	13.18***			0.52
FW	690	−0.27	0.79	−2.38	3.07	79.12***			0.87
POPD	690	5.38	1.27	1.83	8.32	84.25***			0.84
pGDP	690	8.48	0.81	6.69	10.47	99.11***			0.99
INDU	690	3.58	0.38	1.91	4.09	27.55***			0.48
TERT	690	3.59	0.18	3.02	4.34	63.21***			0.64
IEI	690	1.05	0.81	−3.26	2.97	83.63***			0.84
	EC	FQ	FS	FW	POPD	pGDP	INDU	TERT	IEI
Correlations									
EC	1								
FQ	0.93	1							
FS	0.15	—	1						
FW	0.71	—	—	1					
POPD	−0.04	−0.02	0.35	−0.23	1				
pGDP	0.77	0.77	0.22	0.45	0.33	1			
INDU	0.23	0.30	0.34	0.36	0.34	0.23	1		
TERT	0.49	0.39	−0.06	0.13	0.15	0.57	−0.34	1	
IEI	−0.32	−0.33	−0.12	−0.16	−0.39	0.71	−0.20	−0.31	1

Notes EC is energy consumption per capita; FQ, FS and FW are exhaust gas, wastewater and solid waste emissions per capita, respectively; POPD is population density; pGDP is real GDP per capita; IEI is industrial energy intensity; INDU and TERT are proportion of the industry and tertiary industry sectors, respectively. Obs is observations. CD is cross-section independence. Null hypothesis is cross-sectional independence. Abs (corr) is the absolute value correlation coefficient. The Pesaran (2004) CD test is performed using the Stata "xtcd" commend

***Implies statistical significance at the 1% level

density in the western region increased from 120.97 persons per sq. km in 1990 to 133.04 persons per sq. km in 2012. The growth rate of population density in eastern China (54.18%) was faster than that in the western (10.61%) and central (9.98%) regions. By contrast, energy consumption per capita showed a different pattern. The increase in energy use per capita in the western region was greater than that in the other two regions and increased from 0.87 tons of coal equivalent (tce) in 1990 to 4.07 tce in 2012. The energy consumption per capita in the eastern region increased only by 2.58 tce, with an increase from 1.26 tce in 1990 to 3.83 tce in 2012, and in the central region, the increase was also slight from 0.93 tce in 1990 to 2.96 tce in 2012. This result may be related to the implementation of the China's Western Development policy in 2000.

The industrial pollutant emissions per capita in China also changed significantly over the past two decades. The per capita exhaust gas emissions in the western region showed a greater increase than that in the central and eastern areas and increased from 7258.4 m^3 in 1990 to 64,110.99 m^3 in 2012. Similar to energy use, the

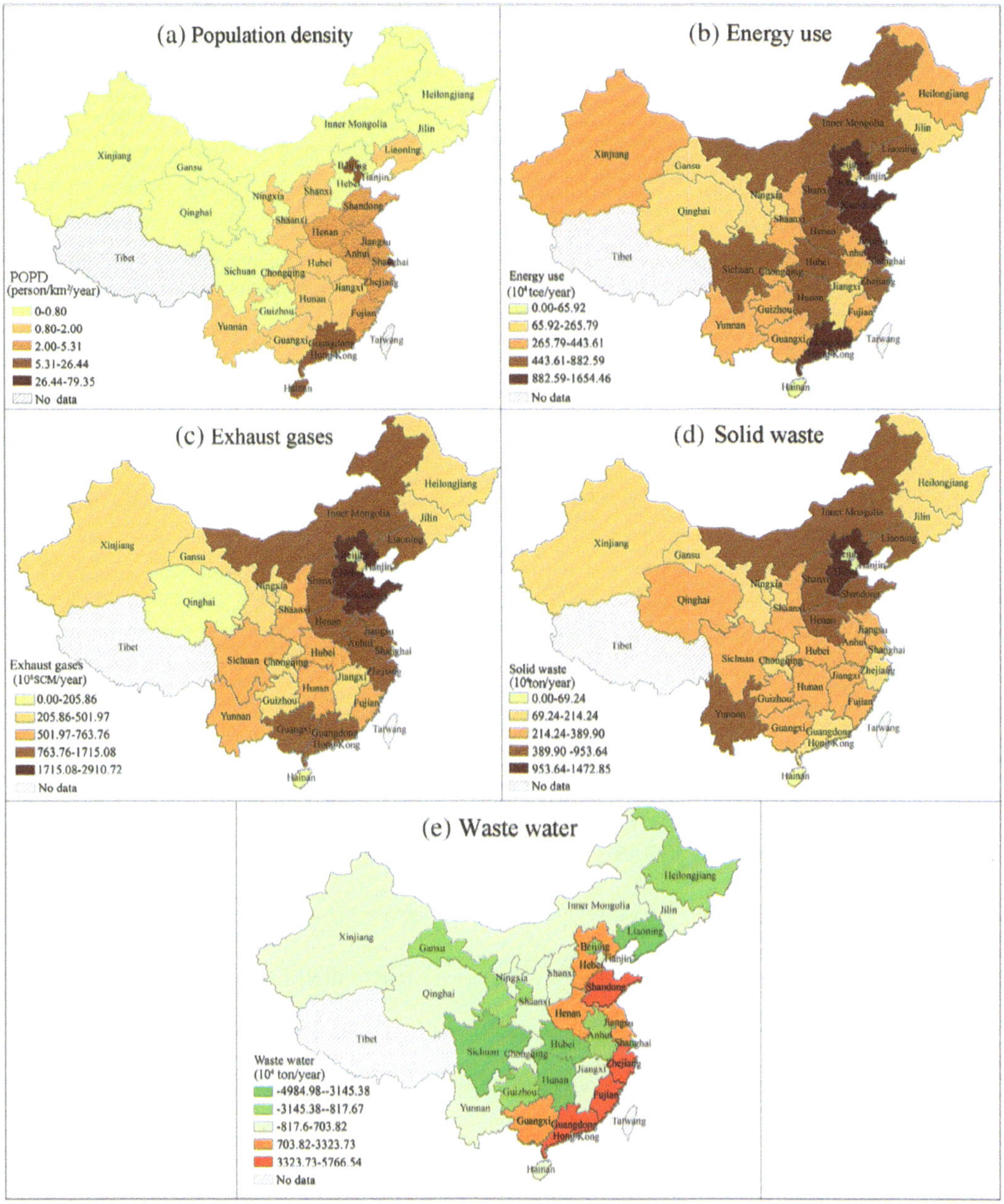

Fig. 7.23 Trend variations in **a** population density, **b** energy use, **c** exhaust gases, **d** solid waste and **e** waste water discharge for 30 provinces in China between 1990 and 2012

increase in solid waste production per capita in the western region was larger than that in the central and eastern regions and increased from 0.49 tons in 1990 to 4.9 tons in 2012. For wastewater discharge, all three regions showed decreases; the decrease in discharge in the eastern area was greater than that in the other two regions.

Table 7.23 Population density, per capita energy consumption and pollutant emissions for China's different regions in 1990 and 2012

Region	POPD		EC		FQ		FS		FW	
	1990	2012	1990	2012	1990	2012	1990	2012	1990	2012
Eastern	636.68	981.66	1.26	3.83	11,833.41	0.57	0.57	19.03	0.012	0.011
Central	276.20	305.51	0.93	2.96	8244.80	0.62	0.62	14.53	0.017	0.013
Western	120.97	133.04	0.87	4.07	7258.74	0.49	0.49	13.10	0.014	0.025

Notes POPD is population density (persons per sq. km); EC is per capita energy consumption (tces per person) and FQ, FS and FW are per capita exhaust gases (10^4 m^3 per person), waste water (tons per person) and solid waste (tons per person), respectively

7.5.3 Empirical Results and Discussions

7.5.3.1 Human Activities and Energy Consumption

Based on the pooled OLS regression with Driscoll-Kraay standard errors (DK), the estimates of the impact of demographic changes, economic development and technology improvement on energy consumption for the whole sample and the eastern, central and western regions are shown in Table 7.24. For the whole sample, all estimated coefficients were statistically significant at the level of 5% or lower. Additionally, the diagnostics are good: the residuals were stationary, and the cross-sectional dependence in the residuals cannot be rejected. The effect of GDP per capita and proportion of

Table 7.24 Estimation results for per capita energy use by DK models, 95% confidence intervals in brackets[a]

Variable	Whole of China	Eastern region	Central region	Western region
Ln (POPD)	−0.211*** [−0.23 −0.20]	−0.065*** [−0.11 −0.02]	−0.239*** [−0.30 −0.18]	−2.18*** [−0.24 −0.20]
Ln (pGDP)	0.554*** [0.51 0.60]	0.456*** [0.34 0.58]	0.279*** [0.16 0.40]	0.423*** [Q27 0.58]
Ln (INDU)	0.524*** [0.46 0.59]	0.948*** [0.65 1.25]	1.854*** [1.47 2.24]	1.456*** [0.70 2.22]
Ln (TERT)	0.968*** [0.751.19]	−11.987*** [0.961.99]	1.160*** [0.50 1.82]	0.838*** [0.06 1.62]
Constant	−8.449** [−9.30 −7.69]	−11.987 [−13.85 −10.13]	−11.614*** [−14.79 −8.43]	−10.30 [−13.99 −6.61]
CIPS	I(0)	I(0)	I(0)	I(0)
Mean rho	0.18	0.15	0.21	0.31
CD (p)	18.65 (0.12)	6.38 (0.34)	8.97 (0.18)	5.79 (0.09)
R^2	0.735	0.802	0.821	0.776
Obs	696	253	184	184

[a]The estimated results based on the FE and FGLS for energy consumption and pollutant emissions models are showed in Tables 7.20, 7.21 and 7.22

Notes The estimated method is pooled OLS. Ln() is natural logarithms; POPD is population density; pGDP is real GDP per capita; INDU and TERT are proportion of industry and tertiary industries, respectively. The stationary of the residuals is determined from the Pesaran (2007) CIPS test and I(0) means stationary. Mean rho is the mean absolute correlation coefficient of the residuals from the Pesaran (2004) CD test. CD is the test statistic from the test along with the corresponding *p*-value in parentheses. The null hypothesis is cross-sectional independence. Obs is observations

**Indicate statistical significance at the 5% level

***Indicate statistical significance at the 1% level

industrial and tertiary industry output on energy consumption were positive, whereas population density negatively affected energy use. The negative elasticity of population density to energy consumption in China was 0.211, which indicates that a 1% increase in population density decreased energy use by 0.211%. The elasticities of GDP per capita, proportion of industrial and tertiary industry output on energy use in China were 0.554, 0.524 and 0.968, respectively. These results suggested that economic growth and changes in industrial structures contributed to the increase in energy consumption in China over the last two decades.

Similar to the national scale, all estimated coefficients of population, income and technology were statistically significantly and negatively correlated with energy consumption in the three regions at the 1% level. The negative effect of population density increases on energy consumption varied across regions, with the greatest impact in the central region, followed by the western and eastern areas (Fig. 7.24). Specifically, a 1% increase in population density decreased energy consumption by 0.239%, 0.218% and 0.065% in the central, western and eastern regions, respectively. This result was consistent with that by Liddle (2004), who found that a 1% increase in the level of population density decreased road energy use by 0.31–0.51% in OECD countries. Due to the lack of basic public services in many cities of the western region, the negative population density elasticity to energy use was likely the result of modernization rather than of scale economies of public infrastructure (Poumanyvong and Kaneko 2010). Due to rapid economic growth and urbanization, eastern China still required many private and public infrastructures to maintain their citizen's lives and economic activities, and more energy resources may be required to construct, maintain and operate the infrastructure. Therefore, the negative elasticity of population density to energy consumption in the eastern region was less than that in the central and western regions.

The influence of other factors on energy use was positive and differed across regions. The elasticity of the GDP per capita to energy use was 0.456 in the eastern region, slightly larger than the elasticity in the western (0.423) and central regions (0.279). The result indicated that economic development may have a strong and positive impact on energy consumption in the three regions, and the impact was the greatest in the eastern region, followed by the western and central regions. The findings by Wang et al. (2012) and Poumanyvong and Kaneko (2010) supported this result. The proportion of industrial output had a positive and significant effect on energy consumption in the three regions. A 1% increase in the proportion of industrial output increased energy consumption by 0.948, 1.854 and 1.456% in the eastern, central and western

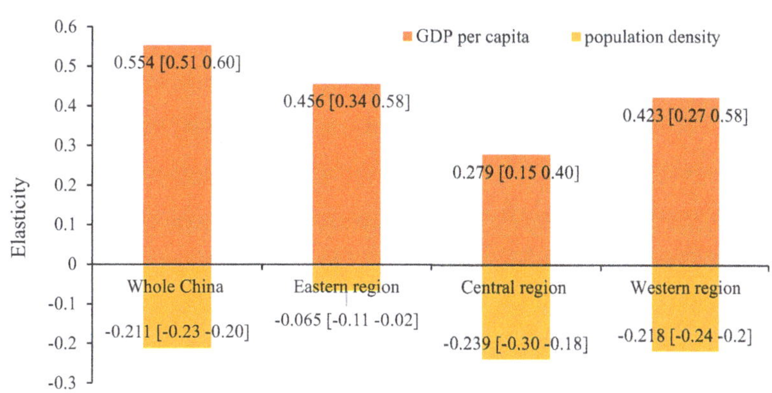

Fig. 7.24 Elasticity of population density and incomes per capita to energy consumption for the whole sample and the three regions (95% confidence intervals in brackets)

regions, respectively. The effect of the proportion of tertiary industry on energy use was also positive and significant in the three regions at the level of 5% or lower. As the eastern region was the first involving in the Reform and Open in China, this region acquired abundant capital due to strong government support and a good socio-economic base. Rapid economic growth, urbanization and industrialization thus increased energy consumption in eastern China. Both the central and western regions have abundant energy resources, especially coal. Coupled with the influence of the Central Rise Strategy and the Western Development Strategy, the economic growth and industrialization of the two regions has accelerated since 2005 and 2000, respectively, contributing to the increase of energy use (Zhang and Xu 2012).

7.5.3.2 Human Activities and Industrial Pollutant Emissions

1. **Industrial exhaust gases**

Table 7.25 provides the estimates of the impact of human activities on industrial exhaust gas emissions for the country and for the eastern, central and western regions. For the whole of China, exhaust gas emissions were statistically significantly and positively associated with the proportion of industrial output and energy intensity, whereas emissions were negatively correlated with population density at the 1% level. A 1% increase in population density decreased exhaust gas emissions by 0.209% in China. When other variables were kept constant, a 1% increase in energy intensity and the

Table 7.25 Estimation results for per capita industrial exhaust gas emission by DK models, 95% confidence intervals in brackets

Variable	Whole of China	Eastern region	Central region	Western region
Ln(POPD)	−0.209*** [−0.24 −0.18]	−0.094*** [−0.15 −0.04]	−0.030 [−0.19 0.13]	−0.035 [−0.09 0.02]
Ln(pGDP)	0.041 [−1.83 1.92]	−0.698 [−3.19 1.79]	−3.326*** [−5.03 −1.62]	−2.373** [−4.70 −0.05]
Ln $(pGDP)^2$	0.058 [−0.05 0.17]	0.105 [−0.03 0.24]	0.235*** [0.14 0.33]	0.223*** [0.09 0.36]
Ln(INDU)	0.558*** [0.44 0.67]	0.855*** [0.60 1.11]	2.063*** [1.27 2.86]	0.439*** [0.02 0.90]
Ln(TERT)	0.191 [−0.09 0.47]	0.271 [−0.12 0.67]	1.284 [0.71 1.86]	0.257 [−0.57 1.08]
Ln(IEI)	0.341*** [0.21 0.47]	0.64*** [0.29 1.00]	0.136*** [0.02 0.25]	0.617*** [0.41 0.83]
Constant	3.117 [−5.03 11.27]	3.527 [−7.19 14.25]	8.622*** [0.49 16.75]	10.826** [0.38 21.27]
CIPS	I(0)	I(0)	I(0)	I(0)
Mean rho	0.16	0.15	0.19	0.24
CD (p)	19.86 (0.11)	5.92 (0.21)	4.91 (0.18)	4.53 (0.07)
R^2	0.787	0.839	0.839	0.840
Obs	690	253	184	184

Notes The estimated method is pooled OLS. Ln() is natural logarithms; POPD is population density; pGDP is real GDP per capita; IEI is industrial energy intensity; INDU and TERT are proportion of industry and tertiary industries, respectively. The stationary of the residuals is determined from the Pesaran (2007) CIPS test and *I*(0) means stationary Mean rho is the mean absolute correlation coefficient of the residuals from the Pesaran (2004) CD test. CD is the test statistic from the test along with the corresponding pvalue in parentheses. The null hypothesis is cross-sectional independence. Obs is observations

**Indicate statistical significance at the 5% level

***Indicate statistical significance at the 1% level

proportion of industrial output increased exhaust gas emissions by 0.341% and 0.558%, respectively. Moreover, the impact of income per capita and proportion of tertiary industrial output on exhaust gas emissions in China were insignificant even at the 10% level.

At the regional scale, the increase in population density had a negative effect on exhaust gas emissions in the eastern region, whereas the impact in the central and western regions was insignificant. A 1% increase in population density decreased emissions by 0.094% in the eastern region (Fig. 7.25). Similar to the whole of China, the proportion of industrial output and energy intensity positively influenced exhaust gas emissions in the three regions. When the industrial energy intensity increased by 1%, exhaust gas emissions increased by 0.643, 0.136 and 0.617% in the eastern, central and western regions, respectively. The results indicated that the impacts of industrial energy intensity on exhaust gas emissions in eastern and western China were greater than that at the central region. The elasticity of exhaust gas emissions to the proportion of industrial output were 0.885, 2.063 and 0.439 in the eastern, central and western regions, respectively. Furthermore, with a quadratic EKC specification, the estimated coefficients on income per capita and its squared value in both the central and western regions were significant with a negative and positive sign, respectively. This could point to the existence of a U-shaped relationship between per capita incomes and exhaust gas emissions in central and western China. However, the impact of economic development on exhaust gas emissions in the eastern region was not statistically significant at the 10% level or higher.

2. **Industrial wastewater**

Table 7.26 presents the estimates of the effects of population, income and technology on wastewater discharge for the country and the three regions. For the whole of China, the estimated coefficients were all statistically significant at the 1% level, except for the coefficient for income per capita and its squared term. The elasticity of industrial wastewater discharge to population density, industrial output and energy intensity in China were 0.127, 0.188 and 0.142, respectively. This indicated that a 1% increase in population density increased wastewater discharge by 0.127% in China (Fig. 7.25). The proportion of

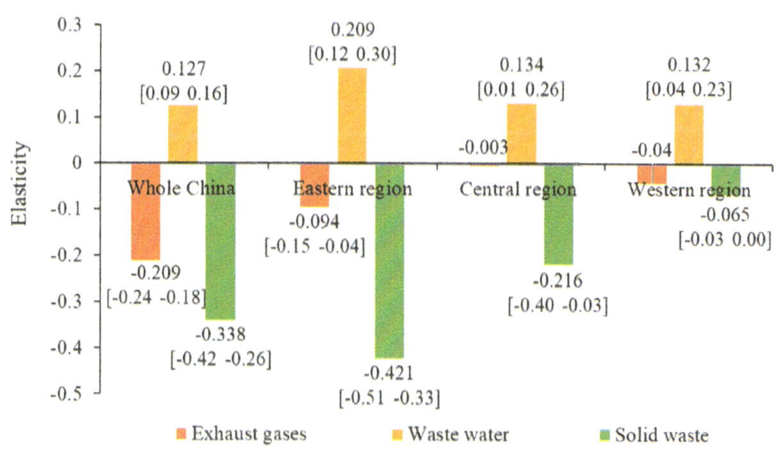

Note: The estimated elasticity of exhaust gases to population density in the central and western regions was listed but it is statistically insignificant at the level 10% or higher.

Fig. 7.25 Elasticity of population density to industrial pollutant emissions (exhaust gases, wastewater and solid waste) for the whole sample and the three regions (95% confidence intervals in brackets)

Table 7.26 Estimation results for per capita industrial wastewater discharge by DK models, 95% confidence intervals in brackets

Variable	Whole of China	Eastern region	Central region	Western region
Ln(POPD)	0.127** [0.09 0.16]	0.209** [0.12 0.30]	0.134** [0.01 0.26]	0.132*** [0.04 0.23]
Ln(pGDP)	0.153 [−1.32 1.63]	1.162 [−0.34 2.67]	−1.455* [−3.15 0.23]	−0.334 [−3.32 2.65]
Ln $(pGDP)^2$	0.004 [−0.08 0.09]	−0.067 [−0.15 0.01]	0.083* [−0.01 0.18]	0.043 [−0.13 0.21]
Ln(INDU)	0.188** [0.11 0.27]	1.046** [0.85 1.25]	−0.419 [−0.86 0.03]	−0.466 [−1.29 0.35]
Ln(TERT)	−0.583*** [−0.99 −0.18]	−0.171 [−0.65 0.30]	−0.506* [−1.02 0.01]	−0.796 [−2.19 0.60]
Ln(IEI)	0.142*** [0.04 0.24]	0.106 [−0.12 0.33]	−0.006 [−0.04 0.02]	0.304** [0.07 0.54]
Constant	1.710 [−5.86 9.28]	−6.542 [−13.62 0.54]	13.027*** [8.40 17.66]	5.806 [−5.40 17.01]
CIPS	I(0)	I(0)	I(0)	I(0)
Mean rho	0.25	0.18	0.16	0.43
CD (p)	3.47 (0.18)	5.97 (0.24)	4.74 (0.26)	4.61 (0.08)
R^2	0.219	0.262	0.839	0.116
Obs	690	253	184	184

Notes The estimated method is pooled OLS. Ln() is natural logarithms; POPD is population density; pGDP is real GDP per capita; IEI is industrial energy intensity; INDU and TERT are proportion of industry and tertiary industries, respectively. The stationary of the residuals is determined from the Pesaran (2007) CIPS test and *I*(0) means stationary. Mean rho is the mean absolute correlation coefficient of the residuals from the Pesaran (2004) CD test. CD is the test statistic from the test along with the corresponding *p*-value in parentheses. The null hypothesis is cross-sectional independence. Obs is observations

*Indicate statistical significance at the 10% level

**Indicate statistical significance at the 5% level

***Indicate statistical significance at the 1% level

tertiary industry output had a negative effect on industrial wastewater discharge in China, and a 1% increase in tertiary industry output resulted in a 0.58% decrease in the wastewater discharge.

At the regional level, population density also positively influenced wastewater discharge. The elasticity of population density to wastewater discharge in eastern, central and western China were 0.209, 0.134 and 0.132, respectively, and a 1% increase in population density resulted in a 0.13%–0.21% increase in wastewater discharge. The effect of energy intensity and economic structure on wastewater discharge was also heterogeneous across regions. The proportion of industrial output was positively correlated with the wastewater discharge in the eastern region, but correlations in the central and western regions were not significant. The significant positive correlation between energy intensity and wastewater discharge was detected only in the western region. Furthermore, the coefficients of income per capita and its squared term to wastewater discharge in the central region were statistically significant at the level of 10% with a negative and positive sign, respectively, indicating a U-shaped relationship. However, the correlations between per capita incomes and wastewater discharge in eastern and western China were statistically insignificant at the 10% level or higher.

3. **Industrial solid waste**

Table 7.27 lists the estimates of the impact of population and other variables on industrial solid

Table 7.27 Estimation results for per capita industrial solid waste production by DK models, 95% confidence intervals in brackets

Variable	Whole of China	Eastern region	Central region	Western region
Ln(POPD)	−0.338*** [−0.42 −0.26]	−0.421*** [−0.51 −0.33]	−0.216** [0.01 0.26]	−0.065** [−0.13 0.00]
Ln(pGDP)	−0.440 [−2.46 1.58]	−3.090** [−5.63 −0.55]	−3.771*** [−6.14 −1.40]	−3.529*** [−4.53 −2.53]
Ln $(pGDP)^2$	0.059 [−0.06 0.17]	0.225*** [0.09 0.36]	0.248*** [0.11 0.39]	0.259*** [0.20 0.32]
Ln(INDU)	0.971*** [0.88 1.06]	1.148*** [0.81 1.49]	1.490*** [0.62 2.36]	1.416*** [0.95 1.89]
Ln(TERT)	0.339 [−0.07 0.75]	0.750** [0.15 1.35]	1.561*** [0.75 2.38]	0.714** [−0.11 1.54]
Ln(IEI)	0.151*** [0.09 0.21]	0.734*** [0.35 1.12]	0.126*** [−0.04 0.02]	0.251*** [0.11 0.39]
Constant	−3.867 [−11.61 3.88]	4.235 [−6.43 14.90]	3.996*** [−6.78 14.77]	3.607** [0.0 17.21]
CIPS	I(0)	I(0)	I(0)	I(0)
Mean rho	0.19	0.17	0.19	0.25
CD (p)	18.99 (0.14)	2.78 (0.25)	1.83 (0.10)	3.02 (0.34)
R^2	0.543	0.620	0.521	0.704
Obs	690	253	184	184

Notes The estimated method is pooled OLS. Ln() is natural logarithms; POPD is population density, pGDP is real GDP per capita; IEI is industrial energy intensity; INDU and TERT are proportion of industry and tertiary industries, respectively. The stationary of the residuals is determined from the Pesaran (2007) CIPS test and *I*(0) means stationary. Mean rho is the mean absolute correlation coefficient of the residuals from the Pesaran (2004) CD test. CD is the test statistic from the test along with the corresponding p-value in parentheses. The null hypothesis is cross-sectional independence. Obs is observations
**Indicate statistical significance at the 5% level
***Indicate statistical significance at the 1% level

waste production for whole China and its three regions. For the whole of China, energy intensity and the proportion of industrial output positively influenced industrial solid waste production, whereas population density negatively affected industrial solid waste production. A 1% increase in population density decreased industrial solid waste production by 0.338%. The elasticities of industrial solid waste production to energy intensity and the proportion of industrial output were 0.971 and 0.151, respectively. This indicated that energy intensity had a strong and positive influence on solid waste production in China. The elasticity of solid waste production to the proportion of tertiary industry output was positive but statistically not significant. Furthermore, the correlation between economic growth and solid waste production was also not significant at the 10% level or higher.

The estimated coefficients were all statistically significant at the 5% level or lower for the three regions. Higher population density had a negative influence on solid waste production at the regional level. With a 1% increase in population density, solid waste production in the eastern, central and western regions decreased by 0.421%, 0.216% and 0.065%, respectively. This result was supported by Liddle (2014), who also found a negative relationship between population density and pollutant emissions. The energy intensity and the proportion of industrial and tertiary industry output had a positive impact on solid waste production in the three regions. Specifically, the elasticity of solid waste production to industrial output in the central region was 1.49, greater than in the western region (1.46) or in the eastern region (1.15). Similar to the industrial output, the impact of the proportion of the tertiary industry output on solid waste production in the central region was greater than that in the eastern and western regions. A 1% increase in industrial energy intensity increased solid waste production by 0.73, 0.13 and 0.25% in eastern central and western China, respectively. Interestingly, both per capita income and its squared value for the three regions were found to be statistically significant at the 5% level or lower. Solid waste production first decreased sharply and then increased with wealth, demonstrating a U-shaped relationship between economic growth and solid waste production at the regional scales.

It can be seen that the effect of the increase in population density on per capita energy consumption and pollutant emissions in China varied across regions (Figs. 7.31 and 7.32). For the entire country, higher population density decreased energy consumption and the emissions of industrial exhaust gases and wastewater but increased solid waste production. At the regional scale, elevated population density decreased energy demand and exhaust gas emissions and solid waste production but increased wastewater discharge in the three regions. Economic growth promoted to the increase in energy consumption in the eastern and western regions, but it decreased exhaust gas emissions and solid waste production in the three regions. Further economic development would increase the emissions of two types of pollutants in the three regions. Economic development first decreased wastewater discharge but then increased discharge in the central region over time. The proportion of industrial output and tertiary industry output had positive influences on energy consumption and solid waste emissions in the three regions, but their impact on wastewater discharge varied across regions. These findings also further confirmed the fact that the impact of population, affluence and technology on pollutant emissions depended on the level of development and the type of pollutants (Fan et al. 2006; Chen et al. 2008). To reduce pollutant emissions, we should exchange the former pattern of high pollution and energy use for economic growth. Meanwhile, we should comprehensively consider the factors affecting the discharge of industrial pollutants and develop effective emission reduction measures to realize the sustainable development of population, resources and environment.

7.5.3.3 The EKC Relationships Between Per Capita Income and Industrial Pollutant Emissions

The relationship between economic growth and environmental quality depended on the type of pollutants in China. The abovementioned DK model estimates suggested that the EKC relationship between economic development and emissions of three industrial wastes were not obvious (Table 7.29). This result was supported by Managi and Kaneko (2009), who suggested that no single relationship existed between environmental quality and income per capita. However, our findings did not in agree with those of a previous study that showed that inverted U-shaped relationships existed for industrial exhaust gases, wastewater and solid waste in China (Song et al. 2008). The possible reason behind this difference is that we also considered additional impact factors (e.g., demographic changes, industrial structures, and industrial energy intensity) on pollutant emissions, in addition to economic growth.

The relationship between economic growth and industrial pollutant emissions were also heterogeneous across regions and depended on the type of pollutants (Table 7.28). In eastern China, a U-shaped relationship was found for solid waste. However, the linkages between per capita incomes and exhaust gases as well as wastewater discharge were not obvious. In central China, there were U-shaped relations between per capita incomes and three industrial wastes. In western China, U-shaped relationships existed between per capita incomes and exhaust gases or solid waste.

7.5.4 Policy Implications

The effects of technology factors on energy consumption and pollutant emissions in China also showed an apparent heterogeneity across regions.

Table 7.28 Relationships between economic growth and three types of industrial pollutant emissions

Sample	Exhaust gases	Waste water	Solid waste
Whole of China	Not obvious[a]	Not obvious	Not obvious
Eastern region	Not obvious	Not obvious	U-shaped
Central region	U-shaped	U-shaped	U-shaped
Western region	U-shaped	Not obvious	U-shaped

Notes The results obtained from the above DK models are based on the estimated coefficients for per capita income and their squared values

[a]Not obvious means that the estimated coefficients for income per capita and their squared terms were not statistically significant at the 10% level or higher

The proportion of industrial output had a significant and positive influence on energy consumption and industrial pollutant emissions in whole of China and its three regions. The proportion of the tertiary industry output was also positively correlated with energy consumption in the three regions of China, but its impact on industrial exhaust gas emissions was insignificant. These results demonstrated that improving energy intensity, changing the energy consumption structure and upgrading industrial structures may still be necessary and useful to mitigate the emission of industrial pollutants in China.

The relationship between economic growth and environmental impact depended on the type of pollutants and differed across regions as well. No strong evidence was found to support the existence of the EKC hypothesis between three industrial waste emissions and incomes per capita at the regional scale. However, U-shaped curves existed for solid waste production in the eastern region, three industrial wastes in the central region, and exhaust gases and solid waste in the western region.

Our empirical results will be of special interest to policy makers and urban planners in China. Urbanization is expected to continue to be one of the primary driving forces for future economic growth in China. Currently, the Chinese government proposes a people-oriented urbanization concept in the National New Type Urbanization Plan (2014–2020). One of the important targets of new-type urbanization is to gradually settle the rural population that has migrated to cities, which will inevitably require additional urban infrastructure and possibly more energy resources and thus results in the emission of more pollutants. It is thus essential for the government of China to take measures to control the pace of the urbanization process, to develop more sustainable infrastructure systems, and to improve energy efficiency as well as upgrade industrial and production structures. In the context of such rapid industrialization and urbanization, policymakers also require more detailed information about the complex linkages between human activities and environmental impact. Moreover, urban planning and management policymakers should take regional pollutant reduction priorities into consideration. Considering the strongest impact of population, income and technology on pollutant emissions, eastern China should focus its efforts on industrial wastewater and exhaust gases, whereas central and western China should pay more attention to solid waste.

7.6 Effects of Urban–Rural Transformation on Energy Consumption and CO_2 Emissions

7.6.1 Theoretical Analysis: Urban–Rural Transformation and Energy and CO_2 Environmental Effects

China's rapid industrialization and urbanization have caused an excessive consumption of resources and environmental degradation.

Existing studies have extensively discussed the impact of urbanization on energy consumption and CO_2 emissions at different spatial scales. Wei et al. (2007) used a consumer lifestyle approach to quantify the direct and indirect impact that the lifestyles of urban and rural residents had on China's energy consumption and CO_2 emissions. Urbanization slowed per capita residential energy consumption growth in China, but it contributed to an increase of residential indirect CO_2 emissions. Using the STIRPAT model, Zhang and Lin (2012) found that urbanization increased energy consumption and CO_2 emissions in China. In addition, Zhou et al. (2013) used provincial panel data to analyze the relationship between industrial structure transformation and CO_2 emissions. Their research shows that the upgrading and optimization of China's industrial structure have inhibited CO_2 emissions, and confirmed that China's urbanization process has increased energy consumption and CO_2 emissions.

These studies provide a scientific basis for effective energy saving and emission reduction strategies for different countries and regions. However, most of the existing studies focus on the impact of urbanization on energy consumption and CO_2 emissions, and have not fully consider other factors in the process of urban and rural development transformation (RUDT), such as employment transformation, industrial structure transformation and land use pattern change. In addition, there are obvious regional differences in natural geographical environment and economic development in China, and there are regional differences in a series of environmental problems brought about by the RUDT. A better understanding of the relationships among RUDT, energy consumption and CO_2 emissions with consideration of regional differences is still necessary and urgent. Thus, the main aim of this study was to examine the impacts of RUDT on energy consumption and CO_2 emissions in China. Based on a balanced panel dataset of 30 provinces in China from 1990 to 2015, this chapter used an improved STIRPAT model to investigate the effects of RUDT on energy use and CO_2 emissions with full consideration of regional differences.

7.6.2 Empirical Methods and Data

1. **Data description**

This paper used a balanced panel dataset of 30 provinces in China over the period 1990–2015 (Hong Kong, Macao, Taiwan and Tibet are not included for lack of data). The data of the provincial population size, GDP, urban population, the added values of secondary and tertiary industry, the percentages of population employed in secondary and tertiary industry, and the per capita disposable income of urban residents and the per capita net income of rural dwellers were obtained from the China Statistical Yearbook and the China Compendium of Statistics. Cultivated land occupied by construction was collected from the China Land and Resources Almanac. The data on energy consumption were obtained from the China Energy Statistical Yearbook. The energy-related CO_2 emissions were calculated according to the formula of CO_2 emissions in the Intergovernmental Panel on Climate Change (IPCC) (Provided by website: http://www.ipcc-nggip.iges.or.jp/). To eliminate the influence of price index, the GDP was calculated at the constant price (1990 prices) and the added values in secondary and tertiary industries were calculated accordingly. Table 7.29 provides a detailed description of all variables used in this study.

2. **Measurement of RUDT**

In this study, measuring RUDT primarily involved the transformation of demographic, employment and industrial structures in rural and urban areas and the land use pattern transitions. The change in the proportion of the urban population relative to the total population (urbanization) was used to reflect the demographic transition (DT) between rural and urban regions. Percentages of the population employed in secondary and tertiary industries were chosen to

Table 7.29 Description of the variables used in the study for the period 1990–2015

Type	Variable	Definition	Unit
Environment impact	Carbon dioxide emissions (CO_2)	Emission from fuel combustion	10^4 tons
	Energy intensity (EI)	Energy consumption divided by GDP	Tec/10^4 yuan
	Energy consumption (EC)	Sum of energy use	10^4 tec
Demographic transformation (DT)	Urbanization (URBA)	Percentage of the urban population in the total population	Percent
Employment structural transformation (EST)	Share of secondary industry employment (EP2)	Percentage of population employed in secondary industry	Percent
	Share of tertiary industry employment (EP3)	Percentage of population employed in tertiary industry	Percent
Industrial structural transformation (IST)	Share of secondary industry in GDP (INDU)	Percentage of the added value of secondary industry to GDP	Percent
	Share of tertiary industry in GDP (TERT)	Percentage of the added value of tertiary industry to GDP	Percent
Land use transformation (LUT)	Farmland conversion (LAND)	Cultivated land occupied by construction	Hectare
Rural–urban disparity	Rural–urban income gap (URI)	Ratio of the per capita disposable income of urban residents to the per capita net income of rural households	Ratio

measure the employment structural transformation (EST), and percentages of the added value of secondary and tertiary industries in GDP were used to measure the industrial structural transformation (IST).

Land use transition (LUT) reflects the changes in land use morphology (including dominant and recessive morphology) of a certain region over a certain time period driven by socio-economic change and innovation (Long 2014). Dominant morphology refers to the quantity, structure and spatial pattern of land, and recessive morphology includes land use features, e.g., quality, property rights, price, and management mode. The dominant form refers to the spatial pattern of quantity, structure and land; the recessive form includes the characteristics of land use, such as quality, property rights, price and management mode. Due to the limitation of data availability, we focused on only one aspect of the changes in dominant morphology, i.e., cultivated land occupied by construction land. In the process of RUDT, rapid urban population growth and economic growth, as well as the construction and maintenance of urban infrastructure, are likely to cause more energy consumption and CO_2 emissions.

3. **STIRPAT model**

The effect of demographic (P), economic (A) and technologic (T) factors on the environment is mainly postulated in the IPAT model (I = PAT), which is firstly proposed by Ehrlich and Holdren (1971). The IPAT model is limited because it analyzes a problem by changing a factor while keeping others constant, resulting in proportionate impacts on the dependent variable. To overcome these weaknesses, Dietz and Rosa (1994) reformulated the IPAT model into a stochastic model (STIRPAT), which can statistically assess non-monotonic or non-proportional impacts of driving forces on the environment. York et al. (2003) refined the STIRPAT model and expanded it by incorporating additional factors, such as quadratic terms or different components of P, A or T. Presently, the STIRPAT model has been successfully utilized to analyze the effects of driving forces on a variety of environmental impacts (Wang et al. 2012, 2013), it can be given as the following equation:

$$I_{it} = \alpha P_{it}^{b} A_{it}^{c} T_{it}^{d} \varepsilon_{it} \quad (7.26)$$

$$Ln(I_{it}) = \alpha + bLnP_{it} + cLnA_{it} + dLnT_{it} + \varepsilon_{it} \quad (7.27)$$

where suffixes *i* and *t* refer to provinces and years, respectively; *P* is population size; *A* is real GDP per capita; *T* represents technology; and the dependent variable *I* stands for pollutant emissions; ε_i is the error term, *a* is the constant term; *b*, *c* and *d* are, respectively, the coefficients of *P*, *A* and *T*.

The STIRPAT model allows factors to be added to assess their impact on the environment. In the STIRPAT model, factors *P* and *A* are decomposable, as is *T*. To further examine the impact of RUDT on energy consumption and CO_2 emissions, we extended the STIRPAT model by incorporating the DT, IST, EST, LUT, and energy intensity factors into the model. The improved STIRPAT model was changed as follows:

$$\begin{aligned} Ln(EC_{it}) = \alpha_0 &+ \alpha_1 Ln(EP2_{it}) + \alpha_2 Ln(EP3_{it}) \\ &+ \alpha_3 Ln(URBA_{it}) + + \alpha_4 Ln(URI_{it}) \\ &+ \alpha_5 Ln(INDU_{it}) + \alpha_6 Ln(TERT_{it}) \\ &+ \alpha_7 Ln(LAND_{it}) + \varepsilon_{1it} \end{aligned} \quad (7.28)$$

$$\begin{aligned} Ln(I_{it}) = b_0 &+ b_1 Ln(EP2_{it}) + b_2 Ln(EP3_{it}) \\ &+ b_3 Ln(URBA_{it}) + b_4 Ln(URI_{it}) \\ &+ b_5 Ln(INDU_{it}) + b_6 Ln(TERT_{it}) \\ &+ b_7 Ln(LAND_{it}) + b_8 Ln(EI_{it}) + \varepsilon_{1it} \end{aligned} \quad (7.29)$$

where EP2 and EP3 represent EST (expressed as the percentages of population employed in secondary and tertiary industries, respectively); URBA represents DT (expressed as the percentage of the population living in urban areas); INDU and TERT represent IST (expressed as shares of the added value of secondary and tertiary industry to the GDP, respectively); URI is the rural–urban income gap; LAND represents LUT (expressed as the cultivated land area occupied by construction); EI is energy intensity (expressed as the energy use per unit of GDP); EC denotes energy consumption; and I represents the environmental impact (proxied by energy-related CO_2 emissions). The energy intensity variable is included in only the emission models because it contains part of the dependent variable.

4. **Model estimation**

First, the LLC, IPS, Fisher-ADF and Fisher-PP statistics were used to test for the presence of the panel unit root (Maddala and Wu 1999; Levin et al. 2002; Im et al. 2003; Choi 2001). The panel unit root test results show that many variables are non-stationary in their levels, but most of them become stationary at the 5% significance level after taking first differences (Table 7.30). Second, we estimated the whole sample without consideration of regional differences. Furthermore, to investigate whether the impact of RUDT on the environment varies across regions in China, the whole sample was divided into three regions, i.e., the eastern region (including Liaoning, Beijing, Tianjin, Hebei, Shandong, Jiangsu, Shanghai, Zhejiang, Guangdong, Fujian and Hainan), the central region (including Heilongjiang, Jilin, Shanxi, Henan, Anhui, Jiangxi, Hubei and Hunan) and the western region (including Xinjiang, Gansu, Qinghai, Inner Mongolia, Ningxia, Shaanxi, Sichuan, Chongqing, Guizhou, Yunnan and Guangxi). The classification by region is common in China, considering China's economic development level and geographic position. The impact of the RUDT on energy use and CO_2 emissions for the whole sample and for the three regions were estimated using three methods: fixed effects (FE), the feasible generalized least squares (FGLS) and the linear regression with Driscoll-Kraay standard errors (DK), generating twenty-four models. Specifically, the FE estimation was applied to alleviate heterogeneity

Table 7.30 Panel unit root tests

Variable	Levels				First differences			
	LLC	IPS	ADF	PP	LLC	IPS	ADF	PP
Ln(EC)	−0.82	−0.13	58.75	116.50***	−10.69***	−9.04***	182.47***	290.77***
Ln(CO_2)	−0.36	−0.13	63.95	44.88	−9.17***	−8.66***	176.63***	235.00***
Ln(EI)	−3.47***	−5.47***	130.56***	88.64***	−12.71***	−11.28***	202.66***	274.02***
Ln(EP2)	−0.30	0.05	65.48	23.52	−7.68***	−9.19***	190.82***	306.73***
Ln(EP3)	−3.12***	−0.85***	99.04***	62.41***	−13.07***	−14.65***	305.67***	426.65***
Ln (URBA)	−0.77	−0.16	54.55	45.43	−16.86***	−12.89***	252.29***	323.56***
Ln(URI)	−2.15***	−4.60***	105.41***	106.32***	−12.92***	−8.95***	187.78***	300.52***
Ln (INDU)	−2.62***	−3.27***	101.68***	53.68	−15.29***	−13.96***	285.85***	410.55***
Ln (TERT)	−6.07***	−6.32***	158.87***	290.73***	−12.35***	−19.64***	393.05***	1552.77***
Ln (LAND)	−6.22***	−6.88***	156.83***	172.48***	−16.04***	−18.42***	373.22***	1360.26***

Notes Exogenous variables: individual effects and individual linear trends. Automatic selection of lags based on AIC: 0 to 4. Newey-West automatic bandwidth selection and Bartlett Kernel. ***denotes rejection of the null hypothesis of nonstationary at the 1% significance level

bias. The autocorrelation in the fixed effect models was verified by the Wooldridge test and the autocorrelation in the fixed effect models was found (Wooldridge 2002). Moreover, we confirmed the presence of groupwise heteroskedasticity in all modes using the modified Wald statistic (Greene 2000). We also detected the existence of cross-sectional dependence by Pesaran test in the fixed effect models (Pesaran 2004). The FGLS estimation was employed to address those issues. However, the FGLS method is infeasible if the panel's time dimension T is smaller than its cross-sectional dimension N, and it tends to produce unacceptably small standard error estimates (Beck and Katz 1995). To solve the issues, the Driscoll-Kraay estimation was applied. Besides being heteroscedasticity consistent, the Driscoll-Kraay standard error estimates are robust to general forms of cross-sectional and temporal dependence (Hoechle 2007). Additionally, the multicollinearity among the explanatory variables in the FD models was tested by the variance inflation factors (VIF). The VIF values are all less than 10, indicating no multicollinearity.

7.6.3 Energy Consumption and CO_2 Emission Effect of Urban–Rural Transition

1. **Transformation of urban and rural development**

RUDT involves population agglomeration to cities, the increase in the number of cities and the population growth of individual cities. As a country with a large population in the world, China has experienced unprecedented urbanization process in recent decades. The urban population increased from 302 million in 1990 to 770 million in 2015, and the urban rate increased from 26.41% in 1990 to 56.1% in 2015. At the same time, China's rural population decreased from 841 million in 1990 to 603 million in 2015, an average annual decrease of 1 percentage point. In 2011, the urban population surpassed the rural population for the first time. The migration between provinces is positively stimulated by the income gap between urban and rural areas, but also negatively affected by the geographical distance from the destination. The total number of interprovincial migrations in China has increased significantly from 10.7 million in 1990–1995 to 55.2 million in 2005–2010. In the past three decades, eastern provinces have been the main destinations for interprovincial migration, with high net population migration, especially to coastal cities and megacities.

The transformation from rural economy to urban economy is a part of RUDT. In the past two decades, great changes have taken place in China's industrial structure, which is reflected in the continuous decline in the proportion of the primary industry and the continuous increase in the proportion of non-agricultural industry output value. In particular, the rapid development of the tertiary industry has promoted the continuous optimization of the proportion of industrial structure. Specifically, from 1990 to 2015, the proportion of added value of the three industries increased from 26.6:41.0:32.4 in 1990 to 9.0:40.5:50.0 in 2015, and the proportion of tertiary industry increased from 32.4% to 50%. RUDT is accompanied by the transformation of agricultural production activities to non-agricultural activities with higher economic benefits. The change of China's economic structure and industrial distribution has led to great changes in China's employment structure, which also means that more and more people are engaged in the secondary and tertiary industries. In 1990, the proportion of the employment structure of the three industries was 60.1:21.4:18.5, and by 2015, the proportion of the employment structure of the three industries reached 28.3:29.3:42.4.

Land use is a mirror of society. Many social, economic and environmental problems brought about by RUDT can be reflected in land use. Since 1998, the Chinese government has implemented strict restrictions and highly centralized policies and regulations on the conversion of cultivated land to non-agricultural construction land. However, in China's rural areas where a large number of cultivated lands is converted to non-agricultural construction land, illegal land use and construction are still common (Tang and Chung 2002). From 1996 to 2008, China's cultivated land converted by construction land increased from 83,800 hectares to 259,000 hectares, with an increase of 7640 hectares per year. The amount of agricultural land occupied by construction increased sharply from 1.3 million hectares in 1990–2000 to more than 2 million hectares in 2000–2010. Among them, from 1990 to 2015, the total cultivated land occupied by construction in eastern China was about 2.25 million hectares, while that in central and Western China was only 1.44 million hectares and 1.31 million hectares. Therefore, economic growth, population transfer and industrial structure adjustment are the main reasons for the rapid transformation to non-agricultural land (Liu et al. 2014b).

2. **Energy consumption and CO_2 emission**

The rapid growth of energy demand made China the world's largest CO_2 emitter in 2007 and the world's largest energy consumer in 2010. From 1990 to 2015, China's total primary energy consumption increased from 969 million tons of standard coal to 4473 million tons of standard coal in 2015, with an average annual growth rate of 6.3 percent (Fig. 7.26a). In 2012, China's energy consumption accounted for 22% of the world's primary energy. It is estimated that by 2030, China's primary energy consumption will be twice that of 2010. Due to the structure of domestic energy reserves, China's energy consumption relies heavily on coal, which has become the main cause of high air pollution in China. China's CO_2 emissions increased from 683 million tons in 1990 to 2.782 billion tons in 2012 (Fig. 7.26a). In the past two decades, China's CO_2 emissions have grown relatively fast. It is estimated that by 2030, China's primary energy demand and CO_2 emissions will reach 22.8 and 31% of the world total respectively (IEA 2012b). To better meet the international restrictions on greenhouse gas emissions, the Chinese government has set a target in the "Twelfth Five-Year Plan", which will reduce energy consumption per unit GDP by 16% and CO_2 emissions by 17% in 2015 compared with 2010 (Wang et al. 2011).

In 1990 and 2015, energy consumption in eastern, central and western China showed an increasing trend (Fig. 7.26b). The average annual

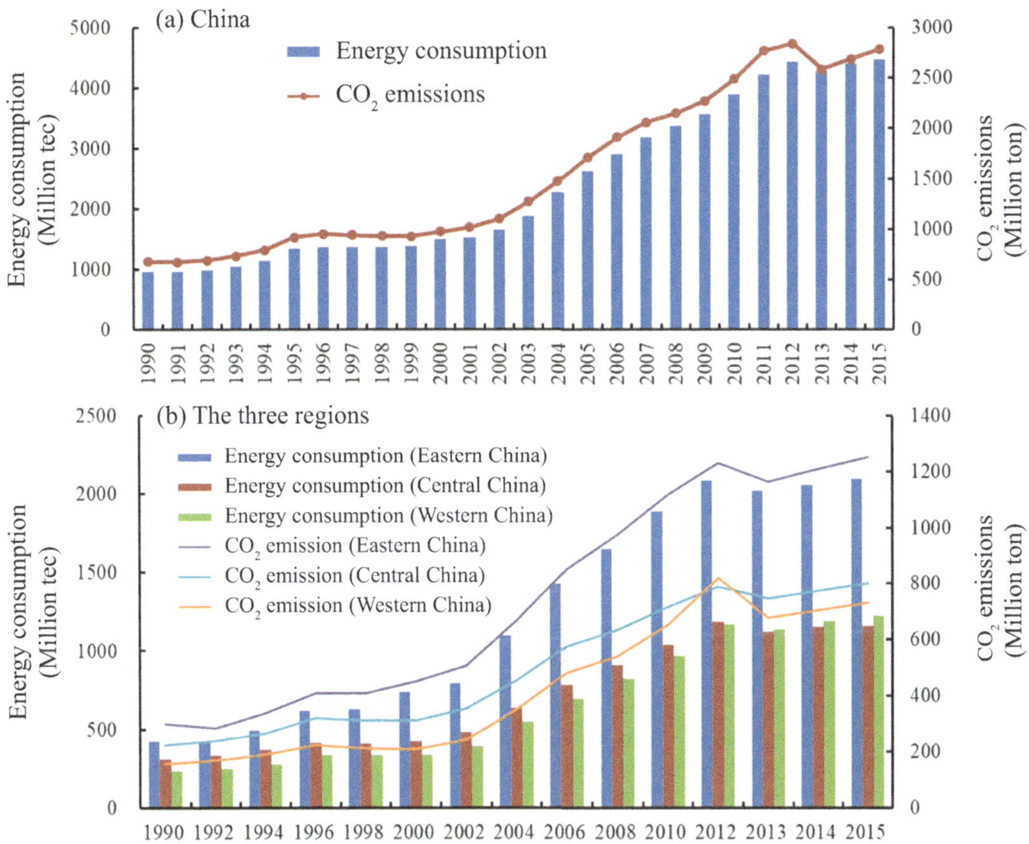

Fig. 7.26 Change in energy consumption and CO_2 emission in different regions

growth rates of energy use in the eastern, central and western regions are 6.59, 5.39 and 6.81%. The results show that in the past 20 years, the growth rate of energy consumption in Western China is higher than that in eastern and central China. Accordingly, from 1990 to 2015, energy-related CO_2 emissions in the eastern region remained at a relatively high level. In 2012, the CO_2 emission in the western region exceeded that in the eastern region, which may be related to the higher growth rate of energy consumption in the western region in recent years.

3. **The Impact of RUDT on energy consumption**

Table 7.31 shows the assessment of the impact of urban–rural transformation on energy consumption at the national and regional scales. The fitting degree of DK model is higher than other models, and the analysis results are based on this model. At the national level, population transformation has a positive impact on China's energy consumption. The energy consumption elasticity of China's urbanization is 0.744, which indicates that energy consumption increases by 744 tons of standard coal for every 1% increase in urbanization level. The elasticity index of employment ratio of tertiary industry to energy consumption is 0.421, which indicates that the increase of non-agricultural employment increases China's energy consumption. Land conversion has a significant impact on China's energy consumption, and the energy consumption increases by 1247 tons of standard coal per unit of cultivated land for construction. Furthermore, the urban–rural income gap will increase energy consumption. For every unit of urban–rural

Table 7.31 Estimated results of energy consumption at national and regional scales

Variable	China			Eastern region		
	FE (1)	FGLS (2)	DK (3)	FE (4)	FGLS (5)	DK (6)
Ln(EP2)	−0.198	0.386***	−0.082	−0.135	−0.674***	−0.941***
Ln(EP3)	0.421***	0.107***	0.180	1.181***	−0.673***	−0.646
Ln(URBA)	0.744***	0.368***	0.379***	0.536***	0.523***	0.488***
Ln(URI)	0.436***	0.277***	0.334**	0.315***	1.101 ***	1.131***
Ln(INDU)	1.247***	1.235***	1.238***	0.852***	2.437***	2.826***
Ln(TERT)	0.855***	0.350***	0.391	0.862***	2.424***	1.962**
Ln(LAND)	0.027*	0.472***	0.504***	−0.64***	0.371***	0.369***
Constant	−3.183***	−2.868***	−3.571**	−3.898***	−6.849***	−11.054***
R^2	0.858	–	0.744	0.904	–	0.789
Autocorrelation test	F(1, 29) = 152.407***			F(1, 10) = 19.33***		
Cross-sectional dependence test	43.657***			15.53***		
Heteroskedasticity test	$x^2(30)$ = 547.07***			$x^2(11)$ = 180.58***		
Observations	780	780	690	286	286	286
Variable	Central region			Western region		
	FE (7)	FGLS (8)	DK (9)	FE (10)	FGLS (11)	DK (12)
Ln(EP2)	−0.379***	0.009	−0.089	−0.224	−0.288***	−0.445**
Ln(EP3)	0.362***	−0.659***	−0.942**	0.251***	0.476***	0.706***
Ln(URBA)	1.134**	0.664***	0.759***	0.538***	0.236***	0.236**
Ln(URI)	0.226**	0.727***	0.736***	0.324***	0.089	0.067
Ln(INDU)	0.387**	1.365***	1.632***	1.929***	1.156***	1.139***
Ln(TERT)	0.045	1.378***	1.640***	0.471***	−0.569***	−0.949**
Ln(LAND)	0.059**	0.145***	0.238***	0.174***	0.468***	0.582***
Constant	2.545***	−3.136***	−5.039***	−4.177***	0.707	0.969
R2	0.891	–	0.796	0.904	–	0.808
Autocorrelation test	F(1, 7) = 62.63***			F(1, 10) = 86.88***		
Cross-sectional dependence test	14.21***			8.09***		
Heteroskedasticity test	$x^2(8)$ = 37.28***			$x^2(11)$ = 10.14***		
Observations	208	208	208	286	286	286

Notes FE (fixed effects); FGLS (feasible generalized least squares); DK (regression with Driscoll-Kraay standard errors). ***, ** and * denote rejection of the null hypothesis of nonstationary at the 1%, 5% and 10% significance level, respectively

income gap, energy consumption will increase by 436 tons of standard coal.

At the regional level, models 6, 9 and 12 show that the impact of RUDT on energy consumption varies with regions. The elasticity coefficients of urbanization to energy consumption in eastern, central and western regions are 0.488, 0.759 and 0.706, respectively. The increase of the employment ratio of the tertiary industry will increase the energy consumption in the western region and reduce the energy consumption in the central region, but the impact on the energy consumption in the eastern region is not obvious; the expansion of the urban–rural income gap will increase the energy consumption in the eastern, central and western regions; the increase in the proportion of the industrial industry will significantly increase the energy consumption of the three regions and the proportion of the output value of the tertiary industry The increase of energy consumption will increase the energy consumption in the eastern and central regions, but the impact on the energy consumption in the western region is not obvious. Among them, the elasticity of energy consumption increased by the added value of the secondary and tertiary industries in the eastern region is 2.826 and 1.962, that of the central region is 1.632 and 1.640, and that of the western region is 1.139, −0.949, which indicates that there are regional differences in the impact of industrial transformation on energy use, that is, the impact on the eastern region is significantly greater than that on the central and western regions. The regression coefficients of urban–rural income gap to energy consumption in eastern and central China are 1.131 and 0.736, respectively, while those in western region are not significant. This difference may be due to the relatively small income gap between urban and rural areas in the eastern region. Compared with the central and western regions, the eastern region needs more energy resources to support and maintain the normal operation of public infrastructure. In addition, the elasticity coefficients of cultivated land occupied by construction in the eastern, central and western regions are 0.369, 0.238 and 0.582 respectively, indicating that the impact of land use transformation on energy consumption in different regions is different.

4. **The impact of RUDT on CO_2 emissions**

Table 7.32 shows the impact of the RUDT on CO_2 emissions at national and three regional scales. According to model 15, the elasticity coefficient of population transformation, industrial transformation and land use transformation for CO_2 emission is positive and significant at the 5% level. The results showed that the CO_2 emission effect of China's urban–rural transformation is significant, and the carbon emission effect of energy consumption intensity is also significant.

The impact of RUDT on CO_2 emissions effects of urban–rural development transition are significantly different in regions. The urbanization process in the eastern and central regions increased CO_2 emissions, but it was not significant in the western regions. For every one percentage point increase in the urbanization rate in the eastern and central regions, CO_2 emissions will increase by 43.1 thousand tons and 0.401 thousand tons respectively. The carbon emission effect of the increase in the proportion of employment in the secondary industry presents different patterns in the eastern, central and western regions; the increase in employment in the tertiary industry will promote the increase of carbon emissions in the western region, but the carbon emission effect in the central and eastern regions is not obvious. In terms of industrial transformation, an increase in the output value of the secondary industry will increase CO_2 emissions in the three regions of the east, central and west, and an increase in the output value of the tertiary industry will increase carbon emissions in the eastern and central regions, but will reduce carbon emissions in the western region. The increase in the output value of the secondary industry has an effect on carbon emission elasticity coefficients of 3.344, 2.036, and 1.044 in the eastern, central, and western regions, respectively, indicating that the increase in the output value of the secondary industry has a significantly higher carbon emission effect on the eastern region than in the central and western regions; Increasing carbon emissions in the eastern, central, and western regions has a higher carbon emission effect in the western region than in the eastern and central regions; the expansion of urban–rural income gap and the increase in

Table 7.32 Estimation result for CO_2 emission models (the whole sample and the eastern region)

Variable	China			Eastern region		
	FE (13)	FGLS (14)	DK (15)	FE (16)	FGLS (17)	DK (18)
Ln(EP2)	−0.248	0.410***	0.415*	−0.094	−1.075***	−1.391***
Ln(EP3)	0.236**	0.071***	0.137	0.878***	−0.495	−0.589
Ln(URBA)	0.796***	0.509***	0.530***	0.459***	0.592***	0.631***
Ln(URI)	0.445***	0.122***	0.166*	0.292**	0.730***	0.728**
Ln(INDU)	1.461***	1.537***	1.509***	1.995***	3.162***	3.344***
Ln(TERT)	0.871***	0.272***	−0.229	1.478***	2.304***	3.180***
Ln(LAND)	−0.001	0.0.475***	0.506***	−0.091	0.265***	0.435***
Ln(EI)	0.022*	0.155***	0.163***	0.038	0.428***	0.502***
Constant	−3.674***	−3.963***	−4.242***	−8.534***	−11.613***	−16.66***
R2	0.818	–	0.706	0.869	–	0.819
Autocorrelation test	F(1,29) = 454.159***			F(1,10) = 115.07***		
Cross-sectional dependence test	39.885***			12.81***		
Heteroskedasticity test	$x^2(30) = 1242.32$***			$x^2(11) = 484.70$***		
Observations	780	780	780	286	286	286
Variable	Central region			Western region		
	FE (19)	FGLS (20)	DK (21)	FE (22)	FGLS (23)	DK (24)
Ln(EP2)	−0.218	−0.001	−0.097	−0.378***	−0.414***	−0.552**
Ln(EP3)	0.329**	0.129	−0.009	0.138	0.269***	0.414**
Ln(URBA)	1.034***	0.348***	0.401**	0.743***	0.396**	−0.534**
Ln(URI)	0.280*	0.439***	0.339**	0.477***	0.043	0.054
Ln(INDU)	0.074	1.791***	2.036***	1.751***	1.089***	1.044**
Ln(TERT)	−0.121	1.454***	1.817***	−0.238***	−0.865**	−1.284**
Ln(LAND)	0.047	0.105***	0.164***	0.162***	0.421***	0.546***
Ln(EI)	−0.093	0.461***	0.458**	−0.105*	−0.113***	−0.061*
Constant	3.880***	−6.460***	−8.482***	−3.188***	2.486***	2.599
R2	0.844	–	0.801	0.856	–	0.706
Autocorrelation test	F(1,7) = 160.75***			F(1,10) = 550.75***		
Cross-sectional dependence test	10.93***			12.87***		
Heteroskedasticity test	$x^2(8) = 243.24$***			$x^2(11) = 89.73$***		
Observations	208	208	208	286	286	286

Notes FE (fixed effects); FGLS (feasible generalized least squares); DK (regression with Driscoll-Kraay standard errors). ***, ** and * denote rejection of the null hypothesis of nonstationary at the 1%, 5% and 10% significance level, respectively energy consumption intensity will also promote carbon emissions in the eastern and central regions.

7.6.4 Policy Implications to Reduce Energy Consumption and CO_2 Emissions

The transformation of China's urban–rural development has caused major changes in the composition of urban–rural elements and spatial linkages, and has promoted the development of urban–rural economic and social integration; on the other hand, rapid population migration and the non-agriculturalization of agricultural land have contributed to urban congestion, serious air and water pollution, Empty waste and inefficient use of rural land. Therefore, revealing the impact of urban–rural transformation on energy consumption and carbon emissions can provide a useful reference for national strategies for urbanization and environmental planning. This study found that urban–rural transformation has a certain impact on energy consumption and carbon emissions; however, the impacts of urbanization, employment transfer, and industrial transformation on different regions are quite different. Taking urbanization as an example, urbanization in the central and eastern regions has significantly promoted the increase in energy consumption and carbon emissions, but the impact in the western region is not significant. Cole and Neumayer (2004) found through a study of 86 countries around the world that the urbanization rate increased by 10%, and the increase in carbon dioxide emissions was 7%. The slowdown in the urbanization process helped reduce energy consumption and pollutant emissions in the Middle East and North Africa. Due to the lack of basic public services in western my country, the negative correlation between urbanization and energy utilization may be the result of modernization, rather than the size of the economy or the scale of public facilities. Urbanization is a complex process involving the shift of population from rural to urban areas and the change of rural lifestyles to urban people. More urban infrastructure construction will inevitably need to support the growing urban population, which may lead to more energy demand and more carbon dioxide emissions (Wang et al. 2013). In most areas of China, urbanization and related income and lifestyle changes are important forces driving carbon dioxide emissions (Feng et al. 2012).

Carbon emissions caused by land-use changes have become an important source second only to fossil fuel combustion, and the change from agricultural land to construction land increases carbon emissions (Qu et al. 2011). From 1850 to 1998, land-use changes contributed about 33% of total emissions in the past 150 years, and 12.5% of total emissions from 1990 to 2010. In the process of rapid urbanization, deforestation and forest degradation are currently estimated to contribute 6–17% of carbon dioxide emissions. By 2030, urban expansion is expected to have a significant impact on global carbon stocks, especially in the subtropics. China's urbanization is characterized by the excessive conversion of agricultural land to non-agricultural land, which has become an important source of carbon emissions in China. From 1950 to 2005, carbon dioxide emissions from land-use changes in China were approximately 10.6 billion tons, accounting for 12% of global carbon emissions from land-use changes over the same period (Dong 2010). The research in this section finds that the impacts of energy consumption and carbon dioxide emissions from land-use changes in different regions are also different. The transformation of land use has increased energy consumption and carbon dioxide emissions in China as a whole and in the three regions of east, central and west, but the degree of impact on the west, east, and central regions has decreased in turn. With the development of the western region, the scale of commercial and transportation activities and urban infrastructure in the western region has further expanded. Coupled with underdeveloped industrial technology, the energy efficiency of the western region is lower than that of the eastern region. Therefore, in the process of changing land use from agricultural land to non-agricultural construction land, the western region will likely need more energy and emit more carbon dioxide. Due to the scarcity of land resources in the eastern region, as the urban population increases, market competition and technological innovation encourage traditional energy sources to switch to renewable energy resources, which will help reduce fossil energy consumption and substitution.

There is a close relationship between urban and rural development transformation, energy consumption and carbon emissions. Related research needs to pay attention to the process of urban and rural development transformation and how the model affects carbon dioxide emissions. It is necessary to strengthen the mutual feedback mechanism between urban and rural development transformation and environmental changes to promote healthy urbanization and coordinated development of urban and rural areas. At present, as the transformation of urban and rural development requires more energy consumption and carbon dioxide emissions, it is necessary for the central and local governments to control the speed of urbanization and expansion of construction land, optimize the land structure, build sustainable infrastructure systems, and improve energy efficiency. Considering the regional differences in the transformation of energy and carbon dioxide emissions from urban and rural development, the development strategy of eastern and central China should focus on controlling the scale of urbanization development, optimizing the industrial structure, and improving the energy efficiency. Western China should pay more attention to the improvement of the employment structure and the protection of cultivated land. Due to regional differences in the impact of human activities on the environment, formulating energy-saving and emission reduction strategies oriented to regional characteristics is an effective measure to achieve sustainable development.

7.7 Analysis of the Impacts of Village Transformation on Resources and Environment

7.7.1 Theoretical Analysis: Village Transformation and Its Resource and Environmental Effects

Village development refers to the benign evolutionary process of agricultural production development, stable economic growth, social harmony and progress, continuous improvement of environment and continuous cultural inheritance in village system under a certain village-town spatial pattern (Li et al. 2013). Village development is a non-linear process, which is characterized by fluctuating or stepped development. This process can be regarded as village transformation, that is, the main body of village development dynamically optimizes, adjusts and innovates the institutional mechanism, operation mode and development strategy of village development based on the changes of the internal and external environment of the village system, thus realizing the transformation from the old development model to the new development model in line with the requirements of the current era. This process is bound to be related to the development, transportation, utilization, production and living emissions of water and soil, mineral resources, energy and other resources, affecting the regional ecological environment and producing a series of resource and environmental effects. In terms of resource utilization, it is mainly reflected in the changes of land resources, water resources and energy utilization modes; in terms of environmental effect, it is mainly reflected in the impacts of agricultural non-point source pollution, livestock and poultry breeding pollution, rural industrial pollution, domestic waste pollution and other endogenous pollution from the village.

The resource and environmental effects of village development are different from those of industrialization and urbanization at national and regional scale. National and regional economic growth is a spatial process of population, factors and industries gathering to cities, the intensity and scale of resource consumption and pollution emissions are larger, and the distribution is more concentrated. Village systems can be regarded as semi-natural artificial system in which resource consumption and pollution emissions tend to be less intense and more dispersed. However, due to the large number of villages and the large population they carry, the total change has a global effect and is worthy of attention. In addition, affected by income, education and other factors, villagers' subjective perception of resource and environmental problems may be different from that of urban residents. Moreover, due to differences in the nature of the community, the practice of villagers' response to resource and environmental problems in "acquaintance society" is often different from that of urban residents. This is one of the important starting points for the study of village scale.

Village development and resources and environment mutually interact and restrict, and their types of coupling are diverse. ① In resource-based village, resources exploitation has become an important driving force for village development, but the over-exploitation may bring serious negative environmental effects. If insufficient attention is paid to resources and environmental effects, or if village development is overly dependent on resource exploitation, it is likely that the "resource curse" will restrict the long-term development of rural industrial economy (Auty 1993), or the village economy collapses and the development level drops because of resource depletion and environmental degradation (Fig. 7.27a). ② In general, few villages can achieve the "big import and big export" of resource factor input and environmental pollution emissions and the long-term economic growth by virtue of rich resource endowment and strong environmental self-purification ability (Fig. 7.27b), unless the resource supply mainly comes from the outside and environmental negative effects are discharged out of the village. ③ However, some villages achieve economic growth with the

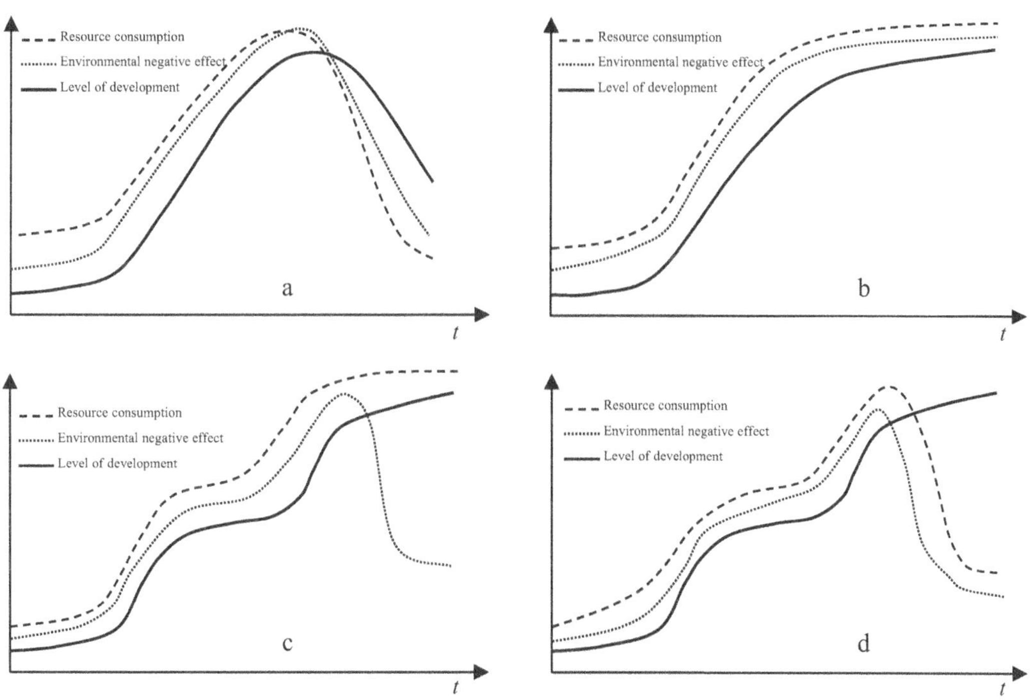

Fig. 7.27 Coupling types of village transformation and its resource and environmental effects

help of external resources. With the improvement of village development level and villagers' demands for environmental health, certain technical measures can be taken to reduce emissions and decrease the negative environmental effects (Fig. 7.27c). ④ The ideal coupling relationship should be that the recycling and cleaner production of various input factors in the process of village development can be realized with the help of technology, economy, system and other relevant measures, the efficiency of resource utilization can be improved, and the negative effect of environmental pollution can be reduced (Fig. 7.27d).

In conclusion, the focus of the research on village transformation and its resource and environmental effects is to analyze the process of transformation and the main types, evolutionary characteristics, potential consequences and internal mechanisms of resource and environmental effects based on multi-disciplinary theories and methods, so as to explore the optimal regulation path and mode. Under the long-term urban–rural dual structure in China, the awareness of resource utilization and environmental pollution control is weak in rural areas, the mechanism is not complete, and the investment is insufficient, resulting in the low efficiency of resource utilization and severe environmental situation. Especially in terms of environmental conditions, the coexistence of point and nonpoint source pollution, the superposition of domestic pollution and industrial pollution, the interweaving of all kinds of new and old pollution and secondary pollution, the transfer of industrial and urban pollution to rural areas, and the increasingly serious soil pollution have become the constraints for the sustainable development of rural economy and society in China. Most of the villages along the eastern coast and in the suburbs of big cities have entered the key stage of transformation and upgrading (Liu 2007), their experience and lessons are of positive reference value for the vast number of villages in the central and western China that are choosing or practicing their own development path, can be used as a pilot area for the study of village transformation and its resource and environmental effects and optimal regulation.

7.7.2 Empirical Methods and Data

1. **Research area**

Beicun village is located in Shunyi District of Beijing, adjacent to National Highway 101 in the east and north sixth Ring Road exit to the south, 30 km from the main city and 2 km from the entrance of Jing-Cheng Expressway. It has a superior geographical location and a good ecological environment, and is located in the green agricultural industrial belt in the suburbs of Beijing. The total area of the village is approximately 4.4 km^2. In 2009, there were 520 households with a population of 1,600, and the per capita net income exceeded 20,000 yuan. Since the early 1990s, the village constantly adjusted development strategy based on its own resource endowment and external market demand, realized the transformation from a traditional agricultural village based on grain production and pig breeding to a modern agricultural industrial village integrated with specialized field crops, flower seedling planting, centralized piglet and pig breeding, park agricultural products processing and brand marketing, vigorously promoted the development of clean production and agricultural leisure tourism, and solved the problems of resources and environment while realizing transformation. The characteristics of the village's transformation process and the measures to optimize and control the resource and environmental effects are of reference value to rural construction in the new period.

2. **Material and methods**

The material used in this study were obtained from four field surveys in December 2010 and December 2011. According to the research objectives, the corresponding data collection methods are adopted. ① The historical information of village development was obtained through the informal discussion of typical farmers, interviews with key figures and collection of village historical data, and the process characteristics of village transformation were discussed. ② The characteristics of land-use change in the village were analyzed based on aerial images in 1967, TM images in 1992 and 1999, and Google Earth images in 2010. Because the accuracy of TM image data did not meet the research needs of the study at village scale, the interpretation of the images in 1992 and 1999 was carried out with the participation of the villagers by referring to the aerial images of 1967 and Google Earth images of 2010. ③ The information about the characteristics and changes of water resources and energy utilization was mainly obtained through interviews, discussions and questionnaires.

In general, research on environmental effects should use the environmental monitoring data of time series, sections or even panels, but such data at village scale are extremely scarce. Therefore, this study assumes that the local villagers' subjective cognition of pollution situation in different periods of village development can reflect the dynamic changes of the environmental effects of village transformation, and attempted to obtain the trend data of environmental changes through villagers' perceptions of environmental pollution conditions. Specifically, combined with the features of village pollution, four sub-indexes (agricultural non-point source pollution index, domestic waste pollution index, livestock and poultry breeding pollution index, and rural industrial pollution index) and one composite index (comprehensive environmental pollution index) were defined. After introducing the connotation of the index, the interviewees were invited to score the four sub-indexes every five years since 1980, and determine the weight of the comprehensive environmental pollution index synthesized by the four sub-indexes (Table 7.33). The participants were composed of 15 local villagers aged 40–60, including village cadres, heads of village enterprise departments and ordinary villagers, all of which have good group representation.

Table 7.33 Types and measurement of environmental pollution in villages

Type	Description	Score range of index	Weight
Agricultural nonpoint source pollution	Pollution caused by the unreasonable application of chemical fertilizers and pesticides to farmland	Very little pollution (0–40 points) Relatively little pollution (40–60 points) Light pollution (60–70 points) Moderate pollution (70–80 points) Relatively heavy pollution (90–100 points) Heavy pollution (90–100 points)	0.13
Domestic waste pollution	Pollution caused by recyclable garbage (such as waste paper, plastic and metal), kitchen waste, harmful garbage (such as waste batteries and expired medicines) and other garbage in daily life		0.16
Pollution of livestock and poultry breeding	Waste residue discharged by various livestock and poultry farms; harm and damage to the environment caused by the sewage and stench produced by cleaning livestock and poultry bodies, breeding grounds and utensils		0.41
Rural industrial pollution	Pollution caused by waste residues, waste gas and waste water discharged from rural industrial enterprises to the environment		0.30
Comprehensive environmental pollution	The comprehensive representation of the environmental pollution obtained by multiplying the sub-index by the corresponding weight		–

Note The weight was the mean of the values determined by the participants; due to the different generation mechanism and regulation path, the exogenous urban domestic pollution and industrial pollution were not considered

7.7.3 The Process of Village Transformation and Its Resource and Environmental Effects

1. **Process characteristics of village transformation**

Based on the participatory investigation, the development process of Beicun village since the reform and opening-up was divided into three stages:

(1) From 1978 to 1992, traditional agriculture drove the slow development of village. The innovation of the household responsibility system and the increase of modern agricultural production factors led to a substantial increase in Beicun's grain output, and pig breeding also began to take shape, effectively solving the problem of food and clothing. Because Beicun is located in the suburbs of Beijing, there were many nonagricultural employment opportunities, and talented people set an example by engaging in diversified businesses such as flour processing and operating a gas station. As a result, the vitality of rural nonagricultural economy made a preliminary appearance. However, constrained by the urban–rural dual system such as the "scissors difference" in the price of industrial and agricultural products, as well as the rapid population growth and limited human capital, the development of Beicun village was still relatively slow.

(2) From 1993 to 1999, innovation and implementation of the joint-stock system to drive the development of the village gradually started. In the new round of reform and opening-up since 1992, Guangdong Nanhai's experience in joint-stock cooperative system has been affirmed by governments at all levels. Drawing lessons from the experience of Nanhai and the management practice of villagers, the village realized the joint-stock transformation of the village industry at the end of 1993 through the extensive mobilization and resources integration by the two village committees, established pig breeding farm, slaughterhouse, flower and tree center, etc. In 1996, Beicun people's own agricultural, industrial and trade group was established, and the intensive breeding community was subsequently built. The imitation and innovation of property rights mechanism and organizational structures played an important role in promoting village development, and the villagers became investors, operators and beneficiaries of village industries, which effectively solved the problems of fund shortage and incentive insufficient.

(3) Since 2000, the industrialization of modern multifunctional agriculture has driven the development and upgrading of the village. Based on its own endowment, market demand and industry-research cooperation, Beicun village took "developing green economy, creating green environment, contributing green products and sharing green life" as its development concept, Beicun village had made great efforts to develop modern efficient agriculture such as flower seedlings, breeding pigs, pig breeding, agricultural products processing, agricultural sightseeing picking, etc., promote the biogas project and rainwater collection project, forming a modern and circular village economic system with the characteristics of suburban villages in big cities. At the same time, Beicun village also attached great importance to the construction of innovation capability, brand building and the fairness of income distribution and welfare treatment, emphasized the comprehensive development of the community, and promoted the multifunctional transformation and upgrading of the village.

The development of Beicun village is a typical process of bringing emerging and high-added-value industries into the village industrial system through overall planning and innovative practices. Therefore, the village resources and their capitalization income are intercepted and even gathered in the village. This kind of development path and mode is representative and leading in the suburbs of big cities.

2. **Changes of resource utilization in the process of village transformation**

(1) Land resources

During the study period, the land-use structure and landscape pattern of Beicun village changed

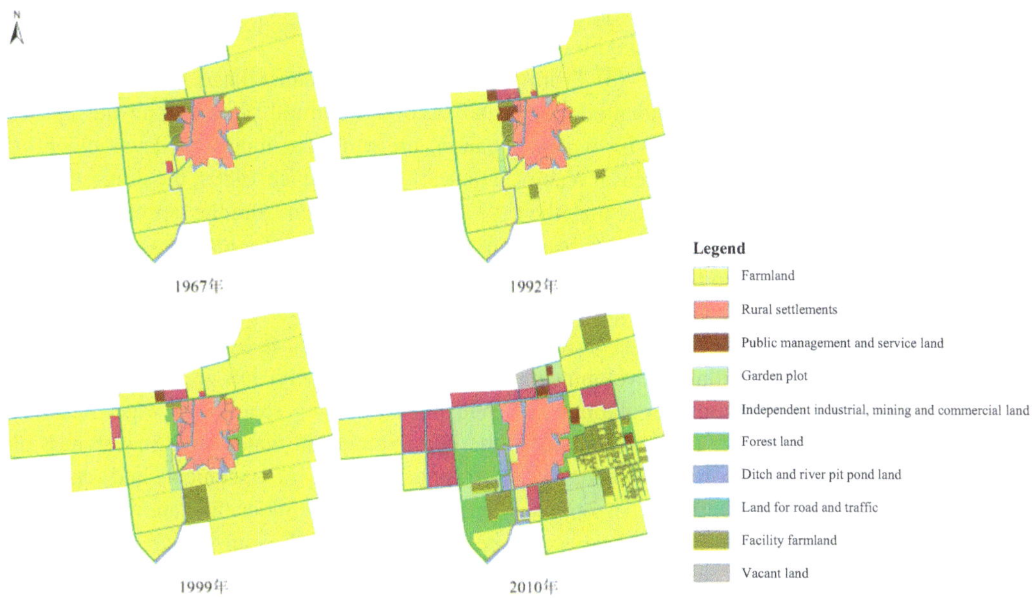

Fig. 7.28 Land use change in Beicun village from 1967 to 2010

greatly (Fig. 7.28): ① In terms of time sequence, the dynamic degree of land-use change in 1967–1992 and 1992–1999 was relatively small, and there was a great change in the last decade. ② In terms of space, it was mainly manifested in the southward expansion of residential land, rapid growth of commercial and service lands along the traffic trunk lines, and the "enclave" expansion of industrial and mining land and facility agricultural land far away from residential areas. ③ In terms of type transformation, a large number of cultivated land had been transformed into facility agricultural land (mainly used for centralized breeding of live pigs and breeding pigs and greenhouse flower planting), independent industrial and mining land and commercial and service land (mainly used for agricultural product processing and the tertiary industry), garden and forest land (mainly used for flower seedling planting) and rural residential land (mainly used for villagers' living), and the degree of land use land use diversification and farmland fragmentation increased significantly. ④ In terms of functional evolution, it had gradually transformed from a traditional village dominated by crop production and concentrated residence to a multifunctional village integrating with ecological cultivation, modern processing, circular production, local services and centralized residences.

The rapid non-grain and non-agriculture of cultivated land are important features of land use change in the transformation process of Beicun village in recent years, while the speed of homestead expansion is lower than that of the change rate of village land in traditional agricultural areas. ① The main performance of non-grain is the change of cultivated land to facility agricultural land, operating garden land and forestland.[3] From 1992 to 2010, the area of facility agricultural land increased by 40.51 hm^2, while the area of operating garden land and forestland increased by 89.64 hm^2. ② The main performance of non-agriculture is the transformation of cultivated land to independent

[3] Taking the breeding community as an example, the village collective and the agriculture, industry and trade group take back the contracted land in the form of leasing, build the pig house, water supply system, sewage system and other basic buildings, and then rent them back to the farmers with the same rent (the average annual rent per mu is about 800 yuan). This method has stimulated the land circulation, promoted the adjustment of agricultural structure, large-scale operation and the improvement of land-use efficiency.

industrial and mining land as well as commercial and service land, From 1992 to 2010, the kind of change had increase by 50.03 hm^2.[4] ③ From 1967 to 2010, the growth of rural residential land was 60.14%, less than half of that of typical villages in traditional agricultural areas (Wang et al. 2010), and the main reason was that the regulations of homestead management have been optimized and adjusted through endogenous institutional arrangements. On the one hand, the boundary of residential expansion was strictly controlled with the help of village development planning; on the other hand, some regulations in the process of homestead application were made to realize the intensive use of homestead.[5] In general, no matter the input intensity or output intensity per unit area, the land use efficiency has been significantly improved and tends to be maximized. To a large extent, this reflects the characteristics of "market guiding, planning control, enterprise leading, villagers participating, clear property-rights, revenue sharing and intensive sustainability", making the land value-added income remain in the village.

(2) Water resource

The water in Beicun village is mainly used for daily life, agricultural planting and nursery irrigation, breeding and industrial development. ① Domestic water. Since 1980s, the domestic water of villagers has been from the tap water extracted from the deep well for a long time. ② Irrigation water. In the 1980s and early 1990s, Beicun village adopted the model of "pumping by motor well, water conveyance by canal system and field flood irrigation", with low utilization efficiency. With the increasing scarcity of water resources, the improvement of water-saving awareness and the enhancement of technical and economic feasibility, the sprinkler irrigation project has been carried out in Beicun village since the mid-1990s, adopting the mode of "pumping by motor well, water conveyance by concealed pipe and field sprinkler irrigation", which greatly improves the utilization efficiency of irrigation water. Irrigation water for flower seedling base mainly comes from the groundwater extracted by motor well, and then the purified reclaimed water is used. ③ Breeding water. In the stage of family free range, this type of water is mainly the groundwater. After the completion and use of the breeding community, the way of "dry manure cleaning", unified water supply and paid use is adopted to realize water-saving breeding. ④ Industrial water. At first, the industrial water was mainly the groundwater extracted from the motor well. In recent years, the village has invested more than 600,000 yuan in installing real-time monitoring remote water meters for the water supply pipelines of 21 enterprises in the village, so as to carry out real-time monitoring and management of enterprise water consumption, thus reducing the total amount of enterprise water consumption by nearly 30%. In particular, to realize the purpose of saving water and increasing efficiency, the village level circulating water work has been effectively carried out since 2005. Two village level sewage treatment stations have been built, and the ponds have been dredged to collect rainwater and reclaimed water from the sewage treatment station. As a result, most of the water for greening and irrigation of flower and

[4]Taking agricultural products processing park as an example, under the guidance of the upper-level land use planning and village development planning, the village collective and the agriculture, industry and trade group shall take back the contracted land within the scope of the planned industrial area with the consent of the villagers, lease it to the enterprises, especially agricultural enterprises, after land leveling as well as the supply of water, electricity, access, communication and drainage; then a certain proportion of the rent (approximately 70%) is returned to the villagers (the annual rent of approximately is 3,000 yuan per acre). As a result, agricultural products processing cluster has been gradually formed, which promotes the adjustment of economic structure in Beicun village and drives the agricultural structure adjustment of surrounding villages. Of course, it is not suitable or necessary for all villages to build parks. It is necessary to make a comprehensive evaluation on the development intention, land use, village and town system, industrial layout and other related planning.

[5]For example, if there is only one son in the family, the family cannot apply for new homestead when he get married as a adult. This institutional arrangement reduces the demand for expansion and empty houses to a large extent, which is in sharp contrast with the low application threshold and lax examination and approval control of most villages in traditional agricultural areas.

seedling bases in the village comes from the accumulated reclaimed water, which realizes the water cycle and reduces the exploitation of groundwater and the cost of water use.

(3) Energy consumption

The energy consumption in Beicun village mainly includes industrial energy and domestic energy. Industrial energy is mainly electricity, and energy conservation is strongly advocated in the industrial park. In terms of domestic energy, ① the cooking energy has transformed from straw to "straw + liquefied gas" and then to biogas, and the degree of cleaning and convenience has been improved. ② Heating energy has evolved from coal-fired to coal-electricity combination since 2000. ③ Electricity has been the main energy for lighting since 1980. ④ The energy consumption for bathing mainly occurred in winter. Before the 1990s, the bathing energy source was mainly straw, and liquefied gas and solar energy were added in the late 1990s. In recent years, the village has established a collective bathroom heated by solar energy, which saved not only money but also water resources. Generally, the extensive and effective use of clean energy such as biogas and solar energy is the main characteristic of the energy consumption transformation in the village since 2000, and the energy utilization structure has been continuously optimized.

3. **Environmental pollution in the process of village transformation**

(1) Agricultural non-point source pollution

In China, the agricultural non-point source pollution mainly occurs in intensive agricultural areas, especially in vegetable-growing areas and rice-growing areas in the south. The environmental pollution caused by agricultural activities in the suburbs of Beijing suburbs mainly came from the unreasonable use of chemical fertilizers and pesticides, especially after the "vegetable basket" project was launched in the late 1980s (Guo et al. 2004). Relatively speaking, the agricultural non-point source pollution in Beicun village is not serious. During the ten years from the mid- and late 1980s to the mid- and late 1990s, the application of pesticides and fertilizers increased year by year, resulting in mild agricultural non-point source pollution. Since 2000, the village has promoted the standardized planting of grain, fruits and vegetables, emphasizing the supervision of the agricultural production process. The application of pesticides and chemical fertilizers has been performed strictly in accordance with the corresponding standardized production procedures, and the proportion of organic manure application was also increased, resulting in a significant reduction in agricultural non-point source pollution.

(2) Domestic waste pollution

Domestic waste is a practical problem that troubles urban–rural development in China. Before the 1990s, there was relatively little domestic waste in Beicun village; since the 1990s, the living standard of villagers has been greatly improved, and the domestic waste increased significantly, resulting in environmental pollution caused by random stacking. In 1996, Beicun village began to build an open-air "three-wall" garbage dump, and the garbage was transported daily to the designated place for landfill by the village committee. Since 2000, roofs were installed because rain often caused waste and sewage to flow across the land. After 2005, the whole village implemented garbage classification. Each household had a clean bucket, which was collected and then dumped in the designated garbage dump site. The cleaners were responsible for cleaning public spaces, and the sanitation vehicles sent by Shunyi District were responsible for garbage collection and transportation every morning. In particular, the village established a paid recycling system for waste plastic bags to encourage the villagers to collect waste plastic bags. At present, it has basically achieved the "community-based collection, specialized transportation, socialized operation and standardized management" of domestic waste, and initially solved the problem of domestic waste pollution in the village. However, the treatment mode of domestic waste is still simple, and the environmental protection performance and scientificity need to be further improved.

(3) Livestock and poultry breeding pollution

Livestock and poultry breeding pollution is different from general agricultural nonpoint source pollution in terms of generation mechanisms, transmission paths and prevention and control measures, and has the characteristics of point source pollution. In addition, the scale of livestock and poultry breeding in the suburbs of large cities is often large, so this pollution is listed separately from agricultural pollution. In general, the pollution of livestock and poultry breeding is related to breeding scale, technology level, process supervision and other factors. The livestock and poultry breeding in Beicun village had reached a relatively large scale since the mid- and late 1980s, with an annual output reaching thousands of head, and all of them were scattered among families. Because the epidemic prevention and control measures and fecal treatment measures were not perfect, the breeding areas were smelly, and the environmental pollution was serious. It peaked in the middle and late 1990s. To expand the scale, reduce pollution, improve the environment, and enhance epidemic prevention and industrial competitiveness, Beicun village had implemented three measures: centralized breeding, ecological treatment and process supervision.

Beicun village has been organizing and promoting the construction of ecological breeding community since 1998. After soliciting public opinions, the village committee designated 1,000 mu of land in the southeast of the village to develop centralized breeding. Each community covered an area of 3 mu, including 1 mu of pigsty, 2 mu of vegetable land, unified and paid water supply. Pig manure is applied into vegetable land by dry cleaning, and pig urine is discharged with the ditch. However, the excretion of pig urine caused serious pollution to the ditch outside the village after 2 years of operation. To reduce breeding pollution, Beicun village collaborated closely with China Agricultural University, Chinese Academy of Agricultural Sciences and other scientific institution. Relying on the scientific and technological projects of large-scale pig farm manure water treatment, the village invested more than 5 million yuan to construct a biogas station in 2002, and adopted the advanced technology of biogas production by biological fermentation for manure water treatment. The biogas generated by fermentation was purified and stored in the biogas tank under pressure and then distributed to each household through special pipes. Through the implementation of this project, it not only realized ecological breeding, but also turned waste into treasure, so that the whole village could use clean fuel. The waste residue separated from biogas slurry was used to produce biological organic fertilizers, part of which was used for farming in the village, while the remainder was sold as a commodity. The discharged wastewater was treated innocuously and reused after reaching the reclaimed water standard. To further improve the efficiency of resource utilization, Beicun village invested 0.57 million yuan in 2007 to establish an organic fertilizer factory, which used biogas residue and excess feces for composting processing to produce organic fertilizers. It is estimated that the average annual cost of biogas for a family of five is about 300 yuan, which is 700 yuan less than coal-fired cooking and 800 yuan less than liquefied gas. Correspondingly, each household can save 1.5 t of coal annually, and 780 t of standard coal can be saved in the whole village every year, thus reducing the emission of 1900 t CO_2, 2500 t CH_4, 6.63 t SO_2 and 5.8 t NO_x. Moreover, the whole cost can be recovered within the design life of 30 years, and the effect of energy saving and emission reduction, cost saving and income increasing is obvious (Duana et al. 2011).

(4) Rural industrial pollution

Rural industrial pollution mainly refers to the environmental pollution caused by waste residues, waste gas and waste water discharged to the environment by rural industrial enterprises.

The enterprises in Beicun village were mainly engaged in agricultural product processing, such as flour processing, pig slaughtering and food processing, with relatively little pollution. Among them, pig slaughtering produced some pollution, mainly sewage discharge, which gradually increased in the 1990s. Since the beginning of 2000, the village began to carry out comprehensive wastewater treatment, and two village-level sewage treatment stations have been built in recent years. The treated water enters the pond for purification and then is used for irrigation of flowers and seedlings. In general, there is little pollution from rural industry.

(5) Local villagers' perception of environmental pollution

By virtue of the local villagers' subjective cognition of the pollution situation in different periods of village development, the dynamic changes of the environmental effects of the village transformation were revealed. During the survey, it was found that the score dispersion of the same index at the same time was low, that is, the villagers' cognition of environmental pollution had a strong consistency, which could basically reflect the trend of environmental effects. Figure 7.29 showed that: ① both the sub pollution index and comprehensive pollution index went through a process of first rising and then falling; ② the mid- and late 1990s, which focused on scale expansion, was a period of relatively serious pollution of all kinds; ③ agricultural nonpoint souece pollution, domestic waste pollution, and livestock and poultry pollution in 2010 were even weaker than those in 1980; ④ the environmental comprehensive pollution index scored basically the same in 1980 and 2010. Generally, in the process of village transformation, the environmental pollution in Beicun village first gradually increased and then gradually decreased, showing an obvious inverted "U" shape.

4. **Mechanism of optimal regulation of resource and environmental effects**

The village is a spatial unit produced by the organic integration of production, life and ecology, and is a typical "ecological-economic-social" complex system. Based on the main line of "analyzing system configuration—dividing evolutionary process—defining participants—clarifying the intention of participants—clarifying functional relationship", the internal mechanism of optimizing and regulating the resource and environmental effects in the process of village transformation is discussed. The village system can be deconstructed into three types of structures: ① natural-biological structure, which is mainly composed of resources, environment,

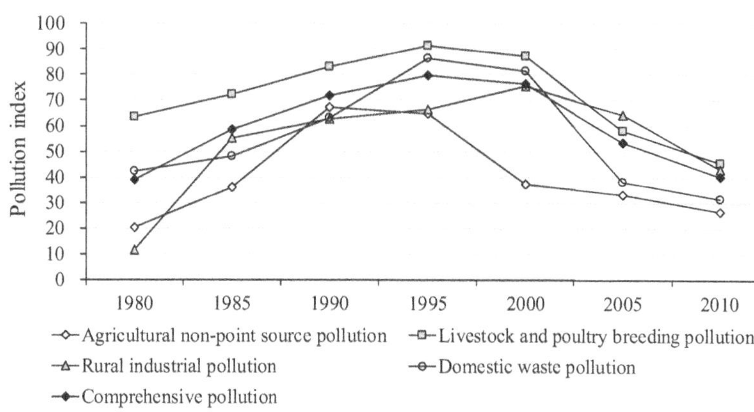

Fig. 7.29 Changes of environmental pollution index in Beicun village since 1980

ecology and other elements; ② technology-economic structure, which is mainly composed of technology, cost, income, market and other factors; and ③ social-institutional structure, which is mainly composed of population and labor force, social culture, laws and regulations, institutions and policies. In general, the resource endowment and environmental bearing capacity of the village are limited, the operation of village system may lead to the shortage or even exhaustion of local resources, environmental pollution or even collapse, resulting in the transformation of the structure of the village system and affecting the benign operation and function of the village system. At this time, the optimization, adjustment and even reconstruction of system elements and structures must be realized to ensure the benign operation of the system. If the adaptive and response mechanism of the system is strong, it may realize automatic optimization through internal action; if it is weak, the appropriate intervention of external forces is needed to achieve system structural optimization and function improvement.

Since the reform and opening up in 1978, Beicun village has experienced three stages, i.e., slow development, gradual start-up and transformation and upgrading. There are also stage differences in the resource and environmental effects, realizing the transformation of resource investment from low-efficiency to high-efficiency and environmental pollution from high-pollution to low-pollution (Fig. 7.30). The initiation of the adaptive mechanism and external intervention mechanism of the village system is a process of multi-agent participation, including villagers, enterprises, village committees, technical institutions, local governments and even the central government. The two mechanisms differ in different stages due to the different needs of the subject. In the initial stage of development, farmers and village enterprise operators focus on the expansion of scale to achieve large-scale economy, but lack of attention is paid to the quality of development, especially resource utilization and environmental pollution, which have certain externalities. In the process of government decision-making, the importance of environmental issues is also significantly weaker than increasing economic output and employment. However, with the scarcity of resources, the aggravation of environmental pollution, and the gradual improvement of income level and education level, the public and the village subject pay more and more attention to the sustainable use of resources and environmental protection. Moreover, the resources and environmental problems are partially transmitted to decision-makers through the market price mechanism, and their demand has changed from a relatively single demand for production efficiency to a comprehensive demand for production efficiency, quality life and livable environment, resulting in the internal driving force of regulating the effect

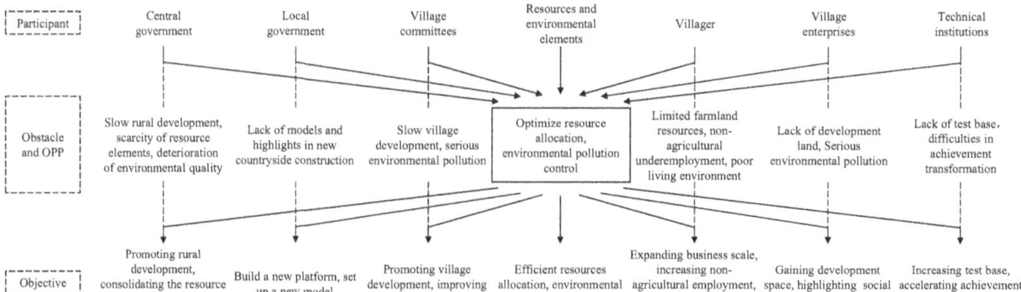

Fig. 7.30 The participant, objective, obstacle and obligatory passage point (OPP) of the optimal regulation of resource and environmental effects in Beicun village

of resources and environment. In addition, the progress of science and technology, the enhancement of economic strength and the increase of social capital improve the acceptability and availability of technology, together with the active guidance and appropriate support from central and local governments, which provides an important premise for the realization of the regulation goal.

According to the investigation, the process of optimizing and controlling the resource and environmental effects of Beicun village includes six parts. ① The emergence of the problems. Resource and environmental problems arise and affect the realization of the main objectives and system functions. ② Observation and evaluation. The participants of village development observe, perceive and evaluate the process, status, causes and consequences of resource and environmental problems, and determine the practical difficulties and key problems to be solved. ③ Stimulation and integration. This mainly refers to that the village cadres, capable people and cooperative organizations stimulate internal demands and integrate external forces to reach a consensus around practical problems and form a joint force of regulation and control. Under the mobilization, organization and coordination of village cadres, capable people and cooperative organizations, all participants believe that resources can be used optimally, environmental pollution can be improved significantly, their core goals can be realized and their interests can be satisfied (Fig. 7.22). ④ Function assignment. Under the comprehensive coordination of market mechanisms, management mechanisms, laws and regulations and emotional elements, the interests, roles, functions and status of relevant participants are redefined, rearranged and recognized (Table 7.8). ⑤ Joint action. The participants perform their duties and show their abilities, promote technological and institutional innovation, and actively promote practices of element renewal and structural optimization. ⑥ System reconstruction. Through the above processes of observation and evaluation, element renewal and structural optimization, the system function reconstruction is successfully realized, and the optimization of resource and environmental effect is achieved. In general, this process is similar to the construction of an actor network in actor network theory (Woods 1998).

Generally, the internal mechanism of realizing the optimization of resources and environmental effects and promoting the village transformation in Beicun village lies in that it takes village cadres, capable people and cooperative organizations as the core, successfully stimulates the endogenous needs of villagers and village enterprises, effectively integrates the external forces of governments and technical units at all levels, and successfully establishes an actor network with clear objectives and functions, feasible technology and good benefits. The optimal regulation practice has the following characteristics (Table 7.34). ① Villagers' subjective cognition and internal demand are the endogenous driving force of optimal regulation. Villagers' subjective cognition of resources and environmental problems is transformed into the internal needs of village transformation and environmental improvement, and becomes an important driving force for optimal regulation. ② Technical and economic feasibility is an important prerequisite for optimal regulation. If there is a lack of relevant technology or the cost of technology is too high, it is bound to be difficult to carry out optimal regulation. ③ Technical cooperative institutions play a scientific and technological supporting role in the construction of village development networks and the optimal regulation of resource and environmental effects. The long-term close cooperation between village enterprises and universities or scientific research institution has realized and ensured the organic combination of scientific research and practical application. ④ The endogenous or exogenous institutional arrangement and innovation is the control power of optimal regulation. Both

Table 7.34 The participants and functional orientation of the optimal regulation of resources and environmental effects in Beicun village

Participants	Main functions
Village committees	They are mainly responsible for the daily organization and management, especially to stimulate internal power and integrate external power, so as to realize the coordination of internal and external power. They are the organizational core of village transformation and the optimal regulation of resources and environmental effects, and the key subjects in constructing village development network.
Villagers	Based on subjective cognition and policy encouragement, they actively participate in the process of efficient resource utilization and environmental protection. Capable people and pig cooperatives can also play the role of stimulation and integration. They are also the important groups of work, supervision and benefit in the practice of village transformation and the optimal regulation of resource and environmental effects.
Village enterprises	Mobilized by the village committees, they actively cooperate and participate in the intensive use of resources and the prevention and control of environmental pollution. They are the market players in the practice of village transformation and the optimal regulation of resources and environmental effects, and the main node of factor flow and transformation. Increasing economic benefits and reflecting social responsibility are its important goals.
Technical institutions	Take Beicun village as the platform, they actively carry out technological innovation and demonstration promotion, such as resource efficient utilization technology and pollution prevention and control technology. They are the main technology supplier in the practice of village transformation and the optimal regulation of resources and environmental effects. Also, strengthening the support of science and technology, increasing economic benefits and fulfilling social responsibility are their important goals.
Local government	They carry out the instructions from the higher-level government and provide certain support to rural development in finance, projects, publicity and mobilization. They are the management subject in the practice of village transformation and the optimal regulation of resources and environmental effects. No dislocation, no offside and no vacancy are the basic requirements of the public.
Central government	Through the formulation of laws and regulations and the introduction of policies, they construct an exogenous system of village development. They set up the institutional and policy background of the practice of village transformation and the optimal regulation of resources and environmental effects, and support and recognize the practice from the aspect of major policies.
Resources and environmental elements	As the key elements of village development, they have the same importance as human elements, and are the direct objects of intensive resource conservation and recycling. They are the target subjects in the practice of village transformation and the optimal regulation of resources and environmental effects.

endogenous variable such as domestic waste treatment mechanism, land transfer mechanism and homestead management mechanism, and exogenous variable such as rural development system and joint-stock system, play an important role in the optimal regulation of resource and environmental effects. ⑤ The guidance and support from the government is an important driving force for optimal regulation. For example, the effective supply of basic public services, support and incentive policies for innovation activities play a strong role in promoting village transformation. ⑥ Village cadres, capable people and cooperative organizations play an important role in inspiring, integrating, mobilizing and organizing. For example, the overall coordination of village cadres, the demonstration of capable persons and the publicity and mobilization of breeding cooperatives have played a prominent role in willingness, strategy formulation and practice promotion. ⑦ The active cooperation and effective participation of local village enterprises are also important factors in the smooth development of optimal regulation.

5. **Thinking on the transformation of resource-saving and environment-friendly village**

The transformation process, operational characteristics and regulation of resources and environmental effects of Beicun village have strong reference value for the development of suburban villages in large cities. In the new era of urban–rural integrated development, the important direction of resource-saving and environment-friendly rural development in the suburbs of large cities is to maintain traditional or modern agricultural production and protect the rural landscape based on the location advantages of neighboring cities and rural resource endowment, thus building an ecological agricultural food chain and providing safe and high-quality agricultural products, sightseeing and leisure, catering and conference services to the urban residents. This development practice can also be regarded as the construction of a network of actors, which is a complex process of multi-agent collaborative participation and interwoven by factors such as cognition, demand, responsibility, science and technology, interest, policy and system. Based on the experience of Beicun village, the specific operations are as follows. ① It can give full play to the organization and coordination role of cooperatives/joint-stock enterprises in the purchase of agricultural materials, daily production, technical services, product sales and other links. ② Standardized production should be adopted to ensure the quality of agricultural products, and the production-education-research cooperation with technical institutions should be emphasized to gradually realize the biological control of diseases and pests by rotation and intercropping. ③ The biogas system should be taken as the core to realize material circulation, energy flow, cost savings and increasing efficiency. ④ Based on common interests and mutual trust, it is necessary to strengthen the relationship between participants such as village cadres and capable people and organization such as manufacturers and government, which are closely related to village development. ⑤ The process links from production base to consumers should be reduced as much as possible to reduce cost and ensure agricultural product safety. ⑥ Agricultural specialization and cleaner production should be promoted to enhance industrial competitiveness, so as to speed up village residential environment and community construction, and realize the comprehensive development with community residents as the core. ⑦ The layout of processing enterprises should consider the current scale and development potential, and moderately concentrates in central villages and towns with relatively convenient transportation and guaranteed water supply.

References

Antweiler W, Copeland BR, Taylor MS. Is free trade good for the environment? Am Econ Rev. 2001;91(4):877–908.

Auty RM. Sustaining development in mineral economics: the resource curse thesis. London: Routledge; 1993.

Bai X, Chen J, Shi P. Landscape urbanization and economic growth in China: positive feedbacks and sustainability dilemmas. Environ Sci Technol. 2012;46(1):132–9.

Batisani N, Yarnal B. Urban expansion in centre county, Pennsylvania: spatial dynamics and landscape transformations. Appl Geogr. 2009;29(2):235–49.

Beck N, Katz JN. What to do (and not to do) with time-series cross-section data. Am Polit Sci Rev. 1995:634–47

Bloom DE, Canning D, Fink G. Urbanization and the wealth of nations. Science. 2008;319(5864):772–5.

Cairns G, Goodwin P, Wright G. A decision-analysis-based framework for analyzing stakeholder behavior in scenario planning. Eur J Oper Res. 2016;249(3):1050–62.

Chen H, Jia B, Lau SSY. Sustainable urban form for Chinese compact cities: challenges of a rapid urbanized economy. Habitat Int. 2008;32(1):28–40.

Chen H, Pontius RG. Sensitivity of a land change model to pixel resolution and precision of the independent variable. Environ Model Assess. 2011;16(1):37–52.

Choi I. Unit root tests for panel data. J Int Money Finance. 2001;20(2):249–72.

Cropper M, Griffiths C. The interaction of population growth and environmental quality. Am Econ Rev. 1994;84(2):250–4.

Cole MA, Neumayer E. Examining the impact of demographic factors on air pollution. Popul Environ. 2004;26(1):5–21.
Dietz T, Rosa EA. Rethinking the environmental impacts of population, affluence and technology. Hum Ecol Rev. 1994;1(2):277–300.
Dong Z. The territorial planning under the concept of low-carbon. Urban Stud. 2010;17:1–5.
Du L, Wei C, Cai S. Economic development and carbon dioxide emissions in China: provincial panel data analysis. China Econ Rev. 2012;23(2):371–84.
Duana N, Lina C, Liu X, et al. Study on the effect of biogas project on the development of low-carbon circular economy—a case study of Beilangzhong eco-village. Procedia Environ Sci. 2011;5:160–6.
Dumitrescu EI, Hurlin C. Testing for Granger non-causality in heterogeneous panels. Econ Model. 2012;29(4):1450–60.
Ehrlich PR, Holdren JP. Impact of population growth. Science. 1971;171(3977):1212–7.
Fan Y, Liu LC, Wu G, et al. Analyzing impact factors of CO2 emissions using the STIRPAT model. Environ Impact Assess Rev. 2006;26(4):377–95.
Feng Z, Yang Y, Zhang Y, et al. Grain-for-green policy and its impacts on grain supply in West China. Land Use Policy. 2005;22(4):301–12.
Feng K, Siu YL, Guan D, et al. Analyzing drivers of regional carbon dioxide emissions for China: a structural decomposition analysis. J Ind Ecol. 2012;16(4):600–11.
Fischer G, Sun L. Model based analysis of future land-use development in China. Agric Ecosyst Environ. 2001;85(1–3):163–76.
Ga R, Huang K. Analysis and enlightenment of the definition of fusion in the western society. Theor Perspect. 2008;2:47–9.
García AM, Santé I, Boullón M, et al. A comparative analysis of cellular automata models for simulation of small urban areas in Galicia, NW Spain. Comput Environ Urban Syst. 2012;36(4):291–301.
Gobin A, Campling P, Feyen J. Logistic modelling to derive agricultural land use determinants: a case study from southeastern Nigeria. Agric Ecosyst Environ. 2002;89(3):213–28.
Gong P, Liang S, Carlton EJ, et al. Urbanisation and health in China. Lancet. 2012;379(9818):843–52.
Gong J, Hu Z, Chen W, et al. Urban expansion dynamics and modes in metropolitan Guangzhou, China. Land Use Policy. 2018;72:100–9.
Greene WH. Econometric analysis. New Jersey: Prentice Hall; 2000.
Grossman GM, Krueger AB. Environmental impacts of a north American free trade agreement. Soc Sci Electron Publishing. 1992;8(2):223–50.
Grossman GM, Krueger AB. Economic growth and the environment. Q J Econ. 1995;110(2):353–77.
Guan D, Li H, Inohae T, et al. Modeling urban land use change by the integration of cellular automaton and Markov model. Ecol Model. 2011;222(20–22):3761–72.
Guan D, Liu Z, Geng Y, et al. The gigatonne gap in China's carbon dioxide inventories. Nat Clim Change. 2012;2(9):672–5.
Guo S, Chen X, Qiu H. Situation of non-point source pollution by agricultural activity and counter-measures in Beijing suburban. Environ Prot. 2004;4:32–6.
Han B, Liu H, Wang R. Urban ecological security assessment for cities in the Beijing–Tianjin–Hebei metropolitan region based on fuzzy and entropy methods. Ecol Model. 2015;318:217–25.
Hoechle D. Robust standard errors for panel regressions with cross-sectional dependence. Stata J. 2007;7(3):281–312.
Huang JC, Lin HX. Spatial evolution analysis and multi-scenarios simulation of Beijing-Tianjin-Hebei urban agglomeration. Geog Res. 2017;36(3):506–17.
Im KS, Pesaran MH, Shin Y. Testing for unit roots in heterogeneous panels. J Econometrics. 2003;115(1):53–74.
International Energy Agency (IEA). Key world energy statistics international energy agency. Paris: Statistics Division; 2012a.
International Energy Agency (IEA). World energy outlook. New Policies Scenario; 2012b.
Jalil A, Mahmud SF. Environment Kuznets curve for CO2 emissions: a cointegration analysis for China. Energy Policy. 2009;37(12):5167–72.
Jenerette GD, Wu J. Analysis and simulation of land-use change in the central Arizona-Phoenix region, USA. Landscape Ecol. 2001;16(7):611–26.
Ji Y, Yan H, Liu J, et al. A MODIS data derived spatial distribution of high-, medium- and low-yield cropland in China. Acta Geographica Sinica. 2015;70(5):766–78.
Lambin EF, Turner BL, Geist HJ, et al. The causes of land-use and land-cover change: moving beyond the myths. Glob Environ Change. 2001;11(4):261–9.
Lambin EF, Geist HJ, Lepers E. Dynamics of land-use and land-cover change in tropical regions. Ann Rev Environ Resour. 2003;28(1):205–41.
Leeuwen V, Eveline S, Nijkamp P. The urban-rural nexus: a study on extended urbanization and the hinterland. Stud Reg Sci. 2006;36:283–303.
Levin A, Lin CF, Chu CSJ. Unit root tests in panel data: asymptotic and finite-sample properties. J Econometrics. 2002;108(1):1–24.
Li F, Dong S, Li X, et al. Energy consumption-economic growth relationship and carbon dioxide emissions in China. Energy policy. 2011;39(2):568–74.
Li Y, Long H, Liu Y. Industrial development and land use/cover change and their effects on local environment: a case study of Changshu, eastern coastal China. Front Environ Sci Eng China. 2010;4(4):438–48.
Li Y, Wu Q. Validation of Kuznets curve for economic growth and cultivated land conversion: evidence from provincial panel data in China. Resour Sci. 2008;30(5):667–72.
Li Y, Liu Y, Long H, et al. Village transformation, resources and environment effects and their optimal regulation in the metropolitan suburbs: the case of

Beicun in Shunyi District, Beijing. Acta Geographica Sinica. 2013;68(6):825–38.
Li Y, Zhang W, Ma L, et al. An analysis of China's grain production: looking back and looking forward. Food Energy Secur. 2014a;3(1):19–32.
Liddle B. Demographic dynamics and per capita environmental impact: using panel regressions and household decompositions to examine population and transport. Popul Environ. 2004;26(1):23–39.
Liddle B. Impact of population, age structure, and urbanization on carbon emissions/energy consumption: evidence from macro-level, cross-country analyses. Popul Environ. 2014;35(3):286–304.
Liddle B. What are the carbon emissions elasticities for income and population? Bridging STIRPAT and EKC via robust heterogeneous panel estimates. Glob Environ Change. 2015;31:62–73.
Liddle B, Lung S. Age-structure, urbanization, and climate change in developed countries: revisiting STIRPAT for disaggregated population and consumption-related environmental impacts. Popul Environ. 2010;31(5):317–43.
Lima ML, Zelaya K, Massone H. Groundwater vulnerability assessment combining the DRASTIC and DYNA-CLUE model in the Argentine Pampas. Environ Manage. 2011;47(5):828–39.
Liu Y. Rural transformation and new countryside construction in eastern coastal area of China. Acta Geographica Sinica. 2007;62(6):563–70.
Liu Y. Introduction to land use and rural sustainability in China. Land Use Policy. 2018;74:1–4.
Liu Y, Chen S. Study on the relationship between economic growth and carbon emission in typical developed countries based on IPAT equation. Ecol Econ. 2009;(11):28–30.
Liu X, Fang C. Progress and prospect of study on carrying capacity of resource and environment of city clusters. Prog Geogr. 2008;27(5):35–42.
Liu J, Kuang W, Zhang Z, et al. Spatiotemporal characteristics, patterns, and causes of land-use changes in China since the late 1980s. J Geogr Sci. 2014b;24(2):195–210.
Liu R, Wang C, Hao J, et al. Measuring environmental carrying capacity. J Basic Sci Engineering. 2009;17(1):49–61.
Liu Y, Yang R. Potential of land consolidation of hollowed villages under different urbanization scenarios in China. J Geogr Sci. 2013;23(3):503–12.
Liu L, Song M, Yokogawa H, et al. Exploring the environmental Kuznets curve hypothesis between economic growth and farmland conversion in China. J Agric Kyushu Univ. 2008a;53(1):321–7.
Liu J, Liu M, Zhuang D, et al. Study on spatial pattern of land-use change in China during 1995–2000. Sci China Ser D Earth Sci. 2003;46(4):373–84.
Liu Y, Wang J, Long H. Analysis of arable land loss and its impact on rural sustainability in Southern Jiangsu Province of China. J Environ Manage. 2010;91(3):646–53.
Liu Y, Long H, Chen Y, et al. Report on China's rural development: rural hollowing and consolidation strategy. Beijing: Science Press; 2011.
Liu M, Wang Y, Li D, et al. Dyna-CLUE model improvement based on exponential smoothing method and land use dynamic simulation. International conference on geo-informatics in resource management and sustainable ecosystem. Berlin, Heidelberg: Springer; 2013. p. 266–277.
Liu Y, Fang F, Li Y. Key issues of land use in China and implications for policy making. Land Use Policy. 2014a;40:6–12.
Liu D, Zheng X, Wang H, et al. Interoperable scenario simulation of land-use policy for Beijing–Tianjin–Hebei region, China. Land Use Policy. 2018a;75:155–65.
Liu Y, Zhang Z, Zhou Y. Efficiency of construction land allocation in China: an econometric analysis of panel data. Land Use Policy. 2018b;74:261–72.
Liu L, Song M, Yokogawa H, et al. Exploring the environmental Kuznets curve hypothesis between economic growth and farmland conversion in China. J Agric Kyushu Univ. 2008;53(1):321.
Liu Y, Wang L, Long H. Spatio-temporal analysis of land-use conversion in the eastern coastal China during 1996–2005. J Geogr Sci. 2008b;18(3):274–82.
Long H. Land use transition and rural transformation. Prog Geogr. 2012;31(2):131–8.
Long H. Land use policy in China: Introduction. Land Use Policy. 2014;40:1–5.
Long H, Li Y, Long H, et al. Accelerated restructuring in rural China fueled by 'increasing vs. decreasing balance' land-use policy for dealing with hollowed villages. Land Use Policy. 2012;29(1):11–22.
López E, Bocco G, Mendoza M, et al. Predicting land-cover and land-use change in the urban fringe: a case in Morelia city, Mexico. Landscape Urban Plann. 2001;55(4):271–85.
Lu D. Function orientation and coordinating development of subregions within the Jing-Jin-Ji urban agglomeration. Prog Geogr. 2015;34(3):265–70.
Lu W, Dai C, Guo H. Land use scenario design and simulation based on Dyna-CLUE model in Dianchi Lake Watershed. Geogr Res. 2015;34(9):1619–29.
Mao H. Innovation of mechanism and regional policy for promoting coordinated development of the Beijing-Tianjin-Hebei region. Prog Geogr. 2017;36(1):2–14.
Mao H, Yu D. Regional carrying cpacity in Bohai Rim. Acta Geogr Sin. 2001;56(3):363–71.
Maddala GS, Wu S. A comparative study of unit root tests with panel data and a new simple test. Oxford Bull Econ Stat. 1999;61(S1):631–52.
Mahadevan R, Asafu-Adjaye J. Energy consumption, economic growth and prices: a reassessment using panel VECM for developed and developing countries. Energy Policy. 2007;35(4):2481–90.
Managi S, Kaneko S. Environmental performance and returns to pollution abatement in China. Ecol Econ. 2009;68(6):1643–51.

McCoskey S, Kao C. Comparing panel data cointegration tests with an application of the twin deficits problem. Mimeo; 1999.
Minx JC, Baiocchi G, Peters GP, et al. A "carbonizing dragon": China's fast growing CO2 emissions revisited. Environ Sci Technol. 2011;45(21):9144–53.
Myint SW, Wang L. Multicriteria decision approach for land use land cover change using Markov chain analysis and a cellular automata approach. Can J Remote Sens. 2006;32(6):390–404.
National Bureau of Statistics (NBS). China statistical yearbook 2011. Beijing: China Statistics Press; 2011.
Ning J, Liu J, Kuang W, et al. Spatiotemporal patterns and characteristics of land-use change in China during 2010–2015. J Geogr Sci. 2018;28(5):547–62.
Niu Z, Yan H, Chen J, et al. Comparison of crop gross primary productivity estimated with VPM model and MOD17 product in field ecosystem of China. Trans Chin Soc Agric Eng. 2016;32(4):191–8.
Panayotou T, Peterson A, Sachs JD. Is the environmental Kuznets curve driven by structural change? What extended time series may imply for developing countries. CAER II Discussion Paper No. 80, 2000.
Pedroni P. Fully modified OLS for heterogeneous cointegrated panels. Nonstationary panels, panel cointegration, and dynamic panels. Emerald Group Publishing Limited; 2001.
Pedroni P. Panel cointegration: asymptotic and finite sample properties of pooled time series tests with an application to the PPP hypothesis. Econometric Theory. 2004:597–625.
Pesaran MH. General diagnostic tests for cross-sectional dependence in panels. Empirical Econ. 2004:1–38.
Pesaran MH. A simple panel unit root test in the presence of cross-section dependence. J Appl Econometrics. 2007;22(2):265–312.
Pontius RG Jr, Schneider LC. Land-cover change model validation by an ROC method for the Ipswich watershed, Massachusetts, USA. Agric Ecosyst Environ. 2001;85(1–3):239–48.
Pontius RG, Boersma W, Castella JC, et al. Comparing the input, output, and validation maps for several models of land change. Ann Reg Sci. 2008;42(1):11–37.
Pontius RG Jr, Peethambaram S, Castella JC. Comparison of three maps at multiple resolutions: a case study of land change simulation in Cho Don District, Vietnam. Ann Assoc Am Geogr. 2011;101(1):45–62.
Poumanyvong P, Kaneko S. Does urbanization lead to less energy use and lower CO2 emissions? A cross-country analysis. Ecol Econ. 2010;70(2):434–44.
Qu F. Research on rural land problems in the process of industrialization and urbanization in China. Beijing: Economic Science Press; 2010.
Qu F, Lu N, Feng S. Effects of land use change on carbon emissions. China Popul Resour Environ. 2011;21(10):82–9.
Santé I, García AM, Miranda D, et al. Cellular automata models for the simulation of real-world urban processes: a review and analysis. Landscape Urban Plann. 2010;96(2):108–22.
Shi A. The impact of population pressure on global carbon dioxide emissions, 1975–1996: evidence from pooled cross-country data. Ecol Econ. 2003;44(1):29–42.
Tan M, Li X, Xie H, et al. Urban land expansion and arable land loss in China: a case study of Beijing–Tianjin–Hebei region. Land Use Policy. 2005;22(3):187–96.
Tang W, Chung H. Rural-urban transition in China: illegal land use and construction. Asia Pac Viewpoint. 2002;43(1):43–62.
Tao S, Zheng T, Lianjun T. An empirical test of the environmental Kuznets curve in China: a panel cointegration approach. China Econ Rev. 2008;19(3):381–92.
The State Forestry Administration of the People's Republic of China. Jointly promote the coordinated development of Beijing Tianjin Hebei Forestry Ecology and take the lead in breaking through the framework agreement. http://www.forestry.gov.cn/portal/ghzj/s/5338/content-987913.html.
Tugcu CT. Tourism and economic growth nexus revisited: a panel causality analysis for the case of the Mediterranean region. Tourism Manage. 2014;42:207–12.
Turner B, Skole D, Sanderson S, et al. Land-use and land-cover change: science/research plan [No source information available]; 1995.
van Vliet J, Bregt AK, Brown DG, et al. A review of current calibration and validation practices in land-change modeling. Environ Model Softw. 2016;82:174–82.
Verburg PH, Overmars KP. Combining top-down and bottom-up dynamics in land use modeling: exploring the future of abandoned farmlands in Europe with the Dyna-CLUE model. Landscape Ecol. 2009;24(9):1167.
Verburg PH, Soepboer W, Veldkamp A, et al. Modeling the spatial dynamics of regional land use: the CLUE-S model. Environ Manage. 2002;30(3):391–405.
Verburg PH, Alexander P, Evans T, et al. Beyond land cover change: towards a new generation of land use models. Curr Opin Environ Sustain. 2019;38:77–85.
Wang J, Liu Y, Chen Y. Spatial expansion pattern and its dynamic mechanism of typical rural settlements in Huang-Huai-Hai plain. Geogr Res. 2010;29(10):1833–40.
Wang S, Zhou D, Zhou P, et al. CO2 emissions, energy consumption and economic growth in China: a panel data analysis. Energy Policy. 2011;39(9):4870–5.
Wang Z, Yin F, Zhang Y, et al. An empirical research on the influencing factors of regional CO2 emissions: evidence from Beijing city, China. Appl Energy. 2012;100:277–84.
Wang P, Wu W, Zhu B, et al. Examining the impact factors of energy-related CO2 emissions using the STIRPAT model in Guangdong Province, China. Appl Energy. 2013;106:65–71.
Wang S, Ma H, Zhao Y. Exploring the relationship between urbanization and the eco-environment: a case

study of Beijing–Tianjin–Hebei region. Ecol Indic. 2014a;45:171–83.
Wei Y, Liu L, Fan Y, et al. The impact of lifestyle on energy use and CO2, emission: an empirical analysis of China's residents. Energy Policy. 2007;35(1):247–57.
Woods M. Researching rural conflicts: hunting, local politics and actor-networks. J Rural Stud. 1998;14 (3):321–40.
Wooldridge JM. Econometric analysis of cross section and panel data. MIT Press; 2002.
Wu C. The core of study of geography: man-land relationship areal system. Econ Geogr. 1991;11(3): 1–6.
Wu F, Webster CJ. Simulation of land development through the integration of cellular automata and multicriteria evaluation. Environ Plann B Plann Des. 1998;25(1):103–26.
Xie H, He Y, Xie X. Exploring the factors influencing ecological land change for China's Beijing–Tianjin–Hebei region using big data. J Cleaner Prod. 2017; 142:677–87.
Xu L, Li Z, Song H, et al. Land-use planning for urban sprawl based on the clue-s model: a Case study of Guangzhou, China. Entropy. 2013;15(9):3490–506.
Yang M, Hens L, Ou X, et al. Tourism: an alternative to development? Reconsidering farming, tourism, and conservation incentives in northwest Yunnan mountain communities. Mt Res Dev. 2009;29(1):75–81.
Yang Y, Zhang S, Liu Y, et al. Analyzing historical land use changes using a historical land use reconstruction model: a case study in Zhenlai County, northeastern China. Sci Rep. 2017;7:41275.
Yao S, Wang C, Zhang L, et al. The influencing factors of resources and environments in the process of urbanization of China. Prog Geogr. 2008;27(3):94–100.
York R, Rosa EA, Dietz T. STIRPAT, IPAT and ImPACT: analytic tools for unpacking the driving forces of environmental impacts. Ecol Econ. 2003;46 (3):351–65.
Zhang C, Lin Y. Panel estimation for urbanization, energy consumption and CO2 emissions: a regional analysis in China. Energy Policy. 2012a;49:488–98.
Zhang C, Xu J. Retesting the causality between energy consumption and GDP in China: evidence from sectoral and regional analyses using dynamic panel data. Energy Econ. 2012b;34(6):1782–9.
Zhang Y, Zhao Y. How to optimize land development in the coordinated development of Beijing, Tianjin and Hebei? China Land. 2016;1:42–4.
Zheng Y, Li Z, Feng S, et al. Biomass energy utilization in rural areas may contribute to alleviating energy crisis and global warming: a case study in a typical agro-village of Shandong, China. Renew Sustain Energy Rev. 2010;14(9):3132–9.
Zhong W, Yuan W, Li SX, et al. The performance evaluation of regional R&D investments in China: an application of DEA based on the first official China economic census data. Omega. 2011;39(4):447–55.
Zhou X, Zhang J, Li J. Industrial structural transformation and carbon dioxide emissions in China. Energy Policy. 2013;57:43–51.
Zhou D, Xu J, Lin Z. Conflict or coordination? Assessing land use multi-functionalization using production-living-ecology analysis. Sci Total Environ. 2017;577: 136–47.
Zhou T, Jiang G, Zhang R, et al. Addressing the rural in situ urbanization (RISU) in the Beijing–Tianjin–Hebei region: spatio-temporal pattern and driving mechanism. Cities. 2018;75:59–71.

8 Optimizing Ideas and Institutional Innovations for Urban–Rural Transformation in China

Abstract

Urban–rural integration is the fundamental goal of urban–rural transformation. Revolutionizing the dualistic urban–rural system, optimizing the urban–rural development pattern, improving the social governance ability, balancing public resources allocation and coordinating the remediation of resource and environment are the main tasks of urban–rural transformation in China. In terms of the goals of urban–rural transformation, its cores are to realize the free flow and equal exchange of elements at the micro scale, the division and coordination of labor of industries at the medium scale and the equalization of urban–rural development strategies and systems at the macro scale, finally promoting the balanced development of urban and rural areas. Based on the optimizing idea of urban–rural transformation, this chapter also puts forward an optimizing framework of urban–rural transformation, including three levels of goal level, operation level and support level, and discusses six localized urban–rural transformation modes, including metropolis-driven type, development zones (DZs)-led type, urban construction-driven type, urban–rural interactive development type, TVEs-driven type, rural characteristic industry-driven type. At the operational level, the mechanism innovation mainly includes the industry-city coordinated mechanism, regional coordinated mechanism of main function orientation, village-town construction and rural development mechanism and multi-level subject overall control mechanism. Additionally, the system guarantee of urban–rural transformation is put forward from household registration system, rural land system, public resources allocation system, governance system, government performance appraisal mechanism and fiscal and taxation system.

8.1 Main Tasks of Urban–Rural Transformation

8.1.1 Revolutionizing the Dualistic Urban–Rural System

China's GDP surpassed US$1000 and US$3000 in 2003 and 2008 respectively, implying that the country has possessed the economic strength of "supporting agriculture with industry and driving the countryside with cities". Actually, the central state has begun to change its long-standing development strategy that gave priority to urban development since 2004. Guided by the annual "No. 1" central documents in 13 consecutive years, the state promoted rural and agricultural development through various policies supporting and benefiting the agriculture (i.e., increasing grain subsidies, abolishing agricultural taxes, and facilitating construction of new countryside). In particular, the urban–rural transformation in the

Y. Liu, *Urban-Rural Transformation Geography*, Sustainable Development Goals Series,
https://doi.org/10.1007/978-981-16-4835-9_8

eastern coast shows the trend of urban–rural interaction and rapid rural transformation. Exemplified by the integrated innovations of "modern new countryside construction" in southern Jiangsu, "renovation of ten thousand villages and the demonstration of thousands of villages" in Zhejiang, construction of ecological civilization villages in Hainan, and "Shan Hai Jing" in Fujian etc., the urban–rural integration has been promoted through vigorously developing the collective economy and accelerating the construction of central villages. Meanwhile, western and central regions in China also speeded up the exploration of urban–rural coordinated development models. For instance, Chongqing and Chengdu were assigned to be the national urban–rural coordinated comprehensive supporting reform pilot zones in 2007, which were relatively completed in terms of supporting policies and achieved remarkable results.

Nevertheless, the urban–rural dual system still exists in China, and the institutional and conceptual inertia still play important roles in impeding urban–rural integrated development, especially in the underdeveloped central and western regions. For example, in the context of the inherent urban–rural dualistic household registration system and the social security system, the phenomenon of "educated in the countryside, working in the city, and flow back to the rural after retirement" has not been fundamentally changed; the construction land supply quotes is obviously urban-biased, while the demands for rural public services, cultural and recreational facilities construction and rural production and development was relatively neglected; rural collective land is grabbed in the process of new rural construction by gathering agriculture in large-scale agricultural parks, farmers in new rural communities, and industry in development zones, namely the way of "three concentrations" (Liu 2008). Moreover, the dual difference between urban and rural investment is also extremely obvious. The proportion of rural fixed asset investment in total social investment declined from 21.86% in 1995 to 1.64% of 2016. Specifically, the nominal amount of rural fixed asset investment in 2016 was 996.5 billion yuan, only 1.67% of that in urban areas (Fig. 8.1).

In the context of urban–rural dualistic system, a large number of rural skilled labors, agricultural capital, and other elements flowed into urban areas, which leads to the following results. Firstly, the ability of main forces in the construction of new countryside is weakened, and the rural development lacks sufficient element support. In addition, the current response mechanism is still facing deeper bottlenecks like insufficient system support, lagging public service facilities, and limited investment capacity, which consequently restricts the implementation performance of the construction of new countryside. Secondly, the phenomenon of "emigration with the housing vacancy" gradually

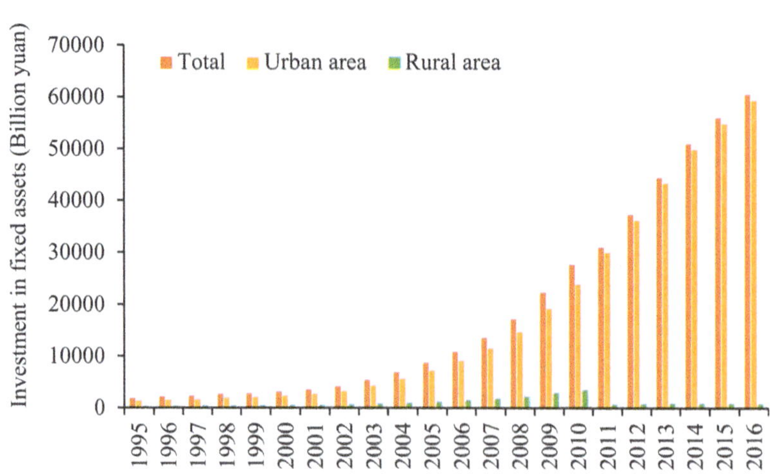

Fig. 8.1 Urban–rural fixed asset investment in China, 1995–2016

transforms from some individual cases to regional issues, and finally leads to the "hollowing" of the whole village and results in waste of land resources, destruction of arable land resources, rural dysfunction and decline in sustainable development capabilities. All these issues have become major obstacles to promoting the construction of beautiful villages. Thirdly, large and megacities developed rapidly, while small and medium-sized cities and small towns are underdeveloped in the process of urbanization. The supply of employment, education, medical care, infrastructure, and other resources is insufficient in small towns, where accelerates outflowing of population. This weakens the county's economic support for rural areas and is not conducive to the coordinated transformation of urban and rural areas. Fourthly, the contradiction between urban construction occupation of cultivated land and the red lines for both cultivated land and ecological land has become increasingly prominent and grabs the land development rights for rural development within the dualistic state-collective owned land system and the unique land acquisition system (Liu et al. 2010; Lin and Xu 2014).

In response to the increasingly serious issue of hollowing village, some regions have carried out practices of rural residential land consolidation and relocation. Given the influence of urban-biased development strategy and local land finance notion, there are still some serious problems in practice. The first is the irrational investment and construction through unauthorized pilots of "increasing vs. decreasing balance" initiative or arbitrarily expansion the scope of projects with the purpose of increasing urban construction land quotes. The second is the excessive pursuit of land finance by demolishing rural houses, which is against farmers' willingness, and the large-scale occupation of rural collective land in construction. The third is to unilaterally recognize the "increasing vs. decreasing balance" initiative as the restructuring of construction land, which results in the preference for the remediation of villages in suburbs or along roads with better location and greater value-added potential rather than that of those "hollow villages" that have been abandoned for many years. The fourth is to simply regard the "increasing vs. decreasing balance" initiative as a land consolidation project, which blindly pursue evacuation of rural residential land and construction of high-rise new settlements with limited consideration of factors such as convenience for agricultural production and rural living (Liu 2010).

Although the macroscopic agricultural and rural development policies have stepped into a new stage, the urban–rural dual structure still exists, and the urban-biased tendency is still prominent in China. It is therefore urgent to realize the benign interaction between urban and rural areas by breaking the dualistic system and improving the integrated market of elements. The key links are the establishment of market-based mechanism for the allocation of resource/elements and the realization of orderly and bidirectional flow of elements, in particular land and labor, between urban and rural areas. Foremost, urban and rural labor markets need to be integrated, the key point of which is to enhance the radiating exemplary role of urban areas and then to schedule the gradual transfer of rural labor, to create an institutional environment that guarantees fair competition and equal employment (equal pay for equal work) along the urban–rural continuum with the help of a public service and vocational training system for both urban and rural workers and an urban–rural integrated employment and labor management system. Coincidently, the dualistic land markets need to be integrated as well. A large amount of arable land has been occupied with rapid economic growth and urbanization, causing an increasingly prominent contradiction between socio-economic development and red lines for arable land and ecological protection in China. Land in the vast rural areas become vacant and abandoned, which makes it difficult for rural land to share the bonus of urbanization. Accordingly, it is necessary to deepen the reform of the rural land system, and to improve the system innovation in agricultural land, homesteads, commercial construction land, land acquisition system, and property rights system. It is also necessary to carry out the comprehensive consolidation of rural land and to build a platform for rural construction and urban–rural integration with the support of the "increase vs. decrease balance" land use policy and under the guidance of the strategy of "three integration" (i.e., the integration of rural organizations, industries, and spaces).

8.1.2 Optimizing the Urban–Rural Development Pattern

Although industrialization and urbanization promote regional socio-economic development, they also trigger changes in regional inequality that exists not only among cities of different sizes, but also between urbanized regions and the agricultural and ecological counterparts. In general, spatial inequality is closely related to regional strategy, economic foundation, industrial structure, and regional functions. Based on the proportion of state-owned economy, private economy, and foreign-invested enterprises in the local economic structure, China can be zoned into different subregions (e.g., the Yangtze River Delta dominated by private economy and the Pearl River Delta dominated by foreign-invested enterprises). In addition, we can also classify functional zones into agriculture-, manufacturing-, business/service-oriented, and balanced development type. According to the theory of comparative advantage, it is expected to achieve the goal of regional and urban–rural integration if each region seeks a scientific industrial division of labor based on its own comparative advantages.

Considering the development trajectories of different rural areas such as suburbs, exurbs, remote agricultural zones, non-agricultural zones, and major crop-producing zones, local practices of urban–rural integration have to some extent combined the regional main functions and urban–rural spatial heterogeneity, and thereafter portrayed suitable blueprints to coordinate urban–rural development and rural construction. However, the following issues have not been clarified yet in the course of policy practice. Firstly, the respective rights and obligations of major crop-producing regions and the non-agricultural counterparts are still confusing. The major crop-producing regions' functions of food security, social stability, and ecological conservation have not yet been properly reflected through agricultural product prices, GDP growth commissions, regional fiscal transfer payments, and any other channels, which to a certain extent led to or aggravated puzzles of "non-grain" agriculture, non-agriculturalization of arable land, and widening regional disparity. Secondly, the rights and obligations of ecological conservation and beneficiary regions are still puzzled. The opportunity cost of ecological conservation zones' protecting the regional ecology has not been well compensated, which to some extent intensifies the vicious circle of "poverty-environmental degradation" in mountainous and hilly areas. The last but not the least important is the respective functional value of the city and the village along the urban–rural continuum. Multiple functions of production, consumption, regional support, and body development carried by the rural can be better implemented through the manifestation of element value and the rational distribution of element value-added benefits. To a certain extent, this leads to or even aggravates issues of village hollowing, labor aging and weakening, and urban–rural gap widening (Chen 2010). As a result, in the context of rapid urbanization and industrialization, there are still large differences among regions, between urban and rural areas, and within rural areas.

Identifying the pattern of regional division of labor and constructing a reasonable mechanism are considered as the basic requirements and important directions for realizing the value of urban and rural areas and facilitating urban–rural integrated development. Foremost, local authorities should make scientific analyses on their advantages and disadvantages and comprehensively define their superior functions based on the demand of both international and domestic markets. The regional division of labor under the constraints of market orientation and development conditions, as well as the functions of urban and rural areas, can be clarified. Secondly, the functional positioning of local development can be clarified at a strategic level by formulating regional dominant function plans. Thereafter, the introduction of fiscal transfer payment and industrial support policies might be favorable to foster regional dominant, characteristic, and advantageous industries. The movement of "one village-one product, one town-one industry, and one county-one specialty" can therefore be carried out by developing rural economies based on local conditions. Finally, the benign interaction mechanism of promoting the construction of new countryside with the development of production and industrial development should be established to facilitate rural development.

8.1.3 Improving the Social Governance Ability

In the context of rapid urbanization and industrialization and the significant regional inequality, China's urban–rural transformation is rooted in the trend of the transformation from rural society to urban society, which originated from the main needs of farmers and was promoted by the top-down governmental intervention. The urban and rural governance system is composed of both governments at all levels and social organizations that manage urban and rural economic growth, social development, and ecological construction. Since the foundation of New China, the country, as a traditional agricultural state and rural society, has formed a parallel governance system, namely the governance of cities by urban governments and the governance of villages by rural collective autonomy organizations. After the reform and opening-up, the household contract responsibility system provided opportunities for rural development. The autonomy governance of village committees was confirmed by the Constitution in 1982. By the end of 2007, there were more than 620,000 village committees across the country. As the focus of reform shifts from rural to urban, the strategic and policy characteristics of industry- and urban-biased become more and more obvious. The urban governance thus has the top priority in urban–rural governance system, yet the rural governance is at the subsidiary position. Since the 1990s, rural society has become increasingly open, mobile, diversified, and complicated. Village-level and township organizations in rural China are confronted with tremendous changes and transformations. Since 2003, the central state has put the coordination of urban–rural economic and social development at the top of the "five coordination" and regarded it as a national strategy. With the agricultural tax exempted in 2006, the central state introduced a series of farmer-beneficial policies that strengthened rural infrastructure construction and social welfare. In practice, with the outflow of rural population and the construction of new countryside, all provinces have carried out experimental work on rural community construction, and gradually moved from the experiment to a new stage of full coverage. Meanwhile, some provinces have also devoted efforts to promoting the urban–rural integration and the comprehensive reform of urban–rural coordination, including land use, household registration, public services, and social security systems, to advance the reform of the township system and explore the township governance system.

Influenced by the synthesized effect of "push–pull" factors, large-scale rural–urban migration has become a prominent feature in the current economic, social, and demographic transition in China. According to the "2016 Migrant Workers Monitoring Survey Report" published in April 2017, the total number of rural migrant workers in 2016 was 281.71 million with an increase of 4.24 million compared to the previous year, of whom 169.34 million were emigrant workers. With regard to the demographic structure of migrant workers, the ratio of male to female is 65.5:34.5. The cohort aged 21–50 accounted for 77.6%, implying that the majority of migrant workers is young and middle-aged rural laborers that have been better educated and skilled. The large number of emigrant workers is a "double-edged sword" for rural development. On the one hand, it may be helpful to realize large-scale operation through land transfer. Emigration can also help to improve the skill level of rural labors and then facilitate the entrepreneurship through the inverse migration. On the other hand, the quality of the labor force left-behind in the countryside is always at a low level, which has led to difficulties in the promotion of agricultural science and technology, slow-down of agricultural transition, faults in the construction of rural grassroots organizations, increased social security problems, heavy education tasks for left-behind children, and insufficient supply of rural public goods. The large-scale rural–urban flow of high-quality young and middle-aged labor has impacted traditional agricultural production methods and rural social structures and has affected the supply of urban household registration management and education services. In summary, an urban–rural governance system that adapts to the mobility of population and social transformation in China has not yet been established. Due to the existing disparity between urban and rural development patterns, there are also obvious regional differences in the urban–rural governance system.

8.1.4 Balancing Public Resources Allocation

Balanced allocation of public resources is an important guarantee for integration of urban and rural development. It involves interests of the state, collectives, and individuals, and is a major issue related to national economy and livelihood of the masses. The long-standing system of "more emphasis on city rather than countryside" has resulted in significant inequality of financial and social investments in urban and rural areas (Figs. 8.2 and 8.3). Since the tax-sharing reform in 1994, China's central and local governments had a certain unreasonable division of financial and administrative powers, decentralizing fiscal responsibilities and intensifying fiscal pressure of local governments. Local municipal governments are now required to raise funding for infrastructure projects and most of its welfare programs by themselves. The financial pressure on local government is therefore growing significantly, bringing about more governmental interventions on economic operations (Guo and Lu 2011). Industrialization and urbanization overly dependent on rural labor and land resources but give limited support to rural social services and infrastructure. The rural taxes and fees reform exempted farmers from the obligation to pay for township planning, so most townships lost their extra-budgetary financial sources. This has greatly affected the general fiscal revenue of towns and townships, whose fiscal function has been weakened as well. Given that there was no independent financial power at the township level, the funds required for the administrative operation and various social development projects completely dependent on the budget of the county. Most county-level finances were also relatively tight, especially in the central and

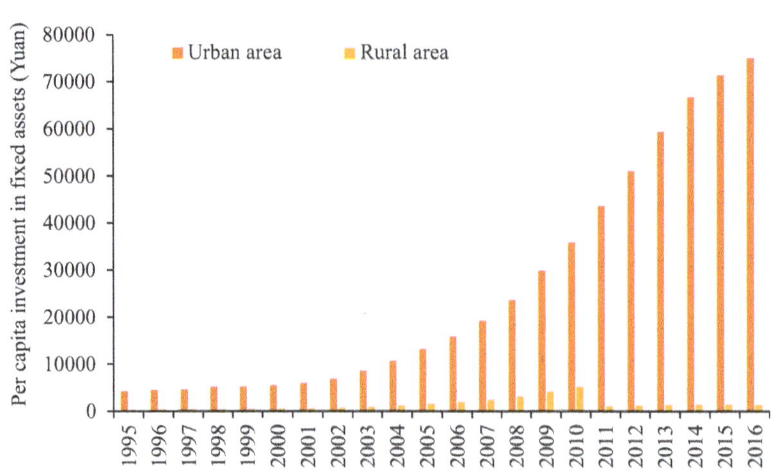

Fig. 8.2 Urban–rural fixed asset investment per capita in China from 1995 to 2016

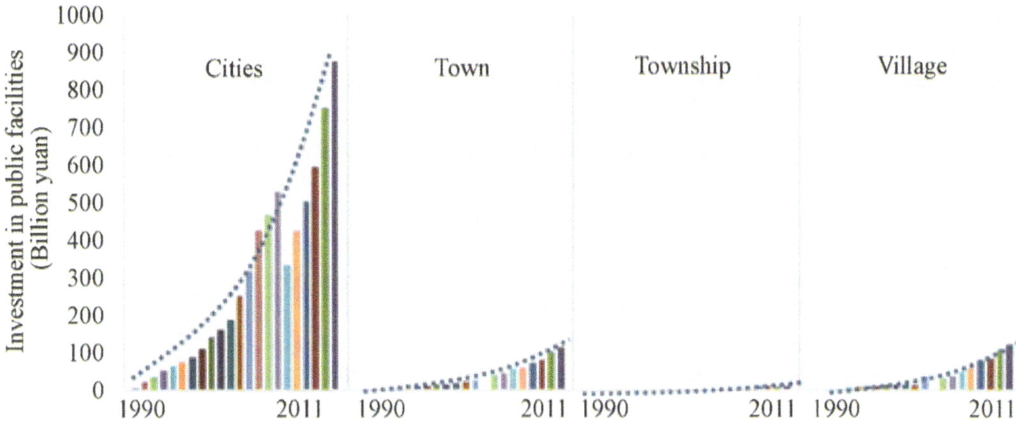

Fig. 8.3 Public utilities investment in cities, towns, township, and countryside from 1990 to 2011

western regions, where the county-level finance supporting capacity was extremely insufficient. These issues directly affected the implementation of village and town construction projects. In addition, many towns/townships had not formulated practical financial management measures. Although some townships have established related measures, they were in vain. Consequently, arbitrary utilization of funds was quite common at the township level. In 2013, most township governments across the country were in debt, with a total volume of about 476 billion yuan. The average debt of each township government was about 25.5 million yuan, leading to higher pressure of debt repayment. In all, the lack of funds, as well as the improper use, planning, and management of these funds became an important problem impeding the construction of villages and towns.

Under the dualistic urban–rural and economic growth-oriented system, rural public infrastructure construction is due for much more investment. Rural production and living conditions are poor. Rural residents have difficulty in transportation, water supply, and medical service. Issues like dirty and chaotic environment and backward public utilities are very serious. To date, only about 50.5% of villages in the China have centralized treatment of garbage and only 11% villages have centralized treatment of sewage. In recent years, some villages and towns have successively built rural public infrastructure such as schools, hospitals, village-level roads, agricultural water conservancy facilities, domestic sewage treatment projects, collective affairs buildings, and rural activity centers, which promoted rural economic and social development and improved farmers' quality of life to some extent. But the problem of "emphasizing construction and neglecting management" happened here and there. A lot of attention was paid to construction of projects by various departments at all levels. However, the management and protection work were gradually ignored once the project was completed, due to the complicated management process and high maintenance costs. For example, a large amount of manpower and financial resources were invested in the construction of canal system. Several years later, some of them were seriously damaged. Especially, small agricultural water conservancy projects had been used and unattended. Pumping stations, irrigation facilities had even been lost and stolen from time to time. In some newly built drinking water source projects, due to the lack of implementation of the management system, the unsustainable utilization resulted into water shortage when drought hit (Yu 2009).

In addition, affected by resource endowments, environmental carrying capacity, and economic foundation, there are significant differences in the level of public resource investment in regions, which are also presented among regions, cities, and groups of dwellers. First of all, there are large differences in government and social investment in infrastructure, education, medical care, and social security in eastern, central and western China. In particular, the economically backward regions in the central and western impoverished mountainous areas and the central plains agricultural areas still insufficiently invest in infrastructure and social services. Secondly, small and medium cities and small towns lack service functions, making it difficult to effectively attract spatial agglomeration of production factors in the surrounding rural areas. According to the Development Report of Floating Population in China 2017, the total number of floating people reached 245 million by 2016. The flow pattern of agglomeration in mega and large cities has not changed. Thirdly, there is an internal imbalance in the allocation of public resources among urban residents, especially between the rural migrant workers and original urban residents. A large number of migrant workers are engaged in working in the "dirty, chaotic, and poor" environments. They can hardly be equally treated as their urban counterparts in terms of children's education, medical and health, family planning, and social assistance. In 2013, the proportion of migrant workers in the manufacturing and construction industries was 31.4% and 22.2%. The proportions of migrant workers who are covered by pension insurance, employment injury insurance, medical insurance, unemployment insurance, and maternity insurance were only 15.7%, 28.5%, 17.6%, 9.1% and 6.6% respectively, which were far lower than urban residents.

Since the CPC Central Committee proposed the strategy of building a new socialist countryside in 2005, governments at all levels have emphasized "government leadership and farmers as the mainstay". In practice, they have achieved gratifying results of rapid growth in farmers' per capita net income, and the emergence of successful development models, including the modern agricultural industrialization in traditional rural areas, multi-functional agriculture in suburbs, village-enterprise integration promoted by rural industrialization. Comparing documents issued by various provinces like "Guiding Opinions on Promoting the Construction of New Socialist Countryside" and "Decision on Comprehensively Promoting the Construction of a New Socialist Countryside", the construction of new countryside in the "11th Five-Year" and "12th Five-Year" still put economic development in the first place. According to our field survey, the government has over intervened and led the construction of new countryside by simplifying the policy system covering "producing development, wealthy life, polite rustic, clean and tidy environment, and democratic management" to an economic policy, and by one-sidedly understanding them as superficial projects like repairing roads, building new houses, and constructing new villages (Ye and Yang 2006). Local policy practices that failed to meet the demands of rural residents and solve the historical problems in rural medical and health, social security, and environmental protection became important bottlenecks hindering rural development. Given the fact that farmers' needs were neglected and rural public goods were still in shortage, it is urgent to establish a long-term mechanism for the supply of rural public goods to increase their effective supply. The Third Plenary Session of the 18th Central Committee clearly stated that farmers should fairly share the benefits of urbanization and land appreciation with the equal exchange of urban and rural elements and the balanced allocation of public resources. And urban investment in rural construction and various undertakings is largely encouraged; the equalization of basic public services in urban and rural areas should be promoted by coordinating urban–rural infrastructure facilities and construction of communities. It has been an important issue to build rural areas, develop agriculture, inherit culture, and maintain the environment in the construction of beautiful villages in the new era. The construction of beautiful villages is a large-scale social construction activity for all farmers, who are the direct participants, ultimate benefits, and the leader of this activity. Effectively solving the key problems that farmers faced in their production and daily life, changing the traditional development strategy of "emphasis on cities over the countryside", optimizing public resource allocation, bringing cities and villages into the same framework, and promoting the balanced allocation of urban and rural public resources are hereby argued as a pathway to address the issues of rural endogenous power lack and deprivation by urban development.

8.1.5 Coordinating the Remediation of Resource and Environment

The occurrence of problems in resource utilization and environment is closely related to the model of economic growth and development stage. Theories of development stage represented by the Kuznets Curve elaborated the correlation well. Under the traditional economic growth and urbanization model, China's economy and society were developing rapidly, which also faced puzzles of inefficient utilization of resources, environmental pollution, and ecological degradation. In the late 1970s, the reform of rural economic development made township and village enterprises (TVEs) become an important force in rural economy, which absorbed a large amount of surplus rural labor. Most TVEs are, however, small-scale with scattered distribution and lack of pollution control facilities, which had become the main reasons for resource depletion and environmental pollution in rural areas. As reported by the "National Survey of Industrial Pollution in Township Areas 2006", the proportions of industrial waste water, waste gas and solid waste emissions in township areas reached 22, 24.9 and 39.4%. Given that economic growth is the main assessment indicator for local authorities, local governments focus on economic growth in the short term, ignoring the protection of resources and the environment, and even sacrifice resources and the environment for pursuing economic growth. With the upgrades of urban industries, enterprises with high energy depletion and heavy pollution gradually transferred part of their production from cities to rural areas. Some industrial wastes and garbage were discarded or discharged into rural areas without treatment, which intensified the rural environmental pollution. The problem, recently reported as "cancer villages", has reflected the deterioration of the rural environment. Since the production and domestic waste generated in cities and industrial parks is difficult to be addressed, it needs to be disposed of by means of landfilling and natural degradation on a larger scale. The country tends to treat urban and rural environmental management and planning separately within the dualistic system. Environmental protection regulations such as the "Environmental Protection Law" and "Environmental Impact Assessment" mainly target urban environmental protection and industrial pollution prevention, and hardly include rural TVEs and environmental protection. The central and local governments have developed more infrastructure and service facilities in cities. And waste and sewage discharge systems have been well developed in urban areas. However, environmental governance in rural areas is seriously lagging. The contradiction between the backward infrastructure in rural areas and the heavy environmental load has become increasingly significant.

Driven by economic growth and construction of central cities, local governments have focused their efforts on urbanization and industrial parks construction. Urban construction land and industrial land occupies a large amount of arable land, resulting in a large amount of land being occupied and not being developed. The mismatch between rural population migration and land nonfarm use is obvious. Rural economic growth and population migration have caused the phenomenon of "expansion outside and empty inside" of rural construction land, and some villages have become hollowing. Due to the gap between urban and rural areas and the imperfect security system, the process of rural population migration to cities and towns did not realize "land follows people". With the purpose of development and income growth, farmers even take some short-sighted behaviors of environmental contamination and resource depletion, such as overgrazing, reclamation of slope land, and deforestation. To increase crop yields, a large number of pesticides and fertilizers are used in agricultural production. The utilization of chemical fertilizers increased by 5% annually from

1985 to 2010. The use of chemical fertilizers per unit area of arable land was close to 4 times the world average level. Unreasonable chemical fertilizers utilization not only caused waste of chemical fertilizers, but also became the main source of agricultural point source pollution (Luan et al. 2013).

In general, urban–rural resource and environmental issues are closely related to the economic model of development and the urban–rural dualistic system in China. The direct influence of the free flow and agglomeration of urban and rural elements (i.e., labor, capital, goods, and information) and the resources as well as environment problems generating in the process of flow and agglomeration were intertwined with each other. The development of non-agriculturalization of industries, urbanization of population and non-agriculturalization of employment during the transition period is fundamentally the flow of factors emerging under China's economic growth mode and dualistic system. The existence of economic models, management systems, and urban–rural gaps has worsened the resource and environmental problems. The carrying capacity of resources and environment in urban areas declined, while problems such as resource depletion, water and soil pollution, and land desolation have emerged in rural areas. In the meanwhile, urban areas transfer the pressure of resources and environment to rural areas through elements flow. Therefore, it is necessary to construct a governance system of urban and rural resources and environment from perspectives of systematic evolution of urban–rural areas and urban–rural integrated development.

8.2 Optimizing Ideas and Pathways Towards Urban–Rural Transformation

8.2.1 Optimizing Ideas of Urban–Rural Transformation

The National New-type Urbanization Plan emphasizes people-oriented urbanization and argues that the coordination of urban–rural development is an important content and strategic fulcrum (Lu and Chen 2015). The integration and equivalence of urban and rural development is the approach to and goal of urban and rural development. The core is to achieve balanced development of urban and rural areas with regard to economic, social and ecological dimensions, of which the focuses is the common prosperity of people and regions. Based on the urban–rural dichotomy notion, it becomes more difficult to adapt China's current system design to actual needs of the integration of urban–rural development in the process of economic growth and urbanization. Urban–rural social governance, sustainable use of land resources, and urban–rural management systems face challenges. At the same time, paths and problems of urban–rural development vary in different regions; regional characteristics should be taken into full consideration in the draft of development strategies and mechanism innovations. The authors hereby argue that the core of urban–rural transformation is to realize the free flow and equal exchange of elements at the micro scale, to realize the division and coordination of labor of industries at the medium scale, to realize the equalization of urban–rural development strategies and systems at the macro scale, and to finally promote the balanced urban–rural development (Fig. 8.4).

1. **Optimizing ideas of the allocation of urban–rural elements**

The idea of optimizing the allocation of urban–rural elements is rooted in both equal exchange and classified regulation of elements. The former means that population, land, technology, capital, information, and any other elements could be developed and employed equally based on market comparative advantages in urban and rural areas. Within China's urban–rural dualistic system, elements can hardly be exchanged equally. Rural space and elements, as well as the values added, are usually deprived by urban construction. Meanwhile, urban capital and residents are also impeded by the rural land system and grassroots governance system when flowing into

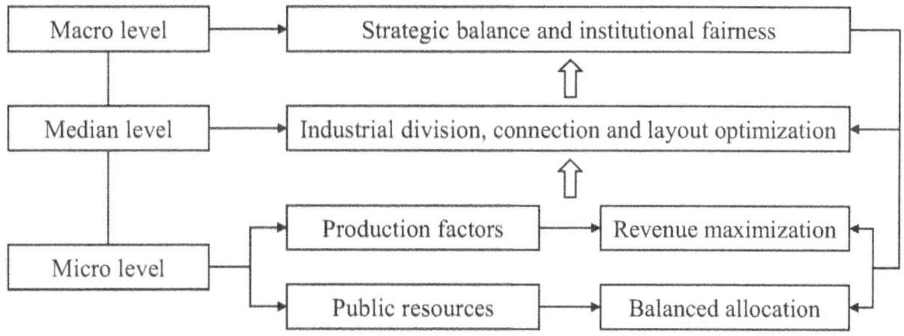

Fig. 8.4 Optimizing ideas of the urban–rural transformation

the rural. Nowadays, China has entered a new normal period of "economic growth slowdown, growth impetus conversion, structural optimization and adjustment". The demographic bonus has disappeared, and land finance is unsustainable. It is necessary to change the low-cost pathway for advancing economic development. The equal exchange of urban–rural elements is an inherent requirement to promote the integration of urban and rural development in the new era, which is presented by a bidirectional flow and equivalent exchange of elements. Therefore, it is necessary to change the development strategy of "emphasizing the city and industry over the rural and agriculture", to build a system and policy platform for the free flow and equal exchange of elements, and to improve the employment environment for migrant workers and the treatment of citizens. Additionally, we suggest actively guiding urban capital and residents to startup commerce, agricultural processing, modern agriculture, and other industries in rural areas, and exploring the formation of rural land markets and rural housing markets.

In general, the attributes of different production factors vary. Besides the economic value, there are also different degrees of non-market values. Factors such as technology and capital are often aimed at maximizing returns and need to be employed in production activities to the largest extent. However, land and water, as non-exclusive public resources, must meet the needs of different stakeholders and functions as the supplement of their economic values. It is necessary to distinguish the attributes of different elements and to adopt different measures to regulate for the coordination of urban–rural development. For the allocation of market-related resources, the role of market mechanism should be strengthened by reforming related systems that hinder the functioning of market mechanisms; with regard to the allocation of those non-market resources, the coordination of economic and non-economic attributes with the goal of maximizing the total welfare of the entire society is the key.

2. **Balance development between urban and rural areas**

In the process of industrialization and urbanization, the geographical separation and functional differentiation of urban and rural areas have general rules. Balanced urban–rural development means a relatively balanced development status in multiple dimensions of economy, living, and ecology along the urban–rural continuum, which is realized by balancing the development relationship based on geographic and functional values. With regard to economic development, it is necessary to innovate the division of labor and linkages between urban and rural industries, to optimize the spatial layout of industrial parks in cities, towns/townships, and villages, to advance moderate scale agricultural operations and intensive processing of agricultural products; to promote the integration of primary, secondary and tertiary industries and the development of rural service industries by taking central towns and rural communities as the hubs and nodes respectively, and thereby to eliminate the income inequality among labors working in different sectors and urban–rural areas. Secondly, it is necessary to comprehensively consider the economic, social, and ecological benefits of urban–rural continuum; to pay more attention to the functions of food production, ecological service, and social security in rural areas as well as protection of the rural natural environment; to maximize the comprehensive revenue of urban and rural areas based on the multi-functional utilization and value feedback on rural areas. Thirdly, it is also suggested to promote the balanced allocation of urban and rural infrastructure and public services, and gradually incorporate the employment, education, medical care, and social security of urban and rural residents into the same system.

3. **Region-specific optimization ideas**

Due to heterogeneity of natural conditions, economic growth, history and culture, the regional contexts of urban–rural transformation vary obviously. The regional characteristics and difficulty of achieving equal exchange of elements and balanced development are therefore varied across regions. As there are intra-regional urban–rural and inter-regional connections, promotion of urban–rural integrated development involves a composite relationship, not only between urban and rural areas but also among regions. The coordination of urban–rural development at the regional scale requires comprehensive consideration of spatial–temporal dimensions. In terms of spatial dimension, comprehensively considering natural conditions, economic society, and institutional environment, analyzing the main conflicts and problems of urban–rural development, and then adopting appropriate countermeasures is the key to form a congruent development path. In terms of time scale, long-term and short-term goals should be combined. In the short-term, integration of urban–rural development is promoted at the county or prefecture levels. In the long-term, it is necessary to promote the integration of urban–rural development between regions at the provincial and inter-provincial levels. Only in this way, we can finally realize the free flow and equal exchange of production elements and public resources nation-wide.

8.2.2 Conceptual Framework of Urban–Rural Transformation

Coordinating urban–rural development is an inevitable precondition in the middle and late stages of industrialization. At present, innovations in China's household registration system, land system, social security system, and agricultural support policies have reconciled the conflicts between urban and rural areas and promoted rural development. However, the problem of divided system design with actual needs still exists. Coordinating the transformation of urban–rural economy, society, and ecosystem is an important feature of urban–rural integration and constructing optimization strategies for urban–rural transformation. Comprehensive design is required from the perspective of system transformation. Based upon the previous analysis of process, pattern, types, and typical regions of urban–rural transformation in China, we hereby build a multi-level and triple dimensional framework through integrating transformation issues and mechanisms. Figure 8.5 portrays the details of the framework, including the goal, operation, and support levels.

1. **Goal level**

Urban–rural development integration is the ultimate goal of urban–rural economic and social evolution and the basis for national and regional strategic regulation. Two premises are required: one is a higher socio-economic development level with strong economic links and clear industrial division between urban and rural areas, the other is a higher urban–rural coordination level with smaller economic gaps, equivalent living standards, coordinated ecological environment, and particularly a balanced allocation of public resources and infrastructure.

2. **Operation level**

Based on the ultimate goal of urban–rural development integration, this level provides optimization ideas and scientific basis for

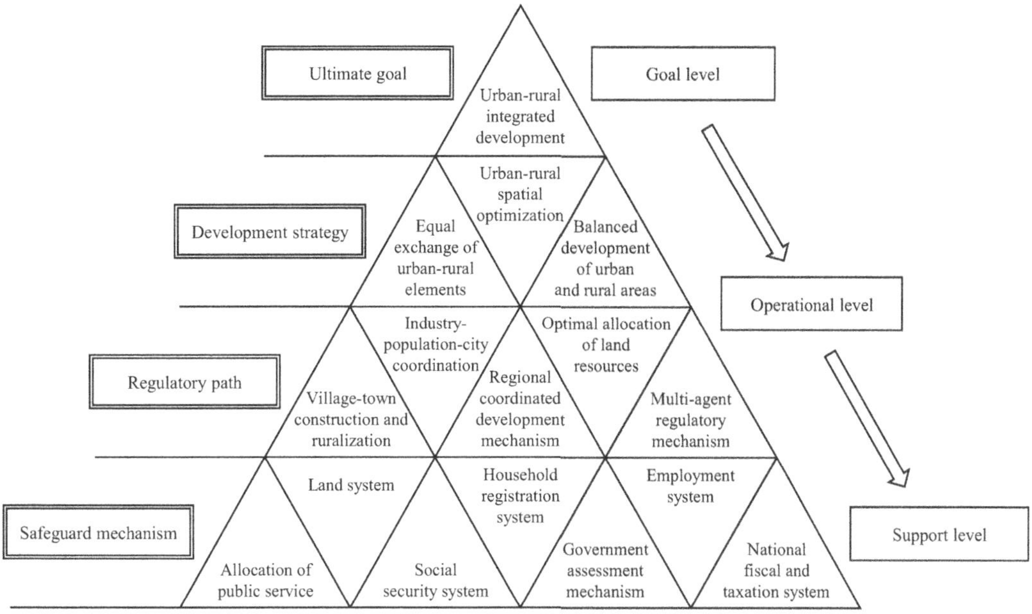

Fig. 8.5 Optimizing framework of coordinating urban–rural development

addressing issues of rural decline with urban expansion, land use contradictions, and poor population transformation, and for central and local governments' adjustment for both strategies and mechanisms of urban–rural coordination. This level can be further divided into two sub-levels, namely development strategy and regulatory path. Regarding the development strategy, we must adhere to the principle of urban–rural coordination and people-oriented and the guideline of "three synergies" (i.e., the synergy of "four modernizations",[1] of economy-society-ecological, and of efficiency and fairness), promote the equal exchange of urban–rural elements and the balanced urban–rural development, and then maximize the benefits of factors allocation and the social welfare in urban and rural areas. With respect to the regulatory path, it is an innovative mechanism to achieve the goals and strategies of urban–rural development, which includes the construction of towns and rural vitalization, the regional coordination mechanism based on regional major functions, the coordination mechanism of population-industry-towns, the optimal mechanism of land resources allocation, and multiple regulatory mechanism. In view of the different types of urban–rural transformation, distinct regulatory paths should be adopted based on the characteristics, problems, and major determinants of urban–rural transformation.

3. **Support level**

As an institutional and policy support, this level provides strategies and mechanisms for urban–rural development. The core systems include household registration system, land system, employment system, social security system, education system, and medical system. Institutional innovation aims to coordinate the promotion of various reforms around the free flow and equal exchange of urban–rural elements, and adapt to the social needs of rural–urban migration for employment, security, and welfare, and the needs of rural economic and social transformation.

The regulatory system for urban–rural development integration is an integrated system, in which the triple levels of goal, operation, and

[1]*Note:* "Four modernization" means industrialization, informatization, urbanization, and agricultural modernization.

support are interconnected and mutually influenced. The goal of urban–rural development integration has elucidated the direction for the formulation of government regulation measures and the improvement of policies and systems. The operating approaches provide an operational direction for promoting the integration of urban–rural development. The policy system fundamentally provides institutional support for realization of goals and path conversion. This framework provides ideas and routes for regional coordination of urban–rural development, and provides a scientific basis for solving urban–rural development problems in various regions and for optimizing the urban–rural development pattern. It plays an important role in guiding development.

8.2.3 Pathways Towards Urban–Rural Transformation

1. **Accelerating the development of small and medium-sized cities with industry-city integration**

Econmic growth and industrial revolution are the basis for the formation of cities, as well as the spatial restructuring and functional evolution. Cities can provide services such as labor, information, and facilities for industrial development, reduce costs of information, transaction, and employment for enterprises, provide a demand market generated by innovation, and finally promote agglomeration of population and industries. Benefiting from the advantage of historical foundation, geographical location, and opening-up policies, eastern coast and large cities in the central and western regions have developed rapidly with agglomeration of industrial enterprises and social public resources. Driven by the effect of scale economy, mega/large cities continue to expand. However, due to the lagging updating of industries, the population living space is separated from industrial development space in metropolitan areas, and migrant workers has to par much more for settling down in cities. The capabilities of infrastructure and public service as well as industrial and population agglomeration in small and medium-sized cities are weak. Consequently, the driving impact for surrounding rural areas is limited. The development model based merely on mega and large cities can hardly promote coordination of urban–rural development and narrow interregional development gaps.

Manufacturing enterprises upgrade the products and relocate based on comprehensive evaluation of market environment, supply–demand relationship, and their cost–benefit. The market regulatory mechanism plays a fundamental role in relocating enterprises. Symbiosis of industrial relocations, urban restructuring, and population migration is an important premise for urban–rural integrated development. Through macro measures such as industrial planning and fiscal and taxation policies, the government makes full use of land rent gap to regulate industrial layout, and promote rural migrants moving between towns and cities. In particular, local governments facilitate the effective transfer of industries and public resources from the urban centers of big cities to suburbs and small towns and coordinate the systems of industrial hierarchy and the urban function. More efforts are needed to advance the capability of infrastructure and public service in small and meddle-sized cities and towns and to promote integration of industrial and urban development by resettling the industries migrated from large cities.

2. **Ensuring economic, social and ecological benefits by optimizing allocation of land resources**

Sustainable use of land resources and balancing multiple demands and limited supply are the key issues to innovate land management system. Economic growth and urbanization have generated a strong demand for land; while in the context of ecological civilization construction, the demand for ecological land is increasing.

Accordingly, the conflict of land demand among food security, economic growth, and ecological security is still a primary problem in land resource management. To coordinate urban and rural development, it is necessary to solve the problems of underutilization of urban and rural construction land, excessive occupation of agricultural land, damage to farmers' rights and interests, and fragmentation of arable lands. To change the innovation of land system that takes a single measure for a single problem, the top-level design of land system innovation should be implemented from a systematic perspective (Liu et al. 2014a). To meet the needs of urban–rural economic and social development, it is necessary to innovate the rural land system, and to develop a reform plan consisting basic, core and strategic domains, as an effective way to deal with the contradiction between land supply and demand and to coordinate urban–rural development (Liu and Zhou 2015). Among them, the core includes the systems of land contract management, homestead withdrawal, operating collective construction land transfer, transaction of rural property rights, and land acquisition.

To realize sustainable use of land resources, it is necessary to take according measures to address the following issues: (1) Clarifying the spaces for residence, industry, agriculture, and ecology by identifying the permanent basic farmland, urban growth boundaries, and ecological red lines. (2) Scheming a plan on the stock construction land in urban and rural areas (Luo and Xu 2015). Confronted with the extensive use of urban industrial and mining construction land, land supply and demand should be determined by comparing the land input and output, which is also the key to optimize the spatial layout of urban construction. In terms of the problem of "building new without dismantling the old" and rural homesteads underutilization, it is recommended to make rural stock land plan, find out the basic situation of rural stock construction land, and explore the potential and reuse methods under the local transformation. (3) Promoting rural land consolidation. Complying with the trend of scale operation of agriculture and rural transformation and combining with the experimental exploration of rural land system reform, it is recommended to formulate master or special plans for consolidating fragmented agricultural land and underutilized homesteads, and then to build platforms for construction of beautiful countryside and development of modern agriculture.

3. **Promoting regional coordinated development based on main functional orientation**

Physical geographic environment is the basis for formation of the terrestrial surface space. Based on the heterogeneity in physical conditions, economic foundations, geographical locations, and institutional contexts, the industrial focus and urbanization mode in urbanizing areas, main crop-producing areas, and ecological areas are determined. Based on regional major functions, different paths of urban–rural transformation are explored. Development in various functional areas is coordinated through regulation of the central state. Areas with rapid urbanization and urban agglomerations will accelerate growth of emerging industries and industrial transformation by taking advantage of economic growth and urbanization, and then promote development of rural areas and traditional agriculture. Consequently, the industrial and spatial systems of urban–rural collaboration are optimized. In the densely populated crop-producing areas, industrialization and urbanization can be used as instruments to accelerate migration of rural population, to promote the scale operation of agriculture and extension of industrial chains, to increase support for infrastructure and public services in rural counties, and to strengthen development of service industries in central towns and rural communities. In ecologically fragile areas, more attention should be paid to carrying capacity of resources and regional environment, cultivation of local special industries, development of ecological agriculture and green industries, and ecological migration of residents in uninhabitable geological disaster areas. In contiguous poverty-stricken areas, under the guidance of the new strategies of poverty alleviation and development, poverty alleviation and development are organically combined with the strategies of new-type urbanization and construction of new countryside. Particularly, vocational training and modern education for labor are strengthened. Rural poverty alleviation is achieved through labor transfer and special industry training. Inter-regional trade rules, financial transfer and official evaluation mechanism in different major functional zones need to be improved. Only in this way, can the urban–rural integration development in national main crop-producing areas and ecologically fragile areas be promoted.

4. **Building platform for urban–rural coordination by the construction of village-town pattern**

With continuous outflow of rural population, decline and aging of rural labors, as well as underutilization of rural land have become an inevitable trend. Rural society urgently needs to build a new construction pattern of villages and towns to promote transformation. The notion of "construction pattern of villages and towns" refers to spatial layout, hierarchical relationships, and governance systems of counties, key towns, central towns, and central villages (communities) in rural areas. It is necessary to build a new main force, new power, and new system of rural development based on the county and township regional space, and then to reconstruct the functions of industrial development, service supporting, ecological conservation, and cultural inheritance (Liu et al. 2014b). Rural transformation and restructuring require the combination of endogenous power and external promotion. In terms of rural endogenous power, it is necessary to develop a variety of industries based on the rural nature, resources, and cultural characteristics, especially to promote agricultural modernization. For the urban external promotion, it is necessary to rely on the driving force of urban capital, industry, and consumption, and to emphasize the role of urban factors in promoting rural development in cognition, strategy, and institution terms. The construction pattern of villages and towns is a fundamental requirement to lay the foundation of rural development and to build a platform for urban–rural coordination.

Based on the trend of urban–rural transformation, villages and towns should be zoned and classified, and corresponding development and rectification measures should be proposed by fully considering the needs of farmers. For villages and towns with good industrial foundation and well-equipped infrastructure, supply of public services and facilities can be increased. Through extension of industrial chains and attracting concentration of elements, development capabilities can be enhanced. Thereafter, the regions are built into the cores. For traditional ancient villages with obvious cultural characteristics, comprehensive protection should be given priority to avoid disappearance of rural culture and traditions. For rural areas that are severely hollow, under the premise of fully respecting farmers' willingness, new communities will be built, and new cultivated land will be reclaimed through rural land consolidation. For the mountainous and hilly areas in the central and western regions with poor habitat environment and inconvenient transportation, the rural production and living space is reconstructed through measures such as gully land consolidation, comprehensive development, and ecological migration. In a word, the construction of villages and towns should be tailored to local conditions in accordance with the times and focus on the spatial and functional integration of industrial cultivation and community construction to avoid the new hollowing phenomenon.

5. **Building a coordinated regulatory system with the multi-level overall planning**

Government administration in China is hierarchal, and government's implementation capacity

Table 8.1 Hierarchy system of coordinating urban–rural development in China

Strategic goals	Regulatory approaches	Key points
Coordinate economic, social, and ecological relations; regional integration, urban–rural coordination, human-earth harmony; equal exchange of urban–rural elements	Macro strategic level: national scale Construct a regional integration mechanism to compensate the main crop-producing and ecological areas; ensure food security and ecological security; innovate urban and rural management systems; equalize the basic public service in urban and rural areas; optimize the layout of productivity and build a spatial pattern of balanced development	Reform the urban–rural dual system to alleviate social contradictions; increases investment in rural agriculture
Enhance economic growth capacity, urban–rural coordination, industry-agriculture integration, urbanization and urban construction, employment transfer and public resource allocation	Middle coordinating level: provincial scale Promote the industrialization of agriculture and establish a modern agricultural production system; establish an orderly pattern of large city-small/medium cities-small towns-central villages-communities/villages; improve the mechanism of urban driving rural and industry-promoting agriculture; narrowing the urban–rural gap, balanced allocation of public resources among different counties	Balanced allocation of public resources among different counties; driven by two wheels of new countryside and urbanization
Improve economic capacity of counties, coordinate agriculture and industry, and modernize agriculture	Micro operating level: county/township scale cultivate the characteristic rural economy and improve the ability of rural self-development; restructure the rural space to form a system of town-central community-village; improve the quality of farmers and cultivate new farmers; promote the construction of rural grassroots organizations and improve rural autonomy and democratic management capabilities	Develop characteristic agricultural industries, cultivate county-level supporting industries, and restructure rural spaces

and scope are closely related to administrative level. The goals, regulatory measures, and institutional innovations vary a lot for governments at different levels in regulating urban–rural transformation. Therefore, coordinating urban–rural development requires cooperation at all levels. From the perspective of administrative functions and spatial scales, coordination of urban–rural development can be divided into three levels, namely the macro-strategic level, middle- coordinating level, and micro-operating level. Correspondingly, it is necessary to promote equal exchange of urban–rural elements and balanced regional development through multi-scalar linkages of regulatory (Table 8.1).

Macro-strategic regulation means the national strategies and policies of urban–rural development released by the central state. At the level of macro-system and policy, central state formulates strategies and institutions that coordinate economic-social-ecological, regions, urban–rural, and human-earth relationships to promote equal exchange of urban–rural elements. Specific approaches include establishing a regional integration mechanism, compensating crop-producing and ecological areas, formulating different development strategies, improving financial transfer payment mechanism, ensuring national food and ecological security, innovatively promoting urban and rural management system reform, urban and rural basic public services equalization, optimizing layout of productivity and building a spatial pattern of balanced development.

Middle-coordinating regulation refers to coordination of the economic development and public resource allocation among counties from provincial/prefectural perspectives, optimization of urban–rural spatial structure, and innovation of "urban driving rural and industry promoting agriculture" mechanism. Specific measures include advancing industrialization of agriculture and establishing a modern agricultural production system, establishing an orderly spatial pattern of large cities-small/medium cities-small towns-central villages-communities/villages, improving the mechanism of "urban driving rural and industry promoting agriculture", narrowing the urban–rural gap and balanced allocation of public resources among counties.

Micro-operating regulation refers to the role of local governments in promoting socio-economy, agricultural industry, and rural reconstruction at the county and town level. Specifically, it includes cultivating characteristic rural economy and improving ability of rural self-development, restructuring space of villages and towns to form a spatial system of towns-central communities-villages, improving quality of farmers and cultivating new farmers, promoting construction of rural grassroots organizations and improving rural autonomy and democratic management capabilities.

6. **Building a sustainable development mechanism by coordinating human-earth relationship**

Coordinating economic, population, resource, and environmental development is the core for optimizing the human-earth relationship in the process of urban–rural transformation. Industrialization and urbanization, as the leading process of China's socio-economic development, directly affect utilization of water and land resources, as well as ecological environment, the degree and trajectory of which are in turn closely related to the transformation process. China's urban–rural resources and environmental problems stem from the development mode, stage, and urban–rural incoordination. Besides reducing emissions and use of fertilizers and pesticides with technological innovation, innovating management approaches, optimizing industrial structure, improving rural development capabilities, and narrowing urban–rural gaps matter for addressing the issues regarding resources and environment in urban–rural transformation. In particular, it is necessary to change the traditional industrialization and urbanization mode that rely excessively on land, water, energy and mineral resources, and promote coordination of human-earth relations through intensive and efficient development.

Coordinating urban–rural economic development is the basis for addressing the issues of resource depletion and environment pollution and is also an important condition for building a sustainable development mechanism. Given the specific urban–rural differences in economy and society, flow and agglomeration of production elements are unidirectional and unbalanced, which has triggered resource and environmental problems. For example, demographic urbanization and industrialization promote expansion of urban construction land. Driven by China's unique land requisition and financial systems, local governments' pursuit for land finance increased occupation of rural land by towns and parks. Due to lack of efficient management, these occupied rural lands tend to be underutilized. The deprivation of rural land development rights has therefore exacerbated urban–rural gaps. At the same time, due to urban–rural differences in systems and governance capabilities, it is difficult for rural areas to effectively control pollution from both point and non-point sources. To solve this problem, we need to effectively improve rural development capabilities and farmers' income and narrow the gaps between urban and rural department.

The sustainable development mechanism requires cooperation of the government, society, organizations, enterprises, and individuals. Each participant assumes corresponding responsibilities in resource utilization and environment protection and helps to establish a coordinating mechanism for market regulation and government control. In the complex multi-agent system, government needs to strengthen institutional regulation and social management with legislation when adjusts the economic structure and development model, restricts resource and environmental constraint for economic development, and then regulates the production behavior of enterprises. By activating all social forces and educating the public, government will guide people to form healthy lifestyle and habits, which is also an important pathway towards solving resource and environmental problems.

8.2.4 Regional Types of Urban–Rural Transformation

The urban–rural transformation model is a summary of regional and stage characteristics and is widely considered as being closely related to its driving forces. To date, the urban–rural transformation model generally stems from urban or rural development. The first is a top-down mode from the perspective of cities and urbanization, which highlights the driving impact of cities and industrial parks on regional development. The second is development of industries in rural areas from the perspective of the construction of small cities and towns, which emphasizes the endogenous model of rural development. The third is to consider the connection and interaction between urban and rural areas, and emphasize the role of small towns, industrial division of labor, and technological innovation in urban and rural development. China has a large population with relatively limited land resources, and the gap among regions and between urban and rural areas is obvious. Urban–rural transformation is restricted by economy, system, and resources. Therefore, it is necessary to fully consider the comprehensive and regional characteristics of urban–rural transformation and explore the implementation of localized types of urban–rural transformation.

1. **Metropolis-driven type**

Since the accelerating stage of urbanization, large and mega cities in China have developed rapidly, and have become the main destination for population immigration and enterprise agglomeration, which prompts the formation of the spatial pattern dominated by large and mega cities with prefecture-level cities as the main body. Depending on the advanced network of transportation and communication infrastructure, compact and economically connected urban agglomerations (e.g., the Yangtze River Delta, the Pearl River Delta, Beijing-Tianjin-Hebei, the middle reaches of Yangtze River, Chengdu-Chongqing, Harbin-Changchun, central and southern Liaoning, Shandong Peninsula) have emerged across the country.

Based on major large and mega cities, the aforementioned agglomerations have driven farmers in the surrounding areas to work in supporting industries such as services and catering in towns or industrial parks through clustering development of machinery, chemicals, petroleum, agro-products processing, and financial services. It has promoted the migration of surplus agricultural labors in rural counties. With the locational advantages, surrounding counties and towns optimize their industrial structure and production environment through infrastructure construction and establishment of supporting industries. Taking advantages of agricultural resources, rural culture, and tourism resources, some others actively develop urban agriculture, rural tourism, agricultural tourism, and leisure tourism. The emergence of new business formats and demands has significantly promoted development of rural areas.

2. **Development zones (DZs)-led type**

Regions with development zones (DZs) are widely considered economically developed and have a good industrial foundation. Construction of DZs further promotes regional socio-economic development and becomes an important spatial carrier for economic growth. The core of DZs-led type is to promote urban–rural economic growth, employment transfer, and restructuring of urban–rural spaces through agglomeration of production elements and adjustment of industrial structure. Since the 1990s, the number and spatial scope of DZs have continuously expanded in China. The types of DZs became more diverse, including special economic zones, economic and technological development zones, high-tech industrial development zones, export bonded zones, financial and trade zones, border economic cooperation zones, new zones, and experimental zones of free-trade. To date, there are 219 national economic and technological development zones, 146 national high-tech industrial development zones, and more than 1,200 provincial-level development zones. From 2010 onward, construction of new urban areas has been accelerated as well.

Governments at different levels of provinces, prefectures and counties regard development zones as spatial carriers for regional economic growth and enterprise agglomeration. To promote economic growth and increase fiscal revenue, local governments would provide convenience and preferential treatment for enterprise development by improving infrastructure and supporting public services in DZs. DZs have therefore become important platforms for agglomeration of urban and rural population and production elements, greatly promoting adjustment of urban and rural industrial structure and triggering adjustment of urban and rural spatial structure. Construction of DZs become an important external driving force for urban–rural transformation, which mainly manifests in the supply of a large number of jobs by inducing enterprises and attracting TVEs and in the payment of land rent to local farmers for occupation of arable land resources.

3. **Urban construction-driven type**

The process of urban–rural transformation driven by urban construction is universal and is closely related to the current acceleration of urbanization in China. The core of this type is employment transition of rural labors and migration to urban areas. Reform of urban land use system, loosening of household registration, and restructuring of urban enterprises have given cities a rapid development momentum. With good location, infrastructure and public service conditions, urban areas not only provide a good production and service environment for enterprises but also attract external funds to expand scale of urban economy. Driven by expansion of urban space, some villages on the fringe of cities are directly incorporated into urban construction areas and transformed into urban areas. Some others accelerated transformation due to industrial connections, population mobility, and diffusion of service functions. On one hand, production elements such as population, land, and capital have gradually shifted from TVEs and agricultural departments to urban areas. Distribution of TVEs have changed from "scattering in the rural" to "concentrating in cities and towns", which accelerated spatial expansion and economic growth in urban areas. On the other hand, rural areas have speeded up renewal of old villages and construction of communities and infrastructure with help of urban market demand and financial support.

4. **Urban–rural interactive development type**

Based on the foundation of agricultural production, agricultural counties in the central and eastern plains tend to accelerate development of industries like characteristic agriculture, agricultural product processing, and township manufacturing and then form development models of "mutual promotion of agriculture and industry" and "urban–rural linkage". The core is establishment of connections between industry and agriculture. Main features are as follows: (1) Construction of specialized agricultural production bases, improvement of agricultural production organization system, building of a new industrial system integrating production, processing, and sales, increase of added value of agricultural production, and modernization of agriculture; (2) Coordination of production of field crops, cash crops, and facility agriculture, innovation of diversified agricultural industrialization models such as "company + farmer", "base + farmer", "base + cooperative organization + farmer". On the basis of the agricultural industry, it is recommended to accelerate development of agricultural products processing, agricultural service, logistics, fertilizer, machinery, and other industries, and to enhance value of agricultural products based on industries of logistics and information. With radiation of industry and capital in surrounding big cities, it is also recommended to build a local industrial supporting system, to share value of the industrial chain, and to accelerate development of tourism based on local characteristic resources and folk cultures.

5. **TVEs-driven type**

The TVEs-driven type refers to development of rural economy by relying on TVEs and in-situ urbanization of rural migrants with construction of small towns. The TVEs-driven type is mainly distributed in areas with better economic development conditions, exemplified by orthodox Su'nan model, Wenzhou model, and Pearl River Delta model. Employing the collective or individual capital accumulation, some TVEs developed labor- and resource-intensive industries in developed eastern coastal areas and rural areas adjacent to large and medium-sized cities or rich in mineral resources. Some others took advantage of the foreign investment under the opening-up policies and with help of local governments. The in-situ migration of rural labors occurred. Led by the construction of DZs and small cities/towns, TVEs have witnessed a trend of gathering in industrial parks. Coincidently, rural residents are resettled in concentrated communities with help of land consolidation.

Within the backgrounds of land restriction of urban expansion, the difficulty of demolition and relocation in rural areas, and the transfer of urban population and capital to rural areas, new trends have been witnessed in development of TVEs. Some enterprises leased rural collective operating construction land and set up factories in rural areas, resulting in a decentralized pattern of industries. However, these enterprises are only able to bring land benefits to rural areas and can hardly address the issue of local labor migration.

6. **Rural characteristic industry-driven type**

Employing regional characteristic climate, natural resources, folk culture, and historical materials, the characteristic industry-driven transformation refers to development of industries with local characteristics, such as characteristic agricultural products, agricultural processing, ethnic cultural tourism, and natural tourism etc. Actually, the characteristic industry-driven type is a model of rural endogenous development, which drives local socio-economic development and meets market demand with market-oriented development of immovable natural and human resources. For instance, tourism development in ancient villages and towns, cultivation of characteristic agricultural products in mountainous and hilly areas, and ethnic customs tourism in ethnic areas attract a large number of urban residents in China and beyond. Southwestern China is vast and sparsely populated with undulating terrain and fragile ecological environment, the majority of which is still dominated by rural landscapes. The development of characteristic industries should be strengthened based on its unique resource endowment.

8.2.5 Land Engineering as a Technical Support

Land capacity is an umbrella term for land potential, productivity and functions, and a comprehensive embodiment of the natural potential, economic values, ecological functions, social security, and technical contributions of the land of a specific region (Liu 2018). Here, land capacity differs from what we used to call land potential or productivity, as it is a systematic and comprehensive supporting service and guarantees capacity. The fundamental goal of land capacity building is to introduce land engineering technique, create high-standard farmlands, guarantee sustainable land use, and promote agricultural security as well as sustainable rural development. Relying on agricultural land engineering, the primary approach of land capacity building is to explore the techniques of cultivating and meliorating organic soil, and to make clear the coupling process of water-soil-climate-biology and the occurrence theory of "three-micro elements" (i.e., microstructure, micromorphology, and micro-mechanism) by conducting field observations and scientific tests of research bases. It also aims to put forward a coordinated strategy of the agricultural ternary structure with food crop, economic crop and forage crop, and an agricultural mode of production-living-ecological integration, thus offering a scientific basis for the planning and decision-making of land use and rural sustainable development in typical cases and major regions and at both national and global scales (Liu 2017; Yang et al. 2018). Land consolidation provides a critical means of realizing land capacity building and it is an indispensable way of spatial restructuring in rural China, which will establish a platform for revitalizing the countryside and realizing urban–rural integrated development (Long 2014; Wu et al. 2019; Zhou et al. 2020). With regard to the problems of inefficient land use and spatial scattering and to achieve the goals of efficient land resource utilization and orderly land space utilization, efforts must be made to scientifically promote the comprehensive consolidation of rural land; to facilitate the integration of organizations, industry and space in rural areas; and to effectively shape the new driving forces and new mechanisms of rural revitalization and urban–rural integrated development in China. Land consolidation aims to raise the efficiency of land use, strengthen production capacity of agriculture and highlight the value of modern agriculture and rural areas. These factors help to enhance multi-functional and high-quality agriculture, while simultaneously promoting in-tensive and ecological rural development. The resources, assets, and capital attributes of land dictate that land has the following features which reveals that the process of land consolidation should also focus on these aspects: (1) Economy—Land is the most fundamental means of production. Its economic characteristics mainly include the scarcity of land supply, the dispersion of land use patterns, the difficulty of land use type change and diminishing marginal returns. (2) Ecology—Land requires conservation to solve ecological problems. In accordance with the requirements of ecological civilization construction, the land ecological consolidation is advocated to thoroughly implement integrated ecological protection, restoration for mountains, rivers, forests, fields, lakes, and strongly build ecological land. (3) Sociality—Land use has the sociality and externality, and land is the key to securing people's well-being. (4) Policy and strategy—Land is the spatial carrier of human main economic and social activities. It is a strategic resource. The core of ecological civilization construction, land spatial optimization and the transformation of economic development mode, is the intensive and economical use of land which is featured by policy-orientation. In 2018, the Ministry of Land and Resources and the Ministry of Finance signaled continuous support would be offered for the in-depth pro-motion of regional major project construction according to national land consolidation planning in impoverished areas, old revolutionary base areas, major grain-producing areas, border areas, minority areas, and other areas with abundant cultivated land reserves. This support had the purpose of pushing forward the targeted poverty alleviation strategy and conducting the rural revitalization strategy as well as other national strategies.

According to the goals of land consolidation and the benefits generated by it, land consolidation plays a "five-guarantee" role for promoting rural revitalization: (1) Resources guarantee—Land consolidation in China is mainly marked by the construction of high-standard farmlands. Relying on land consolidation as well as other projects and techniques, land re-sources can be effectively integrated, thus raising the quantity and quality of cultivated land, while guaranteeing food security. (2) Engineering support—Land consolidation is a systematic engineering construction activity which adopts engineering, technical, biological and other measures to conduct the comprehensive consolidation of disused or degraded land. It could increase the area of effective arable land and the efficiency of land use, and promotes the integration of the production, living and ecological spaces. This project is not merely technical; it is also a social project aiming to benefit peasant households through land consolidation. (3) Technical support—Land consolidation adopts the technical means of remote sensing or un-manned aerial vehicles to conduct dynamic monitoring and all-round surveys in project areas, to acquire basic data through rapid remote sensing interpretation and investigation, and to use a big data platform to provide scientific decision-making for rural revitalization. (4) Industry guarantee—Agriculture is multi-functional. Promoting the integrated development of the primary, secondary, and tertiary industries would constitute a basis for rural revitalization. Based on the increase of cultivated land resources via land consolidation and the development of modern agriculture and emerging industries through cultivation technical system of agricultural experiment or land circulation, the aim is to increase the income and employment of peasants and to realize a sustainable development of rural areas. (5) Institutional guaran-tee—Measures are taken to guide the consolidated land through land transfer and ownership adjustments, to determine land-use rights (instead of land itself), to allocate the rights and interests over newly increased cultivated land. It is important to entitle peasants to more property rights through land consolidation, and to increase the asset income of peasants. The breakthrough of rural land consolidation lies in innovating the rural land system, i.e., introducing a new-type mechanism to promote the development of land consolidation (Wang 2012). The roadmap for promoting rural revitalization through land consolidation is shown in Fig. 8.6.

A diversified range of land consolidation projects has been proposed, such as sandy soil regulation, flood-damaged land governance, barren hilly land consolidation, hollowed village consolidation, and polluted land remediation, etc. The essence of land consolidation is to use engineering technique to convert degraded land into productive cultivated land; to transform fragmented and inefficient cultivated land into high-standard farmlands; and to increase peasants' livelihoods resources, assets, and capital. To further protect the cultivated land and to implement the national strategy of requisition-

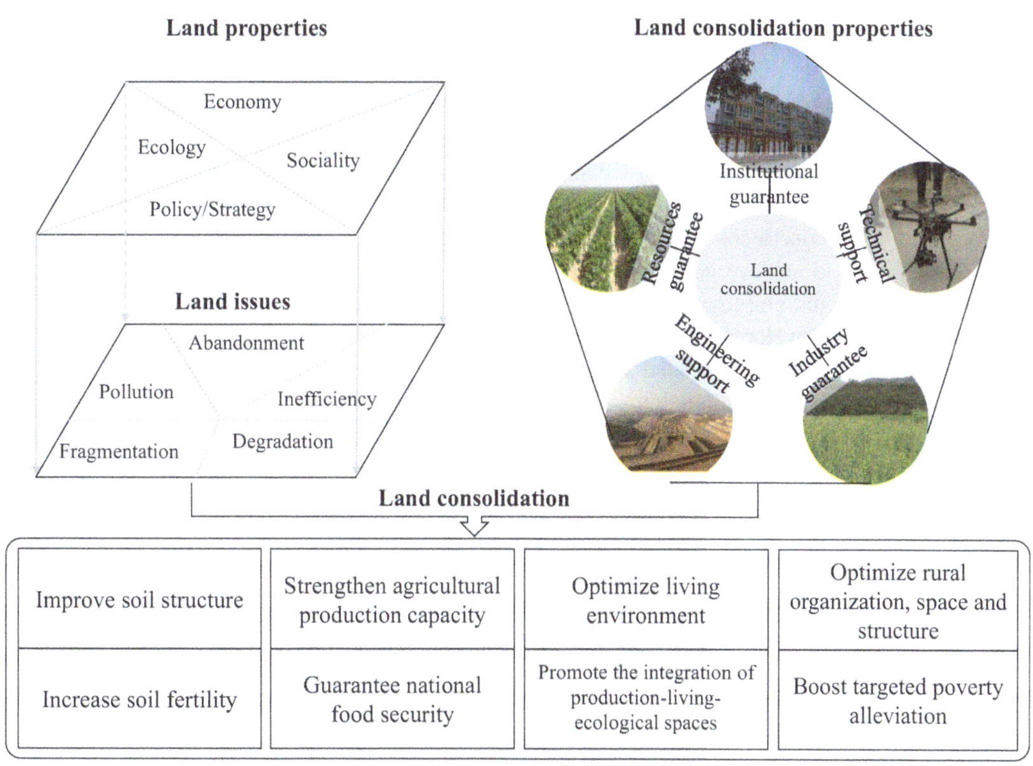

Fig. 8.6 Roadmap for promoting rural revitalization through land consolidation

compensation balance, with quantity increase and quality guarantee, China has systematically launched engineering applications of key techniques related to the comprehensive consolidation of disused or degraded land. The following four main types have been identified in typical areas:

(1) Hollowed village consolidation project: consolidation of disused land in the villages of Yucheng, Shandong Province—By revealing the evolvement course of rural hollowing, this project establishes a technical system of information acquisition, identification survey, potential assessment, and decision-making for disused land of rural villages. The potential of hollowed village consolidation in China was measured at 114 million mu (7.6 million ha) for the first time (Liu et al. 2013). It proposes a "three-integration" mode (i.e., rural spatial restructuring, organizational reconstruction, and industrial reshaping) that can adapt to the development rules of rural transformation and build a spatial restructuring hierarchy for a city-town-central village (community) composition. With regard to the different needs of the government, peasant households, enterprises, and other concerned parties, this project supports to return disused land to farmland and forest after consolidation; it also constructs residential and industrial areas to achieve community-based residence, clustered employment and socialized service. By developing the construction technology of healthy soil related with land cultivation, soil layer compounding, and soil improvement, the problems faced by newly increased cultivated land during village reclamation (such as instable structure and poor fertility) have been solved. Currently, land consolidation in hollowed villages has become one of the most important measures taken by the state towards rural sustainability (Li et al. 2018).

(2) Gully land consolidation project in Yan'an—The "Grain for Green" project in this Loess Plateau has achieved a significant eco-logical construction effect, meanwhile, it has also affected the local household livelihood and food security. Therefore, gully consolidation became a key for strengthening the agricultural production capacity of gullies. Relying on the gully consolidation mode of "grain for green up the hill and gully consolidation down the hill", this project has contributed to the optimization of land use structure and landscape pattern, and help to guarantee the eco-security of watershed (Li et al. 2019). In the gullies of Yan'an, the main cropping pattern is the single cropping of corn at present which does not conform to the macro strategy of China's agricultural structural adjustment. By innovating the modern gully agricultural development pattern based on the production-living-ecology integration, the land consolidation project has introduced forage rape into the agricultural production chain after gully consolidation and promoted the multi-functional agriculture of "planting + breeding + processing + sightseeing". This innovation laid a foundation for realizing the multi-functional gully agriculture pattern from single cropping (corn) into double cropping (corn + rape) (Liu et al. 2019), and from one industry into three industries in the gully regions of the Loess Plateau, while simultaneously promoting the cooperative and professional construction of agricultural bases (Liu et al. 2017; Liu and Wang 2019).

(3) Barren hilly land consolidation project: the land consolidation and circulation project in Fuping County of Hebei Province—The development of mountainous areas of Fuping County is mainly based on a requisition-compensation balance. Damaged construction land and unused land with a slope of less than 25° is developed into cultivated land which is used for a requisition-compensation balance; furthermore, the surplus index of construction land is transferred within the province. The intra-provincial transfer of the land-use index is a market transaction of the land development right and a transaction

between the rural and urban land development rights (Liu and Wang 2019). This transaction not only promotes the bilateral flow and efficient integration of urban–rural resources while increasing the supply of urban construction land but also effectively increases peasants' income from the assets-oriented conversion of land by promoting the capitalization of land resources (Zhou et al. 2020). Furthermore, taking the opportunities of land consolidation and migration relocation and implementing land transfer, the land management right is converted into shares by area and circulated to the village committee. Then, on behalf of peasant households, the village committee contributes in the form of land and engages with agricultural companies in cooperative development and management. As a result, land consolidation creates employment opportunities and widens the employment channels for local farmers to increase their income in the three forms of "share capital, dividend, and wage" (Wu et al. 2019). At the same time, the land consolidation project consolidates geological disaster-stricken land into productive land; effectively reduces landslides, debris flows, and other natural disasters; and efficiently achieves poverty alleviation and disaster reduction.

(4) Sandy soil consolidation: the "double-optimal" project (i.e., optimal formulation of soil layers and optimal selection of improved varieties) for degraded land in Yulin—By analyzing the component structure and evolutionary process of the sandy soil in the Mu Us Sandy Land of Yulin and by relying on the physical complementarity of local sandy soil and red clay, this project experimentally proposed a structural consolidation theory to rehabilitate sandy land into farmland by soil body building, soil layer reconstruction, and soil quality improvement (Wang et al. 2020). An innovative "double-optimal" matching mode was introduced for sandy soil consolidation, whose objective is to understand how to grow optimal crops in particular soils, i.e., how the growth of different crops influences and improves the land, and how different land components affect the growth of crops. Consequently, land suitability and biological fitness are perfectly combined. When applied to sandy soil regulation projects, the composite land-forming technique efficiently transforms sandy soil in wind-blown sand regions into high-productivity farmlands. Land consolidation also in-creases the quality of degraded land and improves its ecological environment.

8.3 Institutional Innovations on Urban–Rural Transformation

8.3.1 Deepening the Reform of Urban–Rural Household Registration (*hukou*) System

Driven by economic growth, population mobility, and integration of urban–rural development, the institutional and policy constraints on the flow of elements have been gradually eliminated. For a long time, the development strategy of "urban/citizens/industry first and rural/farmers/agriculture second" has promoted the formation of urban-biased systems and policy designs, which are finally reflected in the household registration system and others with obvious characteristics of urban–rural disparity. For example, systems of land use, employment, medical, and social security affect the equal exchange and orderly flow of elements along the urban–rural continuum. The household registration system is the core of the dualistic system and the institutional basis of difference in welfare. With the increase of urban–rural migration, the household registration system has witnessed a loosening trend and classifying regulation (Table 8.2). After the reform in 1978, the loosening household registration system provided an institutional environment for farmers with self-catering to migrate to cities. With temporary residence permit, these rural migrants are treated differently from local urban resident. Even so, the quota for

Table 8.2 Evolution of household registration system in China since the reform and opening-up

Time	Institute	Document	Key points
Dec 1981	State Council	Notice on strictly controlling rural labor forces to work in cities and the conversion of agricultural population to non-agricultural population (*nong-zhuan-fei*)	Strictly control recruitment from rural areas and strengthen household registration and grain management
Oct 1984	State Council	Notice on the issue of farmers' resettlement in towns	Allow self-catering farmers to settle down in towns
Sep 1985	NPC Standing Committee	Regulations of resident identity cards in the PRC	The system of "everyone has one ID card" was established
Oct 1989	State Council	Notice on strictly controlling the excessive growth of "*nong-zhuan-fei*"	Strengthen the macro regulation and management of "*nong-zhuan-fei*"
Aug 1992	Ministry of Public Security	Notice on the implementation of the local valid urban household registration	The "local valid *huji*" system is implemented in small towns, special economic zones, economic DZs, and high-tech industrial DZs, but is managed differently from original local *huji*
Aug 1994	Ministry of Labor	Interim provisions on the administration of inter-provincial migration of rural labor	
Jun 1997	State Council	Notice on experimental plan for the household registration reform in small towns and opinions on perfecting rural household registration	Reform the *huji* in small towns while continuing to strictly control the mechanical growth of population in large and medium cities
Jul 1998	State Council	Opinions of the Ministry of Public Security on solving several outstanding issues in current household registration management	New-born babies can voluntarily register with one parent. Household registration policies resolve the separation of couples and allow elderly parents to take refuge in their children
Mar 2001	Central Committee of the Communist Party of China (CPC) and State Council	Notice of State Council approving and transmitting the opinions of the Ministry of Public Security on promoting the reform of household registration system in small towns	Completely reform the *huji* in small cities and towns, abolish urban expansion fees or other similar expenses, and abolish the indicator of "*nong-zhuan-fei*"
Feb 2012	State Council	Notice of General Office of the State Council on actively and steadily promoting the reform of the household registration system	Guide the orderly transfer of non-agricultural industries and rural residents to small and medium-sized cities, gradually meet the needs of eligible rural residents to settle down, and equalize the urban–rural basic public services
Nov 2013	Central Committee of CPC	Decision of the Central Committee of CPC on several major issues of comprehensively deepening reform	Innovate population management, speed up the reform of *huji*, fully liberalize the restrictions on residential registration in towns and small cities, orderly liberalize that in medium-sized cities, reasonably determine that in large cities, and strictly control the population in megacities

(continued)

Table 8.2 (continued)

Time	Institute	Document	Key points
Jul 2014	State Council	Opinions on further promoting the reform of the household registration system	Unify the urban–rural *huji* and cancel the agricultural *huji*; fully liberalize the restrictions on residential registration in towns and small cities, orderly liberalize that in medium-sized cities, reasonably determine that in large cities, and strictly control the population in megacities
Feb 2016	State Council	Several opinions of the State Council on further promoting the construction of new urbanization	Encourage further resettlement conditions relax; allow rural migrants to register in their working places beyond megacities; restrictions like house purchase, investment taxation, and point-based system can only be adopted in megacities
Dec 2019	General Offices of Central Committee of CPC and the State Council	Opinions on the reform of the system and mechanism for promoting the social mobility of labor and talent	Restrictions on settlement in cities with a population of less than 3 million are completely removed, and the condition in large cities with a population of 3–5 million is fully relaxed

farmers' entering cities was limited. The number and trajectory of migration, as well as welfare that migrants have, were strictly controlled by the government. Since 1997, government has strengthened the regulation of urban classification, reformed the household registration system in small towns, gradually cancelled the urban capacity expansion fees or other similar expenses, and cancelled quota restrictions on migration. In 2014, the restrictions on the residential registration were fully liberalized in towns and small cities and sequentially liberalized in medium-sized cities. At present, the residential registration is still strictly controlled in large and mega cities with instruments like score-based residential registration.

Urban economic growth has attracted a large amount of migrant labor without providing them due guarantee in terms of infrastructure and public service supply. Although urban capital and citizens have an increasing demand for rural space and landscapes, migration of urban residents to the countryside is restricted by the household registration system. Equalizing the various benefits of the household registration system is the direction and key to reform of household registration system. Current household registration system is the basis for the dualistic urban–rural structure in China. Uneven development of cities with different scales and in different regions is also an important reason for difficulty of reforming existing household registration system. In 2014, the State Council proposed a migration policy of comprehensively liberalizing the restrictions on residential registration in towns and small cities, sequentially liberalizing the restrictions in medium-sized cities, reasonably determining the conditions in large cities, and strictly controlling the population size of megacities in the document "Opinions on Further Promoting the Reform of Household Registration System". Large and mega cities have obvious advantages in employment, education, medical care, and social services, which is helpful to form a huge attraction for rural population. Small and medium-sized cities and small towns are relatively lagging in supply of employment and social services, making those places an unattractive place to rural population. The monitoring report of floating population in China has coincidently proved that floating population has witnessed a trend of moving to large and mega cities. This inconsistency between directions of population flow and household registration system reform is still the key to future reform. Nowadays, all provinces implemented a unified urban–rural household registration system and canceled the agricultural household registration. Given the slow progress of reforms in rural land and collective economic systems, the urbanizing ability of rural migrants is limited and residents in some areas are concerned about canceling agricultural *hukou*.

Taking the elimination of dualistic urban–rural structure and narrowing the regional gap as directions, the reform of household registration system should further clarify and make breakthroughs in the following aspects. Firstly, the specific direction of the household registration system reform should be clarified. The original functions of the system are identification, demographics census, and social management. Nowadays, various rights and welfares are attached to this system, which is considered as the fundamental and connotation of the reform. The objective of reform should therefore be removing these rights and benefits from the original functions of the system. In this regard, reform of the household registration system should be orderly and steadily advanced with top-level policy design and multi-departmental coordination. Secondly, the path and timetable of reform should be scientifically determined according to local conditions. The current reform is aimed at liberalizing restrictions on registration in small cities and towns and relaxing those restrictions in large and mega cities. The regional differences should be taken into consideration in clarifying the timetable and plans for reform in different regions. Only in this way, can equalization of public resources in urban and rural areas be gradually realized, the rights and benefits be completely removed from the system, and the unified national residence permit or household registration management system be established. Thirdly, the reform of various related supporting policies should be promoted as well. To date, up to 20 rights and benefits are linked to the household registration system, the most obvious of which include employment opportunities, children's education, social security, family planning, and car and housing purchases. To further the reform of household registration system, we hereby suggest that introduction of new policies about the resource allocation and welfare should be decoupled with the household registration system. Various existing policies related to household registration in employment, education, medical care, elderly care, and housing should be gradually removed. The standards should be unified in different regions and along the urban–rural continuum. The reform of household registration system must be carried out in conjunction with related reforms of land use, employment, family planning, education, and social security. A diversified cost-sharing mechanism involving the government, society, enterprises, and individuals is required given that supply of services and facilities is essential to urbanization of migrant workers.

8.3.2 Improving Rural Land System Reform and Reconstruction for Management

Rural land system reform is an important support for rural revitalization. Implementing the rural revitalization strategy requires adhering to the priority development of agriculture and rural areas and establishing a sound policy system for the integrated urban–rural development. Optimization of the rural land system not only provides the premise to effectively solve the "three rural problems" but also constitutes the key toward a thorough implementation of the rural revitalization strategy. Based on the proposed framework of the land-use policy in China (Liu et al. 2014a), we constructed a framework of the rural land-use policy reform in China (Fig. 8.7), which consists of one overall goal, two basic systems, five core systems, and multiple basic guarantee systems. Specifically, the two basic systems are the cultivated land-protection system and the intensive land-use system. In the context of rural revitalization, the rural land system reform mainly covers three fields: the rural

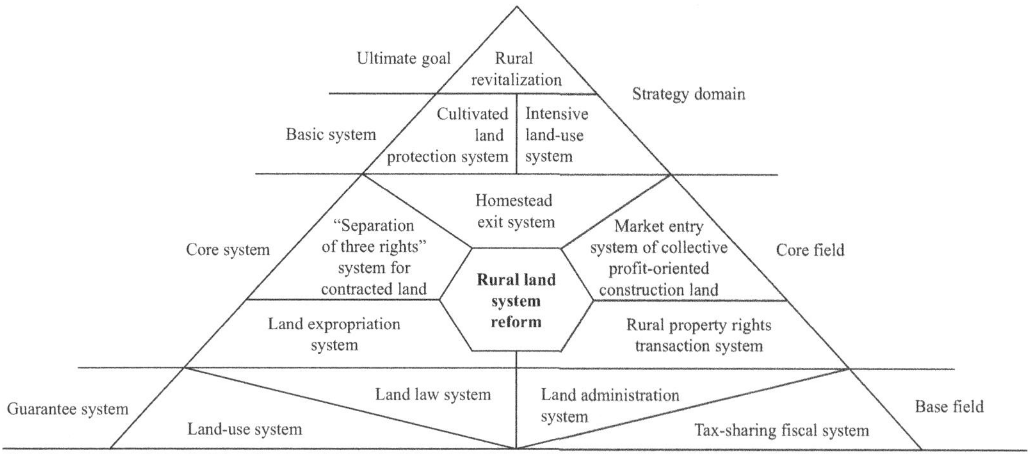

Fig. 8.7 Framework of the rural land-use policy reform in China

contracted land, the home-stead, and the collective construction land. Its five core systems include the "separation of three rights" (i.e., collective ownership, contractual right for peasant households, and management right for operators) system, the homestead exit system, the market entry system of a collective profit-oriented construction land, the land expropriation system, and the rural property rights transaction system.

The demands of rural revitalization require a further consolidation of the "separation of three rights" system of rural land, thus extending the rural homestead system reform and stimulating the endogenous power of the market entry reform of rural collective profit-oriented construction land. In doing so, the innovation of the land expropriation system can be deepened to provide a decision-making reference for improving the adaptability of rural land system reform, and maximally strengthening the optimal utilization of rural land resources. Details are summarized as follows:

(1) Consolidate the "separation of three rights" system for rural land—Efforts should be focused on perfecting the "separation of three rights" system for rural land, and keeping the land contractual relationship stable as well as unchanged on a long-term basis. In the case of no change in the subject of rural collective land ownership, it is necessary to protect contractual right of peasant households, accelerate the enlivening of land management right, and introduce a policy allowing the legitimate use of the management right of rural contracted land as a mortgage for financial institutions and as a contribution toward the industrialized management of agriculture (Wang and Zhang 2017). The current round of land contracts will be extended for a further 30 years upon expiry to provide a long-standing institutional guarantee for rural revitalization. The "separation of three rights" reform of rural land does not affect the collective ownership of rural land but divides the households' contracted management right into two parts, which include nontradable contractual rights of collective members and tradable rural land management rights by land transfer from collective members to agricultural producer (Han 2016). The differences between land contractual right and land management right include the subject of the right, the content of the right and the nature of the right (Xu et al. 2019). This institutional arrangement aims to active the land transfer mechanism, satisfy the demands of modern agribusiness, and ensure rural residents' rights and social security (Li et al. 2018; Xu et al. 2019; Zhou et al. 2020).

(2) Expanding rural homestead system reform—According to the requirements of rural revitalization strategy, the rural homestead is a critical index reflecting "ecological livability,

rural civilization, and effective governance". Efforts should be made to perfect policies related to peasants' idle homesteads and idle houses. It is also necessary to explore the "separation of three rights", i.e., to implement the collective ownership of homesteads, to guarantee the homestead qualification right of peasant households and the house property rights of peasants, and to moderately en-liven peasants' use right over homesteads and houses. With regard to the problems faced by peasant households (such as the difficulty of obtaining a homestead, extensive use, and un-smooth release), the modes of safeguarding and acquiring homestead rights and interests must be perfected. Furthermore, researchers must explore multiple forms of realizing the housing security of peasants in different regions; the possibility of implementing paid use with regard to the "over-standard occupation of the homestead", "one household with multiple homesteads", and other circumstances caused by historical reasons; and the possibility of implementing a voluntary paid release or transfer of homesteads by peasants who have settled down in cities within their collective economic organization. For example, rural residential land use right is suggested to be acquired at a stepped tariff rather than free of charge, especially for those households with more than one residential plot or with the per capita area beyond the prescribed standard (Gao et al. 2020). In 2018, China established an index of newly-increased cultivated land and a mechanism for cross-provincial regulation of the surplus index for linking the amount of urban and rural construction land.

(3) Stimulating the endogenous power of the market entry reform of rural collective profit-oriented construction land—The market entry of rural collective profit-oriented construction land becomes the biggest breakthrough for pilot projects in this new round of land system reform. To meet the rural revitalization strategy, the market entry reform of rural collective construction land should focus on stimulating the endogenous power of the system reform, abide by the basic principles of "same price for same right, smooth circulation, and income sharing", and identify the equilibrium point of interests among the state, the collective, and individuals (Du and Huang 2018). The primary tasks of this reform are to improve the property right system of rural collective construction land, define the scope and way of entering the market, and establish market transaction regulation and service monitoring system (Zhang and Li 2018; NPCC 2018). This strategy aims to deepen the rural collective property rights system reform, safeguard the property rights and interests of peasants, and strengthen the collective economy.

(4) Deepening the innovation of the land expropriation system—The land expropriation system reform aims to realize the "three transformations", i.e., resources into assets, funds into shared capital, and peasants into shareholders. This measure is used to reinvigorate land resources, solve land development problems, and safeguard the rights and interests of peasants during land expropriation. The practice follows the pattern of contribution in the form of land, i.e., peasants contribute in the form of land contracted management rights, while agricultural companies contribute in the form of cash. In this way, peasants can receive an equity dividend from a joint venture they have jointly established. In this practice, land is formally transformed into assets and assumes the credit and debt of the joint venture. This method of land system reform can fully bring into play the resources allocation role of the market, while continuously moving towards the same land, same price and same right for cities and rural areas.

8.3.3 Reforming the System of Public Resources Allocation

Insufficient supply of rural public goods and large urban–rural gap in the supply of public goods are great challenges for rural construction in China. In terms of basic education, the urban basic education funds are included in the urban fiscal budget, which can basically support the demand, while the story in the rural is quite different and basic education funds are seriously insufficient. Development of rural basic education is facing increasingly severe situation. In terms of medical insurance, urban residents can basically enjoy public medical care or purchase insurance within the coverage of the medical insurance system. Although the rural cooperative medical care system has been improved in recent years, the overall rural medical insurance standard is still low. Given the localized payment and management regulation, the proportion of rural residents' medical reimbursements outside their home counties has decreased successively, making it difficult to enjoy better medical and health services for rural residents. In terms of social security, most urban residents can enjoy social security, while farmers who account for about 50% of the total population are basically excluded from the social security system. In recent years, China's social subsidies has covered rural elderly groups, making them that are over 60 eligible to enjoy certain subsidies. In addition, rural production and living infrastructure is relatively backward, and there are huge differences in infrastructure construction investment between urban and rural areas. After abolition of the agricultural tax, the problem of insufficient supply of public goods in rural areas has become more prominent (Kennedy 2007). The increase in both the quantity and quality of public goods in those underdeveloped areas can significantly reduce income gaps (Calderón and Servén 2004). Evidences from developed countries that have completed economic transition, showed that priority of investment would change in a particular period within the transition from a developing country to a developed country. Investment in rural areas must be prioritized and amount of investment must increase rapidly (Timmer 1988). At present, China's per capita GDP has exceeded 7000 U.S. dollars, and the macro-agriculture and rural development policies have stepped into a new stage, which makes it possible and necessary to improve the expression mechanism of farmers' demand for rural public goods and to clearly understand the real consumption of farmers. Establishing a long-term mechanism for supply of rural public goods and increasing investment in rural basic education, medical insurance, social security, and infrastructure construction are the key to realize efficient supply of rural public goods and ultimately realize equalization of basic public services in urban and rural areas.

In general, rural public goods are jointly provided by the central state, local governments, and villages. Special attention should be paid to the enthusiasm of local governments, especially the towns/townships-level governments and village committees. It is necessary to combine the supply of public goods by government and market with the self-supply by using various existing resources in rural communities (Cao and Zhang 2009). Firstly, it is important to improve the decision-making mechanism for the supply of rural public goods to fully reflect opnios of farmers. Secondly, government should increase investment in the supply of rural public goods, and particularly in those underdeveloped rural areas. Thirdly, multiple measures should be taken to attract private investment, encourage the society to pay close attention to supply of rural public goods, enrich the supply structure of public goods, and establish a wide range of financing channels for rural public goods. For example, the corporate social responsibility and encourage enterprises to participate in rural construction should be strengthened. Finally, two methods can be employed in resolutely putting an end to power corruption: one is to regulate and restrict governments' financial behavior through legalization, the other is the supervision and inspection of use of public resources.

8.3.4 Innovating the Collaborative System of Urban–Rural Governance

During the period of 12th five-year plan, the new central government put forward development strategies such as building beautiful villages and improving urban–rural integration, which provided a good system for promoting equal exchange of urban–rural elements and balanced allocation of urban–rural public resources. However, it should be noted that the integration of urban–rural development in China was mainly aimed at the emerging agricultural, rural, and peasant problems. The measures were therefore problem-driven, such as lagging of agricultural modernization, lack of rural economic organization, plight of migrant workers' citizenship, and imperfect rural land system. Moreover, the policy formulation and promotion of measures have shown a top-down trend with defects of "emphasizing on construction and resettlement and neglecting development and entrepreneurship", "construction with lagging plan, model with lagging policy, practice with lagging theory", "the pilot demonstration is eager to achieve success, and is keen on image engineering and icing on the cake." There are still differences in organization and functions of new urban streets, rural communities, and economic organizations that adapt to urban–rural transformation.

In the coming 5–10 years, urbanization will still be the main driving force of urban–rural migration, and centrality of cities will continue to be strengthened. To adapt to the new normal of economic development and the new trends of population mobility, it is necessary to improve urban and rural governance capabilities, coordinate urban and rural governance systems, and enhance synergy between urban and rural construction.

Firstly, we should innovate the governance system of urban residential areas. With continuous increase in the number of rural migrants, the systems of household registration, land use, society, employment, and fiscal and taxation that are divided between urban and rural areas are increasingly difficult to satisfy the needs of a mobile society. As the main destination of the floating population, city is the main area for daily residential service supply and social management. It is necessary to continuously improve the management system of population, employment, education, medical care, and service based on the place of residence. Secondly, we should innovate the rural grassroots administrative system. Rural population decline and aging are inevitable with continuous outflow. However, urban aging population, consumer groups, urban capital and technology will transfer to the countryside driven by the increasing demand for rural landscape, food security, leisure, and entertainment. The restructuring of rural economic and social spaces is an important direction for future rural development. To adapt to changes in population mobility and social needs, it is necessary to innovate the rural grassroots administrative system, promote the integration of multi-stakeholders in the rural and beyond, promote the establishment of a new rural social organization system, straighten out the relationships between the grassroots of CPC at the township and village level, village committees, communities, and collective economic organizations, establish formal and informal institutions and diversified governance agencies and relationship networks, and finally enhance the rural governance capabilities along with the trend of modernization.

8.3.5 Improving the Government Performance Appraisal Mechanism

The implementation of the tax-sharing reform brought two major impacts on regional economic growth and governance. Firstly, local governments directly intervene various economic activities, and have brought about the sustained high growth of the Chinese economy through the competition of economic performance. Secondly, it is easy for the higher-level government to judge the degree of effort of local officials based on comparison of economic performance of various localities, which forms a GDP-oriented official promotion model (Zhou 2007). Therefore, local governments at all levels must compete for resources for local interests, triggering and intensifying competition among local governments. The survey shows that competition at the county levels is most intense and that current administrative performance evaluation and promotion mechanism drives local officials to seek political achievements (Hu and Ma 2010). The negative effects of excessive competition are evident given lack of relevant regulations, such as repeated construction of infrastructure, structural convergence of regional industries, severe market segmentation, and chaotic competition in attracting investment (Zhang 2010). Local economic growth driven by competition has strengthened fiscal expenditure and public goods supply capacity to a certain extent. However, agriculture that lacks economic benefits in the short term and relatively disadvantaged rural areas are still hard to benefit from economic growth, and even suffer from its negative impact. Firstly, local governments cut down land transfer prices, or even implemented zero land prices, to attract investment, which on the other hand accelerate loss of arable land and generate land-lost farmers (Tao et al. 2009). Secondly, polluting industries have opportunities to relocate in rural areas, which brings about serious environmental pollution problems in the move-in place. Thirdly, to highlight political achievements, rural construction is eager for success and keen on "icing on the cake" and doing "face-saving projects", resulting in disproportional investment of limited funds in relatively well-developed villages and towns and therefore marginalizing those underdeveloped villages and towns (Zhang et al. 2007).

Therefore, it is urgent to advance performance evaluation mechanism of local government to provide an important institutional support for promoting integration of urban–rural development. In terms of guiding ideology, the CPC and government management performance evaluation system should reflect the scientific outlook on development and correct political performance. In terms of content, the attention of local governments should be focused on the ability and satisfaction of serving the local public by changing the original practice of taking increase of GDP as main or even the only evaluation index (Zhou 2007). To make the performance evaluation more fair, comprehensive and objective, indexes including the cultivated land protection, environmental quality, ecological protection, income growth rate of low-income groups, the rate of comprehensive rural social and economic development, residents' satisfaction and happiness index, social harmony and resident's sense of security should be added to reflect the content of local government's social management and public service work. Regarding the methods and approaches of evaluation, it is necessary to combine not only specific terms of office of local governments with the local political situation, the public conditions, and the ecological environment, but also the original foundation with subjective efforts, the visible "significant achievements" with temporarily "invisible" "potential achievements", the current performance with long-term performance, to establish a performance evaluation mechanism with triple dimensions of economic growth, social development, and environmental improvement.

8.3.6 Facilitating the Reform of the Fiscal and Taxation System

Since 1978, China's fiscal system had undergone reforms "delineating income and expenditure" in 1980, "division of taxes, approved revenues and expenditures, and hierarchical responsibility" in 1985, "large contract" in 1988, pilot projects of tax sharing system in 1992, and finally established a tax-sharing fiscal system in 1994. The tax-sharing system broke previous fiscal situation of unified revenue and expenditure by the central government and was helpful to gradually realize fiscal decentralization between central and local governments. This has played a positive role in improving efficiency of government administration, promoting regional economic growth, and more effectively providing public products and services. While the fiscal revenue has been significantly concentrated upwards with the tax-sharing system reform, it failed to adjust the expenditure responsibilities among different levels of government. Actual expenditure responsibilities of local governments have increased significantly. Local governments are eager to increase local financial resources and begin to adopt large-scale investment attracting initiatives to compete for budgetary revenues and open new sources of extra-budgetary income through land transfer and various administrative fees (Tao et al. 2009). In particular, the phenomena of "land finance" and "development zone fever" are widespread nationwide and have caused serious problems such as rapid de-agriculturalization of agricultural land, increasing land-lost farmers, and deprivation of rural rights. In the central and western underdeveloped areas, the ability of local governments to provide basic public services and products has further declined after abolition of agricultural taxes (Kennedy 2007; Fujian Provincial Society for Finance and Banking et al. 2007).

Therefore, deepening the reform of fiscal and taxation system and giving local governments, especially those in underdeveloped regions, more controls over finance are important for promoting urban–rural infrastructure construction and balanced allocation of public services. Firstly, it is necessary to gradually adjust the central-local share of fiscal and taxation, increase the local share, and strive to improve the urban–rural structure of local investment in public goods under the premise of strengthening capital supervision and enhance its capital control ability in rural construction. Secondly, it is necessary to gradually establish a fiscal coordination mechanism in specific provinces, strengthen financial transfer payments among regions in a province based on the major functional zoning, such as cultivated land compensation and ecological compensation, and increase financial transfers to underdeveloped main crop-producing areas and ecological conservation areas to promote the construction of new countryside there. For example, from the perspective of regional division of labor, it is encouraged to coordinate the establishment of the financial transfer payment system between the main crop-producing and consumption areas within Shandong and Jiangsu with patterns of "industry in the east and agriculture in the west" and "industry in the south and agriculture in the north", respectively. This may be helpful to realize pattern of "urban driving rural areas and industry supporting agriculture" by investing more capital into rural construction in underdeveloped areas, and promoting the full use of regional functions in various regions to achieve regional integration and urban–rural development coordination. Thirdly, it is important to innovate land-related taxes. After the reform of China's tax and fee system, the basic structure of the real estate tax system has been established. Currently, taxes directly related to land and real estate include urban land use tax, land value-added tax, cultivated land occupation tax, real estate tax, and deed tax. With acceleration of urbanization and increasing economic value of land, land transfer fees charged by local government increased from 120.7 billion yuan in 2002 to 4.26 trillion yuan in 2014, becoming an important source of local government revenue. Due to limited land resources and the phased nature of urbanization, it is difficult to sustain the land-based fiscal model as main source of government revenue. At the same time, the skyrocketing housing price and high credit leverage have increased social and financial risks. Land prices have even exceeded housing prices in megacities. There is an urgent need to innovate land management system and fiscal and taxation system. The land revenue will gradually shift from land increment to land use preservation with new applicable taxes formulated.

References

Calderón C, Servén L. The effects of infrastructure development on growth and income distribution. The World Bank, 2004.

Cao Z, Zhang P. PPP model for the supply of China's western rural infrastructure. J Chang'an Univ (Soc Sci Edn). 2009;11(3):55–8.

Chen Y. Study on the impact assessment and mechanism of production factors' non-agricultural process on rural system development: a case of Shandong Province. Beijing: Graduate University of Chinese Academy of Sciences; 2010.

Du W, Huang M. An analysis on the reform of rural land system in the context of rural revitalization strategy. J Sichuan Norm Univ (Soc Sci). 2018;1: 12–16.

Fujian Provincial Society for Finance and Banking, Wu G, Zheng H, et al. Study on the construction of new countryside and several issues on financial polices. Fujian Finance. 2007;(1):16–21.

Gao J, Liu Y, Chen J. China's initiatives towards rural land system reform. Land Use Policy. 2020;94: 104567.

Guo Q, Lu B. The influence of tax on factor income distribution. Econ Res J. 2011;6:16–30.

Han C. Triple property system of rural land is a new important innovation of China rural reform. Beijing: Guangming Daily. 2016.

Hu X, Ma J. Economic growth momentum or political promotion competition: analysis report of questionnaire survey of contemporary Chinese local government competition. People Tribune. 2010;10:10–5.

Kennedy JJ. From the tax-for-fee reform to the abolition of agricultural taxes: the impact on township governments in north-west China. China Q. 2007;189:43–59.

Li Y, Wu W, Liu Y. Land consolidation for rural sustainability in China: Practical reflections and policy implications. Land Use Policy. 2018;74:137–41.

Li Y, Li Y, Fan P, et al. Impacts of land consolidation on rural human–environment system in typical watershed of the Loess Plateau and implications for rural development policy. Land Use Policy. 2019;86: 339–50.

Lin J, Xu C. Land development rights, space control, and synergetic planning. City Plann Rev. 2014;38(1):26–34.

Liu Y. How to deal with the problems in the construction of new countryside. Science Times, 2008-02-24.

Liu Y. Rural land consolidation should benefit farmers. People's Daily, 2010-11-12.

Liu Y. Land Engineering for Agricultural Sustainability: From China to the World. Abstract Book of the First IGU-AGLE Commission Conference on Global Rural Development and Land Capacity Building (2017-08), 2017.

Liu Y. Introduction to land use and rural sustainability in China. Land Use Policy. 2018;74:1–4.

Liu Y, Wang Y. Rural land engineering and poverty alleviation: Lessons from typical regions in China. J Geog Sci. 2019;29(5):643–57.

Liu Y, Zhou Y. Challenges and countermeasures for beautiful countryside construction in China. J Agric Res Environ. 2015;32(2):97–105.

Liu Y, Wang J, Long H. Analysis of arable land loss and its impact on rural sustainability in Southern Jiangsu Province of China. J Environ Manag. 2010;91(3):646–53.

Liu Y, Yang R, Li Y. Potential of land consolidation of hollowed villages under different urbanization scenarios in China. J Geog Sci. 2013;23(3):503–12.

Liu Y, Fang F, Li Y. Key issues of land use in China and implications for policy making. Land Use Policy. 2014a;40:6–12.

Liu Y, Chen C, Li Y. The town-villages construction pattern under new-type urbanization in China. Areal Res Develop. 2014b;33(6):1–6.

Liu Y, Chen Z, Li Y, et al. The planting technology and industrial development prospects of forage rape in the loess hilly area: a case study of newly-increased cultivated land through gully land consolidation in Yan'an, Shaanxi Province. J Nat Res. 2017;32 (12):2065–74.

Liu Z, Liu Y, Li Y. Extended warm temperate zone and opportunities for cropping system change in the Loess Plateau of China. Int J Climatol. 2019;39(2):658–69.

Long H. Land consolidation: an indispensable way of spatial restructuring in rural China. J Geog Sci. 2014;24(2):211–25.

Lu D, Chen M. Several viewpoints on the background of compiling the "National New Urbanization Planning (2014–2020)'. Acta Geogr Sin. 2015;70(2):179–85.

Luan J, Qiu H, Jing Y, et al. Decomposition of factors contributed to the increase of China's chemical fertilizer use and projections for future fertilizer use in China. J Nat Resour. 2013;28(11):1869–78.

Luo X, Xu X. Rural transformation and planning responses in the 13th five-year plan period. City Plann Rev. 2015;39(3):15–23.

National People's Congress of China (NPCC). Summary Report of the State Council on the Pilot Reform of Rural Land Expropriation, Collective Operational Construction Land Entry into the Market and Housing Land System, 2018. http://www.npc.gov.cn/zgrdw/npc/xinwen/2018-12/23/content_2067609.htm.

Tao R, Lu X, Su F, et al. China's transition and development model under evolving regional competition patterns. Econ Res J. 2009;7:21–33.

Timmer CP. The agricultural transformation. Handb Dev Econ. 1988;1:275–331.

Wang K. Rural land reclamation in Anhui province and land system innovation research: take Huaiyuan as an example. Hefei: Anhui University; 2012.

Wang Y, Li Y, Li Y. Land engineering consolidates degraded sandy land for agricultural development in the largest sandy land of China. Land. 2020;9(6):199.

Wang Q, Zhang X. Three rights separation: China's proposed rural land rights reform and four types of local trials. Land Use Policy. 2017;63:111–21.

Wu Y, Feng W, Zhou Y. Practice of barren hilly land consolidation and its impact: a typical case study from Fuping County, Hebei Province of China. J Geog Sci. 2019;29(5):762–78.

Xu Y, Huang X, Bao H, et al. Rural land rights reform and agro-environmental sustainability: Empirical evidence from China. Land Use Policy. 2019;74:73–87.

Yang Y, Li Y, Long H. Report on the first IGU-AGLE commission conference on global rural development and land capacity building. J Geog Sci. 2018;28 (1):124–9.

Ye J, Yang Z. Ideas on participatory methods and the construction of new countryside. Chin Rural Econ. 2006;7:36–41.

Yu G. Problems in the construction of rural public infrastructure and path selection. J Shandong Agric Administrators' Coll. 2009;25(1):79–80.

Zhang W. Excessive competition has a lethal effect. People Tribune. 2010;10:18–20.

Zhang X, Li Z. Development of the rural land system. In: China's Rural Development Road. Singapore: Springer, 2018;95–106.

Zhang X, Liu H, Zhang C. Nine major problems in the construction of new countryside need to be solved. Renda Jianshe. 2007;8:7–9.

Zhou Y, Li Y, Xu C. Land consolidation and rural revitalization in China: mechanisms and paths. Land Use Policy. 2020;91:104379.

Zhou L. Governing China's local officials: an analysis of promotion tournament model. Econ Res J. 2007;7:36–50.

9 Conclusions and Research Prospects

Abstract

This chapter summarizes the research of the full text and prospects the future research frontier direction of urban–rural transformation geography. In the context of new-type urbanization and rural revitalization, the study of urban–rural transformation geography in the new era needs to establish a global concept, aim at the basic national conditions, summarize the domestic practice mode, seek the breakthrough in technical methods and basic theories at multiple aspects of urban–rural integration, such as organizational mechanism and regional model, constantly improving the system, policy and operation mechanism of urban–rural integrated development and creating China's human-earth system science and rural science, further building a community with a shared future for mankind and supporting the sustainable development of China and even the world.

In the process of economic globalization and integration, urban–rural relationship, urban–rural regional system and urban–rural transformation have become the frontier and strategic issues to re-examine the urbanization mode and reconstruct the rural spatial system, which is also a systematic and comprehensive subject of geographical innovation and in-depth research based on the research of human-earth land relationship and urban–rural relationship. This book focuses on the organic combination of relevant theories and methods of Geography with the scientific issues and decision-making needs of urban and rural development. Focusing on the urban–rural regional system, the book has carried out in-depth research on the basic theory of urban–rural transformation, the analysis of transformation process, the quantitative evaluation of transformation status, the diagnosis of spatial pattern, and the comprehensive research on the resource and environment effects of urban and rural development and transformation. By exploring the economic and social development situation and transformation theory of China's urban and rural areas, combined with the urban–rural dual structure system and the urbanization development process, this paper constructs the theoretical framework and method system of urban–rural transition geography research, and explores the long-term mechanism and feasible ways to scientifically guide the urban–rural transformation, rural revitalization and promote the coordinated development of urban and rural areas, thus contributing China's wisdom to global sustainable development.

Based on China's regional pattern, imbalance of urban and rural development and institutional constraints, taking the Bohai Rim region as a typical region, this paper analyzes the process of population, economy, society and space transformation of urban and rural areas in Bohai Rim region through geographical comprehensive

Y. Liu, *Urban-Rural Transformation Geography*, Sustainable Development Goals Series,
https://doi.org/10.1007/978-981-16-4835-9_9

evaluation, pattern analysis and spatial exploration, and analyzes the coordinated development of regional population-land-industry, new urbanization and rural transformation. Then it explores the spatial differentiation and different pattern of urban–rural transformation. Based on the coupling perspective of human-earth relationship, this paper analyzes the process of urban–rural transformation and its comprehensive impact on land conversion, CO_2 emission and energy consumption, and rural regional resources and environment. It summarizes the main tasks of accelerating the integration of urban and rural development in China, and puts forward the ideas, path selection and institutional innovation of urban–rural transformation in the new period. Systematically promoting the theoretical research, strategic planning and typical practice of urban–rural transformation geography have important reference value and practical significance for the improvement of the scientific system of urban and rural development, the theoretical system of urban–rural transformation, the strategic system of urban–rural overall planning and achieving global sustainable development, thus laying an important foundation for the construction of the theoretical and disciplinary system of urban–rural transformation geography.

9.1 Main Conclusions

(1) Based on the analysis of the problems and stage characteristics of urban and rural development in China, this paper clarifies the research task of urban and rural transformation geography in the new period, and puts forward the concepts of "transformation tree", "transformation law", "transformation threshold" and "integration body" in urban and rural development to deepen the theoretical interpretation of the level, structure and transformation stage of urban–rural transformation. Furthermore, the theoretical system of geography of urban–rural transformation is constructed from the perspective of transformation process, exploration mechanism and transformation effect. It mainly includes urban–rural structure transformation theory, urban–rural interaction theory, urban–rural relationship evolution theory, urban–rural transformation mechanism theory and system regulation theory. Based on the geography research paradigm, this paper puts forward the main research content system of urban–rural transformation, including five parts: the comprehensive process of urban–rural transformation, regional differentiation pattern, driving mechanism, comprehensive effect and optimization countermeasures.

(2) The method system of urban–rural transformation is constructed, which is embodied in the comprehensive integration method of urban and rural regional system evolution geography. The research methods of urban and rural development transition geography mainly include the comprehensive analysis of urban and rural geographical elements, the optimization of industrial structure and layout, the optimal allocation and reconstruction of space, the transformation mode of different types of regions and the functional coordination among regions. Through a comprehensive evaluation, regional division, mechanism exploration, simulation prediction, etc., a system integration platform for urban and rural regional system evolution geography research has been initially formed, which can better serve the decision-making practice of new urbanization and urban–rural development integration by comprehensively using mathematical statistics, quantitative algorithm, spatial exploration, GIS, system simulation and other methods.

(3) The strategic framework system of China's urban–rural transformation is designed, and the key tasks of urban–rural transformation are clarified. Based on the development process of urban and rural areas, this paper defines the concept and criteria of urban–rural transformation, puts forward the development path of the migrant population living and working in peace and contentment, coordination between urbanization and rural areas, highlights the development path of regional differentiation, and scientifically analyzes the role and influence of government and market, policy bias and regional multi-function. The main tasks of urban and rural regional system transformation and reconstruction in economic, social,

spatial and other aspects are clarified, including the construction of agricultural system, urban and rural industrial division, orderly spatial reconstruction and urban and rural system reform.

(4) This paper systematically evaluates and divides the evolution pattern and regional differentiation characteristics of urban–rural transformation in China. Since the reform, China's urban and rural economic, social and spatial structure has undergone systematic changes, but urban and rural development still faces problems such as widening gap and unbalanced regional development. Based on the multi-index comprehensive evaluation method, the evaluation index system of urban–rural transformation is constructed, and the regional pattern of urban–rural transformation in China is analyzed. The level of urban–rural transformation in eastern coastal areas is significantly higher than that in central and western regions, and the economic, social and structural changes of urban and rural areas in Northeast China are slow. Based on the physical and geographical environment, economic and social conditions, structural transformation and urban–rural relationship indicators, this paper divides the regional types of urban–rural transformation. This paper systematically analyzes the regional characteristics, mechanism differences and development orientation of urban–rural transformation in different regional types.

(5) Based on the evolution of population-land-industry elements and process synergy, this paper analyzes the transformation process and pattern of urban–rural transformation in Bohai Rim region. The rapid transformation of urban and rural population, land, industry, employment, etc. in the Bohai rim area, Beijing and Tianjin are undergoing rapid transformation, while Hebei is the slowest. The transformation speed of prefecture-level cities and municipal districts is faster than that of counties and cities. Using the coordination coupling method, this paper analyzes the coupling and coordination of population, land and industry transformation, explores the transformation law of the three and the spatial differentiation pattern of coupling and coordination; quantitatively evaluates urbanization and rural transformation, analyzes the spatial coupling and collaborative evolution law of the two; explores the spatial succession law of urban–rural transformation in Bohai Rim region; and deeply analyzes the population and society in Bohai Rim region The transformation pattern of urban and rural development under the four dimensions of population, society, economy and land.

(6) Identify the dynamic relationship between urban and rural economic and social factors and key resource and environmental indicators, and explore the resource and environmental effects of urban–rural transformation. Based on the analysis of resource and environmental carrying capacity and environmental Kuznitz curve, this paper analyzes the mechanism and interaction forms of the impact of urban–rural transformation on resources and environment. This paper analyzes the resource and environmental effects of urban and rural development and transformation from three aspects: land non-agriculturalization, dynamic simulation of land-use change, CO_2 emissions and energy consumption, and village resources and environmental changes. The research shows that the transformation of urban and rural development has obvious resources and environmental effects, and the upgrading of industrial structure, the transformation of utilization mode and the effective regulation and control of policies help to reduce the negative impact of urban–rural transformation on resources and environment. Therefore, the formulation of appropriate regional policies and the selection of appropriate development models will help to improve the efficiency of resource utilization and optimize environmental conditions.

(7) Summarize and refine the main tasks, optimization ideas and system innovation of the current integration of urban and rural development in China. To promote the integration of urban and rural development, it is urgent to crack the urban–rural dual system, optimize the urban–rural development pattern, enhance the ability of social governance, balance the allocation of public resources, coordinate the treatment of resource and environmental problems, and so on.

Focusing on the comprehensive perspective of population, economy, society, policy and system, this paper puts forward the idea of optimizing the allocation of production factors and promoting regional balanced development from the factor-industry-system level. This paper focuses on seeking the process path of urban–rural transformation and integration from the aspects of production-city system coordination, land optimal allocation, regional function coordination, village and town pattern construction, multi-agent overall planning and improving the ability of sustainable development. Furthermore, this paper deeply discusses the safeguard measures of urban and rural development and transformation from the aspects of household registration system, land management system, public resource allocation system, urban–rural collaborative governance system, local government performance appraisal mechanism and fiscal and taxation system.

9.2 Research Prospects of Urban–Rural Transformation Geography

Driven by global environmental change and human economic activities, the acceleration of urbanization, the integration of urban–rural development, as well as rural transformation and revitalization will become the most significant "pattern of human geography" in many developing countries, including China. To promote urban–rural development to a higher level, it is of great theoretical value and practical significance to innovate and develop human-earth system science, systematically reveal the process, mechanism and pattern of urban–rural transformation, explore the pattern of sustainable urbanization and the law of urban–rural integrated development, deepen the cross study of natural science, social science and technological science, and further strengthen research on global or regional urban-rural transformation, urban-rural regional system, urban agglomeration complex and global rural revitalization.

9.2.1 Future Research Prospects of Human-Earth System Science

With the progress of science and technology and the rapid development of society, anthropedosphere has become the core position of human behavior on the earth's surface, such as resources protection, ecological construction, environmental governance, disaster prevention and epidemic prevention and control (Liu 2020a). Innovating the theory of human-earth areal system and developing human-earth system science can highlight the subjectivity of human beings, the process of human-earth cooperation and the strategic significance of sustainable development on the surface of the earth (Fig. 9.1). Therefore, the research theme of modern human-earth system science should aim at the coupling and sustainable development of human-earth system, give prominence to the concept of system science, coupling process, coordinated development and international cooperation.

1. **The principle, structure and function of anthropedosphere**
   The anthroposphere is a new cognition of the earth's surface system in the era of "Anthropocene". It is urgent to scientifically define its scope and regional structure of urban and rural area, stability and elasticity of the anthroposphere, study the interaction mode, qualitative parameters, interaction force and periodicity of human-earth system, reveal the processes of the transformation of solid, liquid and gas, the cycle of material flow, energy flow and information flow and the integration of ecology, production and life, further explores their mechanism. Meanwhile, more attention should be paid to investigating the impact of new technologies and new factors, especially informatization and intellectualization, on the pattern and function changes of human-earth system, and deepen the comprehensive researches on the system expression, element monitoring, structure simulation, function evaluation, prediction and early warning of anthroposphere system.

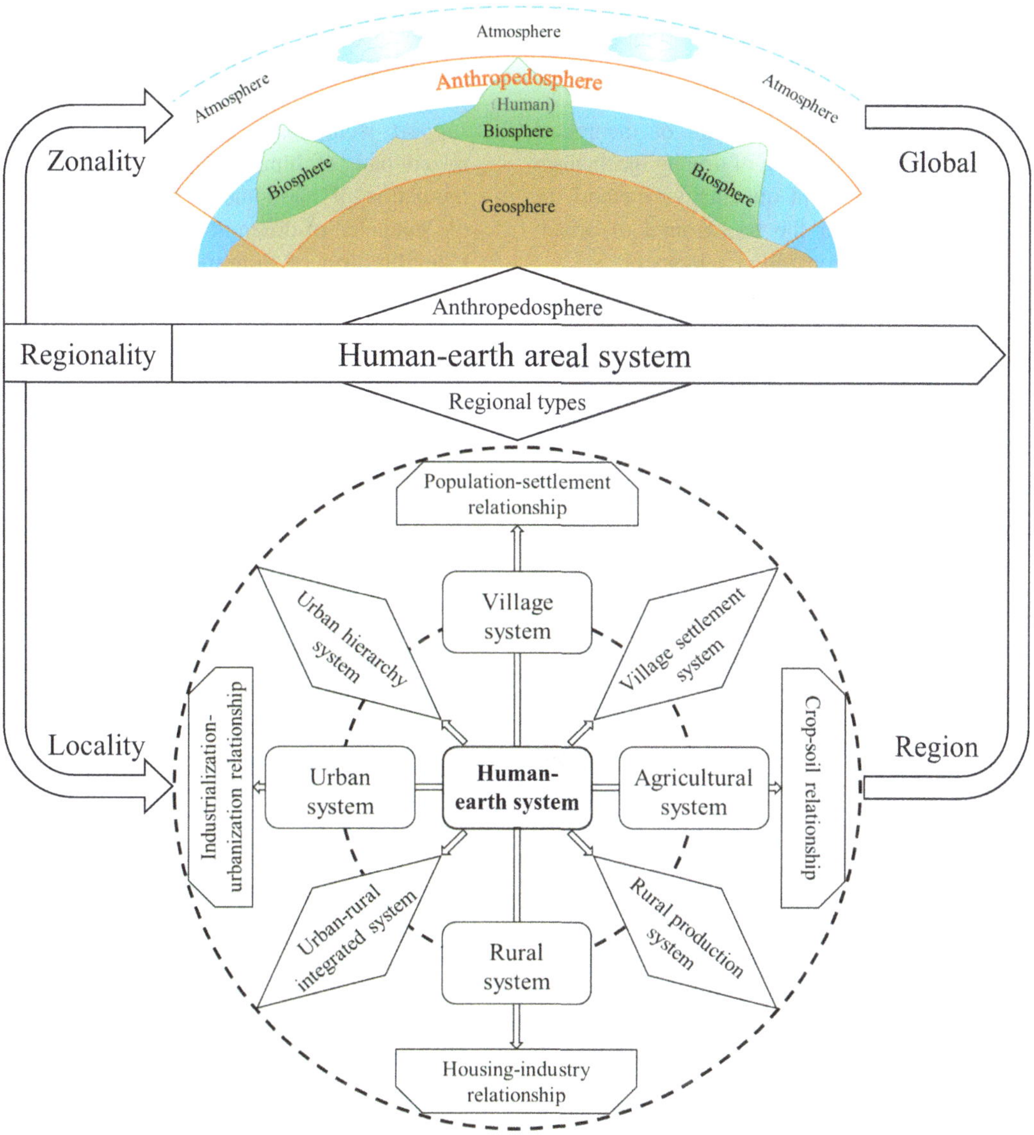

Fig. 9.1 The mode of "double cycle" human-earth system under the background of globalization

2. **The pattern, mechanism and effect of human-earth system**

   In essence, human-earth system is an open system of the "double circulation" and "mutual circulation" between human socioeconomic system and surface natural ecosystem. To explore the pattern, process and mechanism of human-earth system, especially the typical process and influence of the transect on the earth, it is urgent to explore the coupling of natural, economic and technological elements, dynamic structure and interaction mechanism of subsystems of human-earth system, as well as the complex effects of the interaction between different types of human-earth system. For example, population growth and demand for agricultural products are important factors driving the changes of land use system, while global change shapes land use pattern and affects food consumption (Edmonds et al.

2017). To analyze and assess the tremendous pressure of human activities and climate change on land system, it is necessary to clarify the conduction mechanism and path, and further promote the construction of theoretical system, the investigation of driving mechanism and the development quantitative methods of land system under the background of urban-rural transformation and development.

3. **The coupling and dynamic simulation of human-earth system**
   Focusing on the coupling simulation and prediction of complex human-earth system, it is necessary to construct the dynamic model of human-earth system, simulate and analyze the interaction of element coupling and process coupling of human-earth system in different regions, reveal the complexity and dynamics of human-earth system, and explore the mechanism and feedback path of globalization, industrialization, urbanization and other human activities on the anthroposphere. Meanwhile, it is urgent to scientifically diagnose the suitability of human-earth system, define the threshold of coupling and coordinated development of human-earth system, and establish the monitoring and simulation technology system of human-earth system. In particular, more attention should be paid to the researches such as coupling test and pattern simulation of human-earth system in typical urbanizatipn region, including the Loess Plateau, agro-pastoral ecotone, mountainous and hilly areas, plain agricultural area, coastal areas, etc.
4. **Environmental effects and regulation of the high intensity human activities**
   In view of the high-intensity economic activities of human beings and the high-density urban agglomerations, metropolitan areas, industrial belts and other complex human-earth systems, as well as the ecological driving process, various disasters and risks caused by major engineering construction, such as returning farmland to forest and land consolidation, it is urgent to establish the theoretical framework, index system and econometric model of ecological environmental effect assessment of high-intensity human activities, develop regional environmental quality and safety dynamic monitoring technology, major disasters and epidemic early warning and prevention and control support technology, and explore the optimized path of human-earth system and the regulation of harmonious coexistence under the guidance of different regional problems.
5. **Coordination of human-earth system and regional sustainable development**
   Coordination is a kind of social function. As a "villager of the earth", human beings should follow the laws of nature and consciously protect the earth. To adapt to and deal with global change and other issues, it is urgent for human beings to find out the mechanism, dynamic mechanism and conductive path of the conflict and coordination between human and earth system. Through the researches on the state evaluation, simulation and decision support of sustainable development, it is necessary to systematically analyze the evolution law and scenarios of modern human-earth system and the impact process of human activities on the environment, explore the response and adaptation mechanism of human-earth system to global or regional environmental change and increasingly intense human economic activities, develop key technologies for the restoration and reconstruction of degraded environment system, and create a model system and its typical paradigm of "anthropedosphere-regional type-factor flow" to promote the coordination and sustainable development of human-earth system. In particular, it is necessary to deepen the study of special areas such as fragile ecological areas, traditional agricultural areas and major grain-producing areas.

9.2.2 Future Research Prospects of Urban–Rural Transformation

The rural–urban relationship is a kind of mother-child relationship, which has a unique kinship and geographical relationship, and should be intimate, integrated and coordinated development. It can also be said that cities and villages are an organism and a community of destiny. Only when both of them achieve sustainable development can they support each other and evolve together (Liu and Li 2017). Therefore, how to deal with the industry-agriculture relationship and urban-rural relationship determines the success or failure of China's modernization to a certain extent. However, after more than 40 years' development since the reform and opening-up, the urban–rural unbalanced and uncoordinated development has become the most prominent structural contradiction in China's social and economic development, and is also the root of many problems (Liu 2018). The 19th National Congress of the Communist Party of China clearly put forward the strategy of rural revitalization, which aims to establish and improve institutional mechanism and policy system of urban–rural integrated development in accordance with the general requirements of industrial prosperity, ecological livability, rural civilization, effective governance and rich life. Therefore, it will promote rural development to a higher level through better solving the prominent problems such as unbalanced urban and rural development and inadequate rural development, and build an urban–rural integrated pattern.

The urban–rural integrated development is the core of organic reconstruction of urban–rural space, effective allocation of rural elements, and orderly promotion of rural revitalization. Its object is a multi-body system, and aims at realizing multi-level objectives (Liu 2018). Urban–rural integrated geography is devoted to the study of the formation mechanism and evolution process of urban–rural integrated system or urban–rural organism, as well as their evolution law of human-earth relationship, especially the process and dynamic mechanism of urban–rural population flow, industrial interaction and public service equalization, thus promoting the urban–rural relationship from isolation and opposition to integration and realizing the coordinated development of rural revitalization and new-type urbanization (Fig. 9.2). Guided by the theory of human-earth areal system, it is necessary to construct the basic theory and diagnostic technology system of urban–rural integrated system, reveal the mechanism, process, pattern and effect of urban–rural integration, and form a new theoretical cognition of rural–urban system. Moreover, focusing on the gap and imbalance between urban and rural development, it is also of great significance to study the theory of equal exchange of urban–rural elements and public resources allocation, and urban–rural integrated development strategy, guiding the transformation of urban and rural areas.

From a systematic point of view, urban–rural transformation appears in the regional integration system of urban and rural areas, which is no longer limited to isolated urban or rural regional systems, but expanded and deepened based on communication and interaction between urban and rural systems. The research of urban–rural transformation focuses on the internal elements flow and relationship evolution. From the perspective of process, the urban–rural transformation emphasizes dynamic and change, focusing on the interaction of population, resources, environment and other elements between urban and rural areas, and the transformation trend of urban–rural relations. Based on the national strategic needs and current development stage of China's urbanization and industrialization, it is urgent to open new fields and paradigms of geography research from the contradictions and problems in the process of coordination and transformation. From the perspective of pattern, it is necessary to fully exploit the discipline advantages of geography in the study of spatial heterogeneity and regional patterns, and comprehensively use spatial analysis and simulation technology, select research objects based on different spatial scales, and strengthen the comprehensive research on the temporal and spatial evolution, resource and environment effects and optimizing regulation and control approach of urban–rural transformation, so as to enhance the systematicness, scientificity and operability of the research results. In terms of spatial scale, it includes the micro scale of farmers, enterprises and villages, and meso and macro scales of counties, cities and provinces; while in terms of

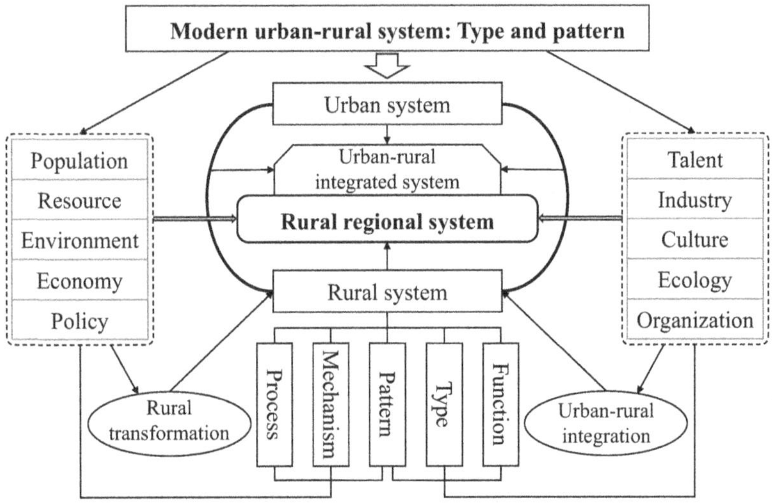

Fig. 9.2 Modern urban–rural system

methods, it includes database construction, model construction, scenario analysis, expert aided decision-making, multi-level problem analysis, multi-scale spatial transformation and multi-scenario simulation. From the perspective of mechanism, it is needed to combine the research of pattern and process to explore the driving factors of the occurrence, development and transformation of the internal elements interaction mode of urban and rural regional system, and reveal the law and mechanism of the development and evolution of urban–rural regional system. This is the basis of deepening the researches on optimization regulation, mode and path of urban–rural transformation. From the perspective of mode, regional resource endowment and economic and social development conditions are different, forming different regional types of urban–rural transformation. In view of the regional pattern, stage characteristics of urban–rural transformation and typical problems of rural resources and environment in different areas, it is urgent to strengthen the systematic research on resource allocation, development efficiency and environmental effect under the background of urban–rural integrated development.

Rural areas are abandoned for reasons that include mobility and technology, poverty, biased policy and inadequate land management. As early as the eighteenth century, Europe's Industrial Revolution saw villages shrivel as towns and cities became packed. The trend of rural depopulation spread to North America in the twentieth century, particularly after the Second World War. Entering the twenty-first century, rural decline has been a global issue (Fig. 9.3) (Liu and Li 2017). It can be seen from Figure 9.3 that the BRICS countries (Brazil, Russia, India, China and South Africa) have the most significant reduction in rural population, becoming the most typical regions of rural population reduction and urban-rural transformation in the process of rapid urbanization in the new era. Against this background, China urgently needs to scientifically promote the implementation of rural revitalization strategy and the integration of urban and rural development in the next 10–20 years, thus promoting socioeconomic sustainable development (Liu 2018, 2020b). To promote urban–rural integrated development, it is urgent to adhere to the principle of sustainable development, and put urban–rural transformation under the overall framework of sustainable science. Looking forward to the future and aiming at the world, we should not only make clear the direction and be full of confidence, but also seize the opportunity and be down-to-earth. Therefore, urban–rural transformation in China urgently needs to establish a global concept, aim at the basic national conditions, summarize the domestic practice mode, seek the breakthrough in technical methods and basic theories at multiple aspects

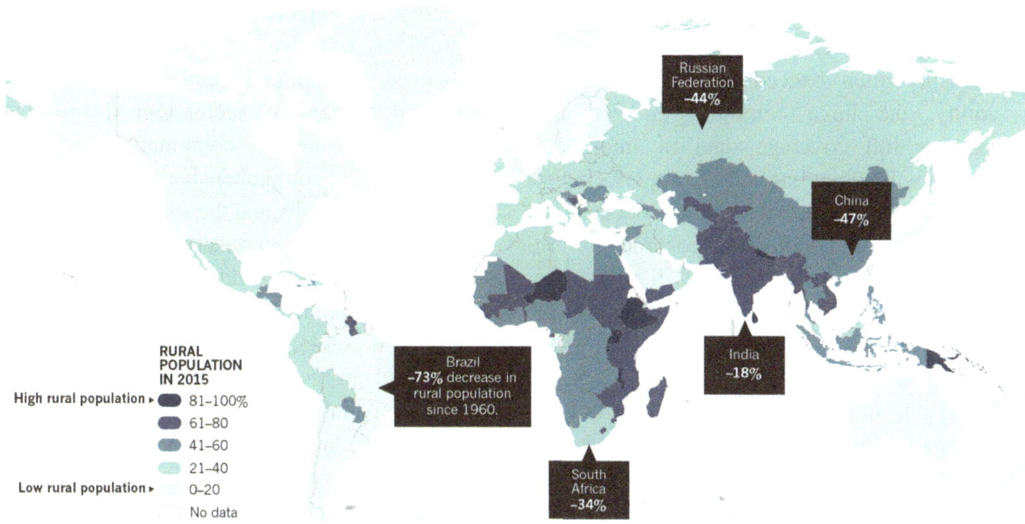

Fig. 9.3 World rural population in 2015 (Liu and Li 2017)

of urban–rural integration through the interdisciplinary of human geography, information geography and physical geography, constantly improving the system, policy and operation mechanism of urban–rural integrated development and creating China's human-earth system science and rural science, further building a community with a shared future for mankind and supporting the sustainable development of China and even the world.

(1) Using new detection technology to strengthen the theoretical cognition of urban–rural transformation. For a long time, the research of geography on urban-rural transformation is mostly based on remote-sensing interpretation, statistical yearbook and field survey. In recent years, the rapid development of information technology promotes the arrival of "big data era", which guides scientific research into the fourth scientific paradigm driven by big data (Zhou et al. 2021). With the development of advanced technologies such as geographic information technology, Internet and cloud computing, all kinds of data, such as geospatial big data and factor flow data, make it possible for people to objectively understand the geographical laws of urban–rural social and economic activities, and provide new ideas and methods for the innovation of urban–rural transformation research. Based on remote sensing (RS), geographic information system (GIS) and global positioning system (GPS), the construction of Beidou satellite navigation system provides the feasibility of establishing a space-sky-earth integrated collaborative observation system based on human-earth system science, modern remote sensing, internet of things, unmanned aerial vehicle (UAV) and other information technology means, which support the analysis of the pattern, process, mechanism and regulation of urban–rural transformation and rural regional system transformation. In addition, the progress of technology promotes the acquisition of multi-source heterogeneous data and its fusion calculation, which makes it possible to innovate geographical research methods, and further deepens the cognition of the pattern, process, mechanism and mode of urban-rural transformation and rural revitalization.

(2) Using artificial intelligence to strengthen technological reform to promote smart urban–rural construction. The rapid progress of modern science and technology promotes the development of computer technology, which makes artificial intelligence technology widely used in various fields. In view of

the complexity and uncertainty of urban–rural system, the changeable relations among elements often exceed human's cognitive ability, the huge computing power and massive full coverage data of intelligent system will help urban–rural complex system to realize self-assessment, automatic optimization and intelligent governance. In general, artificial intelligence will realize the dynamic investigation and monitoring of urban–rural transformation through intelligent means in multi-temporal and spatial scales and multi-dimensional, and simulate the spatial distribution of people, land and industry in urban and rural geographical space, thus building a "classifier", "detector" and "simulator" for urban-rural transformation and rural revitalization. That is, it will help to build a closed-loop research system of "monitoring, evaluation, simulation, prediction, decision-making" for urban–rural transformation, realize the transformation from digital urban–rural areas to smart urban–rural areas, and then to intelligent urban–rural areas, thus solving the problem of unbalanced and inadequate rural development, and promoting the achievement of the "two centenary" goals.

(3) Innovating engineering technology methods to strengthen the response to urban and rural transformation. Under the background of the transformation of the social main contradictions, the problems of urban disease and rural disease faced by China are still severe, resulting in unbalanced urban–rural development, inadequate rural development and uncoordinated regional development, which gives the innovation of modern geographic engineering to comprehensively deepen the new mission of geography serving national strategies. Meanwhile, the urban–rural system is complex. In order to meet the needs of urban–rural integration and rural revitalization, it is urgent to introduce system engineering principles and technical methods, deeply analyze the complexity and differentiation of urban–rural system, form a new understanding of urban–rural transformation theory, and explore new methods to solve the problems of urban–rural transformation. In essence, modern geographic engineering is a natural economic technological process, which is the organic combination and cross fusion of the comprehensiveness and regionality of geography and the systematicness and technicality of engineering. Geographical engineering has a solid foundation for theoretical exploration and practical application in China. It combines the goal orientation with practical problems, and promotes urban and rural spatial reconstruction and system reconstruction through land consolidation project, environmental management project, ecological construction project, etc.

(4) Establishing a global view to guide urban–rural construction practice and optimize urban–rural relationship. China has a historical tradition of "rural China", and the world is also "rural world". It is necessary to further look at the regional pattern of cities and villages in the world from the perspective of the universe, to establish a spatial system of global urban–rural relations, to re-examine the status and nature of rural regional system, and to rediscover the value and function of global rural areas. Nowadays, the focus and strategic core of global environmental change, land degradation, agricultural production, food security, poverty eradication and other issues lie in the rural areas. It is urgent to establish a global rural view and the new rural–urban relationship in theory, and devote to the sustainable development of rural regional system and solve the above regional problems. Urbanization is an economic and social process of rural transformation, non-agricultural elements agglomeration and spatial form reconstruction, the two-way interactive cycle process of promoting urban development with rural development and leading the rural development with urban development is the correct modern rural–urban view and development view. Therefore, the study of urban–rural transition geography has prominent multi-scale characteristics, including global, national, regional and local.

It is of great significance to explore the historical evolution, transformation process and dynamic mechanism of urban–rural relationship in developed countries and emerging economies, and systematically study the urbanization mode, spatial form and regulatory path in low- and middle-income countries, thus optimizing global regional and urban–rural relationships and finally achieving the SDGs by 2030.

(5) Improving human-earth system science to promote the realization of urban-rural integration and global SDGs. In the early 1990s, Mr. Wu Chuanjun, a famous geographer in China, pointed out that the core of geographical research is the human-earth areal relationship. Nowadays, the rapid process of globalization, industrialization and urbanization has or is strongly acting on the human-earth system on the earth's surface, which the coupling of multi-element such as nature, economy, society and technology, as well as the multi-process such as physics, chemistry, biology and humanities more intense. The new surface circle formed by the integration and interaction of human system and physical system is "anthropedosphere". Essentially, urban-rural transformation is the result of human-earth interaction on the earth's surface, which is in the dynamic evolution of coordination-imbalance-re-coordination. Meanwhile, the spatial heterogeneity of human-earth system makes the urban-rural transformation have obvious regional differences. In the Anthropocene, innovating and developing human-earth system or modern human-earth system science based on anthropedosphere will provide a scientific cognitive basis for in-depth analysis of modern human and technological processes, evolution of surface environment and their regional differentiation pattern, thus deepening the cognition of urban-rural transformation and global sustainable development.

(6) Innovating rural–urban relationship to support urban–rural integrated development. In the early stage, rural areas are dominated by natural system and rurality. With the development of the division of labor and the rise of cities, the non-agriculturalization and urbanization in rural areas are developing continuously, and the regional urbanity is increasing and rurality is decreasing. In the middle and late stage of urbanization, regional development has entered a new stage of urban–rural interaction, urban–rural coordination and even urban–rural integration dominated by urbanity. As a result, the countryside is in a subordinate position, forming an urban–rural integrated system. From this point of view, the emphasis on "rural–urban relationship" theoretically has a more rigorous causal chain and logic than the "urban–rural relationship", and the integration and equivalent development of rural-urban system has become the frontier strategy and important direction of new-type urbanization and urban-rural integration in the future. Therefore, it is urgent to change the traditional concept of "urban centrism" in the study of urban-rural relationship and innovate the new cognition of rural–urban relationship. Based on the rural regional system, the global and interactive nature of new-type urbanization and urban-rural integrated development need to be re-examined to deeply understand the internal relationship between rural regional system, urban regional system and urban-rural integrated system, and comprehensively promote the coordination of new-type urbanization and rural revitalization, thus effectively supplementing the shortcomings of rural insufficient development and solving the problem of unbalanced urban-rural development.

References

Edmonds JA, Link R, Waldhoff ST, et al. A global food demand model for the assessment of complex human-earth systems. Clim Change Econ. 2017;8(4): 1750012.

Liu Y. Research on the urban-rural integration and rural revitalization in the new era in China. Acta Geogr Sin. 2018;73(4):637–50.

Liu Y. Modern human-earth relationship and human-earth system science. Sci Geogr Sin. 2020a;40(8):1224–34.

Liu Y. The basic theory and methodology of rural revitalization planning in China. Acta Geogr Sin. 2020b;75(6):1120–33.

Liu Y, Li Y. Revitalize the world's countryside. Nature. 2017;548(7667): 275–77.

Zhou C, Wang H, Wang C, et al. Geoscience knowledge graph in the big data era. Sci China: Earth Sci. 2021;64(7):1105–14.

The manufacturer's authorised representative in the EU is Springer Nature Customer Service Centre GmbH, Europaplatz 3, 69115 Heidelberg, Germany. If you have any concerns regarding our products, please contact ProductSafety@springernature.com

Printed and bound by CPI Group (UK) Ltd, Croydon, CR0 4YY
30/09/2026
02225287-0020